Therapeutic Potential of Vegetables

Vegetables nourish, heal, and protect, a truth recognized for centuries and now reinforced by modern scientific discovery. *Therapeutic Potential of Vegetables* reveals the remarkable bioactive power hidden within everyday vegetables and explores their role as natural allies in disease prevention, management, and holistic well-being.

Combining traditional wisdom with contemporary research, this book uncovers the phytochemicals that make vegetables valuable sources of antioxidants, anti-inflammatory agents, antimicrobials, and nutraceuticals. Through clear scientific discussions, real-world case studies, and emerging innovations such as edible vaccines, nutrigenomics, and vegetable-based pharmaceuticals, it presents a comprehensive view of vegetables as future-ready therapeutic resources.

Ideal for researchers, clinicians, nutritionists, agricultural scientists, industry professionals, and health-conscious readers, this volume in the *Exploring Medicinal Plants* series offers practical insights and a forward-looking perspective on plant-based healthcare. Discover how vegetables can evolve from being simple dietary staples to scientifically validated medicines that contribute to a healthier and more sustainable future.

Exploring Medicinal Plants

Series Editor: Azamal Husen, *Wolaita Sodo University, Ethiopia*

Medicinal plants render a rich source of bioactive compounds used in drug formulation and development; they play a key role in traditional or indigenous health systems. As the demand for herbal medicines increases worldwide, supply is declining as most of the harvest is derived from naturally growing vegetation. Considering global interests and covering several important aspects associated with medicinal plants, the Exploring Medicinal Plants series comprises volumes valuable to academia, practitioners, and researchers interested in medicinal plants. Topics provide information on a range of subjects including diversity, conservation, propagation, cultivation, physiology, molecular biology, growth response under extreme environment, handling, storage, bioactive compounds, secondary metabolites, extraction, therapeutics, mode of action, and healthcare practices.

Led by Azamal Husen, PhD, this series is directed to a broad range of researchers and professionals consisting of topical books exploring information related to medicinal plants. It includes edited volumes, references, and textbooks available for individual print and electronic purchases.

Sustainable Use of Plants and their Products in Neurodegenerative Diseases Management
Edited by Johnson O. Oladele, Ebenezer I. O. Ajayi, María L. Flores-López, and Oluwaseyi E. Okoro

Wild Edibles of Deserts: Food Security and Sustainability
Salman Majeed, Muhammad Zafar, Mushtaq Ahmad, Mohd Kafeel Ahmad Ansari, Shazia Sultana, and Andrea Pieroni

Impact of Climate Change on Medicinal and Herbal Plant micro-RNA
Kanchanlata Tungare, Parul Johri, Sachidanand Singh, and Surojeet Das

Exploring Medicinal Bamboos
Edited by Manu Pant and Azamal Husen

Exploring Herbal Synergies for Optimal Human Health
Edited by Art Devi, Nancy George, Zaved Ahmed Khan, and Vagish Dwibedi

Exploring Medicinal and Commercial Uses of Flowers
Edited by Rakesh Kumar Bachheti, Archana Bachheti, and Azamal Husen

Phenolic Compounds from Medicinal Plants: Pharmaceutical and Health Benefits
Edited by Rakesh Kumar Bachheti, Archana Bachheti, and Azamal Husen

Therapeutic Potential of Vegetables
Edited by Azamal Husen and Megha Barot

Molecular Marker-Assisted Improvement and Production of Medicinal Plan
Edited by Arpan Modi, Megha Barot, Sachidanand Singh, and Azamal Husen

For more information about this series, please visit: www.routledge.com/Exploring-Medicinal-Plants/book-series/CRCEMP

Therapeutic Potential of Vegetables

Edited by
Azamal Husen and Megha Barot

CRC Press
Taylor & Francis Group
Boca Raton London New York

CRC Press is an imprint of the
Taylor & Francis Group, an **informa** business

Designed cover image: Shutterstock

First edition published 2027
by CRC Press
2385 NW Executive Center Drive, Suite 320, Boca Raton FL 33431

and by CRC Press
4 Park Square, Milton Park, Abingdon, Oxon, OX14 4RN

CRC Press is an imprint of Taylor & Francis Group, LLC

ISBN: 978-1-032-99301-0 (hbk)
ISBN: 978-1-032-99919-7 (pbk)
ISBN: 978-1-003-60671-0 (ebk)

DOI: 10.1201/9781003606710

Typeset in Times
by Apex CoVantage, LLC

Contents

Preface

According to Food and Agriculture Organization Corporate Statistical Database, global vegetable production reached an impressive 1808.04 million tonnes in 2023, underscoring both the scale of agricultural achievement and the critical role vegetables play in sustaining human nutrition and food security worldwide. Vegetables have long held an essential place in the human diet, valued not only for their nourishment, flavour, and vitality but also for their deeper contributions to wellness. Across civilizations, traditional knowledge systems have recognized vegetables as powerful natural medicines capable of supporting prevention, healing, and overall well-being. Today, in an era where holistic health, functional foods, and scientifically validated natural therapies are gaining global momentum, this ancient wisdom is being rediscovered and examined through the lens of modern science. Health professionals now increasingly acknowledge that the interplay between drug therapy and nutrition can significantly improve disease prevention and management. Food plants are being actively explored for their prophylactic and therapeutic potential, largely due to their abundance of pharmacologically active compounds. Phytopharmaceuticals, which are standardized and scientifically validated formulations derived from plant bioactive molecules, have emerged as an important bridge between traditional healing practices and contemporary medical innovation. Vegetables, in particular, represent rich reservoirs of antioxidants, flavonoids, alkaloids, carotenoids, glucosinolates, and other phytochemicals known for their anti-inflammatory, antimicrobial, anticancer, cardioprotective, and neuroprotective properties. These bioactive constituents position vegetable-derived compounds as promising candidates for safe, affordable, and sustainable therapeutic applications.

This book under the book series *Exploring Medicinal Plants* explores the multidimensional significance of vegetables as both sources of nourishment and agents of therapy. It examines the phytochemical profiles of widely consumed vegetables such as garlic, turmeric, spinach, onion, broccoli, bitter gourd, and tomato, emphasizing their roles in human health and their potential for development into phytopharmaceutical products. Many plant species, including tea, strawberry, pistachio, potato, and tomato, have become staples in the Indian diet and hold a rich tradition of medicinal use in cultures across the world. Folk medicine, for example, attributes a wide range of therapeutic benefits to beans, using them in remedies for acne, bladder ailments, burns, cardiac issues, diabetes, diarrhoea, dropsy, dysentery, eczema, hiccups, itching, kidney stones, rheumatism, and sciatica. By integrating insights from traditional knowledge with contemporary scientific evidence, this book encourages a deeper appreciation of vegetables as invaluable sources of natural healing. It provides in-depth scientific discussions across diverse themes, including clinical case studies of vegetable-based treatments, challenges and future prospects of vegetable-derived nutraceuticals, nutrigenomics, the potential of vegetables as edible vaccines, industrially important vegetables and their phytochemicals, and the expanding market for vegetable-based phytonutrients. In doing so, it brings together multidisciplinary perspectives spanning biochemistry, nutrition, pharmacology, molecular biology, agricultural science, and public health. In an era marked by rising global health challenges and growing demand for plant-based, low-risk therapeutic options, understanding the medicinal attributes of vegetables is both timely and essential. Our book, *Therapeutic Potential of Vegetables*, aims to offer a robust scientific foundation, practical applications, and forward-looking directions for the study of vegetables as medicinal agents and sources of phytopharmaceutical derivatives. By presenting cutting-edge research along with traditional insights, it seeks to support scientists, healthcare practitioners, nutritionists, policymakers, and industry professionals working toward sustainable, food-based strategies for human health.

We extend our sincere gratitude to all contributors who graciously accepted the invitation to participate in this volume. Their expertise across diverse research domains has greatly enriched the scientific depth of the chapters. We are also grateful for the thoughtful incorporation of editorial suggestions, which has enhanced the clarity, coherence, and overall quality of the book. We hope this book serves as a valuable resource for researchers, educators, clinicians, students, and all those wishing to understand and harness the powerful therapeutic potential of vegetables. As advancements in phytochemistry, nutrigenomics, biotechnology, and food science continue to unfold, we believe that vegetables will play an increasingly significant role in shaping the future of preventive and therapeutic healthcare.

Azamal Husen
Megha Barot

About the Editors

Professor Azamal Husen is an eminent academic and researcher whose career spans over two decades of outstanding contributions to teaching, research, and scientific leadership. His academic journey has been marked by distinction and international engagement, having served as Full Professor and Head of the Department of Biology at the University of Gondar, Ethiopia, and Foreign Delegate at Wolaita Sodo University, Ethiopia. In India, he is currently Adjunct Professor at Graphic Era University and has previously served as Professor at Sankalchand Patel University and Visiting Faculty at several reputed Indian institutions, including the Forest Research Institute, Dehradun. His research has been supported by several prestigious national and international organizations such as the World Bank, National Agricultural Technology Project, Indian Council of Agricultural Research, Indian Council of Forestry Research and Education, and the Japan Bank for International Cooperation. A prolific scholar, Prof. Husen has authored a vast number of research papers, books, and book chapters published by leading international publishers. His commitment to advancing global science is reflected in his editorial and peer review work for numerous high-impact international journals. His expertise has been widely acknowledged through invitations to evaluate book proposals for leading global publishers such as Springer, Taylor & Francis, and Elsevier, as well as to review research projects for prestigious international funding agencies, including the Czech Science Foundation (Czechia) and the National Research, Development and Innovation Office (Hungary). He is a Fellow of the Plantae Group under the American Society of Plant Biologists and a member of several prestigious scientific organizations, including the International Society of Root Research, the Asian Council of Science Editors, and the International Natural Product Sciences Taskforce. Throughout his distinguished career, Prof. Husen has been a keynote and invited speaker at numerous international conferences and symposia. He has played a significant role in academic development, contributing to curriculum design, faculty training, and educational module development across undergraduate, postgraduate, and doctoral programmes. His outstanding achievements in teaching, research, and community service earned him special recognition from the University of Gondar. In the realm of scientific publishing, he serves as Editor-in-Chief of the *American Journal of Plant Physiology* and as Series Editor for several globally recognized academic collections, including *Exploring Medicinal Plants*; *Plant Chemistry, Plant Biology, Sustainability, and Climate Change*; *Smart Nanomaterials Technology*; and the *Handbook Series on Nanotechnology and Its Applications*. Consistently recognized among the World's Top 2% Scientists (2020 onwards) by Stanford University, Prof. Husen has also served as Guest Editor for several influential special issues, including Advances in Plant Metabolomics (*Plant Physiology and Biochemistry*), Exploring Nanotechnology in Medicinal Plants (*Journal of Herbal Medicine*), Integrating Nanotechnology with Plant Compounds (*Frontiers in Plant Science*) and so on. Through his scholarly excellence, editorial leadership, and global collaborations, Prof. Husen continues to advance cutting-edge research at the intersection of plant science, nanotechnology, and sustainability, establishing himself as a leading figure in contemporary biological and environmental research.

Dr Megha Barot is Assistant Professor (Research Cadre) in the Department of Environmental Science, Parul Institute of Applied Sciences, Parul University, Vadodara, Gujarat, India. She has over a decade of experience in environmental science, life science research, and clinical diagnostics. She has served as the National Coordinator of the Biotech Innovation Ignition School (BIIS) at SRISTI, in collaboration with DBT–BIRAC, Government of India, and as Junior Scientist and NABL Internal Auditor at the Neuberg Centre for Genomic Medicine, Ahmedabad, Gujarat, India. A recipient of the Research Contribution Award from GCPC, Govt. of Gujarat, she also attended a fully funded training programme on Marine Environmental Health and Safety at Xiamen University, China. Dr Barot has finalized seven academic books with CRC Press, Springer, and CABI and has numerous research publications and book chapters to her credit.

Contributors

Mostafa Abdelkader
Horticulture Department
Faculty of Agriculture
Sohag University
Sohag, Egypt

Mohd Adnan
Department of Biology
College of Science
University of Ha'il
Ha'il, Saudi Arabia

Adnan Alam
Faculty of Biotechnology
Institute of Biosciences and Technology
Barabanki, Uttar Pradesh, India

Farah Ali
Faculty of Biotechnology
Institute of Biosciences and Technology
Barabanki, Uttar Pradesh, India

Mahmoud A. A. Ali
Horticulture Department
Faculty of Agriculture
Ain Shams University
Cairo, Egypt

Assiya Ansabayeva
Department of Agronomy
Faculty of Agriculture
Akhmet Baitursynuly Kostanay Regional University
Kostanay, Kazakhstan

Iuliana Aprodu
Faculty of Food Science and Engineering
Dunărea de Jos University of Galați
Domneasca, Galați, Romania

Pankaj Barman
Centre for Preclinical Studies
CSIR – North East Institute of Science and Technology
Jorhat, Assam, India
and
Academy of Scientific and Innovative Research (AcSIR), Ghaziabad, Uttar Pradesh, India

Raviraj B. Barot
Department of Biotechnology
Pandit Deendayal Energy University
Gandhinagar, Gujarat, India

Plaksha Budhraja
Faculty of Biotechnology
Institute of Biosciences and Technology
Barabanki, Uttar Pradesh, India

Prity Chakravarty
Centre for Preclinical Studies
CSIR – North East Institute of Science and Technology
Jorhat, Assam, India
and
Academy of Scientific and Innovative Research (AcSIR)
Ghaziabad, India

Ayush Chaurasiya
Faculty of Biotechnology
Institute of Biosciences and Technology
Barabanki, Uttar Pradesh, India

Nwanne Ijere Dike
Department of Public Health
Faculty of Health Sciences
College of Arts, Science and Professional Studies
Aba, Abia State, Nigeria

Hiral Desai
Department of Biotechnology
Nootan Science and Commerce College
Sankalchand Patel University
Visnagar, Gujarat, India

Loredana Dumitrașcu
Faculty of Food Science and Engineering
Dunărea de Jos University of Galați
Domneasca, Galați, Romania

Syeda Bushra Fatima
Department of Clinical Nutrition
College of Applied Medical Science
University of Ha'il
Hail, Saudi Arabia

Mariana Zagalo Fernandes
RISE-Health
University of Beira Interior
Covilhã, Portugal

Pedro Dinis Gaspar
Centre for Mechanical and Aerospace Science and Technologies
Department of Electromechanical Engineering
University of Beira Interior
Covilhã, Portugal

Aadya Gupta
Faculty of Biotechnology
Institute of Biosciences and Technology
Barabanki, Uttar Pradesh, India

Anjali Gupta
Research and Development Cell (RDC)
Parul University, Waghodia
Vadodara, Gujarat, India
and
Department of Biotechnology
Parul Institute of Applied Sciences
Parul University, Waghodia
Vadodara, Gujarat, India

Karim M. Hassan
Horticulture Department
Faculty of Agriculture
Ain Shams University
Cairo, Egypt

Akbar Hossain
Soil Science Division
Bangladesh Wheat and Maize Research Institute
Nashipur, Dinajpur, Bangladesh

Tanvi Jain
School of Engineering and Technology
BML Munjal University
Gurugram, Haryana, India

Komal Janiyani
Research and Development Cell (RDC)
Parul University, Waghodia
Vadodara, Gujarat, India
and
Department of Microbiology
Parul Institute of Applied Sciences
Parul University, Waghodia
Vadodara, Gujarat, India

Md. Abdul Kader
Postharvest Tech and Nutrition Division
Bangladesh Wheat and Maize Research Institute
Nashipur, Dinajpur, Bangladesh

Aditi Kashyap
Faculty of Biotechnology, Institute of Biosciences and Technology
Barabanki, Uttar Pradesh, India

Noopur Khare
Bhai Gurdas Institute of Engineering and Technology
Sangrur, Punjab, India
and
Faculty of Biotechnology
Institute of Biosciences and Technology
Shri Ramswaroop Memorial University
Barabanki, Uttar Pradesh, India

Pragati Khare
BPS Educational Institution of Pharmacy
Etah, Uttar Pradesh, India

Rituraj Konwar
Centre for Preclinical Studies
CSIR – North East Institute of Science and Technology
Jorhat, Assam, India
and
Academy of Scientific and Innovative Research (AcSIR)
Ghaziabad, Uttar Pradesh, India

Tânia M. Lima
Centre for Mechanical and Aerospace Science and Technologies
Department of Electromechanical Engineering
University of Beira Interior
Covilhã, Portugal

Rita Mahapatra
Krishna School of Science
Drs. Kiran & Pallavi Patel Global University
Varnama, Vadodara, India

Saloni Malik
Department of Biotechnology
Pandit Deendayal Energy University
Gandhinagar, Gujarat, India

Sara Morgado Marcelino
C-MAST – Centre for Mechanical and Aerospace Science and Technologies
Department of Electromechanical Engineering
University of Beira Interior
Covilhã, Portugal

Sushmita Mena
Research and Development Cell (RDC)
Parul University, Waghodia
Vadodara, Gujarat, India
and
Department of Microbiology
Parul Institute of Applied Sciences
Parul University, Waghodia
Vadodara, Gujarat, India

Wenapere Chidinma Mercy
Faculty of Basic Medical Sciences
Bayelsa Medical University
Yenagoa, Bayelsa State, Nigeria

Poonam Mishra
Department of Biotechnology
Krishna School of Science, KPGU
Vadodara, Gujarat, India

Arpan Modi
Axenic Agritech Private Limited
Ahmedabad, Gujarat, India

Odangowei Inetiminebi Ogidi
Department of Biochemistry
Faculty of Basic Medical Sciences
Bayelsa Medical University
Yenagoa, Bayelsa State, Nigeria

Arminda Paço
NECE – Research Unit in Business Sciences
Department of Management and Economics
University of Beira Interior
Covilhã, Portugal

Simran Panda
Faculty of Biotechnology
Institute of Biosciences and Technology
Barabanki, Uttar Pradesh, India

Anushka Pandey
Faculty of Biotechnology
Institute of Biosciences and Technology
Barabanki, Uttar Pradesh, India

Anil Patani
Department of Biotechnology
Nootan Science and Commerce College
Sankalchand Patel University
Visnagar, Gujarat, India

Mitesh Patel
Department of Computer Science and Bioscience
Faculty of Engineering and Technology
Marwadi University
Rajkot, Gujarat, India

Swati Patel
Department of Biotechnology
Krishna School of Science, KPGU
Vadodara, Gujarat, India

Ashutosh Pathak
Department of Botany
University of Rajasthan
Jaipur, Rajasthan, India

Mikhail Puchkov
Agronomy Department
Astrakhan State University
Astrakhan, Russia

Kazi N.S. Rafi
Postharvest Tech and Nutrition Division
Bangladesh Wheat and Maize Research Institute
Nashipur, Dinajpur, Bangladesh

Anshika Rai
Faculty of Biotechnology
Institute of Biosciences and Technology
Barabanki, Uttar Pradesh, India

Shristi Rajpal
Faculty of Biotechnology
Institute of Biosciences and Technology
Barabanki, Uttar Pradesh, India

Sparsh Saxena
Faculty of Biotechnology
Institute of Biosciences and Technology
Barabanki, Uttar Pradesh, India

Harvin Selin
Faculty of Biotechnology
Institute of Biosciences and Technology
Barabanki, Uttar Pradesh, India

Pradeep Pushparaj Selvadoss
Department of Biotechnology
Pandit Deendayal Energy University
Gandhinagar, Gujarat, India

Nisha Singh
Department of Biotechnology
Pandit Deendayal Energy University
Gandhinagar, Gujarat, India

Sachidanand Singh
Department of Biotechnology
School of Energy and Technology
Pandit Deendayal Energy University
Gandhinagar, Gujarat, India

Kavita Srivastava
Faculty of Biotechnology
Institute of Biosciences and Technology
Barabanki, Uttar Pradesh, India

Shikha Suman
Biotechnology and Chemical Engineering, School of Engineering
Faculty of Science, Technology and Architecture
Manipal University Jaipur
Jaipur, Rajasthan, India

Malvi Surti
Research and Development Cell (RDC)
Parul University, Waghodia
Vadodara, Gujarat, India
and
Department of Biotechnology
Parul Institute of Applied Sciences
Parul University, Waghodia
Vadodara, Gujarat, India

Tanushree Tandon
Faculty of Biotechnology
Institute of Biosciences and Technology
Barabanki, Uttar Pradesh, India

Vijay Jagdish Upadhye
Research and Development Cell (RDC)
Parul University, Waghodia
Vadodara, Gujarat, India
and
Department of Microbiology
Parul Institute of Applied Sciences
Parul University, Waghodia
Vadodara, Gujarat, India

Disha Vainsh
Krishna School of Science
Drs. Kiran & Pallavi Patel Global University
Varnama, Vadodara, India

1 Introduction to Vegetable-Based Phyto-Nutraceuticals

Aditi Kashyap, Noopur Khare, Plaksha Budhraja, Shristi Rajpal, Farah Ali, Shikha Suman, and Tanvi Jain

1.1 INTRODUCTION

Vegetable-based nutraceuticals are the bioactive compounds that are found in plants, offering numerous health benefits beyond the basic nutrition (Liu, 2013). Vegetables represent most abundant dietary sources of phyto-nutraceuticals – bioactive-plant-derived compounds offering physiological benefits beyond simple nutrition. Although plants make these nonessential nutrients to defend themselves, they can also benefit human health by boosting the immune system, supporting a variety of body processes, and possibly preventing or curing illnesses (Gupta & Prakash, 2014; Husen, 2021, 2022, 2023, 2024a,2024b). Over the time (past ten years), development in clinical nutrition, nutritional biochemistry, and metabolomics has demonstrated the critical role of phytonutrients derived from vegetables, including terpenes, carotenoids, polyphenols, glucosinolates, organosulphur compounds (OSCs), and dietary fibres in regulating human health. These substances can help in treating and preventing chronic illnesses by modifying gut microbiome composition, oxidative stress, lipid metabolism, epigenetic regulation, and detoxification pathways such as diabetes, heart diseases, cancer, and neurological diseases (Cory et al., 2018; James et al., 2021).

Vegetable-based nutraceuticals can also be found as oils, encapsulated formulations, functional foods, extracts, and supplements, and in addition of being ingested in whole foods, they are designed to improve their bioavailability and stability. They are widely available in different forms such as capsules, tablets, and syrups, making them convenient for consumption and incorporation into daily health routines. The global market for nutraceuticals is rapidly growing, and increased demand of consumers for research on plant-based pharmaceutical molecules !into the action mechanism, safety, effectiveness and translational potential of plant based medicinal compounds has accelerated (Masyita et al., 2022).

The main groups of phyto-nutraceuticals generated from plants, their biological traits, mechanism of action, therapeutic relevance, and the emerging concerns affect their clinical efficacy. This chapter deals to know the basic phytochemicals and how they are used in maintaining health in a better manner.

1.2 MAJOR PHYTOCHEMICALS AND THEIR FUNCTIONS IN MAINTAINING HEALTH

Natural substances called phytochemicals are created by plants to assist them fight off illnesses from bacteria, fungus, and plant viruses as well as to prevent insects and other animals from eating them. Alkaloids, glucosinolates, terpenes, carotenoids, polyphenols, phytosterols, and OSCs are the most frequently found in food (Guillamón et al., 2021).

1.2.1 Phenolics

Polyphenols were originally utilized in foods to influence properties such as taste, colour, aroma, and stability (Figure 1.1). Over time, plants evolved to produce these phenolic compounds as a

DOI: 10.1201/9781003606710-1

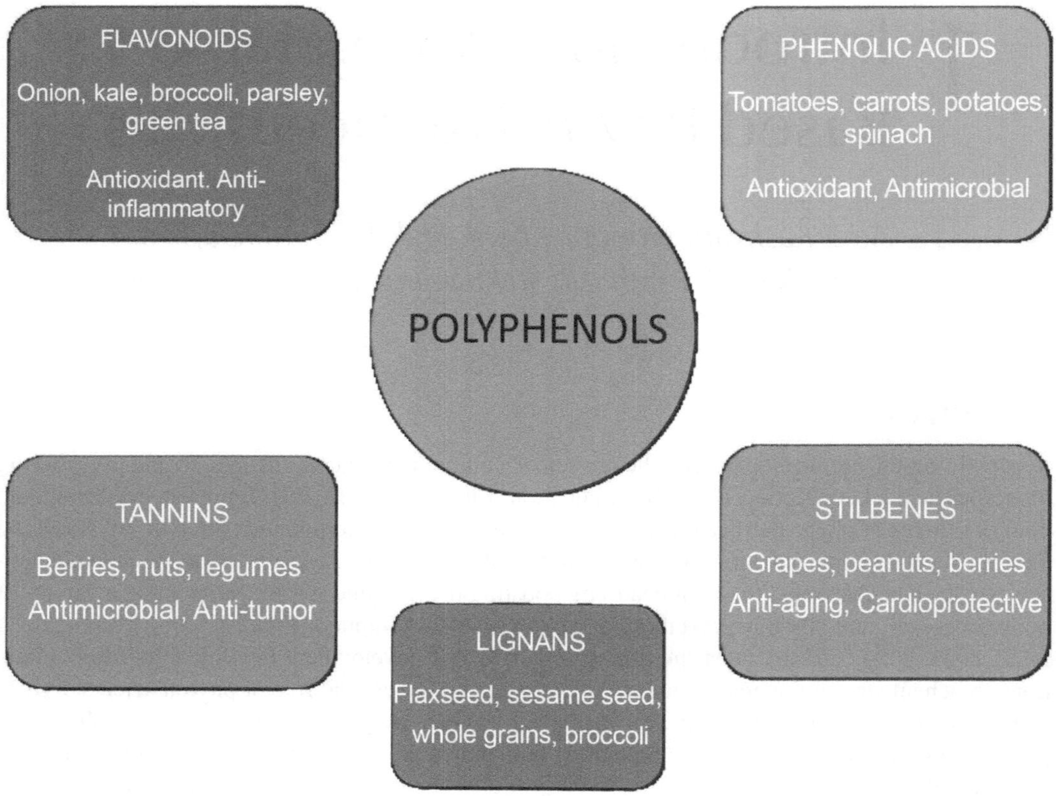

FIGURE 1.1 List of polyphenols.

natural defence mechanism to adapt to changing environmental conditions. Later, extensive epidemiological studies and meta-analyses revealed that regular consumption of plant polyphenols can help protect against several chronic diseases, including diabetes, cancer, osteoporosis, cardiovascular, and neurodegenerative disorders. Based on their chemical structure, polyphenols are categorized into different groups such as phenolic acids, flavonoids, tannins, coumarins, lignans, quinones, stilbenes, and curcuminoids.

Epidemiological studies have shown that consuming foods rich in polyphenols is linked to a lower risk of chronic diseases. Polyphenols act as antioxidants by forming phenoxyl radicals, which interrupt harmful oxidative chain reactions within cells. Research indicates that polyphenols from foods and beverages enhance plasma antioxidant activity by combining with the body's natural antioxidants and aiding in iron absorption. A diet high in polyphenols helps protect lymphocytic DNA from oxidative damage and reduces oxidative stress, thereby lowering the risk of various degenerative diseases (Pandey & Rizvi, 2010; Mutha et al., 2021).

1.2.2 Carotenoids

Carotenoids exert their biological effects through multiple interconnected mechanisms. Additionally, carotenoids engage in cross-talk with the nuclear factor kappa B (NF-κB) signalling pathway, thereby modulating inflammatory responses and contributing to overall cellular homeostasis. These compounds also promote the induction of detoxification pathways and offer potential metabolic benefits, supporting cardiovascular and metabolic health (Fiedor & Burda, 2014). However, human findings on the extent of Nrf2 activation remain inconclusive, as outcomes appear to depend on factors such as dosage, food matrix composition, and biomarker selection (Kaulmann & Bohn, 2014).

It was discovered that the main effects of tomato extract, lutein, and lycopene were pro-oxidation at high doses and antioxidant at low quantities. Additionally, the anti-inflammatory action increases with the concentration of lycopene, tomato, and lutein extract. Tomato extract, lutein, and lycopene can have anti-inflammatory effects by blocking the NF-κB signalling pathway and antioxidant effects, and Nrf2 signalling pathway is activated. Additionally, the expressions of Nrf2 downstream proteins (Nrf2, NQO1, HO-1) and nuclear translocation of Nrf2 were inhibited after treatment with high concentrations of lutein, lycopene, and tomato extract. It was also evident from this that the Nrf2 signalling pathway can function as an antioxidant at low concentrations and as a pro-oxidant at high concentrations. High doses may have an anti-inflammatory effect by preventing NF-κB. It is possible to target the regulation of Nrf2/HO-1 and NF-κB by transforming growth factor-β (TGF-β) and interleukin-10 (IL-10) to produce a low concentration of antioxidants and a high concentration of anti-inflammatory effects (Ba et al., 2023).

1.2.3 Organosulphur Compounds

OSCs comprise a diverse group of bioactive molecules, including glucosinolates present in cruciferous vegetables such as broccoli and cabbage and thiosulphides, which are abundant in Allium vegetables like garlic and onions. These compounds exert multiple biological effects through mechanisms such as inhibition of NF-κB and COX-2 signalling pathways leading to reduced inflammation. Additionally, they exhibit antioxidant properties and contribute to the modulation of lipid metabolism, thereby supporting metabolic health. Their antimicrobial activity also plays an important role in maintaining gut microbial balance and overall gut health (Corzo-Martínez et al., 2007). Evidences from systematic reviews suggest possible cardiometabolic benefits; cancer prevention data are inconsistent across cohorts (Higdon et al., 2007). Allium vegetables, such as garlic and onion, owe their characteristic flavour and medicinal properties to OSCs. When garlic is crushed, the enzyme alliinase converts alliin into allicin that is a highly reactive compound that quickly breaks down into stable OSCs like diallyl disulfide, ajoene, and vinyldithiins, which are primarily responsible for its biological effects (Bazaraliyeva et al., 2022). OSCs exhibit selective antimicrobial activity that suppresses harmful bacteria while promoting beneficial gut microbes. Studies show that allicin and related compounds enhance populations of Bifidobacterium, Lactobacillus, and Lachnospiraceae, improving gut microbiota diversity (Singh et al., 2025). These changes are linked with better lipid metabolism, reduced inflammation, and improved insulin sensitivity in animal models. Additionally, garlic-derived OSCs stimulate beneficial bacteria such as Bacteroides acidifaciens, contributing to metabolic regulation and protection against diet-induced disorders (Chen et al., 2022). OSCs derived from Allium species possess multifaceted immune-modulatory properties that simultaneously attenuate inflammation and enhance host immune responses. Alliin, one of the primary OSCs, suppresses key inflammatory markers in colitis models by suppressing the Mitogen-Activated Protein Kinase (MAPK) and NF-κB signalling pathways (Banerjee et al., 2003). Similarly, allicin strengthens infection defence mechanisms by stimulating $CD4^+$ T cells and Th1 cytokine production while modulating Toll-Like Receptor 4 (TLR4) signalling to prevent excessive inflammation during sepsis (Arreola et al., 2015; Bayan et al., 2014). Alliin has also been shown to inhibit proinflammatory mediators such as nitric oxide (NO), myeloperoxidase (MPO), and inducible nitric oxide synthase (iNOS) in dextran sulphate sodium (DSS)-induced colitis and lipopolysaccharide (LPS)-stimulated macrophages through the suppression of MAPK/NF-κB/AP-1/STAT-1 pathways and activation of peroxisome proliferator-activated receptor gamma (PPAR-γ). Other OSCs, such as propyl propane thiosulphinate (PTS) and propyl propane thiosulphonate (PTSO), further contribute by limiting the secretion of pro inflammatory cytokines like TNF-α, IL-6, and IL-27 and downregulating inflammatory genes (TNF, Nos2, Saa2) in intestinal tissues during parasitic infections (Shi et al., 2017). Allicin also enhances anti-malarial immunity by upregulating $CD4^+$ T cells, promoting dendritic cell maturation, and increasing T helper 1(Th1), cytokines (IFN-γ, TNF), while elevating IL-10 to maintain inflammatory balance (Feng et al., 2012). The gut microbiota plays a crucial role in immune education through metabolites such as short-chain fatty acids (SCFAs), and OSCs modulate this ecosystem by enhancing intestinal barrier integrity and maintaining immune homeostasis, illustrating a significant host–microbe–immune interplay.

The integrated effects of OSCs on gut microbiota and immune signalling offer promising therapeutic implications. In inflammatory bowel diseases (IBDs), alliin and other OSCs alleviate colitis through anti-inflammatory and epithelial-protective mechanisms (Li et al., 2013). In metabolic disorders and obesity, OSC-mediated modulation of gut microbiota and increased SCFA production contribute to improved lipid and glucose metabolism (Wang et al., 2021). Furthermore, in infection control, OSCs enhance Th1-mediated immune defence against infections like malaria while preventing hyperinflammatory tissue damage through TLR regulation (Arreola et al., 2015; Bayan et al., 2014). Despite strong preclinical evidence, translational challenges remain. Most studies are limited to animal models, necessitating well-designed randomized clinical trials to establish human efficacy. Additionally, poor bioavailability of unstable compounds like allicin underscores the need for optimized delivery systems such as encapsulation. Detailed mechanistic mapping of OSC microbiome host signalling pathways and standardization of extraction, species variation, and preparation methods is also essential to ensure reproducibility and clinical reliability. Figure 1.2 illustrates the central role of the gut microbiota in regulating key physiological processes. The microbiota interacts bidirectionally with phytochemicals, influencing their metabolism and bioavailability, while also modulating host metabolism, immune responses, inflammation, and infection resistance. These interconnected pathways highlight how dietary phytochemicals, such as OSCs, support microbial balance and overall host health.

1.2.4 Sulforaphane from Broccoli and Other Cruciferous Vegetables

Sulforaphane (SFN) is an isothiocyanate derived from glucoraphanin, a glucosinolate abundant in cruciferous vegetables such as broccoli, cabbage, kale, and Brussels sprouts. Formed through myrosinase-catalysed hydrolysis during chewing or cutting, SFN is a potent phyto-nutraceutical known for its antioxidant, anti-inflammatory, anticancer, and epigenetic modulatory activities (Traka & Mithen, 2020). Biosynthetically, SFN originates from glucoraphanin upon plant tissue disruption. Even when cooking inactivates myrosinase, gut microbiota can convert glucoraphanin to SFN, preserving its biological effects. Its reactive isothiocyanate (–N=C=S) group allows SFN to interact with cellular macromolecules, influencing gene expression and signalling. Factors such as cooking methods, plant genotype, and storage affect its bioavailability; broccoli sprouts are particularly rich sources, containing 20–50 times more glucoraphanin than mature plants.

Mechanistically, SFN acts through multiple pathways: it activates Nrf2, enhancing phase II detoxification and antioxidant enzymes; inhibits NF-κB, reducing pro-inflammatory cytokines; modulates epigenetic regulators (HDACs and DNMTs); and induces apoptosis, cell cycle arrest, and anti-angiogenic effects in tumour cells (James et al., 2021; Soundararajan & Kim, 2018). These molecular actions collectively contribute to its chemopreventive and epigenetic benefits.

Health-wise, SFN supports cancer prevention, cardiovascular health, neuroprotection, and metabolic regulation. It improves endothelial function, crosses the blood–brain barrier to protect neurons, and enhances insulin sensitivity while reducing inflammation. Furthermore, nanoparticle-based formulations and functional food applications are being developed to enhance its stability and therapeutic potential. Despite these benefits, challenges remain: low bioavailability, loss during cooking, and individual variability in conversion due to differences in gut microbiota limit consistent outcomes. Ongoing research focuses on encapsulation techniques, fermentation processes, and plant breeding strategies to optimize SFN yield and stability (James et al., 2021). Overall, SFN exemplifies how dietary phytochemicals can bridge nutrition and medicine, offering promising avenues in cancer prevention, chronic disease management, and epigenetic therapy. Figure illustrate sources, biosynthesis, mechanism and health benefits of SFN.

1.2.5 Phytosterols

Phytosterols – which include plant sterols and stanols – are naturally occurring steroids that are extensively found in many plant parts, including roots, stems, leaves, flowers, fruits, and entire grasses. They are an important part of plant cell membranes. Every day, humans ingest 100–400 mg of phytosterols, primarily from vegetables, bread, cereals, nuts, and vegetable oils. Clinical applications for phytosterols include preventing cardiovascular diseases (CVDs) and non-alcoholic fatty liver disease (NAFLD) in addition to reducing plasma levels of cholesterol and low-density lipoprotein cholesterol (LDL-C). These days, phytosterols' physiological roles can be shown in a variety of ways, including their antioxidant, anti-inflammatory and antipyretic, and hormone-like properties. Their ability to lower LDL-C concentrations and cholesterol absorption is their most notable feature. Because of their structural resemblance to cholesterol, phytosterols will compete with and prevent the absorption of cholesterol in the intestine, lowering the plasma level of LDL-C (Li et al., 2022).

1.2.6 Terpenes

A single botanical source's bark, buds, flowers, fruits, leaves, peels, roots, seeds, twigs, or even the entire plant can yield essential oils (EOs), which are concentrated liquids made up of complex mixtures of aromatic, hydrophobic, and volatile compounds (Falleh et al., 2020). The major constituents of EOs are terpenes. Terpenes are a diverse group of compounds responsible for the characteristic flavour and aroma of vegetables. These natural products exhibit multiple biological activities through a range of mechanisms, most of which involve interactions with cell membranes and enzyme systems (Singh & Sharma, 2015). Terpenes are also known as isoprenoids, which are characterized by carbon backbones derived from 2-methylbuta-1,3-diene (isoprene units) capable of rearranging into cyclic structures. The number of isoprene units determines the structural diversity of terpenes, forming categories such as hemiterpenes (C5), monoterpenes (C10), sesquiterpenes (C15), diterpenes (C20), triterpenes (C30), and tetraterpenes (C40) (Bhavaniramya et al., 2019). Terpenoids, which are oxygenated derivatives of terpenes, arise through biochemical modifications like methyl group addition or removal.

Over the past few decades, extensive research has highlighted their crucial role in maintaining human health. These compounds, composed of multiple isoprene units, represent the largest class of organic molecules synthesized in plant essential oils. They exhibit diverse therapeutic properties demonstrated through various in-vitro and in-vivo studies, including anticancer, antimicrobial, anti-inflammatory, antioxidant, anti-allergic, neuroprotective, anticoagulant, sedative, and analgesic effects, largely mediated by monoterpenes, sesquiterpenes, diterpenes, triterpenes, tetraterpenes, and glycoside derivatives. Moreover, terpenes serve as vital sources of vitamins A, E, and K, as well as coenzyme Q10, with carotenoids and tocopherols being especially significant for human and animal health. Due to their diverse functional properties, terpenes are extensively utilized in the pharmaceutical, nutraceutical, food and beverage, cosmetic, perfume, synthetic chemical, aroma and flavour, rubber, and biofuel industries (Tetali, 2019; Masyita et al., 2022).

Terpenes demonstrate strong antibacterial properties by disrupting the lipid structure of bacterial cell walls, which causes membrane disintegration, protein degradation, and leakage of cytoplasmic contents, ultimately leading to cell lysis and death. Additionally, due to the presence of conjugated double bonds in their molecular structure, terpenes exhibit a high capacity for scavenging free radicals (Martignago et al., 2023).

1.3 CLASSIFICATION OF PHYTO-NUTRACEUTICALS

Phyto-nutraceuticals can be classified on the basis of their chemical structure, plant part, and nutritional groups that include diverse groups of phytochemicals such as polyphenols, glucosinolates,

TABLE 1.1
Classification of Phyto-Nutraceuticals

Basis of Classification	Category	Examples / Sources	Major Compounds	Role / Function	Key references
Based on Plant Part	**Root vegetables**	Carrot, beetroot	β-carotene, betalains, nitrates	Antioxidant, anti-inflammatory, improves blood flow	Zhai et al. (2021)
	Tuber vegetables	Potato, sweet potato	Alkaloids, carotenoids	Energy source, antioxidant activity	Fiedor and Burda (2014)
	Bulbous vegetables	Onion, garlic	Thiosulphides, flavonoids	Antimicrobial, cholesterol-lowering	Amagase (2006)
	Leafy vegetables	Spinach, kale, cabbage	Lutein, phenolics, glucosinolates	Antioxidant, detoxification, cancer prevention	Higdon et al. (2007)
Based on Nutritional Groups	**Antioxidants**	*Capsicum*, broccoli, spinach, carrot, tomato, kale	Vitamin C, vitamin E, carotenoids (β-carotene, lycopene, lutein)	Protect cells from oxidative stress and chronic diseases	Ba et al. (2023)
	Dietary fibre and prebiotics	Carrots, beetroot, legumes, cabbage, spinach, peas, onion, garlic, asparagus	Soluble fibre, insoluble fibre, inulin, oligosaccharides	Improve gut health, regulate cholesterol, control blood sugar	O'Keefe (2016)
	Phytochemicals/ secondary metabolites	Onion, kale, broccoli, cauliflower, cabbage, garlic	Flavonoids (quercetin, kaempferol), glucosinolates, thiosulphides (allyl sulfides)	Non-nutrient bioactives with disease-preventive roles	Shang et al. (2019)

carotenoids, and OSCs, each linked with protective roles against chronic diseases. Table 1.1 presents the classification of phyto-nutraceuticals based on two main criteria plant part and nutritional groups highlighting their sources, major bioactive compounds, and biological functions. Plant-based phyto-nutraceuticals include a variety of bioactive substances that offer health advantages beyond simple nourishment. Based on their chemical makeup, biological activity, and potential for therapeutic use, they are widely categorized. Polyphenols, which comprise flavonoids, phenolic acids, tannins, and lignans, are one important group. These substances are well known for their cardioprotective, anti-inflammatory, and antioxidant qualities. Fruits, vegetables, and tea all contain flavonoids, such as quercetin, catechins, and anthocyanins, which have a major role in preventing illness. Terpenoids or isoprenoids, which comprise carotenoids, essential oils, and saponins, are another significant class. Carotenoids, such as lutein, lycopene, and β-carotene, have potent antioxidant properties and boost immunity and vision. Aromatic plant-based essential oils have anti-inflammatory and antibacterial properties (Gogoi et al., 2023).

1.3.1 Biological Mechanism of Action of Phyto-Nutraceuticals

Vegetable-based phytonutrients promote health through multiple biological pathways (Williamson, 2017). Their diverse structures allow them to act as antioxidants, anti-inflammatory agents, modulators of hormones and enzymes, regulators of lipid metabolism, and influencers of gut microbiota. These phytochemical compounds protect against oxidative stress (Barot et al., 2015; Ji et al., 2017) and support key physiological functions such as detoxification, immunity, cardiovascular health, and digestion. Phyto-nutraceuticals maintain health and prevent diseases through a range of interrelated molecular and physiological mechanisms. Polyphenols, flavonoids, phytosterols, terpenoids, alkaloids, and carotenoids are examples of plant-derived bioactive substances that alter cellular pathways related to oxidative stress, inflammation, metabolism, and immunological function. Strong antioxidant activity is one of the main mechanisms (Ramya & Patel, 2019). Numerous phyto-nutraceuticals protect cells from oxidative stress-induced damage by scavenging free radicals, increasing endogenous antioxidant enzymes including glutathione peroxidase, catalase, and superoxide dismutase, and lowering lipid peroxidation. By inhibiting important signalling pathways like NF-κB and MAPK and lowering the synthesis of pro-inflammatory cytokines like TNF-α, IL-1β, and IL-6, they also exhibit strong anti-inflammatory effects. Furthermore, a number of phyto-nutraceuticals enhance insulin sensitivity, control lipid profiles, and affect gene expression by activating nuclear receptors like PPARs. By regulating macrophage and lymphocyte activity, encouraging cytokine balance, and improving the integrity of the gut microbiota – which is essential for systemic immunity – some substances improve immune responses (Yoo et al., 2020).

1.3.1.1 Antioxidant Activity

Prevention of Cancer: Antioxidants from vegetables help prevent cancer by reducing DNA damage and protecting cells from harmful oxidative stress (Block et al., 1992).

Cardiovascular Health: Antioxidants prevent the oxidation of LDL cholesterol, which is a key step in the development of atherosclerosis (Steinberg et al., 2008). This lowers the chances of having a heart attack or stroke.

Eye Health: Certain carotenoids, like lutein and zeaxanthin, protect the retina and help prevent conditions such as age-related macular degeneration (AMD) and cataracts.

1.3.1.2 Anti-Inflammatory Effects

Arthritis Relief: By reducing the release of inflammatory substances, these compounds can help ease joint swelling and pain.

Heart Disease Prevention: Chronic inflammation increases the risk of atherosclerosis. Reducing inflammation helps lower the chances of developing heart disease (Galland, 2010).

1.3.1.3 Hormonal and Enzyme Modulation

Cruciferous vegetables can help regulate oestrogen levels, which may reduce the risk of certain cancers that are linked to hormones, including breast, endometrial, and prostate cancer (Higdon et al., 2007).

Liver Health: Enhanced detoxification processes help protect the liver from damage caused by chemicals. One of the main ways that nutraceuticals and bioactive substances affect human health both therapeutically and preventively is through hormonal and enzyme regulation. By interacting with hormone receptors, controlling hormone synthesis, or altering their metabolism, a variety of compounds derived from plants affect endocrine balance. For example phytoestrogens such as isoflavones bind to oestrogen receptors (ER-α and ER-β), producing modest oestrogenic or anti-oestrogenic effects depending on the body's hormonal condition. This promotes cardiovascular health, bone density, and menopausal health. By impacting iodine metabolism, thyroid peroxidase activity, and peripheral conversion of T4 to the more active T3, several nutraceuticals also control thyroid activity and aid in metabolic regulation (Benvenga et al., 2020).

1.3.1.4 Regulation of Lipid Metabolism

Cholesterol Reduction: Phytosterols found in vegetables can help reduce the amount of cholesterol absorbed by the body, lowering LDL cholesterol levels (Gylling et al., 2014). Prevention of Atherosclerosis: Reducing triglyceride and cholesterol levels lowers the chance of artery-clogging plaque. One important way that many nutraceuticals support cardiovascular health and prevent metabolic diseases is through the regulation of lipid metabolism. By focussing on particular enzymes and signalling pathways, a number of phyto-nutraceuticals affect lipid absorption, synthesis, transport, and breakdown. For instance phytosterols structurally resemble cholesterol and lower plasma LDL levels by competitively inhibiting intestinal absorption. By inhibiting the activity of HMG-CoA reductase, the rate-limiting enzyme in cholesterol manufacturing, polyphenols like resveratrol and catechins regulate lipid metabolism and reduce endogenous cholesterol production. By stimulating the breakdown of fatty acids in the liver and improving triglyceride clearance, omega-3 fatty acids promote lipid oxidation by activating peroxisome proliferator-activated receptors (PPAR-α). Furthermore, flavonoids increase HDL and promote reverse cholesterol transfer by enhancing the expression of important transport proteins including ABCA1 (Tan et al., 2021).

1.3.1.5 Modulation of Gut Microbiota

Digestive Health: Fibre from vegetables helps improve bowel function, prevents constipation, and maintains a healthy colon.

Reduced Colon Cancer Risk: When fibre is fermented by gut bacteria, it produces short-chain fatty acids that protect colon cells from cancer (O'Keefe, 2016). Modulation of gut microbiota is an essential strategy by which nutraceuticals boost general health. Numerous plant-based substances, such as dietary fibres, polyphenols, and prebiotics, specifically promote the growth of good bacteria like *Lactobacillus* and *Bifidobacterium* while inhibiting pathogenic microorganisms. These chemicals boost microbial diversity, improve SCFA synthesis, and strengthen the intestinal barrier. Consequently, a balanced microbiome enhances nutritional absorption and metabolic processes, lowers inflammation, and supports immunological modulation. By altering the composition and activity of the gut ecology, nutraceuticals contribute to better digestive health and protection against metabolic and inflammatory illnesses (Caponio et al., 2022).

1.4 BIOACTIVE COMPOUNDS FROM VEGETABLE BY-PRODUCTS: INDUSTRIAL PROSPECTS

One of the main factors that has sped up the investigation of vegetable by-products as a dependable industrial resource is the growing demand for natural, sustainable, and functional ingredients. Rich concentrations of bioactive molecules appropriate for various industries were found in by-products that were thrown away as waste (Figure 1.2). A rich and underutilized source of bioactive chemicals with substantial industrial potential is vegetable by-products, which are produced during processing and post-harvest handling (Caponio et al., 2022). Despite being frequently thrown away as garbage, these residues – such as peels, stems, seeds, leaves, and pulps – contain significant concentrations of beneficial phytochemicals such polyphenols, flavonoids, carotenoids, dietary fibres, vitamins, minerals, and essential oils. In addition to supporting circular economy models, the extraction of these chemicals provides a sustainable method of valuing agro-industrial waste. Bioactive substances derived from vegetable by-products have potential industrial uses in a variety of industries. Because

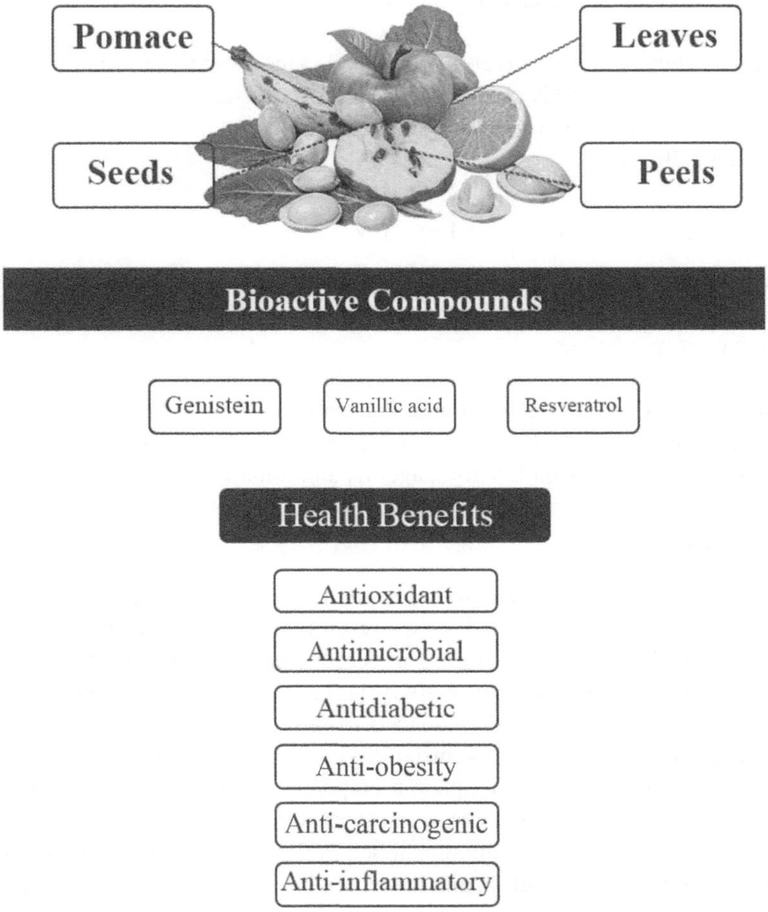

FIGURE 1.2 Bioactive compounds.

of their antioxidant, anti-inflammatory, and metabolic health effects, extracted polyphenols and fibres are added to supplements, fortified foods, and beverages in the nutraceutical and functional food industries. Carotenoids, essential oils, and antioxidants produced from plants are used by the cosmetics industry to create anti-aging, skin-brightening, and antibacterial formulas. Alkaloids, flavonoids, and sulphur compounds from by-products are used by the pharmaceutical industry to create natural remedies, medication precipients, and bioactive components for the treatment of chronic illnesses. Biodegradable films enhanced with antioxidants or antibacterial extracts from vegetable waste improve the shelf life and safety of packed meals in the food packaging industry. Additionally, the fermentation and biotechnology sectors use fibre-rich residues as substrates to make probiotics, organic acids, and enzymes. The efficacy and stability of these bioactive components are further enhanced by technological developments including nanoencapsulation, ultrasound-assisted extraction (UAE), and supercritical fluid extraction (SFE). Vegetable by-products therefore present significant industrial opportunities, making them important resources for high-value product creation, waste reduction, and sustainable innovation (Mateus et al., 2024).

1.4.1 Biomedical and Pharmaceutical Uses

By-product bioactive compounds have demonstrated significant therapeutic potential:

- *Anti-inflammatory agents* (quercetin from onion skins, hesperidin from citrus peels)
- *Anticancer compounds* (SFN precursors from broccoli stems)
- *Antimicrobial extracts* for wound dressings, topical formulations, and natural antiseptics (Mateus et al., 2024).

1.4.2 Personal Care and Cosmetics Industry

Phenolics and pigments derived from by-products are widely used in:

- Creams that prevent ageing.
- Sunscreens with natural antioxidants and skin-brightening serums.
- Hair care products and anti-acne formulations; among many other beneficial qualities for clean beauty formulation, anthocyanins, vitamin C, flavonoids, carotenoids, and plant sterols help shield the skin from UV rays, ageing, and inflammation (Mateus et al., 2024).

1.5 CONCLUSION

Vegetable-derived phyto-nutraceuticals are an expanding class of bioactive compounds with significant potential for human health and disease prevention. Allium vegetable-derived OSCs, leafy green-derived polyphenols, and SFN from cruciferous vegetables illustrate diverse mechanisms from antioxidant and anti-inflammatory activity to immune modulation and epigenetic effects. Although their health benefits are well-evidenced from preclinical and epidemiological research, limitations including poor bioavailability, instability, and absence of standardized clinical validation are important obstacles to their complete therapeutic use. Generally, vegetable-based phyto-nutraceuticals have enormous potential as natural, sustainable therapeutic interventions for the management of chronic disease and promotion of health. Future research should be aimed at clinical translation, new formulation approaches, and incorporation into functional foods and nutraceuticals to overcome the gap between experimental findings and health outcomes in real life. Incorporating these compounds into final commercial applications boosts their value and promotes circular and sustainable bioeconomic practices.

REFERENCES

Amagase, H. (2006). Clarifying the real bioactive constituents of garlic. *Journal of Nutrition*, 136(3 Suppl), 716S–725S. https://doi.org/10.1093/jn/136.3.716S

Arreola, R., Quintero-Fabián, S., López-Roa, R. I., et al. (2015). Immunomodulation and anti-inflammatory effects of garlic compounds. *Journal of Immunology Research*, 2015, 401630. https://doi.org/10.1155/2015/401630

Ba, W., Xu, W., Deng, Z., Zhang, B., Zheng, L., & Li, H. (2023). The antioxidant and anti-inflammatory effects of the main carotenoids from tomatoes via Nrf2 and NF-κB signaling pathways. *Nutrients*, 15(21), 4652. https://doi.org/10.3390/nu15214652

Banerjee, S. K., Mukherjee, P. K., & Maulik, S. K. (2003). Garlic as an antioxidant: The good, the bad and the ugly. *Phytotherapy Research*, 17(2), 97–106.

Barot, M., Ji, N. K., & Kumar, R. N. (2015). Nutraceutical properties of green, red and brown seaweeds from the Eco-sensitive Zone, India as potential dietary and pharmaceutical applications. *Proceedings of the India International Science Festival-Young Scientists' Meet Department of Science and Technology*, Government of India–Dec, 4–8.

Bayan, L., Koulivand, P. H., & Gorji, A. (2014). Garlic: A review of potential therapeutic effects. *Avicenna Journal of Phytomedicine*, 4(1), 1–14.

Bazaraliyeva, A., Moldashov, D., Turgumbayeva, A., Kartbayeva, E., Kalykova, A., Sarsenova, L., & Issayeva, R. (2022). Chemical and biological properties of bioactive compounds from garlic (Allium sativum). *Pharmacia*, 69(4), 955–964.

Benvenga, S., Ferrari, S. M., Elia, G., Ragusa, F., Patrizio, A., Paparo, S. R., . . . & Fallahi, P. (2020). Nutraceuticals in thyroidology: A review of in vitro, and in vivo animal studies. *Nutrients*, 12(5), 1337.

Bhavaniramya, S., Vishnupriya, S., Al-Aboody, M. S., Vijayakumar, R., & Baskaran, D. (2019). Role of essential oils in food safety: Antimicrobial and antioxidant applications. *Grain & Oil Science and Technology*, 2(2), 49–55.

Block, G., Patterson, B., & Subar, A. (1992). Fruit, vegetables, and cancer prevention: A review of the epidemiological evidence. *Nutrition and Cancer*, 18(1), 1–29.

Caponio, G. R., Lippolis, T., Tutino, V., Gigante, I., De Nunzio, V., Milella, R. A., . . . & Notarnicola, M. (2022). Nutraceuticals: Focus on anti-inflammatory, anti-cancer, antioxidant properties in gastrointestinal tract. *Antioxidants*, 11(7), 1274.

Chen, K., Nakasone, Y., Yi, S., Ibrahim, H., Sakao, K., Hossain, M., & Hou, D.-X. (2022). Natural garlic organosulfur compounds prevent metabolic disorder of lipid and glucose by increasing gut commensal Bacteroides acidifaciens. *Journal of Agricultural and Food Chemistry*, 70(18), 5829–5837.

Cory, H., Passarelli, S., Szeto, J., Tamez, M., & Mattei, J. (2018). The role of polyphenols in human health and food systems: A mini-review. *Frontiers in Nutrition*, 5, 87. https://doi.org/10.3389/fnut.2018.00087

Corzo-Martínez, M., Corzo, N., & Villamiel, M. (2007). Biological properties of onions and garlic. *Trends in Food Science & Technology*, 18(12), 609–625.

Falleh, H., Ben Jemaa, M., Saada, M., & Ksouri, R. (2020). Essential oils: A promising eco-friendly food preservative. *Food Chemistry*, 330, 127268.

Feng, Y., Zhu, X., Wang, Q., Jiang, Y., Shang, H., Cui, L., & Cao, Y. (2012). Allicin enhances host proinflammatory immune responses and protects against acute murine malaria infection. *Malaria Journal*, 11, 268.

Fiedor, J., & Burda, K. (2014). Potential role of carotenoids as antioxidants in human health and disease. *Nutrients*, 6(2), 466–488.

Galland, L. (2010). Diet and inflammation. *Nutrition in Clinical Practice*, 25(6), 634–640.

Gogoi, R., Sarma, N., Begum, T., Chanda, S. K., Lekhak, H., Sastry, G. N., & Lal, M. (2023). Agarwood (Aquilaria malaccensis L.) a quality fragrant and medicinally significant plant based essential oil with pharmacological potentials and genotoxicity. *Industrial Crops and Products*, 197, 116535.

Guillamón, E., Andreo-Martínez, P., Mut-Salud, N., Fonollá, J., & Baños, A. (2021). Beneficial effects of organosulfur compounds from Allium cepa on gut health: A systematic review. *Foods*, 10(8), 1680.

Gupta, C., & Prakash, D. (2014). Phytonutrients as therapeutic agents. *Journal of Complementary and Integrative Medicine*, 11(3), 151–169.

Gylling, H., Plat, J., Turley, S., Ginsberg, H. N., Ellegård, L., et al. (2014). Plant sterols and plant stanols in the management of dyslipidaemia and prevention of cardiovascular disease. *Atherosclerosis*, 232(2), 346–360.

Higdon, J. V., Delage, B., Williams, D. E., & Dashwood, R. H. (2007). Cruciferous vegetables and human cancer risk: Epidemiologic evidence and mechanistic basis. *Pharmacological Research*, 55(3), 224–236.

Husen, A. (2021). *Traditional Herbal Therapy for the Human Immune System*. CRC Press. https://doi.org/10.1201/9781003137955

Husen, A. (2022). *Herbs, Shrubs and Trees of Potential Medicinal Benefits*. CRC Press. https://doi.org/10.1201/9781003205067

Husen, A. (2023). *Antidiabetic Medicinal Plants and Herbal Treatments*. CRC Press. https://doi.org/10.1201/b23347

Husen, A. (2024a). *Antimalarial Medicinal Plants*. Taylor & Francis Group. https://doi.org/10.1201/9781003378396

Husen, A. (2024b). *Medicinal Spices and Condiment Crops*. Taylor & Francis Group. https://doi.org/10.1201/9781003387046

James, D., et al. (2021). Sulforaphane: Translational research from laboratory bench to clinic. *Nutrients*, 13(9), 3106.

Ji, N. K., Barot, M., & Kumar, R. N. (2017). Distribution and biochemical constituents of different seaweeds collected from Okha coast, Gujarat, India. *Indian Journal of Geo Marine Sciences*, 46(2), 349–357.

Kaulmann, A., & Bohn, T. (2014). Carotenoids, inflammation, and oxidative stress: Implications of cellular signaling pathways and relation to chronic disease. *Oxidative Medicine and Cellular Longevity*, 2014, 734861.

Li, X., Xin, Y., Mo, Y., Marozik, P., He, T., & Guo, H. (2022). The bioavailability and biological activities of phytosterols as modulators of cholesterol metabolism. *Molecules*, 27(2), 523.

Li, Y., Xu, S., Jiang, B., Cohen, R. A., & Zang, M. (2013). Activation of AMPK-dependent protective autophagy in IBD by garlic compounds. *Free Radical Biology and Medicine*, 64, 1–12.

Liu, R. H. (2013). Health-promoting components of fruits and vegetables in the diet. *Advances in Nutrition*, 4(3), 384S–392S.

Martignago, C. C. S., Soares-Silva, B., Parisi, J. R., et al. (2023). Terpenes extracted from marine sponges with antioxidant activity: A systematic review. *Natural Products and Bioprospecting*, 13(23).

Masyita, A., Mustika Sari, R., Dwi Astuti, A., Yasir, B., Rahma Rumata, N., Emran, T. B., Nainu, F., & Simal-Gandara, J. (2022). Terpenes and terpenoids as main bioactive compounds of essential oils, their roles in human health and potential application as natural food preservatives. *Food Chemistry: X*, 13, 100217.

Mateus, A. R. S., et al. (2024). Unveiling the potential of bioactive compounds in vegetable and fruit by-products: Exploring phytochemical properties, health benefits, and industrial opportunities. *Current Opinion in Green and Sustainable Chemistry*, 48, 100938.

Mutha, R. E., Tatiya, A. U., & Surana, S. J. (2021). Flavonoids as natural phenolic compounds and their role in therapeutics: An overview. *Future Journal of Pharmaceutical Sciences*, 7(1), 25. https://doi.org/10.1186/s43094-020-00161-8

O'Keefe, S. J. (2016). Diet, microorganisms and colon cancer. *Nature Reviews Gastroenterology & Hepatology*, 13(12), 691–706.

Pandey, K. B., & Rizvi, S. I. (2010). Protective effect of resveratrol on markers of oxidative stress in human erythrocytes subjected to in vitro oxidative insult. *Phytotherapy Research*, 24(S1), S11–S14.

Ramya, V., & Patel, P. (2019). Health benefits of vegetables. *International Journal of Chemical Studies*, 7(2), 82–87.

Shang, A., Cao, S. Y., Xu, X. Y., et al. (2019). Bioactive compounds and biological functions of garlic (Allium sativum L.). *Foods*, 8(7), 246.

Shi, L., Lin, Q., Li, X., Nie, Y., Sun, S., Deng, X., Wang, L., Lu, J., Tang, Y., & Luo, F. (2017). Alliin, a garlic organosulfur compound, ameliorates gut inflammation through MAPK–NF-κB/AP-1/STAT-1 inactivation and PPAR-γ activation. *Molecular Nutrition & Food Research*, 61(9)

Singh, A., Kaur, P., Kumar, M. K., Shafi, S., Upadhyay, P., Tiwari, A., Tiwari, V., Rangra, N., Thirunavukkarasu, V., Chacko, S., Roy, D., Ghosh, M., Arora, N., Sharma, N., & Garg, Y. (2025). The role of phytochemicals in modulating the gut microbiota: Implications for health and disease. *Medicine in Microecology*, 24, 100125.

Singh, B., & Sharma, R. A. (2015). Plant terpenes: Defense responses, phylogenetic analysis, and drug development. *Current Drug Targets*, 16(14), 1714–1736.

Soundararajan, P., & Kim, J. S. (2018). Anti-carcinogenic glucosinolates in cruciferous vegetables and their antagonistic effects on prevention of cancers. *Molecules*, 23(11), 2983.

Steinberg, D. (2008). Thematic review series: The pathogenesis of atherosclerosis. An interpretive history of the cholesterol controversy: Part V: The discovery of the statins and the end of the controversy. *Journal of Lipid Research*, 50(Suppl), S179–S185.

Tan, M., Ye, J., Zhao, M., Ke, X., Huang, K., & Liu, H. (2021). Recent developments in the regulation of cholesterol transport by natural molecules. *Phytotherapy Research*, 35(10), 5623–5633.

Tetali, S. D. (2019). Terpenes and isoprenoids: A wealth of compounds for global use. *Planta*, 1–8.

Traka, M., & Mithen, R. (2020). Plant science and human nutrition: Challenges in assessing health-promoting properties of phytochemicals. *Current Opinion in Plant Biology*, 56, 1–9.

Wang, J., Cao, Y., & Sun, B. (2021). Effects of allicin on metabolism and gut microbiota in mice. *Food & Function*, 12(8), 3770–3780.

Williamson, G. (2017). The role of polyphenols in modern nutrition. *Nutrition Bulletin*, 42(3), 226–237.

Yoo, J. Y., Groer, M., Dutra, S. V. O., Sarkar, A., & McSkimming, D. I. (2020). Gut microbiota and immune system interactions. *Microorganisms*, 8(10), 1587.

Zhai, B., Zhang, C., Sheng, Y., & Zhao, C. (2021). Garlic improves intestinal health by enhancing microbiota diversity and SCFA production in mice fed a high-fat diet. *Frontiers in Nutrition*, 8, 683257.

2 Current Scenario of Vegetable-Based Phyto-Nutraceutical Market

Sara Morgado Marcelino, Mariana Zagalo Fernandes, Pedro Dinis Gaspar, Arminda Paço, and Tânia M. Lima

2.1 INTRODUCTION

The emergence of the nutraceutical sector has been driven by public awareness and concerns about healthcare issues, drawing consumers' attention to the potential of food-derived bioactive compounds to improve health conditions and prevent diseases (Tsiaka et al., 2022). In 1989, Stephen DeFelice, founder and chairman of the Foundation for Innovation in Medicine (FIM) in New Jersey, created the concept of 'nutraceuticals' to refer 'any substance that may be considered a food or part of a food that provides medical or health benefits, including the prevention and/or treatment of a disease' (Blaze, 2021; Sharma et al., 2024). The term resulted from a combination of 'nutrition' and 'pharmaceutical', and in its introduction, Stephen DeFelice stated that the nutraceutical revolution would mark a new era of medicine and health, with a similar research orientation to the pharmaceutical industry (Hoti et al., 2022; M. Sharma et al., 2024). Investigations have been suggesting that there is great potential for functional foods and nutraceuticals for chronic disease prevention and to optimize health at both the individual and population levels (Sharma et al., 2024). Among the health benefits of nutraceuticals indicated in formal studies are sleep quality improvement, easing migraines, combating oxidative stress, supporting mental well-being, cognitive function enhancement, managing neurodegenerative diseases, and pain and weight reduction (Takefuji, 2025).

The terms 'nutraceuticals' and 'functional foods' are sometimes used synonymously, as nutraceuticals are derived from foods, but there are some differences. The concept 'functional foods' was coined in Japan in the 1980s, corresponding to 'processed foods containing ingredients that aid specific bodily functions in addition to being nutritious'. Functional foods encompass both processed food products enriched with bioactive components, such as probiotics or plant sterols, and non-processed foods that are naturally high in compounds of health interest, for instance, omega-3 fatty acids or antioxidants (Sharma et al., 2024). Despite being related, nutraceutical is a broader term than functional foods as it covers any health beneficial substance, beyond mere nutrition, whether it may be a food or simply part of a food. Three main segments of the nutraceutical industry are functional foods, dietary supplements, and herbal/botanical products (Chopra et al., 2022; Sawicka et al., 2020). Nutraceutical products are based on bioactive compounds from plants, animals, or microbes that exert reproducible therapeutic effects and can be used in the medicinal form of pills, capsules, or liquids (Frank et al., 2020; Haider, 2023; Shahidi, 2012; Husen & Iqbal, 2021). Figure 2.1 presents visually the connection among the previously defined concepts.

There is an increasing interest in bioactive compounds across a wide range of areas beyond the nutraceutical market, namely, geomedicine, plant science, modern pharmacology, agrochemicals, cosmetics, the food industry, nanobioscience, and related fields (Chaudhary & Garg, 2023). A bioactive compound is a descriptor based on a biological activity and refers to plant, microbial, or animal-derived components with demonstrated activity in biological systems, typically in animals

DOI: 10.1201/9781003606710-2

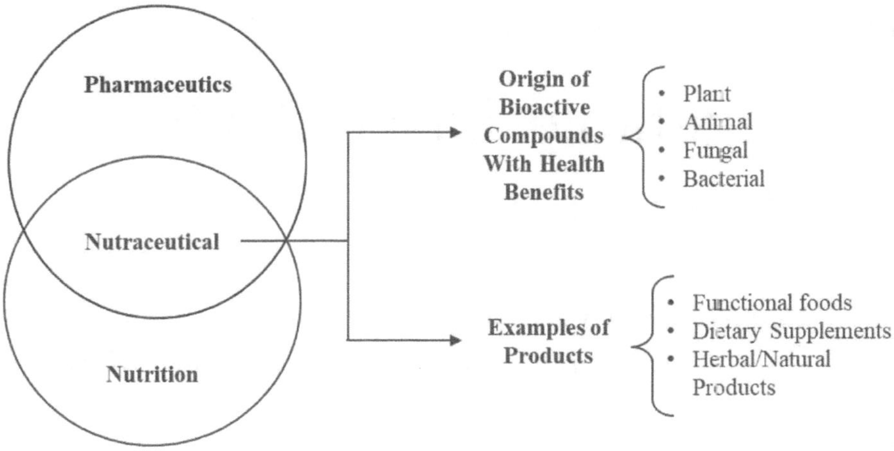

FIGURE 2.1 Conceptual diagram on the overlap of nutraceutical-related terms.

and/or humans (Frank et al., 2020). While bioactive compounds can exert either beneficial or harmful biological effects, nutraceuticals use only bioactive compounds with therapeutic properties that should be reproducible in randomized and controlled animal or human trials (Frank et al., 2020; Sawicka et al., 2020).

This chapter will focus specifically on the phyto-nutraceutical market. The term "phyto" originates from the Greek word for 'plant' (Lalitha, 2024). Phyto-nutraceuticals use bioactive molecules found in plants (fruits, vegetables, legumes, seeds, and nuts), known as phytochemicals or secondary metabolites, which can interact with mammalian biochemical and biotransformation pathways to promote health benefits (Alum & Ugwu, 2023; Delgado et al., 2021; Sajid Khan et al., 2025; Townsend et al., 2023). Unlike primary metabolites produced by plants, which are directly involved in essential metabolic processes, such as normal growth, development, and reproduction (e.g., innate proteins, lipids, and carbohydrates), secondary metabolites confer a selective advantage to the plant organism (Delgado et al., 2021). Plants produce these substances as part of their mechanisms to protect and adapt to stress conditions in unfavourable environments (e.g., water deficit, heat stress, oxygen deficiency, air pollution) (Cisneros-Zevallos, 2003). While plants respond to stress by producing protective secondary metabolites, in contrast, when human cells and tissues are under challenge by stress (e.g., oxidative stress), an inflammatory response can be triggered and, when persistent, can contribute to the development of noncommunicable diseases (Cisneros-Zevallos, 2021).

Plant-derived bioactive compounds contribute to human health by exerting diverse biological effects, including antioxidant, anti-carcinogenic, antiallergenic, anti-inflammatory, antimutagenic, antimicrobial, anti-aggregation, and neurotrophic activities (Husen, 2021, 2022, 2023a,2023b, 2024a,2024b; Dincheva et al., 2025; Korsapati et al., 2025). These properties can have beneficial impacts on various diseases such as autoimmune, inflammatory, cardiovascular, neoplastic, metabolic, and neurodegenerative disorders (Dincheva et al., 2025; Husen & Pant, 2024a,2024b; Pant et al., 2025; Pant & Husen, 2025).

The therapeutic use of plants has deep historical roots in human culture. Ancient civilizations across China, India, Peru, and many other regions recognized the health-promoting potential of botanical resources and used them for that effect (Petrovska, 2012). Today, many of these same plant-derived secondary metabolites are scientifically recognized for their capacity to modulate inflammation and support physiological health and continue to be used traditionally by many cultures around the world (Petrovska, 2012; Sajid Khan et al., 2025; Saleh et al., 2021).

Phytonutrients can influence human health through several mechanisms, including the modulation of the gut microbiota composition and activity. In particular, certain classes of phytochemicals, such as polyphenols, flavonoids, and carotenoids, can promote beneficial microbial taxa, enhance intestinal barrier integrity, and regulate host immune and metabolic responses, which play a pivotal role in promoting an individual's overall health (Chen et al., 2023; Kan et al., 2022; Wang, 2022). The applications of phytochemicals extend beyond human healthcare, as many of them exhibit antimicrobial and antioxidant activities that make them valuable in the food industry, where they can be used, for example, to delay spoilage and oxidative deterioration, inhibit fungal growth on fruits and vegetables after harvest, or serve as preservatives to control postharvest physiological disorders and microbial contaminations (Ahmad et al., 2020; Redondo-blanco et al., 2020; Sawicka et al., 2020).

The growing consumer awareness around the physiological and therapeutic benefits of vegetables has led to a substantial expansion of the vegetable-based phyto-nutraceutical market (Ahmad et al., 2020; Shilpi, 2022). In this market, products such as food supplements, dietary supplements, value-added processed vegetables, as well as non-food formats such as tablets, soft gels, and capsules are included (Ahmad et al., 2020). Vegetables are recognized not only as fundamental components of the human diet but also as functional foods and sources of nutraceuticals, owing to their safety, low caloric value, and high nutritional and therapeutic potential (Femi-Adepoju et al., 2023). Epidemiological evidence indicates that daily consumption of approximately 800 g of fruits and vegetables is associated with a reduced risk of CVD and all-cause mortality, while intakes around 600 g per day are associated with a lower overall cancer risk. The World Health Organization (WHO) recommends a minimum intake of 400 g of edible fruits and vegetables per day to prevent micronutrient deficiencies and support overall health (Arcusa et al., 2021; Prabhakar, 2022). Vegetables represent one of the most accessible and cost-effective sources of essential nutrients and encompass a wide variety of edible plant parts, including roots, tubers, bulbs, leaves, flowers, fruits, seeds, and stems, which collectively contribute to the daily nutritional requirements of humans (Femi-Adepoju et al., 2023; Mahima et al., 2013). Beyond their nutritional value, regular vegetable consumption is associated with a reduced risk for several chronic diseases, including certain cancers, cardiovascular disorders, hypertension, and type 2 diabetes (Ahmad et al., 2020; Prabhakar, 2022). Vegetables provide essential minerals, vitamins, antioxidants, nutrients, phytosterols, and dietary fibre, contributing significantly to human health and disease prevention. Their natural wealth in bioactive medicinal compounds like polyphenols, lycopene, indoles, glucosinolates, carotenoids, anthocyanins, and phytoestrogens is getting more and more consumers' attention (Prabhakar, 2022).

Following the introduction of basic concepts in this first section, the following sections will discuss the current state of the vegetable-based phyto-nutraceutical market, including the evolution of demand, key drivers of market growth, as well as the main challenges and opportunities.

2.2 NUTRACEUTICAL MARKET GROWTH AND MAIN DRIVERS

Understanding market evolution is an important step to know how companies can allocate their resources and in which areas they should focus to be competitive and improve efficiency. The nutraceutical market was predicted to be valued at over $275 billion by 2021, according to the report by the Klynveld Peat Marwick Goerdeler (KPMG) consultant. However, the growth of this market has surpassed expectations. Even in 2019, the nutraceutical market had already worth of $382 billion, $412 billion in 2020 (Chopra et al., 2022), and $412 billion in 2022 (Takefuji, 2025). Another report, published by Allied Market Research, pointed out that the global market size in 2020 was $413.0 billion and is projected to reach $650.5 billion by 2030, registering a Compound Annual Growth Rate (CAGR) of 3.9% from 2021 to 2030 (Allied Market Research, 2022). For the same year, 2030, other market analysis reports point out the values of $919.1 billion (Grand View Research, 2024) and $1025 billion (Takefuji, 2025) in the global nutraceutical market. The global market growth is presented graphically in Figure 2.2 and the market growth per region in Figure 2.3.

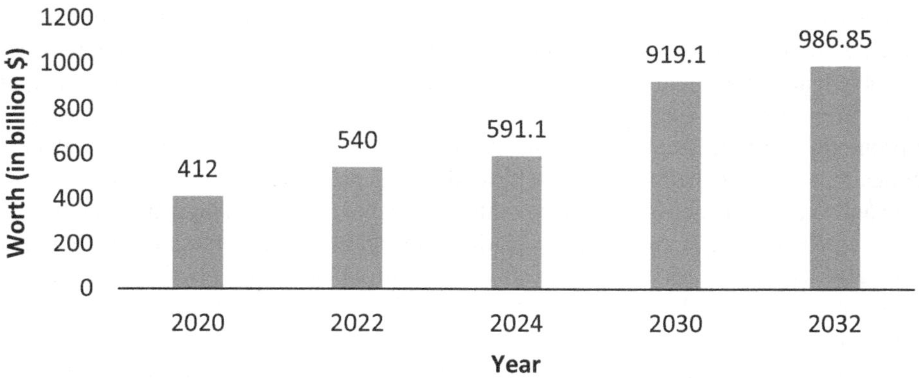

FIGURE 2.2 Global nutraceutical market size.

Source: Based on Allied Market Research (2022), Takefuji (2025), Grand View Research (2024), and Fortune Business Insights,(2025)

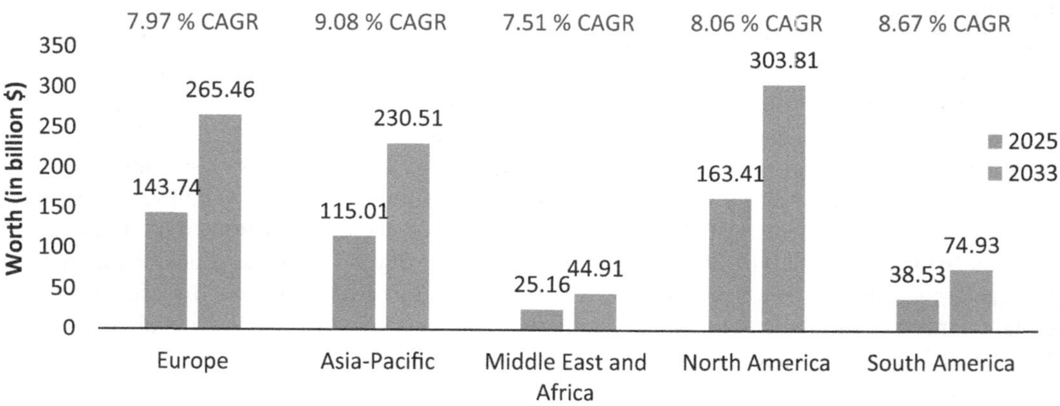

FIGURE 2.3 Nutraceutical market size per region.

Source: Based on Market Data Forecast (2025)

The phytonutrients market has been expanding rapidly, with its global worth significantly increase driven by the growing demand for natural ingredients across the food, cosmetic, pharmaceutical, and nutraceutical industries. Within this expanding sector, the three main types of phytonutrients, namely flavonoids, carotenoids, and glucosinolates, present a predicted market value in 2033 of $11.22 billion, of $4.37 billion, and $1.71 billion, respectively. This market can be broken down into three main sources: fruits, vegetables, and herbs (Consainsights, 2024). In 2023, the fruit segment was valued at approximately $6.49 billion and is expected to nearly double, reaching $11.22 billion by 2033.

The vegetable segment follows, with a market value of $2.52 billion in 2023 and an anticipated increase to $4.37 billion over the same period. Herbal-derived phytonutrients represent a smaller share, estimated at $0.99 billion in 2023 and expected to reach $1.71 billion by 2033. In all three segments, an expansion of market value is expected. For example, the global herbal nutraceuticals market size was valued at $28,329.7 million in 2019 and is projected to reach $48,446.5 million by 2027, registering a CAGR of 7.55% from 2021 to 2027 (Allied Market Research, 2022). Concerning the vegetable extracts market, a report by Mordor Intelligence presents a CAGR of 7.40% from 2020

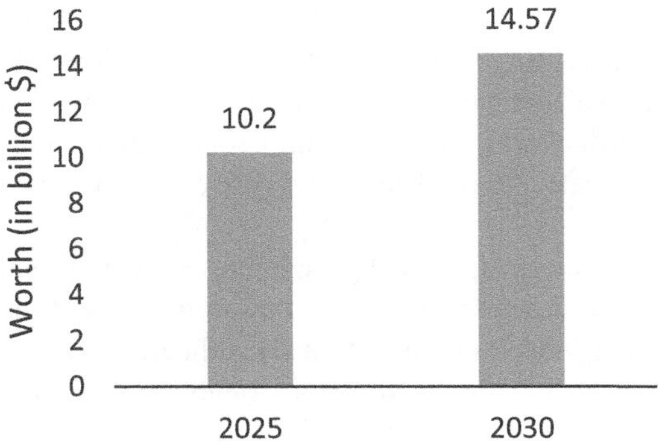

Driver	% Impact on CAGR Forecast
Rising demand for clean-label ingredients	+1.8%
Increasing health consciousness and functional foods	+1.5%
Growth of the nutraceutical and dietary supplement sector	+1.2%

FIGURE 2.4 Vegetable extracts market size and drivers.

Source: Based on Mordor Intelligence (2025)

to 2030. One of the most impactful drivers for this expansion is the growth of the nutraceutical and dietary supplement sector (Mordor Intelligence, 2025), with a quantified impact on the CAGR forecast of 1.2%, as illustrated in Figure 2.4.

By understanding the key market drivers, additional opportunities for growth, expansion, and segmentation can be identified (Chopra et al., 2022). When analysing the nutraceutical sector, several key common drivers were identified in market research and scientific literature, although their impact varies across regions. The main common drivers across the regions presented in Figure 2.3 are discussed below and summarized in Table 2.1 at the end of the section. This summary is not comprehensive, but it is based on data from Market Data Forecast (2025) and Mordor Intelligence (2025) regarding the most representative key market drivers for market growth.

The Asia-Pacific region presents the highest CAGR of 9.08%. Not only China and India but also Brazil are the countries that present the fastest predicted CAGR, corresponding to an above average growth in the nutraceutical market expected over the coming decades (Chopra et al., 2022). One of the key drivers in Asia-Pacific is the ageing population increase, which is projected to more than double by mid-century (Market Data Forecast, 2025). This factor is also impactful in Europe, which has one of the highest proportions of elderly individuals in the world, and it is expected to grow, at least, in the two upcoming decades (Market Data Forecast, 2025). Nutraceuticals have been shown to contribute to the prevention of age-related chronic disorders, having therefore an important role in supporting longevity, minimizing healthcare systems' burden, and enhancing overall well-being among the elderly population (Chopra et al., 2022). In the Middle East and Africa, as well as in North America, an ageing population is also indicated as a relevant driver of market growth.

Another important contribution to the market size expansion, particularly in Europe, Asia-Pacific, the Middle East and Africa, and North America regions, is the increasing digital adoption of e-commerce, allowing brands to reach new submarkets with lower physical retail investment (Market Data Forecast, 2025). Digital marketing approaches, such as email campaigns, social media engagement, mobile applications, e-commerce websites, virtual sales representatives, and restricted online platforms, have become increasingly influential in the promotion and commercialization of nutraceutical and dietary supplement products (Chopra et al., 2022). Broadly, the expansion of digital marketing strategies, including the use of business-to-business and business-to-consumer channels, is accelerating growth in the nutraceutical market. Multilevel marketing, in which independent distributors sell products directly to consumers from home, online, or via telecommunications and receive rewards both for sales and for recruiting new distributors, further contributes to this market expansion (Chopra et al., 2022). Digital platforms also exert a duality effect on the nutraceutical market, as reported in the North America case (Market Data Forecast, 2025). On the one hand, telehealth consultations have been notably increasing nutraceutical recommendations and consumption. However, on digital platforms and social media, strict regulatory oversight combined with the spread of misinformation and unverified health claims constrains market growth by eroding consumer trust and perceptions of product quality, particularly deterring risk-averse consumers (Market Data Forecast, 2025).

With the second largest CAGR of 8.67%, the South American market growth has been amplified due to the increase in the cost of healthcare, nutraceutical entrepreneurs, and awareness about the importance of managing lifestyle-related diseases and inadequate nutrition. Healthcare costs are increasing both in developing and developed countries as healthcare systems around the world are dealing with a rise in demand for medical services, as the global population keeps growing, as well as the percentage of elderly people in many parts of the world grows (Shrivastava & Ali, 2025). This demographic condition results in augmented costs, prompting some consumers to seek more affordable solutions to prescription medications and health products (Blaze, 2021). Plants or herbs, particularly fruits and vegetables, are the most accessible and cost-effective sources of nutrients, namely carbohydrates, protein, vitamins, minerals, and essential amino acids (Mahima et al., 2013). This nutritional abundance can support the expansion of the vegetable-based phyto-nutraceutical market in response to rising healthcare costs.

North America shows the third largest CAGR of 8.06%. This notable increase in the nutraceutical market growth has been especially driven by the escalating prevalence of chronic lifestyle-related diseases. According to data from the Centres for Disease Control and Prevention, it is estimated that 60% of American adults live with at least one chronic disease, such as cardiovascular conditions, diabetes, and obesity (Market Data Forecast, 2025). Additionally, in Asia-Pacific markets, the growing adoption of urban lifestyles and Western dietary patterns has led to widespread health issues, particularly diabetes, CVDs, and metabolic disorders (Mordor Intelligence, 2025). This driver also occurs significantly in the Middle East and Africa territories, where lifestyle-related chronic conditions have been rising healthcare system costs, encouraging both citizens and governments to adopt preventive nutrition policies (Market Data Forecast, 2025). Urbanization and lifestyle changes, including increased sedentary behaviour and poor dietary patterns, have contributed to a decline in nutritional status and, consequently, immune function (Ahmad, 2024; Sajid Khan et al., 2025). In this context, vegetable compounds of interest can contribute to mitigating the social and economic burden of chronic lifestyle-related conditions, including neurodegenerative diseases (Arcusa et al., 2021). In all the different analysed regions, the increasing awareness about preventive health has been one of the key drivers for the nutraceutical market growth.

TABLE 2.1
Key Drivers for Nutraceutical Market Growth Per Region

	Europe	Asia-Pacific	Middle East and Africa	North America	South America
Ageing population	✓	✓	✓	✓	
E-commerce	✓	✓	✓	✓	
Increasing healthcare costs					✓
Increasing prevalence of lifestyle-related diseases		✓	✓	✓	
Rising health consciousness and awareness about preventive health	✓	✓	✓	✓	✓

Source: Based on Market Data Forecast (2025) and Mordor Intelligence (2025)

2.3 EMERGENT PARADIGMS IN HEALTHCARE CONDITIONING THE NUTRACEUTICAL MARKET

In the previous section, we explored some key drivers of market growth in the nutraceutical market mostly from a consumer-centric optic, but emerging trends in healthcare service provision can also play an important role in shifting market evolution. Understanding consumers' motives and social conditions is crucial for guiding stakeholders and decision-makers in the health sector and inclusively in the nutraceutical sector (Takefuji, 2025).

In several areas of knowledge, researchers analysed nature to find inspiration to improve efficiency and develop more sustainable and innovative technologies. This approach, known as biomimetics, involves creating technologies, processes, or methods inspired by biological processes and structures (Kohsaka et al., 2019). Nature constitutes a vast reservoir of bioactive compounds that can be harnessed to promote human and animal health and to enhance food system sustainability (Kussmann & Cunha, 2022). In this sense, problems such as national services crisis, increase in population, unemployment, loss of ethical values, increase in the incidences of pathogens, development of resistance to antibacterial, and other phenomena have forced scientists and researchers to study applications of natural or alternative approaches to medicine (Mahima et al., 2013).

The growing global population, the increasing proportion of elderly individuals, and the rising prevalence of chronic and lifestyle-related diseases are placing significant pressure on the demand for healthcare products and services, resulting in augmented costs. To address these challenges, both individuals and governments must prioritize preventive care, integrate technological innovations, strengthen healthcare workforce capacity, and improve service delivery efficiency to ensure long-term sustainability (Shrivastava & Ali, 2025). There is evidence that numerous diseases can be prevented by phytochemicals, including tumours; heart disease; diabetes mellitus; elevated blood pressure; inflammation; microbial, viral, and parasitical infections; mental health issues; spasmodic diseases; and ulcerated patches (Sajid Khan et al., 2025).

Although modern medical drug development is partly based on plant secondary metabolites, they are derived from modifications of the original natural compounds to enhance their efficacy and enable intellectual property protection (Cisneros-Zevallos, 2021). However, their efficacy in the short term comes with a price. Beyond the augmented costs, concerns about the adverse effects of medical drugs and the risk of toxicity are fostering a worldwide attention to nutraceuticals, considered a powerful instrument to facilitate lifespan, optimal health, and quality of life (Hoti et al., 2022).

Pharmaceutical practices rely on the use of *a substance or a combination of substances presented as having properties for treating or preventing disease in humans* (Witte & Müller, 2025). In the nineteenth century, the emergence of infectious diseases and severe epidemics, such as cholera, tuberculosis, diphtheria, and typhoid fever, spreading in Europe, America, and colonial territories, led to the development of scientific pharmacology and research on the effects of drugs on humans. Throughout that century, the sale of drugs became the economic foundation of the healthcare system (Polak et al., 2025). Since then, with the widespread use of pharmaceuticals, unrelated or unexpected adverse reactions to approved drugs, even under normal conditions of use and dosage, have increasingly become a matter of public concern (Yan et al., 2021). Conventional pharmacological practices are largely designed to deliver individual and short-term therapeutic benefits, often neglecting the broader external 'health costs' imposed on global populations and future generations (Witte & Müller, 2025). The persistence and bioaccumulation of pharmaceutical compounds and their metabolites in the environment raise significant ecological and health concerns. By compromising health and well-being and diminishing the effectiveness of disease treatment, the conventional use of medical drugs conflicts with the concept of sustainable development, defined by the Brundtland Commission as *development that meets the needs of the present without compromising the ability of future generations to meet their own needs* (WCED, 1987). Furthermore, the use of allopathic drugs conflicts with some fundamental principles of Hippocratic medicine.

Hippocrates of Kos (c. 460 to c. 370 BCE) separated medicine from the study of philosophy and changed medical practice by introducing an approach based on direct observation and experience, arguing that medical progress relied on both empirical evidence and logical reasoning (Polak et al., 2025). However, his contribution is not limited to experimental medicine. Hippocrates also established fundamental medical ethics based on the philosophy of 'the health of the people should be the supreme law' (*salus aegroti suprema lex*) and the precautionary principle 'first, do no harm' (*primum non nocere*) (Polak et al., 2025; Witte & Müller, 2025). Hippocrates also acknowledged the importance of diet in the prevention and management of diseases, a concept often summarized by the phrase attributed to him, 'let food be thy medicine, and medicine be thy food' (Wegener, 2014). The frequent preference for medical drugs, despite their adverse effects on human health and the environment, over holistic approaches that consider lifestyle and diet, exposes a clear misalignment with the basilary experimental medicine principles. The divergence between Hippocratic principles with the prevailing approach in modern medicinal drug use highlights the need to prioritize solutions that do not compromise human health and well-being or reduce the effectiveness of disease treatment.

One of the paradigmatic shifts proposed in the literature for a more sustainable healthcare treatment is focusing on treating the root cause, whether it may be on a molecular level or external body factors of illness, instead of directing towards symptom management (Wynendaele et al., 2021). There is a growing interest in integrative medicine, a paradigm that shifts the focus of medical practice from a disease-centred model to one oriented toward healing and the whole person (Seetharaman et al., 2021). Whereas the disease-based approach seeks primarily to counteract the manifestations of pathological processes in an effort to cure, the healing-based perspective emphasizes understanding illness within a broader, holistic context (González et al., 2000). Integrative medicine does not correspond to a rejection of allopathic medicine. Instead, it combines evidence-based complementary and alternative therapies with modern medicine, aiming to promote health, prevent disease, and enhance patient well-being through a personalized and patient-centred approach (Sharma et al., 2023). In search of a more holistic approach to health and in response to the rising costs of allopathic medicine and the numerous side effects associated with conventional drugs, an increasing number of people are turning toward lifestyle modifications and use of plant-based products to prevent various diseases and enhance overall health (Ahmad et al., 2020), including nutraceutical products (Pathak & Kumar, 2022). The use of nutraceuticals may help reduce or even eliminate the need for conventional drugs, while also mitigating the adverse effects linked to conventional food consumption (Sawicka et al., 2020). The rapidly expanding application of phytochemicals in functional foods and nutraceuticals poses new challenges for both the food and pharmaceutical industries (Sajid Khan et al., 2025) and underscores the need for an evolution in contemporary medical practices.

Another strategy for achieving more sustainable healthcare, and specifically for reducing the number of pharmaceuticals released into the environment, as well as minimizing adverse human and ecological side effects, is to enhance bioavailability, defined as the proportion of a substance that, once ingested, is absorbed and effectively utilized by the organism. Bioavailability is influenced by a patient's physical and genetic predisposition and previous diagnosis, and, under a personalized medicine approach, professionals consider these factors to optimize treatment prescriptions, avoiding unnecessary adverse effects on the patient and, indirectly, on the environment (Witte & Müller, 2025). For a wider personalized medicine practice, better strategies to educate and train healthcare professionals must be developed (Goetz & Schork, 2018). For instance, in North America, it is reported that the growing tendency of personalized nutrition to tailor nutraceutical regimens to genetic, metabolic, and lifestyle data, namely diet and symptoms, has been enhancing adherence to nutraceutical products (Market Data Forecast, 2025). The growing demand for nutraceutical products, together with increasing scientific evidence of their potential to prevent and manage numerous health conditions, has created the need for health professionals to be aware of both the advantages and limitations of nutraceutical use, as well as their potential synergistic effects with medications in improving metabolic parameters (Caponio et al., 2022).

Healthcare providers are also shifting into a more participatory approach. While in the past, there was an orientation towards healthcare providers, the spotlight is becoming directed to consumers. The increasing availability of interactive information that is accessible to the masses, through the media and internet, and the rising awareness about preventive care encourage consumers to assume more responsibility for their health (Adams, 2020). This shift in the healthcare industry aligns with the fifth industrial revolution. Within the Healthcare 5.0 paradigm, the importance of patient empowerment is recognized, enabling individuals to take an active role in their own care, participate in decision-making processes, and promote a cost-effective approach aimed at ensuring universal access to healthcare services (Raisa et al., 2025). Given the growing consumer interest in more natural and plant-based healthcare solutions, the trend toward a participatory approach has been driving the increase in the nutraceutical market's value.

Currently, we are assisting a set of emerging approaches to healthcare management, towards preventive and more sustainable healthcare strategies, rather than treatment and disease management (Chopra et al., 2022). While this opens opportunities for the continued growth of the nutraceutical market, professionals and researchers in the field should ensure that this evolution advances towards a truly patient-centred approach, prioritizing human health preservation over short-term socioeconomic interests.

2.4 CHALLENGES

The nutraceutical sector faces multiple interrelated challenges that hinder its scientific validation, regulatory harmonization, and consumer trust. One of the major challenges faced by the nutraceutical sector lies in the lack of consensus on the nutraceutical definition. The definition introduced by Dr. Stephen DeFelice is currently not recognized by any regulatory authority in any country, posing an ongoing legislative challenge (Blaze, 2021; Chopra et al., 2022). The absence of a unified global definition and classification of nutraceuticals results in divergent regulatory standards, generating confusion among consumers and complicating international trade (Komala et al., 2023). Although initiatives such as the U.S. Food and Drug Administration's Good Manufacturing Practice (GMP) requirements have strengthened product quality oversight, many markets still lack effective systems to ensure safety, efficacy, and authenticity (Blaze, 2021). The nutraceutical sector continues to suffer from inadequate quality control, which contributes to cases of adulteration and contamination during production (Chopra et al., 2022). Furthermore, divergent national legislations often result in inconsistent product classifications and varying interpretations of what constitutes a nutraceutical, which can lead to consumer confusion, market fragmentation, and even potential misuse (Ivanovska et al., 2020).

A uniformly developed regulation system is important not only for the identification and classification of nutraceuticals at an international level, but also to ensure a reliable assessment of quality, efficacy, mechanism of action, and safety, potentially benefiting both the consumers and the industry (Ivanovska et al., 2020). Such a framework would also help minimize cases of adulteration and contamination in nutraceutical products, which remain a critical concern. Nutraceuticals can also pose toxic risks to consumers if contaminated with heavy metals, metalloids, mycotoxins, allergens, fertilizers, pesticides, or residues of toxic plants during production (Chopra et al., 2022). Adulteration and various other forms of impurity may go undetected simply due to the near absence of specific and standardized quality control protocols (Shilpi, 2022).

Another concern for the nutraceutical industry is the lack of robust scientific evidence supporting many of the health claims attributed to these products. Most medicinal or nutritional claims remain unsubstantiated due to insufficient knowledge of their mechanisms of action and the limited number of randomized clinical trials confirming their efficacy and safety in specific pathological conditions (Chopra et al., 2022; Sawicka et al., 2020). Although numerous in-vitro studies demonstrate promising biological activity of food-derived bioactive compounds, human studies are still scarce. Key parameters such as bioavailability, metabolism, optimal dosage, and potential toxicity remain poorly understood, which limits the credibility and clinical integration of nutraceuticals (Sawicka et al., 2020).

In addition to the lack of clinical validation, safety concerns and potential drug–nutraceutical interactions represent a growing challenge. Adverse pharmacokinetic and pharmacodynamic interactions between nutraceuticals and conventional therapeutics can lead to severe complications, including cardiac arrhythmias, hyperthermia, cerebral haemorrhage, or hypertensive crises. This risk is exacerbated by the fact that patients frequently fail to inform healthcare professionals about their use of supplements (Chopra et al., 2022; Takefuji, 2025). Furthermore, excessive intake of certain nutraceuticals, such as antioxidants, may have paradoxical effects – acting as pro-oxidants or interfering with pharmacological treatments (Sawicka et al., 2020).

Financial and structural asymmetries between the pharmaceutical and nutraceutical industries also exacerbate the difficulties in quality and efficacy control. While conventional drug research receives substantial international funding, research into nutraceuticals remains significantly underfunded, despite their widespread consumption by millions of individuals worldwide (Chopra et al., 2022). Greater investment in clinical studies could enhance the scientific credibility of nutraceuticals, improve consumer confidence, and increase market value.

Finally, consumer education remains an essential yet underdeveloped area. Although the awareness of nutraceuticals is increasing, many consumers still lack accurate information regarding their appropriate use, potential benefits, and safety risks (Ahmad, 2024). Strengthening educational initiatives and fostering collaboration among industry stakeholders, healthcare professionals, and regulatory agencies are crucial steps to promote the safe, effective, and sustainable integration of nutraceuticals into healthcare systems (Komala et al., 2023; Takefuji, 2025).

2.5 OPPORTUNITIES

The nutraceutical market is undergoing a period of profound transformation, driven by increasing consumer demand for natural, sustainable, and scientifically validated health solutions. This evolving landscape presents significant opportunities for innovation across the entire value chain, from raw material sourcing to product development, marketing strategies, and regulatory frameworks.

From a technological standpoint, innovation in formulations and delivery systems represents one of the most promising avenues for market expansion. Advances in nanotechnology and smart materials are reshaping the sector by enabling controlled release, enhanced bioavailability, and targeted delivery of bioactive compounds. Such innovations not only optimize therapeutic outcomes but also minimize adverse effects. For example, nanoemulsions and emulsomes have demonstrated the ability to improve the absorption and stability of compounds such as curcumin and piperine, thereby enhancing their effectiveness in applications ranging from anti-inflammatory to anticancer therapies (Ashrafpour & Ashrafpour, 2025). In parallel, artificial-intelligence-driven approaches are emerging as powerful tools to accelerate the discovery and translation of natural bioactives, which may be applied to nutraceutical formulation. By enabling the computational prediction of structure, function, and natural sources, artificial intelligence supports a design-for-benefit strategy that can significantly shorten the path from concept to product, overcoming current limitations linked to the complexity and inefficiency of natural product research (Kussmann & Cunha, 2022).

In terms of product development, several key trends reflect evolving consumer priorities, including personalized nutrition, immune support, cognitive health, and healthy ageing (Anushi et al., 2023). Advances in genetics, biomarker analysis, and digital health technologies now enable the creation of nutraceutical products tailored to individual genetic, metabolic, and lifestyle profiles. This evolution marks a paradigm shift from generalized supplementation to personalized prevention, reinforcing the role of nutraceuticals in integrative and precision healthcare (Anushi et al., 2023).

Sustainability and ethical sourcing have also become central pillars for innovation in the nutraceutical sector. The future of nutraceuticals will increasingly depend on regenerative agriculture, agroforestry, and circular economy practices, which minimize environmental impact and promote social responsibility throughout the supply chain (Anushi et al., 2023). Improving the health benefit properties of fresh produce represents an important opportunity to enhance the value of fresh produce and to expand access to health-oriented markets. Monitoring plant responses to environmental stressors, such as water deficit, temperature extremes, oxygen deficiency, and air pollution, is essential, as these factors can restrict the plant's ability to fully express its genetic potential, ultimately impacting the concentration and stability of nutraceutical compounds. Understanding and managing these responses on an agricultural production management level can contribute to the later production of higher quality, more nutraceutical-enriched products. Achieving this objective, however, requires the development and implementation of technologies capable of ensuring the consistent delivery of high-quality products enriched in the desired bioactive compounds (Cisneros-Zevallos, 2003).

One of the most promising approaches involves the valorisation of agro-industrial by-products and food waste, such as fruit peels, seeds, and pulps, which are often rich in polyphenols and other health-promoting phytochemicals (Sawicka et al., 2020; Šubarić & Jokić, 2022; Čižmárová et al., 2023). These by-products, traditionally regarded as waste, can be reintroduced into the food system as functional ingredients with both nutritional and environmental benefits. However, achieving this goal requires the development of eco-friendly and high-efficiency extraction techniques, which preserve the integrity and stability of bioactive compounds (Pai et al., 2022; Sawicka et al., 2020).

At the same time, marketing strategies play a decisive role in strengthening consumer trust and market differentiation. Effective strategies should emphasize the health benefits, safety, quality, and transparency of nutraceutical products, responding to the growing demand for clean-label and sustainably sourced formulations (Anushi et al., 2023). Brands that communicate evidence-based claims and prioritize transparency in labelling and ingredient sourcing are better positioned to build lasting consumer loyalty and credibility.

Finally, the sustainable growth of the nutraceutical sector depends on cross-sector collaboration among industry stakeholders, research institutions, and regulatory bodies (Ahmad, 2024). Such collaboration is essential to foster technological innovation, ensure regulatory coherence, and promote evidence-based product development. Integrating scientific research, technological advancement, and ethical governance will be key to unlocking the full potential of nutraceuticals as safe, effective, and sustainable contributors to global health and well-being.

2.6 CONCLUSIONS

Nature provides an immense reservoir of bioactive compounds that can be harnessed to promote human and animal health while supporting food system sustainability (Kussmann & Cunha, 2022). Extensive evidence indicates that phytochemicals play a preventive role against a wide range of diseases, including cancer, cardiovascular disorders, diabetes mellitus, hypertension, inflammatory conditions, and mental health disorders. Vegetables, in particular, serve as key precursors for a new generation of diet and lifestyle-based interventions aimed at disease prevention and health promotion (Carrillo et al., 2021). Consequently, vegetable-derived phyto-nutraceuticals represent a promising avenue toward a more cost-effective and nature-based healthcare model.

The global nutraceutical market has been expanding rapidly, driven by the ageing population, the rise of e-commerce, increasing healthcare expenditures, the growing prevalence of lifestyle-related diseases, and heightened health awareness among consumers. Despite its notable growth potential, without adequate funding for research, the nutraceutical industry will remain constrained. In the absence of robust scientific evidence, resistance from healthcare professionals toward nutraceutical-based interventions is likely to persist. Nevertheless, the market continues to expand significantly, even in the absence of comprehensive regulation. Therefore, both consumers and healthcare providers must be well informed about the potential benefits and limitations of integrating these products into therapeutic protocols. Establishing robust regulatory frameworks will be essential to minimize cases of adulteration or contamination, which can have serious health consequences. Beyond national efforts, regulators should work toward a coordinated and harmonized international approach to facilitate the safe globalization of nutraceuticals, prevent consumer confusion and misuse, and ensure product quality and consistency across markets. Developing a unified system for the identification and classification of these products would further allow for reliable assessment of their quality, efficacy, mechanisms of action, and safety, safeguarding consumers while fostering the sustainable growth and credibility of the nutraceutical industry. From a business perspective, the findings presented here underscore the importance of educational initiatives to adopt new health management approaches and technology use to benefit consumers. The industry is shifting toward a patient-centric model, as consumers become increasingly informed about their health and are more demanding in seeking natural solutions. Consequently, healthcare professionals should ground their practice in robust evidence, adhere to the precautionary principle, and adopt a mindset of continuous learning to develop and apply innovative methods, which address the root causes of medical conditions rather than merely suppress symptoms. As consumer awareness of the benefits and risks of nutraceuticals grows, businesses that prioritize transparency and implement strong quality assurance mechanisms, from the ethical sourcing of phytonutrients based on agroecological practices to the use of advanced technologies, such as nanotechnology and artificial intelligence, combined with honest marketing strategies, can achieve positive differentiation and build a lasting competitive advantage. Based on the evidence here presented, collaboration among patients, healthcare providers, governments, and researchers is fundamental to developing safer, more ethical, and effective solutions, which prevent and manage medical conditions while promoting overall health outcomes.

ACKNOWLEDGEMENTS

This work was supported by the Fundação para a Ciência e Tecnologia (FCT), Portugal, with Grant reference: 2024.02723.BDANA, the NECE - Research Centre for Business Sciences funded by the Multiannual Funding Program of R&D Centers of FCT, under Grant number UID/04630/2025, DOI: 10.54499/UID/04630/2025 and the Research Unit of Centre for Mechanical and Aerospace Science and Technologies (CMAST-UBI), through the Projects references UID/00151/2025 (https://doi.org/10.54499/UID/00151/2025), UID/PRR/00151/2025 (https://doi.org/10.54499/UID/PRR/00151/2025) and UID/PRR2/00151/2025 (https://doi.org/10.54499/UID/PRR2/00151/2025), funded by FCT, IP/MECI. CMAST team also acknowledges the support provided by LITecS - Laboratory of Innovation and Technologies for Sustainability (www.litecs.ubi.pt).

This research is partly within the results of VITAL project, Reference: CENTRO2030-FEDER-02584000, funded by CENTRO2030, PORTUGAL2030, and the European Union. "The Funds Closer to You."

REFERENCES

Adams, A. L. (2020). Consumer health information. *Public Services Quarterly*, *16*(4), 234–241. https://doi.org/10.1080/15228959.2020.1818665

Ahmad, M., Mailk, A. A., Hussain, K., Malik, G., Wani, S. A., Rather, A. M., & Magaray, M. (2020). Nutraceutical and bioactive healthy compounds in vegetable crops. *International Journal of Chemical Studies*, *8*(4), 3078–3086. https://doi.org/10.22271/chemi.2020.v8.i4ak.10120

Ahmad, S. (2024). India's growing demand for nutraceuticals. *Asian Journal of Pharmaceutical Research and Development*, *12*(1), 126–132. https://doi.org/10.22270/ajprd.v12i1.1368

Allied Market Research (2022). *Nutraceuticals Market* [online]. Available at: https://www.alliedmarketresearch.com/press-release/nutraceuticals-market.html [Accessed 28 October 2025].

Alum, E. U., & Ugwu, O. P. C. (2023). Beyond nutrients: Exploring the potential of phytochemicals for human health. *IAA Journal of Applied Sciences*, *10*(3), 1–7. https://doi.org/10.59298/iaajas/2023/4.1.3211

Anushi, A., Kumar, D., & Ranjan, A. R. (2023). Exploring the growing interest in the medicinal properties of fruits and the development of nutraceuticals. *Journal of Plant Biota*, *3*(1), 13–17. https://doi.org/10.51470/jpb.2024.3.1.13

Arcusa, R., Carrillo, J. Á., Xandri-Martínez, R., Cerdá, B., Villano, D., Marhuenda, J., & Zafrilla, M. P. (2021). Effects of a fruit and vegetable-based nutraceutical on biomarkers of inflammation and oxidative status in the plasma of a healthy population: A placebo-controlled, double-blind, and randomized clinical trial. *Molecules*, *26*(12). https://doi.org/10.3390/molecules26123604

Ashrafpour, S., & Ashrafpour, M. (2025). The double-edged sword of nutraceuticals: comprehensive review of protective agents and their hidden risks. *Frontiers in Nutrition, 12*(March). https://doi.org/10.3389/fnut.2025.1524627

Blaze, J. (2021). A comparison of current regulatory frameworks for nutraceuticals in Australia, Canada, Japan, and the United States. *Innovations in Pharmacy, 12*(2), 8. https://doi.org/10.24926/iip.v12i2.3694

Caponio, G. R., Lippolis, T., Tutino, V., Gigante, I., De Nunzio, V., Milella, R. A., Gasparro, M., & Notarnicola, M. (2022). Nutraceuticals: Focus on anti-inflammatory, anti-cancer, antioxidant properties in the gastrointestinal tract. *Antioxidants, 11*(7), 1–15. https://doi.org/10.3390/antiox11071274

Carrillo, J. Á., Arcusa, R., Zafrilla, M. P., & Marhuenda, J. (2021). Effects of fruit and vegetable-based nutraceutical on cognitive function in a healthy population: Placebo-controlled, double-blind, and randomized clinical trial. *Antioxidants, 10*(1), 1–12. https://doi.org/10.3390/antiox10010116

Chaudhary, M., & Garg, A. P. (2023). Bioactive compounds from plants and animals: A review. *International Journal of Advanced Biochemistry Research, 7*(1), 105–117. https://doi.org/10.33545/26174693.2023.v7.i1b.175

Chen, Y., Liu, X., Ma, Y., He, J., He, Y., Zheng, C., Gao, W., & Ma, C. (2023). Variability analysis and the conservation capacity of soil water storage under different vegetation types in arid regions. *Catena, 230*(May 2022). https://doi.org/10.1016/j.catena.2023.107269

Chopra, A. S., Lordan, R., Horbańczuk, O. K., Atanasov, A. G., Chopra, I., Horbańczuk, J. O., Jóźwik, A., Huang, L., Pirgozliev, V., Banach, M., Battino, M., & Arkells, N. (2022). The current use and evolving landscape of nutraceuticals. *Pharmacological Research, 175*(November 2021). https://doi.org/10.1016/j.phrs.2021.106001

Cisneros-Zevallos, L. (2003). The use of controlled postharvest abiotic stresses as a tool for enhancing the nutraceutical content and adding-value. *Concise Reviews and Hypotheses in Food Science, 68*(5), 1560–1565.

Cisneros-Zevallos, L. (2021). The power of plants: How fruit and vegetables work as a source of nutraceuticals and supplements. *International Journal of Food Sciences and Nutrition, 72*(5), 660–664. https://doi.org/10.1080/09637486.2020.1852194

Čižmárová, B., Hubková, B., Tomečková, V., & Birková, A. (2023). Flavonoids as promising natural compounds in the prevention and treatment of selected skin diseases. *International Journal of Molecular Sciences, 24*(7). https://doi.org/10.3390/ijms24076324

Consainsights (2024). *Phytonutrients Market Report* [online]. Available at: https://www.consainsights.com/reports/phytonutrients-market [Accessed 28 October 2025].

Delgado, Y., Cassé, C., Ferrer-Acosta, Y., Suárez-Arroyo, I. J., Rodríguez-Zayas, J., Torres, A.-R., Torres-Martínez, Z., Pérez, D., González, M., Velázquez-Aponte, R. A., Andino, J., Correa-Rodríguez, C., Franco, J. C., Milán, W., Rosario, G., Velázquez, E., Vega, J., Colón, J., & Batista, C. (2021). Biomedical effects of popular phytonutrients and their phytochemicals: A comprehensive review. https://doi.org/10.20944/PREPRINTS202106.0371.V1

Dincheva, I., Badjakov, I., & Galunska, B. (2025). New insights in the research on bioactive compounds from plant origins with nutraceutical and pharmaceutical potential II. *Plants, 14*(4). https://doi.org/10.3390/plants14040500

Femi-Adepoju, A. G., Olufemi Adepoju, A., Jalloh, A., & Faiyah Bah, A. R. (2023). From natural compounds to disease treatment and control: A review of nutraceuticals vegetable crops and their roles in disease control. *Scientific Review, 9*(91), 1–9. https://doi.org/10.32861/sr.91.1.9

Fortune Business Insights (2025). *Nutraceuticals Market Report* [online]. Available at: https://www.fortunebusinessinsights.com/nutraceuticals-market-102530 [Accessed 28 October 2025].

Frank, J., Fukagawa, N. K., Bilia, A. R., Johnson, E. J., Kwon, O., Prakash, V., Miyazawa, T., Clifford, M. N., Kay, C. D., Crozier, A., Erdman, J. W., Shao, A., & Williamson, G. (2020). Terms and nomenclature used for plant-derived components in nutrition and related research: Efforts toward harmonization. *Nutrition Reviews, 78*(6), 451–458. https://doi.org/10.1093/nutrit/nuz081

Goetz, L. H., & Schork, N. J. (2018). Personalized medicine: motivation, challenges, and progress. *Fertility and Sterility, 109*(6), 952–963. https://doi.org/10.1016/j.fertnstert.2018.05.006

González, M. J., Miranda-Massari, J. R., Mora, E. M., Cruzado, N. A., Jiménez, I., Rosa, M., Matos Vera, M. I., Santiago, C., Román-Eyxarch, M. I., Rodríguez, J. R., Pérez Cortés, C., Riordan, N. H., Riordan, H. D., & Ricart, C. M. (2000). Integrative medicine: A paradigm shift in medical education and practice. *Puerto Rico Health Sciences Journal, 19*(4), 389–392.

Grand View Research (2024). *Nutraceuticals Market (2025–2030)* [online]. Available at: https://www.grandviewresearch.com/industry-analysis/nutraceuticals-market [Accessed 28 October 2025].

Haider, R. (2023). Understanding the concept of "nutraceutical". *Global Journal of Pharmacy & Pharmaceutical Sciences, 11*(1). https://doi.org/10.19080/gjpps.2023.11.555805

Hoti, G., Matencio, A., Pedrazzo, A. R., Cecone, C., Appleton, S. L., Monfared, Y. K., Caldera, F., & Trotta, F. (2022). Nutraceutical concepts and dextrin-based delivery systems. *International Journal of Molecular Sciences*, *23*(8). https://doi.org/10.3390/ijms23084102

Husen, A. (2021). *Traditional Herbal Therapy for the Human Immune System*. CRC Press. https://doi.org/10.1201/9781003137955

Husen, A. (2022). *Herbs, Shrubs and Trees of Potential Medicinal Benefits*. CRC Press. https://doi.org/10.1201/9781003205067

Husen, A. (2023a). *Exploring Poisonous Plants: Medicinal Values, Toxicity Response and Other Potential Uses*. CRC Press. https://doi.org/10.1201/b23017

Husen, A. (2023b). *Antidiabetic Medicinal Plants and Herbal Treatments*. CRC Press. https://doi.org/10.1201/b23347

Husen, A. (2024a). *Antimalarial Medicinal Plants*. Taylor & Francis Group. https://doi.org/10.1201/9781003378396

Husen, A. (2024b). *Medicinal Spices and Condiment Crops*. Taylor & Francis Group. https://doi.org/10.1201/9781003387046

Husen, A., & Iqbal, M. (2021). Plant-based potential nutraceuticals for improving the human immune system. In: *Traditional Herbal Therapy for the Human Immune System* (Ed. Husen, A.). CRC Press, pp. 1–11. https://doi.org/10.1201/9781003137955-1

Husen, A., & Pant, M. (2024a). *Exploring Medicinal Orchids*. Taylor & Francis Group. https://doi.org/10.1201/b23375

Husen, A., & Pant, M. (2024b). *Tissue Culture Techniques and Medicinal Plants: Enhancing Propagation and Production*. Taylor & Francis Group. https://doi.org/10.1201/b23374

Ivanovska, T., Zhivikj, Z., & Petrushevska-Tozi, L. (2020). Regulatory status of nutraceuticals – is it rationale to rethink? *Macedonian Pharmaceutical Bulletin*, *66*(3), 15–16. https://doi.org/10.33320/maced.pharm.bull.2020.66.03.007

Kan, J., Wu, F., Wang, F., Zheng, J., Cheng, J., Li, Y., Yang, Y., & Du, J. (2022). Phytonutrients: Sources, bioavailability, interaction with gut microbiota, and their impacts on human health. *Frontiers in Nutrition*, *9*. https://doi.org/10.3389/fnut.2022.960309

Khan, S., Solanki, T. V., Bakal, R. L., & Hatwar, P. R. (2025). Nutraceuticals and their benefits for health: A review. *GSC Biological and Pharmaceutical Sciences*, *31*(3), 171–178. https://doi.org/10.30574/gscbps.2025.31.3.0224

Kohsaka, R., Fujihira, Y., & Uchiyama, Y. (2019). Biomimetics for business? Industry perceptions and patent application. *Journal of Science and Technology Policy Management*, *10*(3), 597–616. https://doi.org/10.1108/JSTPM-05-2018-0052

Komala, M. G., Ong, S. G., Qadri, M. U., Elshafie, L. M., Pollock, C. A., & Saad, S. (2023). Investigating the regulatory process, safety, efficacy and product transparency for nutraceuticals in the USA, Europe and Australia. *Foods*, *12*(2), 1–13.

Korsapati, A., Korsapati, H. R., & Peddireddy, V. (2025). Neuroprotective phytochemicals in functional foods: Mechanistic insights, translational challenges, and nano-enabled delivery strategies. *Trends in Food Science and Technology*, *165*(October), 105355. https://doi.org/10.1016/j.tifs.2025.105355

Kussmann, M., & Cunha, D. H. A. (2022). Nature has the answers: Discovering and validating natural bioactives for human health. *Bioactive Compounds in Health and Disease*, *5*(10), 222–234. https://doi.org/10.31989/bchd.v5i10.1000

Lalitha, A. (2024). Phytochemicals and health benefits. *Annals of Geriatric Education and Medical Sciences*, *11*(1), 29–31. https://doi.org/10.18231/j.agems.2024.007

Mahima, Verma, A. K., Tiwari, R., Karthik, K., Chakraborty, S., Deb, R., & Dhama, K. (2013). Nutraceuticals from fruits and vegetables at a glance: A review. *Journal of Biological Sciences*, *13*(2), 38–47. https://doi.org/10.3923/jbs.2013.38.47

Market Data Forecast (2026). *Nutraceuticals Market Report* [online]. Available at: https://www.marketdataforecast.com/market-reports/global-nutraceuticals-market [Accessed 28 October 2025].

Mordor Intelligence (2025). *Vegetable Extracts Market* [online]. Available at: https://www.mordorintelligence.com/industry-reports/vegetable-extracts-market [Accessed 28 October 2025].

Pai, S., Hebbar, A., & Selvaraj, S. (2022). A critical look at challenges and future scopes of bioactive compounds and their incorporations in the food, energy, and pharmaceutical sector. *Environmental Science and Pollution Research*, *29*(24), 35518–35541. https://doi.org/10.1007/s11356-022-19423-4

Pant, M., & Husen, A. (2025). *Exploring Medicinal Bamboos*. Taylor & Francis Group. https://doi.org/10.1201/9781003485933

Pant, M., Sohrab, S. S., & Husen, A. (2025). *Genetic Improvement and Conservation Practice of Medicinal Plants*. Taylor & Francis Group. https://doi.org/10.1201/9781003482659

Pathak, N., & Kumar, S. (2022). Nutraceuticals and their role in health promotion. In A. Sharma, G. Malik, & S. Kumar (Eds.), *Medicinal Plants: Ethnomedicine, Pharmacognosy and Therapeutic Values*, pp. 170–181. https://doi.org/10.31995/book.ab220-a22.chapter10

Petrovska, B. B. (2012). Historical review of medicinal plants' usage. *Pharmacognosy Reviews*, *6*(11), 1–5. https://doi.org/10.4103/0973-7847.95849

Polak, A. M., Karska, K., & Drozd, M. (2025). The discovery of medicines: Drug testing on humans and the development of medical ethics before the 20th century. *Drug Discovery Today*, *30*(8), 104418. https://doi.org/10.1016/j.drudis.2025.104418

Prabhakar, B. N. (2022). Vegetables as nutraceuticals and future plant drugs. *Annals of Phytomedicine: An International Journal*, *11*(1), 1–6. https://doi.org/10.54085/ap.2022.11.1.1

Raisa, J. F., Rahman, M. S., Mahmud, I., Kaiser, M. S., & Han, D. S. (2025). Transition toward Healthcare 5.0: Impact of COVID-19 in the healthcare industry. *ICT Express*, *11*(3), 371–389. https://doi.org/10.1016/j.icte.2025.04.002

Redondo-Blanco, S., Fernández, J., López-Ibáñez, S., Miguélez, E. M., Villar, C. J., & Lombó, F. (2020). Plant phytochemicals in food preservation: Antifungal bioactivity: A review. *Journal of Food Protection*, *83*(1), 163–171. https://doi.org/10.4315/0362-028X.JFP-19-163

Saleh, H. A., Yousef, M. H., & Abdelnaser, A. (2021). The anti-inflammatory properties of phytochemicals and their effects on epigenetic mechanisms involved in TLR4 / NF-k B-mediated inflammation. *Frontiers in Immunology*, *12*(March), 1–29. https://doi.org/10.3389/fimmu.2021.606069

Sawicka, B. H., Ziarati, P., Krochmal-Marczak, B., & Skiba, D. (2020). Nutraceuticals in food and pharmacy. A review. *Agronomy Science*, *74*(4), 7–31. https://doi.org/10.24326/as.2019.4.1

Seetharaman, M., Krishnan, G., & Schneider, R. H. (2021). The future of medicine: Frontiers in integrative health and medicine. *Medicina (Lithuania)*, *57*(12), 1–7. https://doi.org/10.3390/medicina57121303

Shahidi, F. (2012). Nutraceuticals, functional foods and dietary supplements in health and disease. *Journal of Food and Drug Analysis*, *20*(Suppl.1), 226–230. https://doi.org/10.38212/2224-6614.2144

Sharma, M., Vidhya C. S., Ojha, K., Yashwanth B. S., Singh, B., Gupta, S., & Pandey, S. K. (2024). The role of functional foods and nutraceuticals in disease prevention and health promotion. *European Journal of Nutrition & Food Safety*, *16*(2), 61–83. https://doi.org/10.9734/ejnfs/2024/v16i21388

Sharma, S. K., Aijaz, M., Rawat, S. L., Ahmad, M., Devi, S., Mehra, M., Kaur, J., & Goswami, M. (2023). Beyond pharmaceuticals: A review on integrative medicine and complementary therapies in modern healthcare. *European Chemical Bulletin*, *12*(S3), 6866–6873. https://doi.org/10.31838/ecb/2023.12.s3.769

Shilpi, S. (2022). Nutraceuticals: Use of additives, applications and regulatory aspects. *Advance Pharmaceutical Journal*, *7*(2), 32–37. https://doi.org/10.31024/apj.2022.7.2.1

Shrivastava, P., & Ali, S. S. A. S. S. (2025). Analyzing the impact of aging populations on healthcare system sustainability. *Journal of Neonatal Surgery*, *14*(1), 393–397. https://doi.org/10.52783/jns.v14.1554

Šubarić, D., & Jokić, S. (2022). Food industry by-products as a sources of phytochemical compounds. *Foods*, *11*(12), 10–11. https://doi.org/10.3390/foods11121724

Takefuji, Y. (2025). Nutraceuticals: A promising, yet unregulated Frontier in Healthcare. *European Journal of Clinical Nutrition*, *79*(5), 413–418. https://doi.org/10.1038/s41430-024-01557-y

Townsend, J. R., Kirby, T. O., Sapp, P. A., Gonzalez, A. M., Marshall, T. M., & Esposito, R. (2023). Nutrient synergy: Definition, evidence, and future directions. *Frontiers in Nutrition*, *10*(October). https://doi.org/10.3389/fnut.2023.1279925

Tsiaka, T., Kritsi, E., Tsiantas, K., Christodoulou, P., Sinanoglou, V. J., & Zoumpoulakis, P. (2022). Design and development of novel nutraceuticals: Current trends and methodologies. *Nutraceuticals*, *2*(2), 71–90. https://doi.org/10.3390/nutraceuticals2020006

Wang, X., Qi, Y., & Zheng, H. (2022). Dietary polyphenol, gut microbiota, and health benefits. *Antioxidants*, *11*(6), 1212. https://doi.org/10.3390/antiox11061212.

WCED. (1987). The Brundtland report: "Our common future." *Environmental Ethics*, *4*(1), 17–25. https://doi.org/10.1080/07488008808408783

Wegener, G. (2014). Let food be thy medicine, and medicine be thy food: Hippocrates revisited. *Acta Neuropsychiatrica*, *26*(1), 1–3. https://doi.org/10.1017/neu.2014.3

Witte, K., & Müller, M. (2025). Sustainable pharmacy – a guiding principle. *Sustainable Chemistry and Pharmacy*, *44*(December 2024), 101897. https://doi.org/10.1016/j.scp.2024.101897

Wynendaele, E., Furman, C., Wielgomas, B., Larsson, P., Hak, E., Block, T., Van Calenbergh, S., Willand, N., Markuszewski, M., Odell, L. R., Poelarends, G. J., & De Spiegeleer, B. (2021). Sustainability in drug discovery. *Medicine in Drug Discovery*, *12*(August), 100107. https://doi.org/10.1016/j.medidd.2021.100107

Yan, Z., Feng, Z., Jiao, Z., Chen, C., Wang, G., & Feng, D. (2021). The severity of adverse drug reactions and their influencing factors based on the ADR monitoring center of Henan Province. *Scientific Reports*, *11*(1), 1–9. https://doi.org/10.1038/s41598-021-99908-3

3 Isolation and Characterization of Vegetable-Based Phyto-Nutraceuticals

Sushmita Mena, Mohd Adnan, Mitesh Patel, Vijay Jagdish Upadhye, and Malvi Surti

3.1 INTRODUCTION

The word 'phyto-nutraceuticals', which is a combination of the term 'phytochemicals' and 'nutraceuticals', describes naturally occurring bioactive substances that are present in plants and offer health advantages beyond simple nutrition (Kalra, 2003). By utilizing the therapeutic potential of molecules produced from plants, phyto-nutraceuticals, in contrast to manufactured medications, provide a preventive approach to health management (Shahidi, 2009; Husen & Iqbal, 2021). In addition to adding to the diet's nutritious content, these substances are essential for reducing the chance of developing chronic illnesses such diabetes, cancer, heart disease, neurological problems, and so on (Kris-Etherton et al., 2002; Liu, 2004; Husen, 2021, 2022, 2023a,2023b, 2024a,2024b; Husen & Pant, 2024a,2024b; Pant et al., 2025; Pant & Husen, 2025). Vegetables play a major part in the wide range of plant sources because of their varied phytochemical profiles, high consumption rates, and widespread availability in cuisines around the world (Slavin & Lloyd, 2012; Vallejo et al., 2003). Numerous structurally varied bioactive substances, such as polyphenols, flavonoids, carotenoids, glucosinolates, alkaloids, and OSCs, can be found in abundance in vegetables. These phytochemicals have been demonstrated to exhibit substantial bioactivity in human physiology, despite being produced as a component of the plant's defence systems against environmental stresses (Rodríguez-Negrete et al., 2024). For instance the glucosinolates found in cruciferous vegetables like kale and broccoli are converted into isothiocyanates, which have been shown to have anticancer effects (Soundararajan & Kim, 2018). Likewise, leafy greens and vibrantly coloured vegetables such as tomatoes, spinach, and carrots are rich in carotenoids and polyphenols, which have anti-inflammatory and antioxidant properties (Sharma et al., 2021). These vegetables are widely consumed, which makes them perfect candidates for the creation of nutraceutical products and functional foods that improve public health (Gul et al., 2016).

Several epidemiological and clinical investigations have provided substantial evidence for the therapeutic potential of vegetable-based phyto-nutraceuticals, consistently linking a high vegetable consumption to a lower incidence of non-communicable diseases (NCDs) (Smith et al., 2022; Woodside et al., 2023). These substances work via a number of biological processes such as regulating cellular signalling involved in metabolism and apoptosis, enhancing immunological responses, reducing oxidative stress, and modulating inflammatory pathways (Gariballa et al., 2023; Hosseini et al., 2018). However, the extraction process, level of purity, and structural integrity of these compounds all have a significant impact on their bio-efficacy, making exact methods for their isolation and characterization necessary (A. Kumar et al., 2023; Mungwari et al., 2025).

The objective of this chapter is to present a thorough summary of the scientific methods and techniques currently used in the separation and characterization of phyto-nutraceuticals obtained from vegetables. It looks at both established and cutting-edge green extraction methods as well

DOI: 10.1201/9781003606710-3

as sophisticated analytical methods for identifying and quantifying these bioactive components (Darwin et al., 2025).

3.2 OVERVIEW OF VEGETABLE-BASED PHYTO-NUTRACEUTICALS

A broad family of naturally occurring bioactive substances, vegetable-based phyto-nutraceuticals are essential for preventative nutrition and have a major positive impact on health. Numerous classes of these substances exist, each with distinct pharmacological characteristics, including polyphenols, glucosinolates, carotenoids, OSCs, and saponins (Rodríguez-Negrete et al., 2024) (Figure 3.1). For instance broccoli is a well-known source of glucosinolates, especially glucoraphanin, which yields SFN, an isothiocyanate with strong anticancer and detoxifying effects, when hydrolysed by enzymes (Kaiser et al., 2021). Carotenoids like lutein and zeaxanthin, as well as polyphenols like ferulic acid and flavonoids like kaempferol, are abundant in spinach and help to support its antioxidant potential

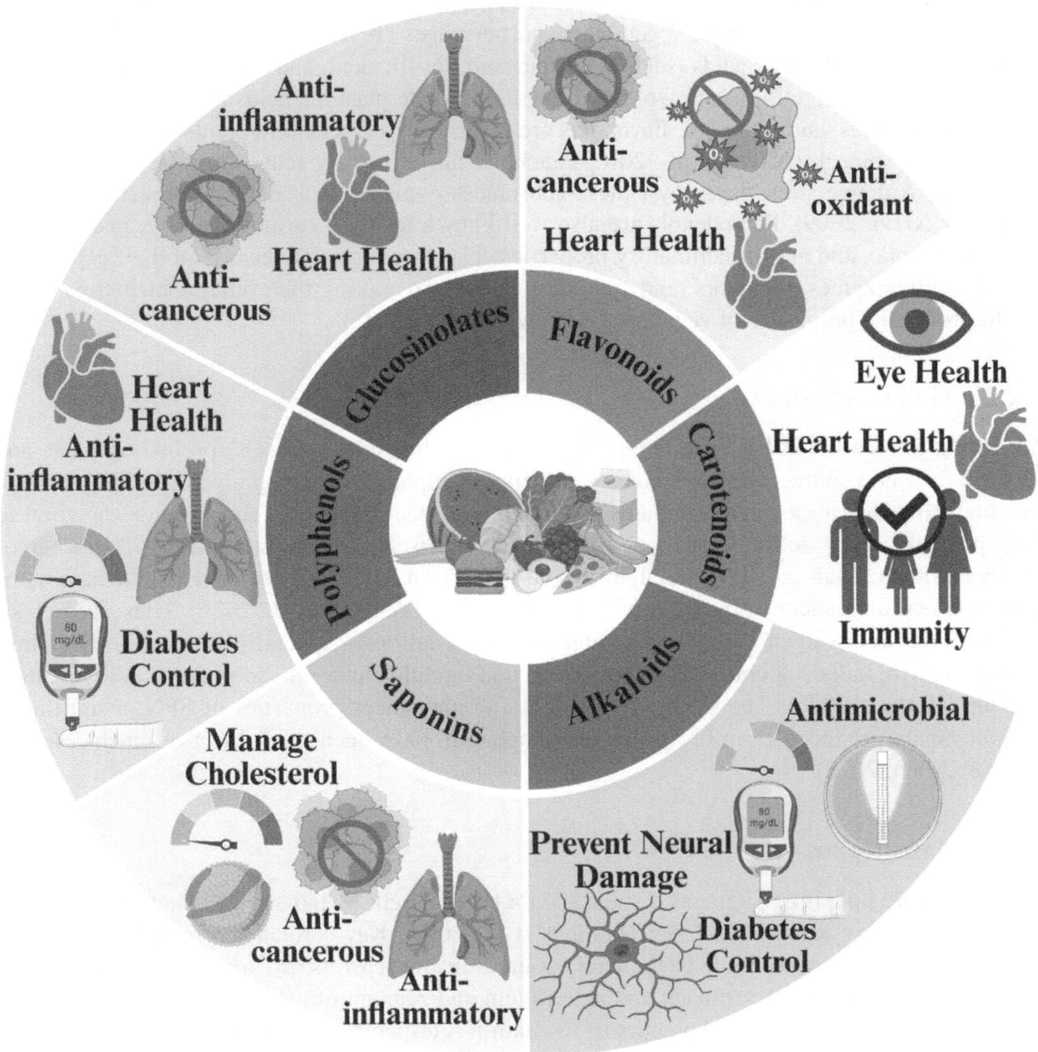

FIGURE 3.1 Several types of phyto-nutraceuticals and their health benefits.

Source: https://BioRender.com

and cardiovascular and visual health benefits (Pandjaitan et al., 2005). Lycopene, a carotenoid with a potent single oxygen-quenching ability and proven benefits in lowering the risk of atherosclerosis and prostate cancer, is abundant in tomatoes (Arab & Steck, 2000). OSCs, such as alliin and its derivative allicin, which are found in garlic, have antibacterial, antihypertensive, and lipid-lowering properties by modifying oxidative stress and enzyme activity (Bayan et al., 2014). These vegetables provide a complex matrix of phytochemicals that work together to boost immune function, lower inflammation, and guard against oxidative cellular damage in addition to providing vital vitamins, minerals, and dietary fibre (Kalogerakou & Antoniadou, 2024). By emphasizing the dual roles of vegetables as food and medicine, their incorporation into regular diets is consistent with the concepts of preventative healthcare and functional nutrition (Table 3.1).

3.2.1 Polyphenols

Polyphenols is a large family of secondary metabolites found in plants that are engaged in defence mechanisms against pathogens and UV radiation. They are naturally occurring substances that are mostly present in cereals, fruits, vegetables, and beverages (K. B. Pandey & Rizvi, 2009). They are separated into phenolic acids (such as caffeic acid, ferulic acid, and chlorogenic acid) and flavonoids (such as quercetin, kaempferol, and luteolin). While phenolic acids are frequently present in carrots, potatoes, and tomatoes, flavonoids are extensively dispersed in plants like onions, kale, broccoli, and spinach (Akyol et al., 2016; Panche et al., 2016). By scavenging free radicals and modifying oxidative stress pathways, these substances exhibit potent antioxidant activity (K. B. Pandey & Rizvi, 2009). Polyphenols are also well known for their cardioprotective, neuroprotective, antidiabetic, and anti-inflammatory properties. They are essential because of their capacity to interact with enzymes, receptors, and signalling pathways, making them vital contributors to the health-promoting properties of vegetables (Hedayati et al., 2023).

3.2.2 Glucosinolates

According to (Clarke, 2010), cruciferous vegetables like broccoli, Brussels sprouts, cabbage, and kale are the main source of glucosinolates, which are sulphur-containing glycosides. Myrosinase enzymatically hydrolyses glucosinolates during cellular disruption (e.g. chopping or chewing) to yield physiologically active substances like indoles and isothiocyanates. Products of glucosinolate breakdown, such as 3,3′-diindolylmethane (DIM) and indole-3-carbinol (I3C), have anti-inflammatory, anticancer, and cardioprotective properties (Baldelli et al., 2025). These hydrolysis products are well known for their chemopreventive qualities, especially their ability to limit tumour growth, induce apoptosis in cancer cells, and modulate phase I and phase II detoxification enzymes (Esteve, 2020). The strong antioxidant and anticancer properties of SFN, a significant isothiocyanate produced from broccoli's glucoraphanin, have been the subject of much research (Otoo & Allen, 2023).

3.2.3 Carotenoids

Carotenoids are lipophilic pigments responsible for the red, yellow, and orange colouration in many vegetables (Bartley & Scolnik, 1995; Hammer et al., 2009). Key carotenoids include β-carotene, lycopene, lutein, and zeaxanthin. Rich dietary sources include carrots (β-carotene), tomatoes (lycopene), and leafy greens like spinach and kale (lutein and zeaxanthin) (Elvira-Torales et al., 2019). Carotenoids are antioxidants and precursors to vitamin A (especially β-carotene). They also support the immune system, promote eye health, and protect against CVD and some types of cancer (Bohn et al., 2023). Their bioavailability is increased by dietary fats, and their health effects are frequently associated with their interaction with cellular oxidative and inflammatory pathways (Rocha et al., 2023).

TABLE 3.1
Comprehensive Table on Phyto-Nutraceuticals: An Insight into Biological Aspects

Phyto-Nutraceutical	Absorption site	Bioavailability (%)	Health Benefits	Mechanism	Stability	Storage Condition	Shelf-life	Key References
Sulforaphane	• Small intestine (mainly jejunum and ileum) • Some absorption in colon after microbial conversion	~40–70%	• Anticancer (induces phase II detox enzymes) • Anti-inflammatory • Neuroprotective • Cardioprotective • Antioxidant	• Inhibits NF-κB inflammatory pathway • Induces apoptosis in cancer cells	• Unstable at high temperatures and alkaline pH	• Store in cool, dark place • Keep extracts frozen or refrigerated	• Fresh cruciferous vegetables: few days to 1 week (refrigerated) • Pure sulforaphane powder/extracts: 6–12 months under proper storage	Kaiser et al. (2021), Otoo and Allen (2023)
Lycopene	• Small intestine (mostly duodenum and jejunum) via passive diffusion	~10–30% from raw sources; can increase up to 50% from processed (heated/oil-based) forms (e.g. tomato paste, juice)	• Antioxidant • Cardioprotective • Anticancer (especially prostate) • Skin protection • Eye and bone health	• Inhibits oxidative stress via singlet oxygen quenching • Modulates inflammatory pathways • Inhibits LDL oxidation	• Sensitive to heat, light, and oxygen • More stable in oil-rich matrices • Trans-lycopene is dominant in raw food; cis-isomers formed upon heating (more bioavailable)	• Store in airtight, light-resistant containers • Refrigeration improves stability for oils/ supplements	• Fresh tomatoes: ~1 week (refrigerated) • Tomato paste (sealed): Up to 12–18 months • Lycopene extract (encapsulated): 1–2 years if stored in cool, dry, dark conditions	Arab and Steck (2000), A. Kumar et al. (2025), Madia et al. (2021)
Lutein	• Small intestine (mainly jejunum)	~ 10–30 % depending on food matrix, fat content, and processing	• Protects against age-related macular degeneration (AMD) • Filters blue light • Antioxidant in retina	• Accumulates in macula lutea of the eye • Neutralizes reactive oxygen species (ROS) • Absorbs blue light	• Sensitive to heat, light, and oxygen • More stable when embedded in fat or microencapsulated forms	• Store in cool, dark, dry place • Use airtight, light-resistant packaging	• Pure form or encapsulated: ~1–2 years • Fortified foods: 6–12 months under proper storage	Mitra et al. (2021)

(Continued)

TABLE 3.1 *(Continued)*
Comprehensive Table on Phyto-Nutraceuticals: An Insight into Biological Aspects

Phyto-Nutraceutical	Absorption site	Bioavailability (%)	Health Benefits	Mechanism	Stability	Storage Condition	Shelf-life	Key References
Zeaxanthin	• Small intestine (mainly jejunum)	~15–40%, generally higher than lutein in some studies due to better micelle incorporation	• Similar to lutein: macular protection • Supports cognitive and skin health	• Localizes in central macula • Inhibits oxidative stress & lipid peroxidation • Modulates inflammation	• Similar profile as lutein; slightly more thermally stable than lutein	• Similar like lutein; refrigeration helps extend shelf-life in formulations	• Similar to lutein shelf-life: 1–2 years in extract/ supplement form	Bouyahya et al. (2021)
Allicin	• Stomach and small intestine (duodenum)	Low and variable (~5–15%), due to instability in acidic pH and rapid degradation	• Antibacterial and antifungal • Antioxidant • Cardioprotective • Anti-inflammatory • Potential anticancer effects	• Reacts with thiol groups in proteins → affects enzyme function • Inhibits lipid peroxidation • Stimulates antioxidant enzymes • Modulates nitric oxide (NO) pathways	• Highly unstable in heat, light, and acidic environments • Decomposes quickly into other sulphur compounds (e.g. ajoene, diallyl disulfide) • Not stable in cooked garlic	• Best stored in dry, cool, and dark conditions • Freshly crushed garlic should be consumed immediately for maximum allicin yield • Freeze-dried garlic may retain more stable precursors	• Pure allicin: Only a few hours unless stabilized • Garlic powder/ extracts: 6–12 months under airtight, light-protected conditions	Chan et al. (2013), Deng et al. (2023)
β-Carotene	• Small intestine (primarily duodenum and jejunum) through passive diffusion; enhanced by dietary fat and bile salts	Highly variable (5–50%); higher in processed or cooked vegetables with oil due to better solubilization and release from cell walls	• Provitamin A activity (converted to retinal) • Antioxidant (neutralizes free radicals) • Supports eye health, immune function, skin health	• Acts as an antioxidant and quencher of singlet oxygen • Serves as a precursor to vitamin A (retinol) • Regulates gene expression involved in immune response and epithelial integrity	• Sensitive to light, oxygen, and heat • Degrades rapidly when exposed to UV or air • More stable in oil-based or encapsulated formulations	• Store in cool, dark, and airtight containers • Refrigeration preferred for long-term storage	• Fresh vegetables: Few days to a week (refrigerated) • β-Carotene extract/ capsules: 12–24 months when properly sealed and stored	Ge et al. (2025)

Quercetin	• Mainly absorbed in the small intestine (jejunum and ileum); some absorption occurs in the colon after microbial metabolism	Low (2–20%) due to poor water solubility and extensive first-pass metabolism; enhanced by co-administration with fat or other flavonoids (e.g. vitamin C)	• Antioxidant (scavenges free radicals) • Anti-inflammatory • Antiviral and antibacterial • Supports cardiovascular and immune health • May protect against neurodegenerative diseases	• Inhibits pro-inflammatory enzymes • Chelates metal ions • Reduces oxidative stress and inhibits apoptosis in healthy cells	• Sensitive to pH, heat, and light • Degrades in alkaline environments and during extended cooking • More stable in dry, encapsulated, or complexed form	• Store in cool, dry, and dark conditions • Protect from moisture, light, and oxygen • Use airtight containers	• Pure quercetin powder or capsules: ~1–2 years under optimal storage	Aghababaei and Hadidi (2023)
Betalains	• Small intestine	Low to moderate (~2–12%); depends on structure, food matrix, and gastrointestinal conditions	• Antioxidant and anti-inflammatory • Hepatoprotective • May reduce oxidative stress and DNA damage • Supports cardiovascular health • Potential anticancer activity	• Scavenges reactive oxygen/nitrogen species • Modulates inflammatory enzymes • Inhibits lipid peroxidation	• Highly sensitive to heat, light, oxygen, and alkaline pH • Degradation leads to loss of colour and bioactivity	• Store at low temperatures (<4°C), protected from light and air • Freeze-drying enhances stability in extracts and powders	• Fresh sources (e.g. beet juice): ~2–5 days refrigerated • Freeze-dried powder/extract: up to 12–18 months if stored in dark, airtight containers	Sadowska-Bartosz and Bartosz (2021)
Polyphenols	• Primarily in the small intestine; a significant portion reaches the colon where it is metabolized by gut microbiota into bioactive metabolites	Highly variable (5–40%), depending on the polyphenol type, food matrix, and gut microbiota activity	• Potent antioxidant and anti-inflammatory activity • Supports cardiovascular health • May protect against neurodegeneration, diabetes, and cancer • Modulates gut microbiota	• Scavenges reactive oxygen species (ROS) • Modulates signalling pathways • Regulates gene expression, enzyme activity, and cell signalling	• Sensitive to heat, light, pH changes, and oxidation • Some are stable in dry and encapsulated forms	• Store in cool, dark, dry places • Polyphenol-rich foods (e.g. tea, berries) should be kept away from light and moisture	• Polyphenol supplements: ~12–24 months if sealed and protected from heat/moisture	Hedayati et al. (2023), K. B. Pandey and Rizvi (2009)

3.2.4 Saponins

A hydrophobic aglycone called sapogenin is joined to one or more sugar chains to form saponins, which are glycosides with typical foaming properties (Cheok et al., 2014). Saponins, which are found in legumes (like beans and peas) and vegetables like spinach and asparagus, are known to have anti-inflammatory, anticancer, and cholesterol-lowering properties (Khan et al., 2022). In the digestive system, they can combine with cholesterol to generate complexes that decrease absorption and increase excretion (Jesch & Carr, 2017). Additionally, saponins can cause cancer cells to undergo apoptosis and alter immunological responses. Their medicinal promise in nutraceutical development is widely recognized, despite their slightly bitter taste and possible anti-nutritional effects in large doses (Kareem et al., 2022).

3.2.5 Alkaloids

Alkaloids are substances that contain nitrogen and have a variety of pharmacological characteristics. Some vegetables, such garlic (alliin derivatives), tomatoes (tomatine), and eggplants (solanine), contain significant alkaloid chemicals; however, they are more common in medicinal plants (Kukula-Koch & Widelski, 2017). These compounds have neuroprotective, analgesic, antidiabetic, and antibacterial properties. For instance allicin and other organosulphur alkaloids found in garlic have been shown to have antibacterial and cardiovascular properties (Lanzotti et al., 2012). Vegetable alkaloids add to the total bioactivity and therapeutic value of plant-based diets, notwithstanding the possibility of toxicity at large dosages.

3.3 ISOLATION TECHNIQUES

Effective sample preparation and pre-processing are essential for the successful isolation of vegetable-based phyto-nutraceuticals in order to maintain the integrity and bioactivity of the compounds (El-Nakhel et al., 2020). Substances like carotenoids and polyphenols are protected from deterioration by proper post-harvest storage, which includes regulating temperature, humidity, and light (Barkat et al., 2018). Samples are stabilized, and moisture is reduced by drying techniques such oven, freeze, and air drying (Langston et al., 2023). To improve solvent penetration, surface area is increased through grinding and homogenization (Câmara et al., 2022). The polarity of the target phytochemicals determines which solvent – methanol, ethanol, acetone, or water – is the best. To improve yield and preserve bioactivity, extraction methods – from straightforward maceration and Soxhlet to environmentally friendly approaches like ultrasonic or microwave-assisted extraction (MAE) – must optimize the solvent-to-solid ratio, temperature, duration, and pH (A. Kumar et al., 2023b) (Table 3.2).

3.3.1 Conventional Methods

3.3.1.1 Solvent Extraction

Because of its ease of use, effectiveness, and scalability, solvent extraction is a popular conventional technique for separating phyto-nutraceuticals from plants. It uses solvents such as water, ethanol, methanol, acetone, or hexane, either separately or in combination, and operates on the basis of differential solubility (Wang & Weller, 2006). In order to improve extraction, the procedure usually entails maceration or refluxing in which ground plant material is immersed in solvent under controlled temperature, duration, and agitation. Important variables that impact yield and extract composition include temperature, extraction time, and solvent-to-solid ratio (Savic & Savic Gajic, 2023). However, there are drawbacks to solvent extraction such as lengthy processing periods, significant solvent consumption, and the possibility of heat-sensitive chemicals degrading. In addition to posing safety and environmental risks, the use of organic solvents necessitates efficient solvent recovery and waste management (A. Kumar et al., 2023b).

3.3.1.2 Steam Distillation and Hydro-Distillation

Traditional, efficient techniques for removing volatile bioactive substances from vegetables like garlic, onions, and leafy greens include steam distillation and hydro-distillation. These techniques also provide essential oils and OSCs (L. Wang & Weller, 2006). Through the process of steam distillation, volatile components in plant material are vaporized, condensed, and gathered. By immersing the plant material in water and heating it, vapor is produced during the hydro-distillation process. Both use essential oils' water immiscibility and volatility to extract thermally stable volatile phytochemicals such as derivatives of allicin that have medicinal properties (Chi et al., 2023). To improve yield and prevent thermal degradation, important parameters like steam rate, particle size, and distillation time must be tuned. These economical, solvent-free techniques work well for heat-stable volatiles but not for non-volatile or heat-sensitive substances (Marčac et al., 2023).

3.3.2 Advanced Techniques

3.3.2.1 Supercritical Fluid Extraction (SFE)

SFE is a cutting-edge, environmentally friendly method for separating beneficial phyto-nutraceuticals from plants. It combines gas-like diffusivity with liquid-like solvating power by using supercritical fluids, primarily carbon dioxide, above their critical temperature and pressure (Anklam et al., 1998). This makes it possible for plant material to be efficiently penetrated and for non-polar to moderately polar chemicals, such as sterols, carotenoids, and essential oils, to be extracted without leaving harmful residues (Akanda et al., 2012). Selective extraction, low temperatures that shield heat-sensitive chemicals, quick processing, and little environmental impact are some advantages of SFE. Polar phytochemicals become more soluble when co-solvents such as ethanol are added. The amount of co-solvent, temperature, pressure, and flow rate are some of the variables that affect extraction efficiency and can be adjusted for the target chemicals. SFE is widely used in the functional food and nutraceutical industries due to its accuracy and scalability, despite the higher equipment costs. It ensures excellent purity and bioactivity by successfully removing bioactives like lutein, lycopene, and glucosinolates from vegetables including spinach, tomatoes, and broccoli (L. Li et al., 2022).

3.3.2.2 Ultrasound-Assisted Extraction (UAE)

UAE is a productive, environmentally responsible method for removing phyto-nutraceuticals from crops using high-frequency ultrasonic waves. Cavitation bubbles produced by ultrasonic waves burst close to plant cells, causing shear pressures that break down cell walls and improve mass transfer and solvent penetration. Compared to conventional techniques, this improves extraction efficiency by hastening the release of bioactives such as polyphenols, flavonoids, and carotenoids (Won & Kwon, 2024). UAE protects heat-sensitive chemicals by using less energy and solvent, cutting down on extraction time, and operating at low temperatures (Bhadange et al., 2024). Ultrasound power, frequency, time, solvent type, and solid-to-liquid ratio are important variables that impact UAE (K. Kumar et al., 2021). UAE preserves the activity of chemicals like lycopene, lutein, and glucosinolates while increasing their yields when applied to vegetables like tomatoes, spinach, and broccoli (Badea et al., 2025).

3.3.2.3 Microwave-Assisted Extraction (MAE)

Using microwave energy to heat the sample and solvent, MAE is a quick and energy-efficient technique for removing phyto-nutraceuticals from vegetables. The release of bioactives such as polyphenols, flavonoids, carotenoids, and glucosinolates is accelerated by this internal heating, which also breaks down plant cell walls and improves solvent penetration (N. Wang et al., 2025). Shorter extraction times, less solvent use, and increased yields are some benefits of MAE over conventional

TABLE 3.2
Phyto-Nutraceuticals: Sources, Extraction, Purification, Analytical Techniques, Yield, and Purity

Vegetable	Major Phyto-Nutraceutical	Chemical class	Extraction Method	Purification Method	Analytical Techniques	Solvent used	Yield (%)	Purity (%)	Key References
• **Broccoli** • **Cabbage** • **Brussels Sprouts** • **Cauliflower**	Sulforaphane, Indole-3-carbinol	Glucosinolates	• Myrosinase-assisted extraction • Ultrasound-Assisted Extraction (UAE) • Microwave-Assisted Extraction (MAE) • Solvent extraction • Supercritical Fluid Extraction (SFE)	• Liquid-Liquid Extraction (LLE) • Solid-Phase Extraction (SPE) • Column chromatography (silica or C18) • Drying	• HPLC • LC-MS/MS • GC-MS • NMR	• Ethyl acetate • Dichloromethane • Methanol • Hexane (optional wash) • Ethanol (food-grade) • Chloroform (lab use)	60–85%	70–98%	Hadi Al-Taai et al. (n.d.), Liang et al. (2005), Men et al. (2023), Lv et al. (2021)
• **Tomato**	Lycopene	Carotenoids	• Solvent extraction • Ultrasound-Assisted Extraction (UAE) • Supercritical Fluid Extraction (SFE) • Microwave-Assisted Extraction (MAE)	• Filtration & centrifugation • Column chromatography (silica or C18) • Liquid-Liquid Extraction (LLE)	• UV-vis spectroscopy • HPLC • NMR/FTIR • LC-MS/MS	• Hexane, petroleum ether • Acetone • Ethanol • Dichloromethane	70–95%	80–98%	Ranveer et al. (2020), Zuorro (2020), Naviglio et al. (2008), Madia et al. (2021)
• **Spinach** • **Kale** • **Red/yellow pepper** • **Green peas**	Lutein, Zeaxanthin	Carotenoids	• Solvent extraction • Ultrasound-Assisted Extraction (UAE) • Supercritical Fluid Extraction (SFE) • Microwave-Assisted Extraction (MAE) • Enzyme-assisted extraction (EAE)	• Silica gel chromatography • Liquid-Liquid Extraction (LLE) • Centrifugation • saponification	• UV-vis spectroscopy • HPLC • NMR/FTIR • LC-MS/MS	• Hexane • Acetone • Ethanol • Petroleum ether • Dichloromethane • Methanol • Ethyl acetate	70–95% (lutein), 60–90% (zeaxanthin)	85–98% (lutein), 80–97% (zeaxanthin)	S. Pandey et al. (2025), Arbab et al. (2024), L. Wang et al. (2019), J. Li and Engelberth (2018)
• **Garlic**	Allicin	Organosulphur compounds	• Aqueous enzymatic extraction • Solvent extraction • Ultrasound-Assisted Extraction (UAE) • Supercritical Fluid Extraction (SFE)	• Liquid-Liquid Extraction (LLE) • Solid-Phase Extraction (SPE) • Silica column chromatography	• UV-vis spectroscopy • HPLC • NMR/FTIR • LC-MS/MS • GC-MS	• Water • Ethanol • Ethyl acetate • Dichloromethane • Methanol	50–70%	70–95%	Dana et al. (2024), Kumar Manikyam and Mani (2024), F. Li et al. (2017), Bose et al. (2014)

• **Carrot** • **Pumpkin**	β-Carotene	Carotenoids	• Solvent extraction (most common) • Ultrasonic-Assisted Extraction (UAE) • Supercritical fluid extraction (SFE) • Microwave-assisted extraction (MAE) • Enzyme-assisted extraction (EAE)	• Filtration • Centrifugation • Concentration via rotary evaporation • Column chromatography	• UV-vis spectroscopy (max absorption at ~450 nm) • HPLC • LC-MS/MS	• Hexane • Acetone • Ethanol • Petroleum ether • Mixtures (e.g. hexane:aceton:ethanol = 2:1:1)	70–90%	80–95%	Ge et al. (2025), Hagos et al. (2022), Bogacz-Radomska and Harasym (2018), Machmudah and Goto (2013)
• **Onion**	Quercetin, sulphur compounds	Flavonoids	• Solvent extraction • Ultrasonic-Assisted Extraction (UAE) • Supercritical Fluid Extraction (SFE) • Microwave-Assisted Extraction (MAE) • Enzyme-assisted extraction (EAE)	• Solid-Phase Extraction (SPE) • Liquid-Liquid Extraction (LLE) • Column chromatography	• HPLC (C18 column) • UV-vis spectroscopy • NMR • FTIR	• Methanol • Ethanol • Acetone (sometimes acidified with HCl or acetic acid)	1–2% (w/w) from fresh vegetable material, ~30–50 mg/g from dried plant extracts (depending on source)	85–99%	Shergujri et al. (2024), Ciardi et al. (2021), Ko et al. (2011), (a method of extracting quercetin from onion, 2009)
• **Beetroot**	Betalains, nitrates, polyphenols	Alkaloids	• Aqueous extraction • Ethanolic extraction • Ultrasound-Assisted Extraction (UAE)	• Centrifugation + filtration • Solid-Phase Extraction (SPE) • Ion-exchange chromatography	• UV-vis spectroscopy • HPLC • LC-MS/MS • FTIR • NMR	• Water (pH 4–6) • Ethanol (50–80%) • Acidifiers (citric acid, ascorbic acid – to stabilize betalains and prevent degradation)	0.5–1.5% (w/w) from beetroot (crude pigment extract), ~30–70 mg/g from dry matter	80–95%	Gonalves et al. (2012), Eyshi et al. (2024), Shakir and Simone (2024), Da Silva et al. (2023)

techniques, particularly for chemicals that are water soluble and heat sensitive. It is an environmentally beneficial method because it works very well with green solvents like ethanol or water. Factors such as temperature, solvent type, time, microwave power, and solid-to-solvent ratio affect extraction efficiency (Dhotre, 2025). MAE enhances the recovery of important nutrients including β-carotene, lycopene, and phenolic acids when applied to vegetables like carrots, tomatoes, and leafy greens. MAE is a useful tool in the food and nutraceutical industries for creating functional goods rich in phytochemicals because of its scalability, automation possibilities, and compatibility with continuous systems (Sharma et al., 2025).

3.4 CHARACTERIZATION TECHNIQUES

3.4.1 Spectroscopic Techniques

In order to analyse the molecular structure and content of phyto-nutraceuticals, spectroscopic techniques are crucial. Functional groups, molecular weight, and chemical bonding are all thoroughly described by techniques such as UV-visible, infrared, NMR, and mass spectrometry (MS). With their high sensitivity and specificity, these methods enhance quality control and research in vegetable-derived nutraceuticals by enabling the identification, quantification, and purity assessment of bioactive components (Figure 3.2).

3.4.1.1 UV-Vis Spectroscopy

UV-Visible (UV-Vis) spectroscopy is a popular method for the initial characterization and measurement of phyto-nutraceuticals as it is easy to use, quick, and affordable. According to Farooq

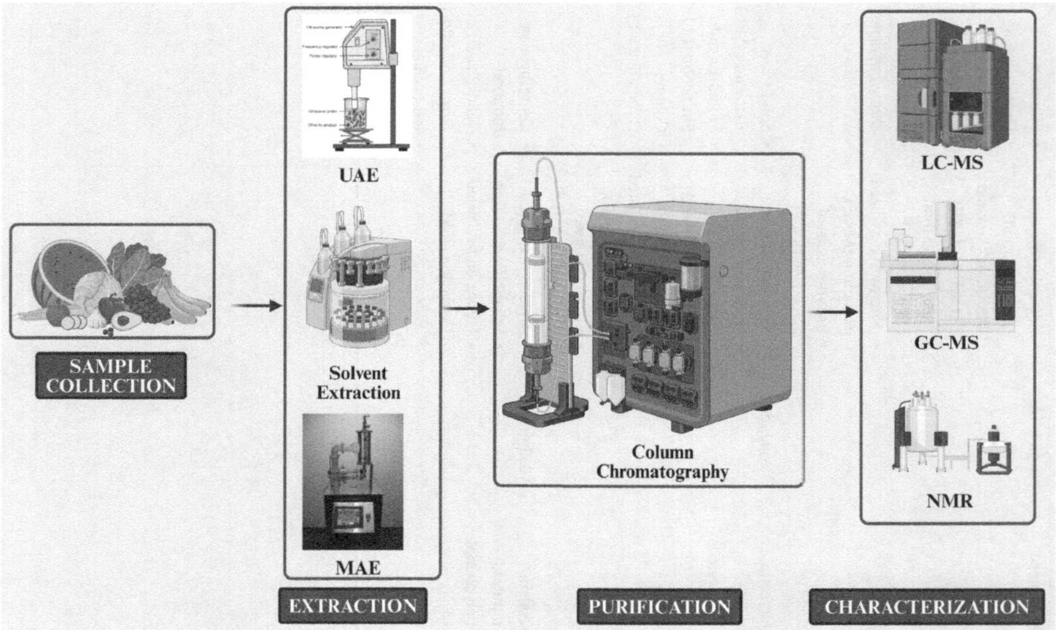

FIGURE 3.2 A schematic representation of extraction, purification, and characterization of phyto-nutraceuticals.

Source: https://BioRender.com

Rather et al. (2024), it detects the absorption of light in the visible (400–800 nm) and ultraviolet (200–400 nm) ranges, which correspond to electronic changes in chromophoric groups within phytochemicals. Numerous substances produced from vegetables, including polyphenols, flavonoids, carotenoids, and anthocyanins, can be identified according to their distinctive absorbance peaks (Bhardwaj et al., 2025), while carotenoids like β-carotene and lycopene absorb around 450–500 nm (A. Kumar et al., 2025), useful for pigment analysis in vegetables like carrots and tomatoes. Colorimetric tests, such as Folin-Ciocalteu, can be used to quantify the total phenolic content (TPC); absorbance is usually measured at 765 nm (Shruthi, n.d.). UV-Vis is nonetheless useful for routine quality control, stability investigations, and high-throughput screening even if it lacks the selectivity and structural detail of methods like NMR or MS.

3.4.1.2 Infrared (IR) Spectroscopy

Another prominent non-destructive method for identifying and characterizing phyto-nutraceuticals made from plants is IR spectroscopy (Losacco et al., 2022). It detects when infrared light is absorbed as a result of molecular vibrations which happen when chemical bonds are stretching and bending. According to Ali Redha et al. (2024), every functional group in a molecule absorbs infrared light at particular frequencies, resulting in a distinct spectral fingerprint that aids in compound identification. Olyphenols, carotenoids, glucosinolates, and saponins are examples of vegetable-based phyto-nutraceuticals that show clear absorption bands in the infrared spectrum. For example glucosinolates exhibit distinctive S=O and C–N vibrations in the fingerprint area (600–1,500 cm^{-1}), whereas polyphenols exhibit broad –OH stretching bands (3,200–3,600 cm^{-1}) and aromatic C=C stretching (~1,600 cm^{-1}) (Ali Redha et al., 2024; Pharmawati & Wrasiati, 2020).

By collecting entire spectra in real time, Fourier-transform infrared (FTIR) spectroscopy increases the speed and sensitivity of traditional IR spectroscopy. Direct analysis of solid or semi-solid samples with little pre-treatment is made possible by methods such as Attenuated Total Reflectance (ATR). IR is especially helpful for keeping an eye on the authenticity, stability, and purity of formulations made from vegetables while they are being processed and stored (Rohman et al., 2020).

3.4.1.3 Nuclear Magnetic Resonance (NMR)

A powerful, non-destructive method for determining the precise molecular structure and measuring phyto-nutraceuticals made from plants is NMR spectroscopy. Chemical surroundings and molecular architecture are revealed by NMR, which measures how specific atomic nuclei (such as ^{1}H and ^{13}C) react to electromagnetic radiation in a magnetic field based on their magnetic characteristics. Complex phytochemicals including flavonoids, carotenoids, glucosinolates, and saponins can be analysed with its help (Mannina et al., 2012). By providing information on carbon-hydrogen frameworks, stereochemistry, and functional groups, NMR makes it possible to distinguish structurally identical substances (such as quercetin and kaempferol) (Şöhretoğlu, 2021). NMR is perfect for identifying novel compounds as it enables direct quantification without reference standards, unlike spectrophotometric techniques. Characterizing lycopene from tomatoes, SFN from broccoli, and saponins from legumes are a few examples of its applications (Valentino et al., 2020).

3.4.1.4 Mass Spectrometry (MS)

By assessing the mass-to-charge (m/z) ratios of phyto-nutraceuticals, MS is a potent analytical method for their identification and quantification. It provides great sensitivity and specificity by ionizing substances and identifying their molecular weights and structural fragments. MS improves the analysis of complex bioactives obtained from vegetables, including flavonoids, phenolic acids, carotenoids, and glucosinolates. It is frequently used in conjunction with chromatographic techniques such as LC-MS or GC-MS. MS is particularly useful for identifying trace substances and assessing their metabolism and bioavailability, such as SFN, lycopene, or allicin (De Vos et al., 2012).

Essential oils, OSCs, and phenolics are examples of volatile, thermally stable phyto-nutraceuticals that can be analysed using GC-MS. Compounds are separated using gas chromatography, and their mass fragmentation patterns are used to identify them. High resolution, sensitivity, and structural insights are offered by GC-MS, which is frequently utilized for sulphur metabolites and aroma chemicals, and for the application of derivatization to non-volatile phytochemicals (Y. Q. Li et al., 2013).

For the analysis of non-volatile, heat-sensitive phyto-nutraceuticals, LC-MS/MS provides excellent sensitivity, selectivity, and precision. Complex chemicals including polyphenols, glucosinolates, and saponins are perfect for profiling with it. It makes it possible to identify and quantify structures by producing product ion spectra. LC-MS/MS is widely used in metabolomics, pharmacokinetics, and bioavailability investigations. It is crucial for quality control and nutraceutical research as it is excellent at detecting trace-level bioactives (Wu et al., 2021).

3.5 APPLICATIONS IN FUNCTIONAL FOODS AND NUTRACEUTICALS

In recognition of their health-promoting qualities, phyto-nutraceuticals made from vegetables are being used more and more in functional foods and dietary supplements (Martirosyan & Singh, 2015). While formulation approaches include encapsulation, nanoemulsions, and polymer-based carriers to improve stability and targeted administration, fortification tactics entail enhancing staple foods with concentrated bioactives like polyphenols, carotenoids, or glucosinolates (K. B. Pandey & Rizvi, 2009), (Saini & Keum, 2018). Significant obstacles still exist in spite of their potential, such as low water solubility, deterioration during processing and storage, and restricted bioavailability after ingestion. Stabilization technology and sophisticated delivery methods are needed to overcome these problems. Furthermore, different countries have different regulatory frameworks, and phyto-nutraceuticals frequently fall somewhere between food and medicine. Consumer protection and market acceptance depend on adherence to safety regulations such as GRAS status (Generally Recognized as Safe), labelling, and dose recommendations (A. Kumar et al., 2023b; Langston et al., 2023).

3.6 CONCLUSION AND FUTURE PERSPECTIVES

The isolation and characterization of phyto-nutraceuticals from vegetables involve a combination of optimized extraction, purification, and analytical techniques to ensure the quality, efficacy, and safety of bioactive compounds. Advances in green technologies and integrated analytical methods continue to enhance yield, purity, and sustainability. Looking ahead, research should focus on standardization, bioavailability, and novel delivery systems. With their rich phytochemical profile, vegetables remain central to the development of safe, effective, and natural nutraceutical products.

Accurate profiling of complex bioactive matrices is hampered by the specificity, sensitivity, and scalability issues that plague the current isolation and characterization techniques for phyto-nutraceuticals derived from vegetables. New technologies that promise to improve chemical discovery, sustainability, and environmental compliance include metabolomics, integrative omics approaches, and green extraction methods (e.g. subcritical water extraction, deep eutectic solvents) (Usman et al., 2023). Cost-effective scale-up, batch-to-batch consistency, and sustainable raw material procurement are still major obstacles, though. To guarantee quality, efficacy, and regulatory approval in the growing field of functional foods and nutraceuticals, future developments must concentrate on environmentally friendly, high-throughput technologies and strong standardization processes (M. Kumar et al., 2023).

REFERENCES

Aghababaei, F., & Hadidi, M. (2023). Recent advances in potential health benefits of quercetin. *Pharmaceuticals*, *16*(7), 1020. https://doi.org/10.3390/PH16071020

Akanda, M. J. H., Sarker, M. Z. I., Ferdosh, S., Manap, M. Y. A., Rahman, N. N. N. A., & Kadir, M. O. A. (2012). Applications of supercritical fluid extraction (SFE) of palm oil and oil from natural sources. *Molecules*, *17*(2), 1764–1794. https://doi.org/10.3390/MOLECULES17021764

Akyol, H., Riciputi, Y., Capanoglu, E., Caboni, M. F., & Verardo, V. (2016). Phenolic compounds in the potato and its byproducts: An overview. *International Journal of Molecular Sciences*, *17*(6), 835. https://doi.org/10.3390/IJMS17060835

Ali Redha, A., Torquati, L., Langston, F., Nash, G. R., Gidley, M. J., & Cozzolino, D. (2024). Determination of glucosinolates and isothiocyanates in glucosinolate-rich vegetables and oilseeds using infrared spectroscopy: A systematic review. *Critical Reviews in Food Science and Nutrition*, *64*(23), 8248–8264. https://doi.org/10.1080/10408398.2023.2198015

Anklam, E., Berg, H., Mathiasson, L., Sharman, M., & Ulberth, F. (1998). Supercritical fluid extraction (SFE) in food analysis: A review. *Food Additives and Contaminants*, *15*(6), 729–750. https://doi.org/10.1080/02652039809374703/ASSET//CMS/ASSET/D01C66C4-27F6-4598-BA32-FA3EAF56EACC/02652039809374703.FP.PNG

Arab, L., & Steck, S. (2000). Lycopene and cardiovascular disease. *The American Journal of Clinical Nutrition*, *71*(6), 1691S–1695S. https://doi.org/10.1093/AJCN/71.6.1691S

Arbab, S., Shafi, A., Farooq, U., Ahmad, S., Hayat, K., Raza, E., & Arbab, Z. (2024). Extraction and characterization of lutein from plant sources. *AJ Life Science*, *2024*(1), 40–53. https://doi.org/10.34091/AJLS.7.1.5

Badea, G. I., Gatea, F., Litescu-Filipescu, S. C., Alecu, A., Chira, A., Damian, C. M., & Radu, G. L. (2025). Optimization of green ultrasound-assisted extraction of carotenoids and tocopherol from tomato waste using NADESs. *Molecules*, *30*(3), 591. https://doi.org/10.3390/MOLECULES30030591

Baldelli, S., Lombardo, M., D'Amato, A., Karav, S., Tripodi, G., & Aiello, G. (2025). Glucosinolates in human health: Metabolic pathways, bioavailability, and potential in chronic disease prevention. *Foods*, *14*(6), 912. https://doi.org/10.3390/FOODS14060912

Barkat, N., Singh, J., Jayaprakasha, G. K., & Patil, B. S. (2018). Effect of harvest time on the levels of phytochemicals, free radical-scavenging activity, α-amylase inhibition and bile acid-binding capacity of spinach (Spinacia oleracea). *Journal of the Science of Food and Agriculture*, *98*(9), 3468–3477. https://doi.org/10.1002/JSFA.8862

Bartley, G. E., & Scolnik, P. A. (1995). Plant carotenoids: Pigments for photoprotection, visual attraction, and human health. *The Plant Cell*, *7*(7), 1027. https://doi.org/10.1105/TPC.7.7.1027

Bayan, L., Koulivand, P. H., & Gorji, A. (2014). Garlic: A review of potential therapeutic effects. *Avicenna Journal of Phytomedicine*, *4*(1), 1. https://pmc.ncbi.nlm.nih.gov/articles/PMC4103721/

Bhadange, Y. A., Carpenter, J., & Saharan, V. K. (2024). A comprehensive review on advanced extraction techniques for retrieving bioactive components from natural sources. *ACS Omega*, *9*(29), 31274–31297. https://doi.org/10.1021/ACSOMEGA.4C02718/ASSET/IMAGES/LARGE/AO4C02718_0007.JPEG

Bhardwaj, C. P., Sharma, P., Gaur, S., & Napit, M. (2025). Isolation, characterization and structural elucidation of a bioactive flavonoid from *Schinus polygama* leaves using chromatographic and spectroscopic techniques. *Pharmacognosy Research*, *17*(2), 627–637. https://doi.org/10.5530/pres.20252098

Bogacz-Radomska, L., & Harasym, J. (2018). β-Carotene – properties and production methods. *Food Quality and Safety*, *2*(2), 69–74. https://doi.org/10.1093/FQSAFE/FYY004

Bohn, T., Balbuena, E., Ulus, H., Iddir, M., Wang, G., Crook, N., & Eroglu, A. (2023). Carotenoids in health as studied by omics-related endpoints. *Advances in Nutrition*, *14*(6), 1538–1578. https://doi.org/10.1016/J.ADVNUT.2023.09.002

Bose, S., Laha, B., & Banerjee, S. (2014). Quantification of allicin by high performance liquid chromatography-ultraviolet analysis with effect of post-ultrasonic sound and microwave radiation on fresh garlic cloves. *Pharmacognosy Magazine*, *10*(Suppl 2), S288. https://doi.org/10.4103/0973-1296.133279

Bouyahya, A., El Omari, N., Hakkur, M., El Hachlafi, N., Charfi, S., Balahbib, A., Guaouguaou, F. E., Rebezov, M., Maksimiuk, N., Shariati, M. A., Zengin, G., El Menyiy, N., Chamkhi, I., & Bakrim, S. (2021). Sources, health benefits, and biological properties of zeaxanthin. *Trends in Food Science & Technology*, *118*, 519–538. https://doi.org/10.1016/J.TIFS.2021.10.017

Câmara, J. S., Perestrelo, R., Berenguer, C. V., Andrade, C. F. P., Gomes, T. M., Olayanju, B., Kabir, A., Rocha, C. M. R., Teixeira, J. A., & Pereira, J. A. M. (2022). Green extraction techniques as advanced sample preparation approaches in biological, food, and environmental matrices: A review. *Molecules*, *27*(9), 2953. https://doi.org/10.3390/MOLECULES27092953/S1

Chan, J. Y. Y., Yuen, A. C. Y., Chan, R. Y. K., & Chan, S. W. (2013). A review of the cardiovascular benefits and antioxidant properties of allicin. *Phytotherapy Research*, *27*(5), 637–646. https://doi.org/10.1002/PTR.4796

Cheok, C. Y., Salman, H. A. K., & Sulaiman, R. (2014). Extraction and quantification of saponins: A review. *Food Research International*, *59*, 16–40. https://doi.org/10.1016/J.FOODRES.2014.01.057

Chi, W., Mak, K., & Wck, M. (2023). Comparisons between hydro and steam distillation processes to extract *Prunella vulgaris* volatile compounds, and their anti-oxidative activities. *Journal of Drug Delivery and Therapeutics*, *13*(3), 51–57. https://doi.org/10.22270/JDDT.V13I3.5955

Ciardi, M., Ianni, F., Sardella, R., Di Bona, S., Cossignani, L., Germani, R., Tiecco, M., & Clementi, C. (2021). Effective and selective extraction of quercetin from onion (Allium cepa l.) skin waste using water dilutions of acid-based deep eutectic solvents. *Materials*, *14*(21), 6465. https://doi.org/10.3390/MA14216465/S1

Clarke, D. B. (2010). Glucosinolates, structures and analysis in food. *Analytical Methods*, *2*(4), 310–325. https://doi.org/10.1039/B9AY00280D

Dana, T., Pathan, A. K., & Ahmad, S. (2024). Optimized extraction and characterization of allicin from snow mountain garlic and its anticancer potential via allicin-loaded CuO nanoparticles. *International Journal of Pharmaceutical Quality Assurance*, *15*(2), 931–942. https://doi.org/10.25258/ijpqa.15.2.61

Darwin, R., Valmon, R., Chithanna, S., Galla, H., Syed, S. H., Ajeed, A., Billah, M., Teja, K., Reddy, K., Varuna, U., & Arjun, N. V. (2025). PV Vaithiyalingam Rd, Velan Nagar, Krishna puram. *Technology and Advanced Studies (VISTAS)*, *600*(117). https://doi.org/10.48309/CHEMM.2025.504050.1892

Da Silva, D. V. T., Baião, D. dos S., Magalhães, A., Almeida, N. F., Conte, C. A., & Paschoalin, V. M. F. (2023). Combining conventional organic solvent extraction, ultrasound-assisted extraction, and chromatographic techniques to obtain pure betanin from beetroot for clinical purposes. *Antioxidants*, *12*(10), 1823. https://doi.org/10.3390/ANTIOX12101823/S1

Deng, Y., Ho, C. T., Lan, Y., Xiao, J., & Lu, M. (2023). Bioavailability, health benefits, and delivery systems of allicin: A review. *Journal of Agricultural and Food Chemistry*, *71*(49), 19207–19220. https://doi.org/10.1021/ACS.JAFC.3C05602/ASSET/IMAGES/MEDIUM/JF3C05602_0006.GIF

De Vos, R. C. H., Schipper, B., & Hall, R. D. (2012). High-performance liquid chromatography-mass spectrometry analysis of plant metabolites in brassicaceae. *Methods in Molecular Biology*, *860*, 111–128. https://doi.org/10.1007/978-1-61779-594-7_8

Dhotre, I. (2025). A comprehensive review on progression and innovations in microwave assisted extraction technology for essential oils. *Journal of Chemical Technology & Biotechnology*, *100*(5), 894–907. https://doi.org/10.1002/JCTB.7823

El-Nakhel, C., Pannico, A., Graziani, G., Kyriacou, M. C., Giordano, M., Ritieni, A., De Pascale, S., & Rouphael, Y. (2020). Variation in macronutrient content, phytochemical constitution and in vitro antioxidant capacity of green and red butterhead lettuce dictated by different developmental stages of harvest maturity. *Antioxidants*, *9*(4), 300. https://doi.org/10.3390/ANTIOX9040300

Elvira-Torales, L. I., García-Alonso, J., & Periago-Castón, M. J. (2019). Nutritional importance of carotenoids and their effect on liver health: A review. *Antioxidants*, *8*(7), 229. https://doi.org/10.3390/ANTIOX8070229

Esteve, M. (2020). Mechanisms underlying biological effects of cruciferous glucosinolate-derived isothiocyanates/indoles: A focus on metabolic syndrome. *Frontiers in Nutrition*, *7*, 111. https://doi.org/10.3389/FNUT.2020.00111

Eyshi, S., Ghareaghajlou, N., Afshar Mogaddam, M. R., & Ghasempour, Z. (2024). Red beet betalains extraction process: A comprehensive review of methods, applications, and physicochemical properties. *Food Science & Nutrition*, *12*(11), 8540. https://doi.org/10.1002/FSN3.4458

Farooq Rather, M., Un Nisa, B., Mushtaq, H., Gulzar, A., Mushtaq, H., Ashraf, R., Rashid, A., Parveen, M., Mansoor, G., Gulaam, S., & Bano, S. (2024). Nanotechnology-enhanced phytonutraceuticals: Innovations in characterization, formulation, and therapeutic applications. *Journal of Advances in Biology & Biotechnology*, *27*(11), 635–646. https://doi.org/10.9734/jabb/2024/v27i111647

Gariballa, S., Al-Bluwi, G. S. M., & Yasin, J. (2023). Increased fruit and vegetable consumption mitigates oxidative damage and associated inflammatory response in obese subjects independent of body weight change. *Nutrients*, *15*(7), 1638. https://doi.org/10.3390/NU15071638

Ge, H., Yang, L., Yao, Q., Gu, Y., Zhang, C., Liu, X., Huang, F., & Zhang, Y. (2025). Carotenes from pumpkin flesh and peel: A review on their physicochemical properties, preparation techniques, detection methods and antioxidant potential. *ACS Food Science and Technology*. https://doi.org/10.1021/ACSFOODSCITECH.4C00964/ASSET/IMAGES/MEDIUM/FS4C00964_0001.GIF

Gonalves, L. C. P., Trassi, M. A. D. S., Lopes, N. B., Dörr, F. A., Santos, M. T. Dos, Baader, W. J., Oliveira, V. X., & Bastos, E. L. (2012). A comparative study of the purification of betanin. *Food Chemistry*, *131*(1), 231–238. https://doi.org/10.1016/J.FOODCHEM.2011.08.067

Gul, K., Singh, A. K., & Jabeen, R. (2016). Nutraceuticals and functional foods: The foods for the future world. *Critical Reviews in Food Science and Nutrition*, *56*(16), 2617–2627. https://doi.org/10.1080/10408398.2014.903384;WGROUP:STRING:PUBLICATION

Hadi Al-Taai, I., Falih Al-Fekaiki, D., & Salim Issa, S. (n.d.). *Extraction, Purification and Identification of The Bioactive Compound Sulforaphane from Broccoli (Brassica Oleracea Var. Italica) By HPLC.*

Hagos, M., Redi-Abshiro, M., Chandravanshi, B. S., & Yaya, E. E. (2022). Development of analytical methods for determination of β-carotene in pumpkin (*Cucurbita maxima*) flesh, peel, and seed powder samples. *International Journal of Analytical Chemistry*, *2022*, 9363692. https://doi.org/10.1155/2022/9363692

Hammer, G. P., Du Prel, J. B., & Blettner, M. (2009). Vermeidung verzerrter ergebnisse in beobachtungsstudien: Teil 8 der serie zur bewertung wissenschaftlicher publikationen. *Deutsches Arzteblatt*, *106*(41), 664–668. https://doi.org/10.3238/ARZTEBL.2009.0664

Hedayati, N., Yaghoobi, A., Salami, M., Gholinezhad, Y., Aghadavood, F., Eshraghi, R., Aarabi, M. H., Homayoonfal, M., Asemi, Z., Mirzaei, H., Hajijafari, M., Mafi, A., & Rezaee, M. (2023). Impact of polyphenols on heart failure and cardiac hypertrophy: Clinical effects and molecular mechanisms. *Frontiers in Cardiovascular Medicine*, *10*, 1174816. https://doi.org/10.3389/FCVM.2023.1174816/XML/NLM

Hosseini, B., Berthon, B. S., Saedisomeolia, A., Starkey, M. R., Collison, A., Wark, P. A. B., & Wood, L. G. (2018). Effects of fruit and vegetable consumption on inflammatory biomarkers and immune cell populations: A systematic literature review and meta-analysis. *The American Journal of Clinical Nutrition*, *108*(1), 136–155. https://doi.org/10.1093/AJCN/NQY082

Husen, A. (2021). *Traditional Herbal Therapy for the Human Immune System*. CRC Press. https://doi.org/10.1201/9781003137955

Husen, A. (2022). *Herbs, Shrubs and Trees of Potential Medicinal Benefits*. CRC Press. https://doi.org/10.1201/9781003205067

Husen, A. (2023a). *Exploring Poisonous Plants: Medicinal Values, Toxicity Response and Other Potential Uses*. CRC Press. https://doi.org/10.1201/b23017

Husen, A. (2023b). *Antidiabetic Medicinal Plants and Herbal Treatments*. CRC Press. https://doi.org/10.1201/b23347

Husen, A. (2024a). *Antimalarial Medicinal Plants*. Taylor & Francis Group. https://doi.org/10.1201/9781003378396

Husen, A. (2024b). *Medicinal Spices and Condiment Crops*. Taylor & Francis Group. https://doi.org/10.1201/9781003387046

Husen, A., & Iqbal, M. (2021). Plant-based potential nutraceuticals for improving the human immune system. In: *Traditional Herbal Therapy for the Human Immune System* (Ed. Husen, A.). CRC Press, pp. 1–11. https://doi.org/10.1201/9781003137955-1

Husen, H., & Pant, M. (2024a). *Exploring Medicinal Orchids*. Taylor & Francis Group. https://doi.org/10.1201/b23375

Husen, H., & Pant, M. (2024b). *Tissue Culture Techniques and Medicinal Plants: Enhancing Propagation and Production*. Taylor & Francis Group. https://doi.org/10.1201/b23374

Jesch, E. D., & Carr, T. P. (2017). Food ingredients that inhibit cholesterol absorption. *Preventive Nutrition and Food Science*, *22*(2), 67. https://doi.org/10.3746/PNF.2017.22.2.67

Kaiser, A. E., Baniasadi, M., Giansiracusa, D., Giansiracusa, M., Garcia, M., Fryda, Z., Wong, T. L., & Bishayee, A. (2021). Sulforaphane: A broccoli bioactive phytocompound with cancer preventive potential. *Cancers*, *13*(19), 4796. https://doi.org/10.3390/CANCERS13194796

Kalogerakou, T., & Antoniadou, M. (2024). The role of dietary antioxidants, food supplements and functional foods for energy enhancement in healthcare professionals. *Antioxidants*, *13*(12), 1508. https://doi.org/10.3390/ANTIOX13121508

Kalra, E. K. (2003). Nutraceutical – definition and introduction. *AAPS PharmSci*, *5*(3), 27–28. https://doi.org/10.1208/PS050325/METRICS

Kareem, O., Ali, T., Dar, L. A., Mir, S. A., Rashid, R., Nazli, N., Gulzar, T., & Bader, G. N. (2022). Positive health benefits of saponins from edible legumes: Phytochemistry and pharmacology. *Edible Plants in Health and Diseases: Volume II: Phytochemical and Pharmacological Properties*, *2*, 279–298. https://doi.org/10.1007/978-981-16-4959-2_8

Khan, M. I., Karima, G., Khan, M. Z., Shin, J. H., & Kim, J. D. (2022). Therapeutic effects of saponins for the prevention and treatment of cancer by ameliorating inflammation and angiogenesis and inducing antioxidant and apoptotic effects in human cells. *International Journal of Molecular Sciences*, *23*(18), 10665. https://doi.org/10.3390/IJMS231810665

Ko, M. J., Cheigh, C. I., Cho, S. W., & Chung, M. S. (2011). Subcritical water extraction of flavonol quercetin from onion skin. *Journal of Food Engineering*, *102*(4), 327–333. https://doi.org/10.1016/J.JFOODENG.2010.09.008

Kris-Etherton, P. M., Hecker, K. D., Bonanome, A., Coval, S. M., Binkoski, A. E., Hilpert, K. F., Griel, A. E., & Etherton, T. D. (2002). Bioactive compounds in foods: Their role in the prevention of cardiovascular disease and cancer. *The American Journal of Medicine*, *113*(9), 71–88. https://doi.org/10.1016/S0002-9343(01)00995-0

Kukula-Koch, W. A., & Widelski, J. (2017). Alkaloids. In: *Pharmacognosy: Fundamentals, Applications and Strategy*, pp. 163–198. https://doi.org/10.1016/B978-0-12-802104-0.00009-3

Kumar, A., Bhawsar, N., Manekar, S., Pendram, B., Pal, P., Ali, D., Alarifi, S., Wanale, S. G., Singh, S., Katare, S., Srivastava, G. M., Niazi, P., Pareek, S., & Yadav, V. K. (2025). Extraction and optimization of lycopene from selected fruits and their assessment as an ultraviolet ray protectant for escherichia coli. *Food Science and Nutrition*, *13*(4), e70090. https://doi.org/10.1002/FSN3.70090;SUBPAGE:STRING:FULL

Kumar, A., Nirmal, P., Kumar, M., Jose, A., Tomer, V., Oz, E., Proestos, C., Zeng, M., Elobeid, T., Sneha, V., & Oz, F. (2023). Major phytochemicals: Recent advances in health benefits and extraction method. *Molecules*, *28*(2), 887. https://doi.org/10.3390/MOLECULES28020887

Kumar, K., Srivastav, S., & Sharanagat, V. S. (2021). Ultrasound assisted extraction (UAE) of bioactive compounds from fruit and vegetable processing by-products: A review. *Ultrasonics Sonochemistry*, *70*, 105325. https://doi.org/10.1016/J.ULTSONCH.2020.105325

Kumar, M., Barbhai, M. D., Puranik, S., Radha, Natta, S., Senapathy, M., Dhumal, S., Singh, S., Kumar, S., Deshmukh, V. P., Anitha, T., Pandiselvam, R., & Lorenzo, J. M. (2023). Combination of green extraction techniques and smart solvents for bioactives recovery. *TrAC Trends in Analytical Chemistry*, *169*, 117286. https://doi.org/10.1016/J.TRAC.2023.117286

Kumar Manikyam, H., & Mani, J. (2024). Cryoethanolic extraction, purification of allicin from *Allium sativum* and HPTLC analysis. *International Journal of Biochemistry Research & Review*, *33*(6), 43–51. https://doi.org/10.9734/ijbcrr/2024/v33i6887

Langston, F. M. A., Nash, G. R., & Bows, J. R. (2023). The retention and bioavailability of phytochemicals in the manufacturing of baked snacks. *Critical Reviews in Food Science and Nutrition*, *63*(14), 2141–2177. https://doi.org/10.1080/10408398.2021.1971944;SUBPAGE:STRING:FULL

Lanzotti, V., Barile, E., Antignani, V., Bonanomi, G., & Scala, F. (2012). Antifungal saponins from bulbs of garlic, Allium sativum L. var. Voghiera. *Phytochemistry*, *78*, 126–134. https://doi.org/10.1016/J.PHYTOCHEM.2012.03.009

Li, F., Li, Q., Wu, S., & Tan, Z. (2017). Salting-out extraction of allicin from garlic (Allium sativum L.) based on ethanol/ammonium sulfate in laboratory and pilot scale. *Food Chemistry*, *217*, 91–97. https://doi.org/10.1016/J.FOODCHEM.2016.08.092

Li, J., & Engelberth, A. S. (2018). Quantification and purification of lutein and zeaxanthin recovered from distillers dried grains with solubles (DDGS). *Bioresources and Bioprocessing*, *5*(1), 1–10. https://doi.org/10.1186/S40643-018-0219-3/FIGURES/3

Li, L., Chen, M., Zeng, Y., & Liu, G. (2022). Application and perspectives of supercritical fluid technology in the nutraceutical industry. *Advanced Sustainable Systems*, *6*(7), 2200055. https://doi.org/10.1002/ADSU.202200055

Li, Y. Q., Kong, D. X., & Wu, H. (2013). Analysis and evaluation of essential oil components of cinnamon barks using GC–MS and FTIR spectroscopy. *Industrial Crops and Products*, *41*(1), 269–278. https://doi.org/10.1016/J.INDCROP.2012.04.056

Liang, H., Yuan, Q., & Xiao, Q. (2005). Purification of sulforaphane from Brassica oleracea seed meal using low-pressure column chromatography. *Journal of Chromatography B: Analytical Technologies in the Biomedical and Life Sciences*, *828*(1–2), 91–96. https://doi.org/10.1016/J.JCHROMB.2005.09.041

Liu, R. H. (2004). Potential synergy of phytochemicals in cancer prevention: Mechanism of action. *The Journal of Nutrition*, *134*(12), 3479S–3485S. https://doi.org/10.1093/JN/134.12.3479S

Losacco, D., Campanale, C., Tumolo, M., Ancona, V., Massarelli, C., & Uricchio, V. F. (2022). Evaluating the influence of nitrogen fertilizers and biochar on *Brassica oleracea* L. var. botrytis by the use of Fourier transform infrared (FTIR) spectroscopy. *Sustainability*, *14*(19), 11985. https://doi.org/10.3390/SU141911985

Lv, C., Zhang, Y., Zou, L., Sun, J., Song, X., Mao, J., & Wu, Y. (2021). Simultaneous hydrolysis and extraction increased erucin yield from broccoli seeds. *ACS Omega*, *6*(9), 6385–6392. https://doi.org/10.1021/ACSOMEGA.0C06319/ASSET/IMAGES/LARGE/AO0C06319_0004.JPEG

Machmudah, S., & Goto, M. (2013). Methods for extraction and analysis of carotenoids. In: *Natural Products: Phytochemistry, Botany and Metabolism of Alkaloids, Phenolics and Terpenes*, pp. 3367–3411. https://doi.org/10.1007/978-3-642-22144-6_145

Madia, V. N., De Vita, D., Ialongo, D., Tudino, V., De Leo, A., Scipione, L., Di Santo, R., Costi, R., & Messore, A. (2021). Recent advances in recovery of lycopene from tomato waste: A potent antioxidant with endless benefits. *Molecules*, *26*(15), 4495. https://doi.org/10.3390/MOLECULES26154495

Mannina, L., Sobolev, A. P., & Viel, S. (2012). Liquid state 1H high field NMR in food analysis. *Progress in Nuclear Magnetic Resonance Spectroscopy*, *66*, 1–39. https://doi.org/10.1016/J.PNMRS.2012.02.001

Marčac, N., Balbino, S., Tonković, P., Medved, A. M., Cegledi, E., Dragović, S., Dragović-Uzelac, V., & Repajić, M. (2023). Hydrodistillation and steam distillation of fennel seeds essential oil: Parameter optimization and application of cryomilling pretreatment. *Processes*, *11*(8), 2354. https://doi.org/10.3390/PR11082354

Martirosyan, D. M., & Singh, J. (2015). A new definition of functional food by FFC: What makes a new definition unique? *Functional Foods in Health and Disease*, *5*(6), 209–223. https://doi.org/10.31989/FFHD.V5I6.183

Men, X., Han, X., Oh, G., Im, J. H., Lim, J. S., Cho, G. hee, Choi, S. I., & Lee, O. H. (2023). Plant sources, extraction techniques, analytical methods, bioactivity, and bioavailability of sulforaphane: A review. *Food Science and Biotechnology*, *33*(3), 539. https://doi.org/10.1007/S10068-023-01434-7

Mitra, S., Rauf, A., Tareq, A. M., Jahan, S., Emran, T. Bin, Shahriar, T. G., Dhama, K., Alhumaydhi, F. A., Aljohani, A. S. M., Rebezov, M., Uddin, M. S., Jeandet, P., Shah, Z. A., Shariati, M. A., & Rengasamy, K. R. (2021). Potential health benefits of carotenoid lutein: An updated review. *Food and Chemical Toxicology*, *154*, 112328. https://doi.org/10.1016/J.FCT.2021.112328

Mungwari, C. P., King'ondu, C. K., Sigauke, P., & Obadele, B. A. (2025). Conventional and modern techniques for bioactive compounds recovery from plants: Review. *Scientific African*, *27*, e02509. https://doi.org/10.1016/J.SCIAF.2024.E02509

Naviglio, D., Pizzolongo, F., Ferrara, L., Aragòn, A., & Santini, A. (2008). Extraction of pure lycopene from industrial tomato by-products in water using a new high-pressure process. *Journal of the Science of Food and Agriculture*, *88*(14), 2414–2420. https://doi.org/10.1002/JSFA.3334;WGROUP:STRING:PUBLICATION

Otoo, R. A., & Allen, A. R. (2023). Sulforaphane's multifaceted potential: From neuroprotection to anticancer action. *Molecules*, *28*(19), 6902. https://doi.org/10.3390/MOLECULES28196902

Panche, A. N., Diwan, A. D., & Chandra, S. R. (2016). Flavonoids: An overview. *Journal of Nutritional Science*, *5*, e47. https://doi.org/10.1017/JNS.2016.41

Pandey, K. B., & Rizvi, S. I. (2009). Plant polyphenols as dietary antioxidants in human health and disease. *Oxidative Medicine and Cellular Longevity*, *2*(5), 270. https://doi.org/10.4161/OXIM.2.5.9498

Pandey, S., Gupta, A., Mahato, D. K., Paul, V., Tripathi, A. D., Rasane, P., Kumar, P., Kamle, M., & Haque, S. (2025). Lutein and zeaxanthin: Source, extraction, stability, bioactivity, and functional food applications. *Current Pharmaceutical Biotechnology*, *26*. https://doi.org/10.2174/0113892010334209241206113640

Pandjaitan, N., Howard, L. R., Morelock, T., & Gil, M. I. (2005). Antioxidant capacity and phenolic content of spinach as affected by genetics and maturation. *Journal of Agricultural and Food Chemistry*, *53*(22), 8618–8623. https://doi.org/10.1021/JF052077I

Pant, M., & Husen, A. (2025). *Exploring Medicinal Bamboos*. Taylor & Francis Group. https://doi.org/10.1201/9781003485933

Pant, M., Sohrab, S. S., & Husen, A. (2025). *Genetic Improvement and Conservation Practice of Medicinal Plants*. Taylor & Francis Group. https://doi.org/10.1201/9781003482659

Pharmawati, M., & Wrasiati, L. P. (2020). Phytochemical screening and FTIR spectroscopy on crude extract from *Enhalus acoroides* leaves (Saringan Fitokimia dan Spektroskopi FTIR Ekstrak Mentah Daun Enhalus acoroides). *Malaysian Journal of Analytical Sciences*, *24*.

Ranveer, R. C., Rathod, N. B., Debaje, P. P., & Sahoo, A. K. (2020). Extraction, purification and characterization of lycopene from tomato (*Cv Vijeta*) processing industry waste. *European Journal of Nutrition & Food Safety*, 83–90. https://doi.org/10.9734/EJNFS/2020/V12I1230337

Rocha, H. R., Coelho, M. C., Gomes, A. M., & Pintado, M. E. (2023). Carotenoids diet: Digestion, gut microbiota modulation, and inflammatory diseases. *Nutrients*, *15*(10), 2265. https://doi.org/10.3390/NU15102265

Rodríguez-Negrete, E. V., Morales-González, Á., Madrigal-Santillán, E. O., Sánchez-Reyes, K., Álvarez-González, I., Madrigal-Bujaidar, E., Valadez-Vega, C., Chamorro-Cevallos, G., Garcia-Melo, L. F., & Morales-González, J. A. (2024). Phytochemicals and their usefulness in the maintenance of health. *Plants*, *13*(4), 523. https://doi.org/10.3390/PLANTS13040523

Rohman, A., Ghazali, M. A. B., Windarsih, A., Irnawati, Riyanto, S., Yusof, F. M., & Mustafa, S. (2020). Comprehensive review on application of FTIR spectroscopy coupled with chemometrics for authentication analysis of fats and oils in the food products. *Molecules*, *25*(22). https://doi.org/10.3390/MOLECULES25225485

Sadowska-Bartosz, I., & Bartosz, G. (2021). Biological properties and applications of betalains. *Molecules*, *26*(9), 2520. https://doi.org/10.3390/MOLECULES26092520

Saini, R. K., & Keum, Y. S. (2018). Carotenoid extraction methods: A review of recent developments. *Food Chemistry*, *240*, 90–103. https://doi.org/10.1016/J.FOODCHEM.2017.07.099

Savic, I. M., & Savic Gajic, I. M. (2023). Development of the sustainable extraction procedures of bioactive compounds from industrial food wastes and their application in the products for human uses. *Sustainability*, *15*(3), 2102. https://doi.org/10.3390/SU15032102

Shahidi, F. (2009). Nutraceuticals and functional foods: Whole versus processed foods. *Trends in Food Science & Technology*, *20*(9), 376–387. https://doi.org/10.1016/J.TIFS.2008.08.004

Shakir, B. K., & Simone, V. (2024). Estimation of betalain content in beetroot peel powder. *Italian Journal of Food Science*, *36*(1), 53–57. https://doi.org/10.15586/IJFS.V36I1.2438

Sharma, S., Dedha, A., Gupta, M. M., Singh, N., Gautam, A., & Kumari, A. (2025). Green and sustainable technologies for extraction of carotenoids from natural sources: A comprehensive review. *Preparative Biochemistry and Biotechnology*, *55*(3), 245–277. https://doi.org/10.1080/10826068.2024.2402905;PAGE:STRING:ARTICLE/CHAPTER

Sharma, S., Katoch, V., Kumar, S., & Chatterjee, S. (2021). Functional relationship of vegetable colors and bioactive compounds: Implications in human health. *The Journal of Nutritional Biochemistry*, *92*, 108615. https://doi.org/10.1016/J.JNUTBIO.2021.108615

Shergujri, M. A., Bhatt, D., Chadha, A., & Bhaduri, G. A. (2024). Single-step process for isolation of pure quercetin from aqueous extract of waste onion peels. *ACS Food Science and Technology*. https://doi.org/10.1021/ACSFOODSCITECH.4C00580/SUPPL_FILE/FS4C00580_SI_001.PDF

Shruthi, G. (n.d.). Phytochemical estimation and antibacterial activity of acorus calamus. *International Journal of Advanced Research in Medical & Pharmaceutical Sciences*, *10*. Retrieved July 28, 2025, from www.ijarmps.org

Slavin, J. L., & Lloyd, B. (2012). Health benefits of fruits and vegetables. *Advances in Nutrition*, *3*(4), 506–516. https://doi.org/10.3945/AN.112.002154

Smith, L., López Sánchez, G. F., Veronese, N., Soysal, P., Oh, H., Barnett, Y., Keyes, H., Butler, L., Allen, P., Kostev, K., Jacob, L., Shin, J. I., & Koyanagi, A. (2022). Fruit and vegetable intake and non-communicable diseases among adults aged ≥50 years in low- and middle-income countries. *Journal of Nutrition, Health and Aging*, *26*(11), 1003–1009. https://doi.org/10.1007/S12603-022-1855-Z

Şöhretoğlu, D. (2021). Advances in NMR techniques applicable to phytochemical analysis. *Phytochemical Analysis*, *32*(1), 5–6. https://doi.org/10.1002/PCA.2983

Soundararajan, P., & Kim, J. S. (2018). Anti-carcinogenic glucosinolates in cruciferous vegetables and their antagonistic effects on prevention of cancers. *Molecules*, *23*(11), 2983. https://doi.org/10.3390/MOLECULES23112983

Usman, M., Nakagawa, M., & Cheng, S. (2023). Emerging trends in green extraction techniques for bioactive natural products. *Processes*, *11*(12), 3444. https://doi.org/10.3390/PR11123444

Valentino, G., Graziani, V., D'Abrosca, B., Pacifico, S., Fiorentino, A., & Scognamiglio, M. (2020). NMR-based plant metabolomics in nutraceutical research: An overview. *Molecules*, *25*(6), 1444. https://doi.org/10.3390/MOLECULES25061444

Vallejo, F., Tomás-Barberán, F. A., & García-Viguera, C. (2003). Phenolic compound contents in edible parts of broccoli inflorescences after domestic cooking. *Journal of the Science of Food and Agriculture*, *83*(14), 1511–1516. https://doi.org/10.1002/JSFA.1585;WGROUP:STRING:PUBLICATION

Wang, L., Lu, W., Li, J., Hu, J., Ding, R., Lv, M., & Wang, Q. (2019). Optimization of ultrasonic-assisted extraction and purification of zeaxanthin and lutein in corn gluten meal. *Molecules*, *24*(16). https://doi.org/10.3390/MOLECULES24162994

Wang, L., & Weller, C. L. (2006). Recent advances in extraction of nutraceuticals from plants. *Trends in Food Science & Technology*, *17*(6), 300–312. https://doi.org/10.1016/J.TIFS.2005.12.004

Wang, N., Zhu, H., Wang, M., Zhao, S., Sun, G., & Li, Z. (2025). Recent advancements in microwave-assisted extraction of flavonoids: A review. *Food and Bioprocess Technology*, *18*(3), 2083–2100. https://doi.org/10.1007/S11947-024-03574-Y/METRICS

Won, S., & Kwon, K. H. (2024). Green technology for extraction of bioactive compounds from edible plants. *Journal of Oleo Science*, *73*(10), 1249–1265. https://doi.org/10.5650/JOS.ESS24162

Woodside, J. V., Nugent, A. P., Moore, R. E., & Mckinley, M. C. (2023). Fruit and vegetable consumption as a preventative strategy for non-communicable diseases. *Proceedings of the Nutrition Society*, *82*(2). https://doi.org/10.1017/S0029665123002161,

Wu, W., Chen, J., Yu, D., Chen, S., Ye, X., & Zhang, Z. (2021). Analysis of processing effects on glucosinolate profiles in red cabbage by LC-MS/MS in multiple reaction monitoring mode. *Molecules*, *26*(17), 5171. https://doi.org/10.3390/MOLECULES26175171

Zuorro, A. (2020). Enhanced lycopene extraction from tomato peels by optimized mixed-polarity solvent mixtures. *Molecules*, *25*(9), 2038. https://doi.org/10.3390/MOLECULES25092038

4 Vegetable-Based Phyto-Nutraceuticals and Their Health Benefits

Odangowei Inetiminebi Ogidi, Nwanne Ijere Dike, and Wenapere Chidinma Mercy

4.1 INTRODUCTION

Plant-based diets are recognized for their health benefits as well as reduced ecological footprint on food production. These plants include both terrestrial and diversified aquatic plants (Viroli et al., 2023; Barot et al., 2015; Husen, 2021, 2022, 2023a,2023b, 2024a,2024b). Among them, the health benefits and illness prevention potential of vegetables have made them an integral part of healthy diet plans. Beyond their basic nutritional value, vegetables are rich in bioactive compounds known as phytonutrients, which include polyphenols, carotenoids, flavonoids, glucosinolates, and dietary fibre. These compounds, collectively referred to as phyto-nutraceuticals, have attracted significant attention in recent years for their therapeutic potential. Phyto-nutraceuticals derived from vegetables are being increasingly incorporated into functional foods, dietary supplements, and nutraceutical products aimed at addressing chronic diseases and enhancing overall well-being (Barwant et al., 2025a, 2025b; Boeing et al., 2012). The health benefits of vegetable-based phyto-nutraceuticals are attributed to their ability to combat oxidative stress, reduce inflammation, and modulate cellular and molecular pathways involved in disease progression. One such chemical is SFN, which is abundant in cruciferous vegetables and has antioxidant and anticancer effects (Traka & Mithen, 2009). Similarly, carotenoids like beta-carotene from carrots and lycopene from tomatoes are associated with improved cardiovascular and eye health, owing to their antioxidative and anti-inflammatory actions (Yasmeen et al., 2025). In addition to supporting eye health, carotenoids found in foods like carrots and leafy greens also have anticancer effects through regulating immune responses and preventing tumour growth (Krinsky & Johnson, 2005). Furthermore, phytonutrients are essential for metabolic process regulation. The cruciferous vegetable family, which includes broccoli, Brussels sprouts, and SFN, is one such example. It activates phase II detoxification enzymes, which help eliminate carcinogens and other harmful compounds. In a similar vein, the heart-healthy anthocyanins in red and purple veggies improve vascular function, lower blood pressure, and lower levels of bad cholesterol (LDL) (Liu, 2021). Additionally, dietary fibre found in leafy greens and root vegetables supports gut health by promoting beneficial microbiota and enhancing metabolic regulation (Ogwu et al., 2025; Slavin, 2013). Diabetes, obesity, CVD, and cancer are just a few of the chronic diseases that can be helped and managed with the use of phyto-nutraceuticals derived from vegetables.

Vegetable phytonutrients impact gut health by enhancing intestinal barrier function and encouraging the growth of beneficial gut bacteria. This will lead to better digestion, less inflammation throughout the body, and fewer gastrointestinal problems. According to new evidence, phytonutrient consumption may be connected with better cognitive function and a lower risk of age-related neurodegenerative disorders including Parkinson's and Alzheimer's (Khan et al., 2020). In addition to their medicinal uses, phytonutrients provide a great opportunity for healthcare prevention,

DOI: 10.1201/9781003606710-4

especially in light of the worldwide epidemic of lifestyle-related illnesses. The 'food as medicine' movement has been growing in popularity, with proponents arguing that a diet abundant in various coloured vegetables is the best way to receive all the phytonutrients your body needs. In addition to the obvious positive effects on health, phytonutrients help keep farming techniques sustainable as they come from plants that grow well in a variety of environments and don't need as many artificial inputs like fertilizers and pesticides (Liu, 2021). Epidemiological studies consistently show that higher vegetable consumption is linked to reduced morbidity and mortality from these diseases (Boeing et al., 2012). Moreover, the integration of phytonutrients into functional foods and nutraceuticals provides a practical approach to delivering health benefits in populations with limited access to fresh vegetables. This chapter highlights the emerging field of vegetable-based phyto-nutraceuticals, emphasizing their potential to revolutionize preventive healthcare.

4.2 MECHANISM OF VEGETABLE-BASED PHYTO-NUTRACEUTICALS AND THEIR HEALTH BENEFITS

The phyto-nutraceuticals found in vegetables are bioactive substances that have several uses beyond just nourishment. Incorporating dietary fibres, polyphenols, flavonoids, carotenoids, and glucosinolates into your diet can improve your health in many ways. Their principal function is to shield cells from oxidative damage, inflammation, and long-term health problems like cancer, heart disease, and neurological disorders.

Akhigbe et al. (2025) found that people whose diets included these plant-based chemicals had superior immunological function, metabolic health, and cognitive performance.

Phyto-nutraceuticals derived from vegetables work primarily through their antioxidant activity, which helps fight free radical-induced oxidative stress. Several long-term health problems and ageing are associated with oxidative stress. Vegetables including berries, onions, and spinach contain polyphenols, which protect cells from free radical damage. Atherosclerosis and neurological illnesses can be lessened by consuming carotenoids, which include lycopene in tomatoes and beta-carotene in carrots. These carotenoids are powerful antioxidants which prevent the oxidation of lipids. By increasing levels of antioxidant enzymes such as catalase and superoxide dismutase, these chemicals also improve cellular defence systems (Ogidi et al., 2024a,2024b).

Another critical function of vegetable-based phytonutrients is their anti-inflammatory properties, which help regulate immune responses and prevent chronic inflammation-related diseases. Flavonoids, present in vegetables like kale and broccoli, inhibit pro-inflammatory cytokines such as TNF-alpha and interleukins. SFN, a compound found in cruciferous vegetables like broccoli and Brussels sprouts, modulates the NF-κB signalling pathway, reducing inflammation at the molecular level. By controlling inflammation, these phytonutrients help lower the risk of autoimmune disorders, CVD, and metabolic syndromes such as obesity and diabetes (Azimova et al., 2025).

Phyto-nutraceuticals derived from vegetables also have an important function in the fight against cancer and prevention of cancer. Metabolized from glucosinolates in cruciferous vegetables are isothiocyanates, which protect healthy cells against cancer by inducing apoptosis (programmed cell death) in malignant ones. Another phytochemical found in cabbage and broccoli, indole-3-carbinol, aids in the detoxification of oestrogen metabolites which are carcinogenic and lowers the incidence of hormone-dependent malignancies including prostate and breast cancer. Further to their ability to fight cancer, these chemicals also impede angiogenesis, the process of creating new blood vessels to support tumour growth, and they prevent DNA alterations (Ogidi et al., 2024c,2024d).

Protecting the cardiovascular system is another important way in which phytonutrients help the body. Vegetables high in nitrate, such as spinach, beets, and lettuce, increase the body's synthesis of nitric oxide, which in turn causes vasodilation and better blood circulation. Because of this impact, blood pressure is lowered, and the risk of CVD is decreased. Vegetables high in dietary fibre, such as carrots, beans, and Brussels sprouts, aid in cholesterol reduction by inhibiting the reabsorption of bile acids and increasing their excretion. Ogidi (2024) found that a lower incidence of CVD and strokes was associated with a diet high in fibre and antioxidant substances.

In addition to improving overall health, phytonutrients found in vegetables also help the digestive system. According to Ogidi and Felagha (2024), prebiotic fibres, which are present in vegetables such as leeks, onions, and asparagus, support a healthy microbiome by feeding good bacteria in the stomach. Metabolites produced by gut microorganisms on polyphenols boost metabolic health and immunological function. Digestive function, food absorption, and defence against harmful bacteria are all dependent on a balanced gut microbiome. Important for metabolic health are SCFAs generated in the gut by fibre fermentation; these SCFAs have anti-inflammatory and insulin-sensitivity-enhancing properties (Ogidi & Felagha, 2024).

The phyto-nutraceuticals found in vegetables have multiple health benefits, including those for the nervous system and brain function. Ogidi and Ajoko (2024) demonstrated that flavonoids like curcumin (turmeric) and luteolin (celery) can lower neuroinflammation by crossing the blood-brain barrier. Being important chemicals in neurogenesis and cognitive function, these drugs also increase BDNF synthesis. Ogidi and Emaikwu (2024) found that people who regularly eat vegetables that are rich in flavonoids have a lower risk of developing neurodegenerative diseases such as Alzheimer's and Parkinson's. Phyto-nutraceuticals derived from vegetables have several health benefits. They can reduce inflammation, defend against cancer, aid the cardiovascular system, modulate the gut flora, and even shield the nervous system from damage. These compounds offer a natural and effective way to prevent and manage chronic diseases while promoting overall well-being. By incorporating a variety of phytonutrient-rich vegetables into daily meals, individuals can harness these powerful health benefits and support long-term health (Ogidi et al., 2024a).

4.3 HEALTH-PROMOTING PHYTO-NUTRACEUTICALS IN VEGETABLES

All chemical compounds with health-promoting characteristics which are derived from plants are called phyto-nutraceuticals. Vegetables contain the bulk of these chemicals, although at much lower concentrations. Nevertheless, they play a considerable role in safeguarding live cells from chronic diseases when ingested in enough numbers. Below, we will go over the most essential phyto-nutraceuticals present in vegetables and know how they can biologically battle chronic diseases. Table 4.1 shows phyto-nutraceuticals from different vegetables, with their experimental findings.

4.3.1 Terpenoids

Although terpenoids are most usually found in plants, some of them have been extracted from other sources. The fragrant oils found in plants and flowers are termed terpenoids. Higher plants, conifers, citrus, and eucalyptus all have them in their leaves and fruits (Breitmaier, 2008). The volatile liquid turpentine was collected from pine trees, and the components it contained were subsequently recognized to be terpenes. Essential oils generated from specific plant and tree tissues and sap are predominantly made of mono- and sesquiterpene, which are simpler hydrocarbons (Croteau et al., 2000). Steam is not a solvent for the di and tri terpenoids. Their source is the gums and resins found in plants and trees. Tetraterpenoids belong to a specific category of compounds referred to as carotenoids. More commonly, terpenoids are used; this word encompasses hydrocarbons and oxygenated versions of them. But the word terpene is already being used to describe terpenoids by some writers (Maggi et al., 2009). In the context of interactions between stationary plants and other species, such as defence, reproduction, or symbiosis, different terpenoids are thought to perform separate but complementary roles (Gershenzon & Dudareva, 2007). The chemical structure of terpenoids is depicted in Figure 4.1b.

TABLE 4.1
Experimental Data on Health-Promoting Phyto-Nutraceuticals in Vegetables

Vegetable	Phyto-Nutraceutical Compound	Health Benefits	Experimental Findings	Reference
Broccoli	Sulforaphane	Anticancer, anti-inflammatory	Sulforaphane inhibits cancer cell proliferation and induces apoptosis in prostate cancer cells.	Zhang et al. (2020)
Carrots	Beta-carotene	Antioxidant, vision health	Beta-carotene reduces oxidative stress and supports eye health by preventing macular degeneration.	Smith and Brown (2019)
Spinach	Lutein, zeaxanthin	Cognitive function, eye health	Lutein supplementation improves cognitive performance and reduces inflammation in elderly individuals.	Miller et al. (2021)
Tomatoes	Lycopene	Cardiovascular health, cancer prevention	Lycopene intake is associated with reduced risk of cardiovascular disease and lower LDL oxidation.	Garcia et al. (2022)
Garlic	Allicin	Antimicrobial, immune-boosting	Allicin exhibits strong antibacterial effects against drug-resistant pathogens.	Lee and Kim (2023)
Onions	Quercetin	Anti-inflammatory, cardiovascular protection	Quercetin reduces blood pressure and enhances endothelial function in hypertensive patients.	Thompson et al. (2020)
Turmeric	Curcumin	Neuroprotection, anti-inflammatory	Curcumin improves memory and reduces neuroinflammation in Alzheimer's patients.	Patel et al. (2018)
Beetroot	Nitrates	Vasodilation, athletic performance	Dietary nitrates improve endurance and lower blood pressure through nitric oxide production.	Robinson and Harris (2021)
Red Cabbage	Anthocyanins	Anti-inflammatory, brain health	Anthocyanins improve cognitive function and reduce oxidative stress in ageing populations.	Davis et al. (2021)
Brussels Sprouts	Indole-3-Carbinol	Detoxification, hormone regulation	Indole-3-carbinol modulates oestrogen metabolism, reducing the risk of hormone-related cancers.	Parker et al. (2020)
Bell Peppers	Capsaicin	Metabolic health, pain relief	Capsaicin promotes fat oxidation and reduces neuropathic pain.	Williams et al. (2019)
Sweet Potatoes	Anthocyanins, beta-carotene	Skin health, immune support	Anthocyanin-rich sweet potatoes enhance skin elasticity and provide anti-aging benefits.	Hernandez et al. (2022)
Cabbage	Glucosinolates	Anticancer, gut health	Glucosinolates support detoxification pathways and gut microbiome diversity.	Foster et al. (2021)
Kale	Flavonoids, vitamin K	Bone health, cardiovascular protection	Kale consumption improves bone mineral density and reduces arterial stiffness.	Turner et al. (2022)
Celery	Apigenin	Anti-inflammatory, neuroprotection	Apigenin modulates inflammatory pathways and reduces risk of neurodegenerative diseases.	Anderson et al. (2020)

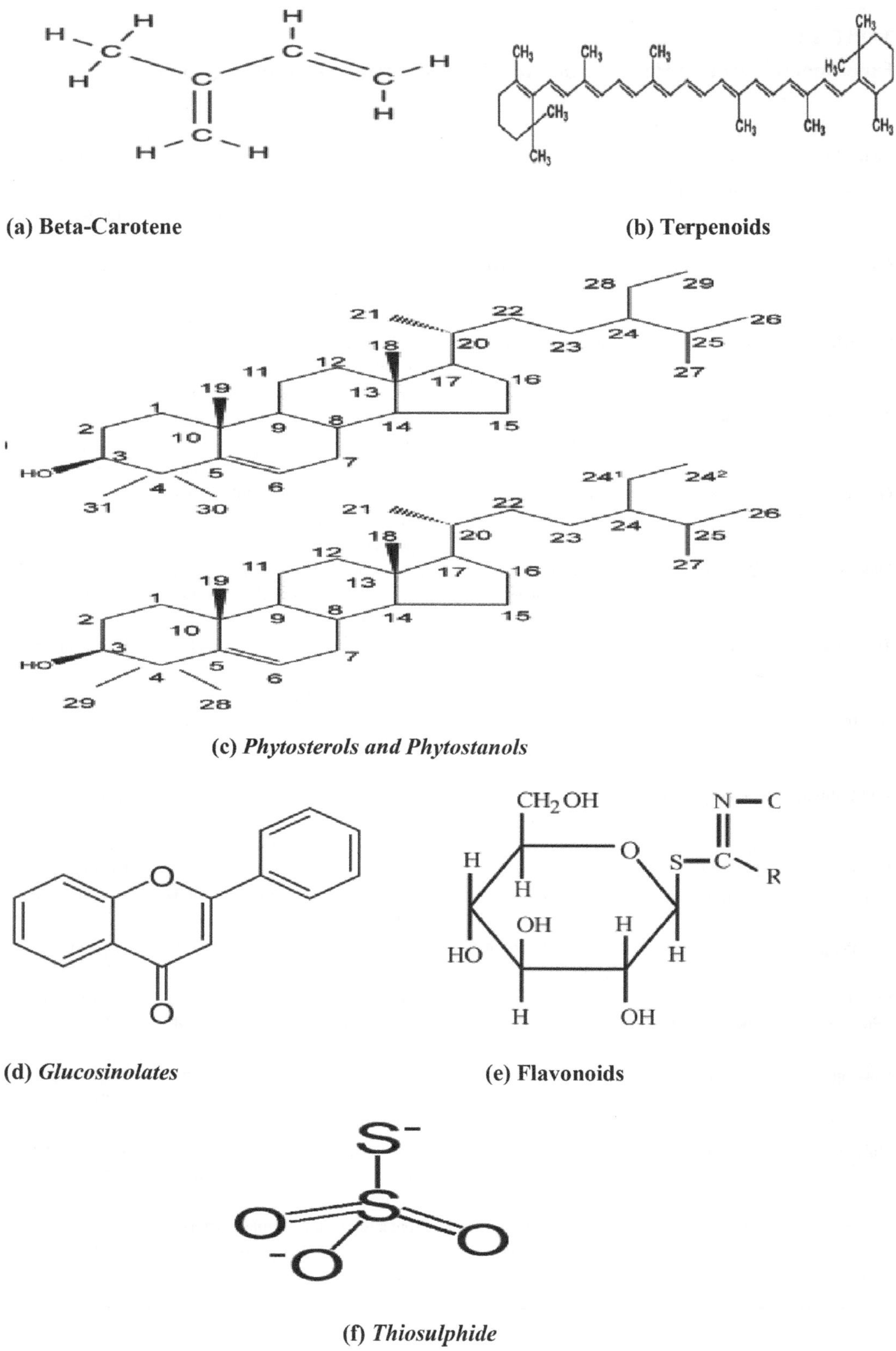

FIGURE 4.1 Chemical structures of phyto-nutraceuticals in vegetables.

4.3.1.1 Potential Health Benefits

Photosynthetic reaction centres, comprising terpenoids, bacteriochlorophylls, rhodopsins, and carotenoids, transform light into chemical energy (Manach et al., 2005). These centres are crucial for the development and sustenance of photosynthetic organisms. Terpenoids have crucial roles in stress response and defence processes, among other recognized functions of plants (Tholl, 2006). Terpenoids are utilized in various domains, including cosmetics, agrichemicals, functional foods, colourants, fragrances, flavourings, and disinfectants. Their biological functions are varied and extensive. One example is the terpenoid artemisinin, which has antimalarial properties (Ogidi et al., 2025a; Van et al., 1999). There are many terpenoids that are very valuable in the business world. Some chemicals, like menthol and sclareol, are used in food and cosmetics as flavourings and fragrances. The quality of agricultural goods is greatly affected by compounds like linalool, which affect things like the flavour of fruits and the scent of flowers (Harbourne et al., 2013). The unique antioxidant properties of terpenes stem from their elongated carbon side chains, enabling them to interact with free radicals through partitioning into lipid membranes. Tocotrienols and tocopherols have been extensively studied for their antioxidant effects among terpenes (Dillard & German, 2000).

4.3.2 Carotenoids

The fat-soluble carotenoids are pigments that exist naturally. It was in the nineteenth century that carotenoids were first recognized. Plant chloroplasts contain carotenoids, which are organic pigments that can have three colours: red, yellow, and orange. The carotenoids help protect against cancer and heart disease as they are strong antioxidants. Precursors to vitamin A are carotenoids. The hydrocarbon molecules known as carotenoids are responsible for the colour. With 40 carbon atoms and a complex conjugated double bond structure, these compounds are tetraterpenes. An increase in the quantity of conjugated double bonds results in a pale yellow, orange, or red hue. Plants, algae, fungi, and bacteria possess the capability to biosynthesize carotenoids (Aizawa & Inakuma, 2007). Figure 4.1a shows the chemical structure of beta-carotene.

4.3.2.1 Potential Health Benefits of Carotenoids

Carotenoids are particularly beneficial to human health due to their anti-inflammatory and antioxidant effects. The ability of these substances to neutralize free radicals offers numerous benefits, including a reduced risk of oxidative stress and chronic illnesses such as CVD and specific cancer types. Lutein and zeaxanthin, two carotenoids, possess antioxidant qualities that aid in the prevention of AMD and cataracts. Beta-carotene, a form of vitamin A, supports both the immune system and ocular health. Research has shown that lycopene helps lower blood pressure and LDL cholesterol, which in turn improves cardiovascular health and lowers the risk of prostate cancer. Furthermore, carotenoids improve skin suppleness and shield skin from UV-induced damage, all of which contribute to better skin health. Arthritis and metabolic syndrome are inflammatory conditions, but their anti-inflammatory properties may make them less likely to develop. Eating foods that are rich in carotenoids, such as sweet potatoes, spinach, tomatoes, and carrots, can help keep you healthy and stave off diseases.

4.3.3 Phytosterols and Phytostanols

Phytosterols are a diverse array of sterols present in plants. The term originates from the Greek words 'phyton', meaning 'plant', and 'stereos', meaning 'solid'. Harbourne et al. (2013) identified more than 250 unique phytosterols and associated compounds in diverse marine and botanical sources. The predominant source of phytosterols in human diets is plant-derived foods such as fruits, cereals, and vegetable oils. Corn, wheat, rye and rice are common dietary sources of phytostanols, which occur in nature in lesser quantities than phytosterols. Phytosterols are present in five common kinds throughout all plant tissues: (a) Phytosteryl hydroxycinnamic acid esters (HSE), (b) acylated steryl glycosides (ASG), (c) steryl esters (SE), (d) steryl glycosides (SG), and (e) Free sterols (FS) (Pascal & Segal,

2006). Similar to cholesterol in humans, they serve a similar purpose in plants. The human body does not produce phytosterols and phytostanols; instead, these compounds must be consumed via food. Vegetable oils, including corn, rapeseed, soybean, and sunflower oils, along with nuts, seeds, and cereals, possess a high concentration of phytosterols and phytostanols (Harbourne et al., 2013). Kardong et al. (2013) identify campestanol, sitostanol, campesterol, and sitosterol as the phytosterols and phytostanols most commonly consumed by people. Phytosterols and phytostanols do not supply the body with energy, but they do have other beneficial effects such as decreasing cholesterol, protecting against cancer, regulating the immune system, and protecting the skin (Ogidi et al., 2025b). Chemical structures of *phytosterols and phytostanols* are shown in Figure 4.1c.

4.3.3.1 Potential Health Benefits of Phytosterols and Phytostanols

The chemicals phytosterols and phytostanols, which come from plants and have a structure similar to cholesterol, are well known for their many health benefits, especially when it comes to maintaining heart health. One way they help lower cholesterol levels is by preventing the intestines from absorbing cholesterol from food. This can lower bad LDL cholesterol without influencing good HDL cholesterol. Because of this, they are useful in lowering the danger of atherosclerosis and coronary heart disease.

Phytosterols and phytostanols not only reduce cholesterol levels, but they also have antioxidant and anti-inflammatory activities, which could further protect against metabolic syndrome and type 2 diabetes. Some cancer cells, including those in the breast, colon, and prostate, may be susceptible to these chemicals' ability to modulate cell growth and induce apoptosis, according to new research. Additionally, phytosterols help modulate the immune system, which improves health and resistance to infections. Vegetable oils, nuts, seeds, and fortified foods (such as yoghurt and margarine) are rich in phytosterols, which can improve cardiovascular health and lower the risk of other chronic diseases.

4.3.4 Flavonoids

Plant phenols, predominantly flavonoids, are low molecular weight compounds with widespread dispersion (Dai & Mumper, 2010). Their inhibitory activity and antioxidant characteristics have recently been the focus of research into cancer progression in animals. The chemical structure of flavonoids is shown in Figure 4.1e. Two other categories of flavonoids are namely, (a) anthocyanins and (b) anthoxanthins, categorized as follows: (a) Anthocyanins – These pigments impart red, blue, and purple colours to vegetables (Clifford, 2000). Furthermore, anthocyanins exhibit anti-inflammatory and anti-obesity properties (Tsuda, 2008; Sasaki et al., 2007). (b) Anthoxanthins – These yellow or colourless compounds are categorized into five primary types: isoflavones, flavonols, flavones, and flavanols. Moreover, among the various types of anthoxanthins and flavonoids, flavonols and flavones are the two most prevalent in plant-based meals.

4.3.4.1 Potential Health Benefits of Flavonoids

The powerful antioxidant, anti-inflammatory, and antimicrobial characteristics of flavonoids – a varied class of polyphenolic chemicals originating from plants – offer a plethora of health advantages. Flavonoids are antioxidants that can be found in a variety of foods, including fruits, vegetables, tea, chocolate, and wine. By scavenging free radicals, they help lower oxidative stress and the danger of conditions like diabetes, CVD, and even some cancers. By improving endothelial function, lowering blood pressure, lowering levels of harmful LDL cholesterol, and boosting levels of good HDL cholesterol, these chemicals support heart health. Flavonoids such as quercetin and kaempferol may have anti-inflammatory properties, which could help with metabolic syndrome and arthritis. The flavonoids included in green tea, like catechins, have the potential to aid with weight loss and metabolic wellness.

Improved cognitive performance and a reduced risk of neurodegenerative illnesses like Alzheimer's and Parkinson's have been linked to flavonoids such as anthocyanins and flavanols. Through their ability to cross the blood-brain barrier and shield neurones, they facilitate memory and learning. In general, flavonoids improve immunity because they affect immune responses and prevent infections from progressing. Consuming a diet higher in flavonoid-rich foods is a simple and effective approach to enhance overall health and reduce the risk of age-related and chronic diseases.

4.3.5 Glucosinolates

Cruciferous plants, especially Brassica species, possess sulphur-containing glucosides known as glucosinolates (GLS) within their cell vacuoles. Cabbage, broccoli, rapeseed, mustard seed, and many condiments contain them in substantial amounts; they are also present in oilseeds (Rodriguez et al., 2006). Chen and Andreasson (2001) identify around 120 unique glucosinolates. The quantity and composition of these chemicals may vary based on the plant species, agricultural practices, and climatic conditions. Most studies have focused on the broccoli-derived isothiocyanate SFN because of its role as an inducer of a specific class of phase II enzymes which have strong anticancer effects. Rodriguez et al. (2006) showed that glucoraphanin concentrations in three-day-old broccoli sprouts were 10–100 times higher than in fully grown plants. This indicates that glucosinolate levels persist in rising post-harvest. The therapeutic importance of an individual phytochemical has been challenged; nonetheless, the significance of an overall dietary pattern in mitigating cancer risk is acknowledged (Oguoma et al., 2025). Figure 4.1d depicts the chemical structure of *glucosinolates.*

4.3.5.1 Potential Health Benefits of Glucosinolates

Cruciferous vegetables, which contain sulphur-containing chemicals called glucosinolates, have several beneficial health effects because of the bioactive by-products of their breakdown, which include indoles and isothiocyanates. Contributing to general health and illness prevention, these chemicals display powerful antioxidant, anti-inflammatory, and anticancer capabilities (Okara et al., 2025; Ogidi & Omu, 2025). Glucosinolates' ability to stave against cancer is their best-documented advantage. In particular, studies on colorectal, breast, and prostate cancers have demonstrated that isothiocyanates like SFN and indoles like indole-3-carbinol can decrease tumour growth and spread by encouraging apoptosis and detoxification of carcinogens. Glucosinolates also help the liver eliminate toxins by increasing the activity of phase II detoxification enzymes.

Glucosinolates have the potential to alleviate oxidative stress and inflammation, two factors linked to long-term health problems like diabetes and CVD. In addition to bolstering the immune system and restoring harmony to the intestinal microbiota, their antibacterial qualities promote gut health. A simple and effective way to improve health, detoxify better, and lower the risk of chronic diseases is to eat more glucosinolate-rich veggies (Ogidi & Adigwe, 2025).

4.3.6 Thiosulphide

Organosulphur compounds, including thiosulphides, are found in alliums (Desjardins, 2008). The species of Allium, farming techniques, and processing methods all influence the composition of OSCs. Some thiosulphides require mechanical exposure, such as cutting or chewing, for formation, which is absent in the bulbs. Crushed garlic has 23 OSCs, but whole garlic bulbs possess 16 OSCs (Verma & Verma, 2008). A separate investigation identified 33 OSCs in fresh garlic, according to Kalra et al. (2006). In the cytoplasm of mature Allium tissues, the predominant thiosulphides discovered are glutamylcysteine, S-methylcysteine sulfoxide, and alliin (Verma & Verma, 2008; Rose et al., 2005). Alongside ethiin, another minor thiosulphide is S-propenyl-cysteine sulfoxide, commonly referred to as isoalliin (Verma & Verma, 2008; Rubec et al., 1999). Aside from methiin, identified in cabbage leaves and certain Brassicaceae, no thiosulphides documented in alliums have been observed in any other plants (Thomas & Parkin, 1994). In recent years, numerous studies have quantified Allium thiosulphide levels (Verma & Verma, 2008; Rose et al., 2005; Yin et al., 2002). The diverse and inconsistent biosynthetic pathways of thiosulphides in alliums have resulted in significant shortcomings. Alliin is highly unstable, while being the most prevalent thiosulphide in garlic. The pungent smell and unique aroma of garlic come from substances that are soluble in lipids. Both the kind and amount of thiosulphides produced by alliums were affected by the substrate's properties and the amount of alliinase activity. Two varieties of thiosulphides, S-propenyl-cystein sulfoxide and S-methylcysteine sulfoxide, have been recognized in onions (Rose et al., 2005; Thomas & Parkin, 1994; Yoo & Pike, 1998). Various Allium species exhibit distinct scents and flavours due to differing amounts of volatile sulphur compounds. Both raw and cooked vegetable alliums contain over 80 identified volatile compounds belonging to this category (Brewster, 2008). They play a crucial role in imparting bitterness to certain onions as well (Randle & Lancaster, 2002). Figure 4.1f shows the chemical structure of *thiosulphides*.

4.3.6.1 Potential Health Benefits of Thiosulphides

The sulphur-containing chemicals called thiosulphides are abundant in Allium plants like garlic, onions, leeks, and shallots. These veggies have anti-inflammatory, antibacterial, and antioxidant effects which are really beneficial to your health. Reduced risk of atherosclerosis and coronary heart disease is a result of the lowering of blood pressure, reduction of LDL (bad) cholesterol, and prevention of platelet aggregation, all of which are contributed to by these bioactive substances, which include diallyl disulphide and allicin.

Particularly effective against gastrointestinal and breast malignancies, thiosulphides promote detoxification, induce apoptosis in cancer cells, and prevent tumour growth and spread. They reduce the likelihood of developing chronic disorders including metabolic syndrome and arthritis by reducing inflammation. In addition to warding off microbial, fungal, and viral infections, thiosulphides boost immunological function by activating immune cells. In addition to promoting liver health, these chemicals help remove toxins by increasing the activity of detoxification enzymes. A natural and effective approach to improve general health and decrease the risk of infectious and chronic diseases is to regularly consume foods rich in thiosulphides (Ogidi & Oguoma, 2025).

4.4 PHYTO-NUTRACEUTICAL VEGETABLES AND HEALTH BENEFITS

An increasing number of individuals are recognizing that consuming an abundance of vegetables is an excellent method to obtain essential vitamins, minerals, fibre, and phytochemicals required by the body. All veggies may assist individuals in preventing the onset of chronic diseases. Crucifers and alliums are the only vegetables which include glycosinolates and thiosulphides. The primary characteristic is that the phyto-nutraceutical combinations and quantities present in various vegetable groups differ from each other and from other vegetables within the same group. Crucifers, alliums, cucurbits, legumes, and solanaceous vegetables (Asteraceae, Apiaceae, Chenopodiaceae) are extensively studied and consumed vegetables, and their health benefits will be emphasized herein. Table 4.2 shows list of various vegetables, their key phytonutrients, and associated health benefits.

4.4.1 Crucifers

The predominant source of glucosinolates in human diets is cruciferous vegetables, belonging to the Brassicaceae family. Many other types of vegetables are included in this, such as cabbage, broccoli, cauliflower, kale, and many more. There are ample number of glucosinolates in crucifers which include kale, cabbage, and broccoli. Various types of malignancies, including those caused by chemicals, can be prevented by eating these vegetables, according to research (Ambrosone et al., 2004; Brennan et al., 2005; Kirsh et al., 2007; Traka, 2010). The risk of certain cancers can be lowered, according to epidemiological research, by eating 10 g of cruciferous vegetables each day (Ambrosone et al., 2004; Brennan et al., 2005; Seow et al., 2002). The glucosinolate profiles of different accessions within each crucifer, as well as different parts of the plant, stages of growth, agronomic methods, and climates, fluctuate significantly in both quantity and quality (Fahey et al., 2001). The main glucosinolate, according to Kushad et al. (1999), is glucoraphanin, and the concentration difference between the two varieties of broccoli is more than 27 times. The highest concentration was seen in

TABLE 4.2
Vegetables, Key Phytonutrients, and Their Health Benefits

Vegetable	Key Phytonutrient(s)	Health Benefits	Additional Information	Reference
Broccoli	Sulforaphane, Vitamin C, Fibre	Anticancer, immune-boosting, anti-inflammatory	Sulforaphane activates detoxification enzymes and reduces oxidative stress.	Traka and Mithen (2009), Zhang et al. (2020), Vallejo et al. (2002)
Carrots	Beta-carotene, Lutein	Vision health, antioxidant protection, skin health	Beta-carotene is converted into vitamin A, essential for eye health.	Smith and Brown (2019), Rao and Rao (2007)
Spinach	Lutein, Zeaxanthin, Iron	Cognitive function, eye health, bone strength	Rich in iron, which supports red blood cell formation and prevents anaemia.	Slavin (2013), Scott and Hart (1995)
Tomatoes	Lycopene, Vitamin C	Heart health, cancer prevention, skin protection	Lycopene helps reduce oxidative damage and protects against UV-induced skin ageing.	Rao (2006), Stewart et al. (2000)
Garlic	Allicin, Sulphur Compounds	Antimicrobial, immune support, cardiovascular health	Allicin has strong antibacterial and antifungal properties.	Rose et al. (2005), Verma and Verma (2008), Yin et al. (2002)
Onions	Quercetin, Sulphur Compounds	Anti-inflammatory, blood pressure regulation	Quercetin has antihypertensive properties and protects against heart disease.	Randle and Lancaster (2002), Yoo and Pike (1998)
Turmeric	Curcumin	Neuroprotection, anti-inflammatory, joint health	Curcumin improves brain function and reduces inflammation in arthritis.	Patel et al. (2018)
Beetroot	Nitrates, Betalains	Athletic performance, blood pressure control	Nitrates enhance oxygen efficiency, improving endurance in athletes.	Robinson and Harris (2021)
Red Cabbage	Anthocyanins, Vitamin K	Brain health, anti-aging, digestive support	Anthocyanins reduce oxidative stress and support memory function.	Tsuda (2008), Rao and Rao (2007)
Brussels Sprouts	Indole-3-Carbinol, Fibre	Detoxification, hormone balance, digestive health	Indole-3-carbinol supports liver detox and oestrogen metabolism.	Parker et al. (2020), Seow et al. (2002)
Bell Peppers	Capsaicin, Vitamin C	Metabolic health, immune support, skin health	High vitamin C content boosts collagen production for skin health.	Williams et al. (2019)

(Continued)

TABLE 4.2 ***(Continued)***
Vegetables, Key Phytonutrients, and Their Health Benefits

Vegetable	Key Phytonutrient(s)	Health Benefits	Additional Information	Reference
Sweet Potatoes	Anthocyanins, Beta-carotene	Skin health, anti-aging, blood sugar regulation	Anthocyanins act as natural anti-inflammatory agents, beneficial for skin.	Tsuda (2008), Rao and Rao (2007)
Cabbage	Glucosinolates, Fibre	Gut health, cancer prevention, digestion	Rich in fibre, aiding digestion and promoting a healthy gut microbiome.	Traka and Mithen (2009), Vallejo et al. (2003)
Kale	Flavonoids, Vitamin K, Iron	Bone health, heart health, anti-inflammatory effects	Vitamin K is essential for bone mineralization and blood clotting.	Turner et al. (2022)
Celery	Apigenin, Polyphenols	Anti-inflammatory, digestive health, hydration	Apigenin has anticancer properties and helps regulate blood sugar.	Slavin (2013)
Asparagus	Saponins, Folate	Brain function, heart health, diuretic properties	High folate content supports foetal development and cardiovascular health.	Scholz-Ahrens et al. (2001)
Eggplant	Nasunin, Fibre	Brain health, cholesterol reduction, antioxidant	Nasunin protects brain cells from oxidative stress and inflammation.	Tsuda (2008)
Mushrooms	Ergothioneine, Beta-glucans	Immune boosting, anti-inflammatory, anti-aging	Beta-glucans enhance immune function and support gut health.	Ogidi and Adigwe (2025), Ogidi and Oguoma (2025)
Cauliflower	Glucosinolates, Choline	Detoxification, brain health, anti-inflammatory	Choline is essential for brain development and liver function.	Traka (2010), Traka and Mithen (2009)
Green Beans	Polyphenols, Vitamin K	Bone health, heart health, digestive support	Rich in fibre and antioxidants, aiding digestion and reducing inflammation.	Slavin (2013)
Zucchini	Lutein, Zeaxanthin, Vitamin A	Eye health, skin hydration, digestive support	High water content supports hydration and promotes healthy skin.	Rao and Rao (2007)
Pumpkin	Beta-carotene, Fibre	Heart health, immune support, skin regeneration	Beta-carotene boosts collagen production and supports wound healing.	Rao and Rao (2007), Smith and Brown (2019)
Cucumber	Flavonoids, Vitamin K, Water	Hydration, skin health, digestion	High water content keeps the body hydrated and aids digestion.	Slavin (2013)
Radish	Anthocyanins, Glucosinolates	Liver detoxification, digestion, anti-inflammatory	Supports liver function and promotes bile production.	Traka and Mithen (2009), Tsuda (2008)

the 'Brigadier' variety, while the lowest was in the 'EV6–1' variety. Studies have shown that glucoraphanin levels in red cabbage are much higher than in white cabbage. There was a lot of diversity among the several red cabbage varieties. According to Fahey et al. (2001), the amount of glucosinolates in broccoli sprouts is 20–50 times higher than in fully grown plant material. According to Wang et al. (2004), Vallejo et al. (2002, 2003), Nilsson et al. (2006), and Borkowski et al. (2008), the main glucosinolates present in broccoli heads are glucoraphanin, glucobrassicin, progoitrin, and gluconasturtiin. Sinigrin, progoitrin, and glucobrassicin are the main glucosinolates found in bell peppers, cabbage, Brussels sprouts, kale, tronchuda, and collard (Cartea et al., 2008). Glucoerucin, glucoraphanin, and glucobrassicin are the main glucosinolates found in turnips and rutabagas (Cartea et al., 2008). Radish has three main glucosinolates: glucoerucin, glucoraphanin, and glucobrassicin (Ciska et al., 2000). Other, less common glucosinolates are present in all of these crucifers. Out of 22 vegetables analysed, cruciferous veggies like kale came in at number two for vitamin content. The vitamin content of broccoli and Brussels sprouts has been highly praised because of their abundance of beta-carotene, vitamins C and E, and other nutrients (Cao et al., 1996). All five of the crucifers studied had vitamin C as their primary nutrient (Kurilich et al., 1999). Cruciferous veggies are rich in folate as well. Broccoli (70 to 90 µg/100 g) and Brussels sprouts (110 to 135 µg/100 g) are the top vegetable suppliers of folate (Kurilich et al., 1999). Crucifers include a lot of beneficial minerals and fibre. According to Kurilich et al. (1999), nutritious fibre makes up around 40% of cauliflower's dry weight, or around 5% of its fresh weight, and is composed of nonstarch polysaccharides. Brussels sprouts had cellulose concentrations of 36% and lignin concentrations of 14.5% of total dry matter, respectively, while cauliflower had values of about 16% and 13%, respectively (Rahn et al., 1999). There is no shortage of calcium-rich cruciferous vegetables; some examples are kale, tronchuda cabbage, pak-choy, kailan, rutabaga, turnip, and Brussels sprouts.

Although broccoli is classified as a Brassica vegetable, Galega kale surpasses it in protein, fibre, calcium, and sulphur content. Crucifers possess minimal selenium concentrations intrinsically, although they can collect substantial quantities when grown in selenium-rich soils. Broccoli, as reported by Pasquali et al. (2008), accumulated sevenfold the selenium content compared to cabbage, collards, and Swiss chard when cultivated in selenium-rich soil. The selenium concentration in edible florets of broccoli plants fertilized with sodium selenate and selenite following outdoor cultivation in a growth medium composed of sphagnum peat moss and vermiculite. Selennocysteine is the selenium variant found in broccoli. Numerous animal studies have shown that selenium-enriched broccoli offers a preventive effect against colon and breast malignancies (Ramakrishn & Jailkhani, 2008; Michikawa et al., 2009). Flavonoids represent an additional category of antioxidants present in cruciferous vegetables.

4.4.2 Alliums

The Allium family (Alliaceae) includes several vegetables such as garlic, leeks, chives, and Welsh onion. They possess several thiosulphides, which have been linked to a reduction in some chronic diseases. Thiosulphides in alliums exhibit significant variation in both kind and quantity, akin to the glucosinolates found in cruciferous vegetables. Among the several phytonutrients found in alliums, flavonoids rank second in importance. Red onions contain anthocyanins, while most yellow fresh varieties of onions contain flavonols including kaempferol and quercetin. Lila (2004) found that among 62 vegetables, the overall flavonoid content was highest in onion leaves. In terms of flavonoids per kilogramme of dry weight, onion leaves contained approximately 2.7 g/kg, whereas garlic had about 1.0 g/kg (Miean & Mohamed, 2001). Of the total flavonoids found in onion leaves, luteolin makes about 14%, kaempferol 31%, and quercetin about 55%. Of the flavonoids found in onion bulbs, quercetin makes up over 95%, whereas kaempferol is present in very small amounts. According to research, the majority of the quercetin in red onion cultivars is concentrated in the outer scales, whereas white onion cultivars have a far lower quercetin content. In terms of total flavonols per gram of fresh weight, red onions had about 1,350 g/g, whereas white onions had only 10 g/g.

The majority of the quercetin in onions was found in the free aglycone form. Of the flavonoids found in garlic cloves, myricetin makes up 72%, apigenin 23%, and quercetin 5% (Miean & Mohamed, 2001). The most common flavonoid in leeks, garlic chives, and chives is kaempferol. One can get as much as 10% of their recommended daily allowance of calcium, potassium, and manganese from onions and garlic. Onions and garlic have trace amounts of selenium, but when grown in soils high in selenium, these plants can collect selenium and produce selenocysteine and selenoproteins. Ip and Lisk (1994) state that as much as 0.02 ppm of selenium is present in 'Stuttgart' onion bulbs and 0.06 ppm in 'Valencia' topset garlic cloves. There is a lot of chromium in onions, making them a great source of this mineral. In diabetic individuals, chromium has been found to enhance glucose tolerance, lower insulin levels, and lower fasting glucose levels. Two hundred grams of onions have enough chromium to meet up to 20% of your daily needs. Onions and other high-fibre vegetables contain the polyfructosan inulin. Extensive study spanning more than a decade has confirmed that inulin-type fructans have positive effects on human health (DGAC, 2005). It changes the gut bacterial mycoflora, reduces the prevalence of dangerous bacteria, and imparts prebiotic qualities. Another fructan included in onions, neokestose, has now been shown to successfully encourage the growth of good bacteria (Kilian et al., 2002). Fruits may help prevent osteoporosis (Scholz-Ahrens et al., 2001), and they also aid in calcium absorption. Research shows that middle-aged people whose diets are heavy in fructose have lower concentrations of insulin, cholesterol, triacylglycerol, phospholipids, and glucose in their blood. Flavonoids and inulin, a prebiotic polysaccharide, are found in onions and may have antidiabetic effects (Srinivasan, 2005). Digestion enzymes are not effective in breaking down inulin.

4.4.2.1 Solanaceous Vegetables

Significant vegetables in the Solanaceae family encompass eggplant, peppers, tomatoes, and potatoes. We shall examine each of these solanaceous vegetables individually due to significant variations in their phytonutrient content.

4.4.3 Tomato

Tomatoes include carotenoids, the most important of which are neurosporene, phytoene, lycopene, and carotenes (Clinton, 1998). There is no food on this planet that compares to tomatoes and tomato-based dishes in terms of lycopene concentration. The average concentration of lycopene in tomatoes, measured on a fresh weight basis, is 35 mg/kg (Scott & Hart, 1995). Lycopene concentrations range from 5 mg/kg in yellow cultivars to 90 mg/kg in red ones. Sauces, pastes, liquids, and ketchup made with tomatoes have two to four times the lycopene concentration of fresh tomatoes (Srinivasan, 2005; Gerster, 1997). In the American diet, tomatoes are a significant source of provitamin A and vitamin A due to their 0.6 to 2.0 mg/kg concentrations of α-, β-, γ-, and σ-carotene. Research from a variety of sources, including studies (Leonardi et al., 2000; Albushita et al., 2000; Arab & Steck, 2000), indicate that diets which are high in tomato and tomato-based products and are rich in the carotenoid lycopene have a decreased chance of getting stomach and rectal cancers (Giovannucci, 1999). Both potassium and lycopene, which are abundant in tomatoes, are staples in the diets of industrialized countries like the United States (DGAC, 2005). Tomato fruits contain a high quantity of ascorbic acid (around 200 mg/kg), making them the second-most abundant source of vitamin C, behind citrus fruits (Rao & Rao, 2007). Tomatoes contain trace levels of lutein, conjugated flavonoids, and α-, β-, and γ-tocopherols (1–2 mg/kg) (Leonardi et al., 2000; Albushita et al., 2000). While fresh tomatoes only contain conjugated flavonoids like quercetin and kaempferol, processed tomatoes have a lot of free flavonoids. Both culture and cultivar determine the amount of flavonoids present. For instance compared to ordinary or meaty tomato varieties, cherry tomatoes contain a substantially higher concentration of flavonoids, and tomatoes produced in the field have a higher concentration of flavonoids than tomatoes cultivated in a greenhouse (Crozier et al., 2000). Numerous tomato cultivars with the high crimson gene contain two to four times the typical amount of lycopene, while Doublerich contains twice the normal amount of vitamin C, and 97L97 contains 40 times the normal amount of vitamin A. Tomatoes with elevated anthocyanin concentrations are purple in hue.

4.4.4 Potato

Despite potatoes' reputation as a carbohydrate bomb, they're actually a great provider of many important amino acids. Potatoes may have better protein quality and better protein utilization due to their high lysine and other metabolite levels. Individuals who consume substantial amounts of cereal but little lysine might consider incorporating potatoes into their diet. Potatoes include high-quality proteins, vitamins, minerals, and phytochemicals such as phenolics, phytoalexins, and protease inhibitors. Almost all naturally occurring phenols in potato tubers consist of chlorogenic acid. In addition, 3,4-dicaffeoylquinic, 3,5-dicaffeoylquinic, 4-O-caffeoylquinic, and neo-chlorogenic acid are all present in potatoes (Friedman, 1997). The phytochemical contribution from potatoes is modest yet noteworthy. There are now other types available that are red, yellow, and purple and contain a lot of phytochemicals. Both the skin and the flesh of a purple potato can change colour to blue when cooked. Mutations which occur at the P locus of cultivars produce the antioxidant anthocyanin (Giovannucci, 1999). There are between half a gram and 1.5 kilograms of total phenolics in potato tubers. According to Al-Saikhan et al. (1995), total phenolics vary greatly across cultivars, yet there is no association between flesh colour and total phenolics. The peel and surrounding tissue of potatoes contain almost half of the total phenolic chemicals; however, this ratio reduces as one travels towards the centre of the tuber (Friedman, 1997).

4.4.5 Peppers

Peppers can be either sweet or hot. A profusion of carotenoids, flavonoids, vitamin K, and vitamin C can be found in any fresh sweet pepper (Bosland, 1996). The antioxidant lycopene, contained in red bell peppers as well, helps stave against cancer and CVD (Agarwal & Rao, 2000; Rao, 2006). There is some evidence that bell peppers can prevent cataracts, which may be due to the vitamin C and beta carotene they contain. One such carotenoid present in red peppers, beta-cryptoxanthin, shows potential in the fight against smoking and second-hand smoke-related lung cancer. In addition to their high fibre content and plethora of powerful phytochemicals, peppers are a fantastic source of nourishment. Vitamin C concentrations exhibited considerable variation among cultivars, although they were not specific to any particular species. The vitamin C content of fruits is typically 1–2 g/kg, which is equivalent to 200–300% of the recommended dietary intake (RDI) for adults (male and female combined) (Howard et al., 2000). Variegated cultivars display varying quantities of carotenoids, especially provitamin A. While certain chilli pepper varieties may contain undetectable amounts of carotenoids, others can contain as much as 12 mg/kg (Howard et al., 2000). Among pepper flavonoids, quercetin and luteolin are the most important. Different cultivars contain conjugated forms of these chemicals in concentrations ranging from 800 mg/kg (which is undetectable) to undetectable (very low) (Lee et al., 1995). The B-complex vitamins, comprising thiamine, riboflavin, niacin, and pyridoxine (vitamin B6), are advantageous when ingested alongside chiles. The body needs specific vitamins from external sources to replenish them, rendering them essential. Chilli peppers contain several minerals such as magnesium, potassium, manganese, and iron. Capsaicinoids are the primary phytochemicals present in hot peppers, including chilli peppers. Chilli peppers comprise more than 20 capsaicinoids, categorized into capsaicin and dihydrocapsaicin groups. Studies showing the capsaicin in chilli peppers can lower blood triglyceride and cholesterol levels, boost immunity, and lower the risk of stomach ulcers. Along with being analgesic, capsaicin has antidiabetic and antibacterial properties. It can treat bladder hypersensitivity, vasomotor rhinitis, and spinal cord hyperreflexia as well (Howard et al., 2000).

4.4.6 Eggplant

Phytochemicals found in aubergine have antioxidant properties, and the fruit is rich in vitamins and minerals as well. Phytochemicals found in aubergine include phenolic substances such chlorogenic acid and flavonoids like nasunin. Nasunin prevents hydroxyl radical generation. Of all the phenolic compounds detected in aubergine, chlorogenic acid is by far the most abundant and effective free radical scavenger in all of the cultivars which were studied (Matsuzoe et al., 1999). Supposedly, chlorogenic acid fights cancer, germs, viruses, and LDL cholesterol. Seven domestically grown U.S. eggplant cultivars and a wide variety of foreign and wild eggplants were the subjects of research (Matsuzoe et al., 1999). Although chlorogenic acid was the most abundant phenolic component in the commercial cultivars, 13 additional phenolic acids were detected, with somewhat varying concentrations. The antioxidant phenolic levels were approximately three times higher in the commercial cultivar 'Black Magic' than in any of the other aubergine kinds tested. The US market is characterized by this style. Aubergine fruits have extra antioxidants including kaempferol and myricetin and carotenoids like lycopene, lutein, and β-carotene (Ben-Amos & Fishler, 1998). In terms of overall antioxidant activity, aubergine ranks first among 22 popular vegetables, with an estimated 190 mol Trolox equivalent per 40 g serving size. In addition to dietary fibre, which aids digestion, eggplants are a good source of manganese, which helps build bones. Potassium, which supports cardiovascular health, and molybdenum, an enzyme catalyst, are abundant in aubergine. Besides magnesium and vitamin K, which are good for your bones, aubergine has vitamin B6, copper, niacin, vitamin C, and folate which are good for your heart.

4.5 FUTURE PROSPECTS

Conventional Indian medical systems such as Ayurveda, homoeopathy, and Unani medicine have historically depended on phytotherapeutic solutions for many chronic health conditions. Numerous phytochemicals present in nature exhibit fascinating biological activity and medicinal properties, rendering nature a plentiful and varied source of phytochemicals. Establishing exact functional and structural-activity linkages regarding phytochemical effects in biological systems is difficult. The intricacy of physiological reactions and the plethora of phytochemicals with similar molecular configurations significantly contribute to this phenomenon. Moreover, given the quantity of phytochemicals found to date, it is plausible that nature harbours many more. Scientists currently confront a significant difficulty in exploring nutritional research prospects which examine the relationship between foods or food components and illness prevention. Health claims associated with functional foods, together with assertions regarding the stability of certain phytochemicals throughout preparation and storage, must be supported by robust, well-established scientific data. To obtain the unique functional health advantages associated with these phytochemicals, further research is required to utilize them directly as functional meals or nutraceuticals with high retention (Noda et al., 2000).

4.6 CONCLUSIONS

Vegetables include phytonutrients which protect the body from a number of chronic diseases; hence, it is obvious that a diet which is abundant in vegetables is advantageous. A complex and mostly unexplained process underlies vegetables' ability to reduce the risk of illness. The health advantages of a balanced diet are expected to be amplified by several components. A wide variety of phytonutrients have antioxidant effects, which can manifest either directly as a neutralization of free radicals or indirectly through the activation of redox-sensitive cell signalling pathways. Blood pressure regulation is influenced by nutrients like potassium. Vegetables, which are abundant in dietary fibre, provide a wide range of potential opportunities for health benefits. One of these is the ability of the human gastrointestinal tract to transfer many minerals and phytochemicals which are attached to the fibre matrix. Other benefits include improving bowel transit, lowering cholesterol levels, assisting with blood glucose management, and removing cholesterol from the body. The purpose of this chapter is to provide a comprehensive overview of the phytonutrients which are found in the vegetables that have been the subject of the most research and are consumed the most frequently, as well as their biological efficacy in treating long-term health concerns. To fully obtain the health benefits of a plant-based diet, and due to the unique composition of phytonutrients in each vegetable, it is advisable for individuals to consume a diverse array of vegetables. Individuals can integrate a diverse array of phytonutrients that enhance health into their diets due to the year-round availability of various vegetables and their awareness of the health benefits associated with them.

REFERENCES

Agarwal, S., & Rao, A. K. (2000). Tomato lycopene and its role in human health and chronic diseases. *Canadian Medical Association Journal*, *163*(6), 739–744.

Aizawa, K., & Inakuma, T. (2007). Quantitation of carotenoids in commonly consumed vegetables in Japan. *Food Science and Technology Research*, *13*(3), 247–252.

Akhigbe, E. O., Izah, S. C., Ogidi, O. I., Iyingiala, A. A., & Raimi, M. O. (2025). The role of edible mushrooms in immune support. In S. C. Izah, M. C. Ogwu, & M. Akram (Eds.), *Bioactive Compounds in Edible Mushrooms* (Reference Series in Phytochemistry). Cham: Springer. https://doi.org/10.1007/978-3-031-52642-8_21-1

Albushita, A. A., Daood, H. G., & Biacs, P. A. (2000). Change in carotenoids and antioxidant vitamins in tomato as a function of varietal and technological factors. *Journal of Agricultural and Food Chemistry*, *48*(5), 2075–2081. https://doi.org/10.1021/jf990715p

Al-Saikhan, M. S., Howard, L. R., & Miller, J. C. (1995). Antioxidant activity and total phenolics in different genotypes of potato (*Solanum tuberosum* L.). *Journal of Food Science*, *60*(2), 341–344. https://doi.org/10.1111/j.1365-2621.1995.tb05668.x

Ambrosone, C. B., McCann, S. E., Freudenheim, J. L., Marshall, J. R., Zhang, Y., & Shields, P. G. (2004). Breast cancer risk in premenopausal women is inversely associated with consumption of broccoli, a source of isothiocyanates, but is not modified by GST genotype. *Journal of Nutrition*, *134*(5), 1134–1138.

Anderson, H., Martinez, R., & Lin, K. (2020). The neuroprotective effects of apigenin from celery. *Journal of Neuroscience Research*, *38*(2), 199–213.

Arab, L., & Steck, S. (2000). Lycopene and cardiovascular disease. *American Journal of Clinical Nutrition*, *71*(6 Suppl), 1691S–1695S.

Azimova, B., Xusniddinova, A., Izah, S. C., Sharipov, A., & Ogidi, O. I. (2025). Assessment of the bioactive compounds and metabolites in the edible mushroom used in management of rheumatic disorders. In S. C. Izah, M. C. Ogwu, & M. Akram (Eds.), *Bioactive Compounds in Edible Mushrooms* (Reference Series in Phytochemistry). Cham: Springer. https://doi.org/10.1007/978-3-031-52642-8_48-1

Barot, M., Kumar, J., & Kumar, N. R. (2015). Seaweed species diversity in relation to hydrochemical characters of Okha Coast Western India. *International Journal of Recent Research and Review*, *8*(3), 16–28.

Barwant, M. M., Ogidi, O. I., Singh, S., & Singh, B. (2025a). Novel drug delivery methods for herbal medicine. In A. Farid (Ed.), *Herbal Pharmacopeia: Nanotechnology and Advancing Drug Discovery* (1st ed.). CRC Press. https://doi.org/10.1201/9781003544418

Barwant, M. M., Singh, S., Singh, B., Sharma, A., & Ogidi, O. I. (2025b). Nanotechnology-enhanced delivery of herbal extract: Formulation, preclinical and clinical efficiency. In Farid, A. (Ed.), *Herbal Pharmacopeia: Nanotechnology and Advancing Drug Discovery* (1st ed.). CRC Press. https://doi.org/10.1201/9781003544418

Ben-Amos, A., & Fishler, R. (1998). Analysis of carotenoids with emphasis on 9-cis-carotene in vegetables and fruits commonly consumed in Israel. *Food Chemistry*, *62*(4), 515–520. https://doi.org/10.1016/S0308-8146(97)00196-9

Boeing, H., Bechthold, A., Bub, A., Ellinger, S., Haller, D., Kroke, A., & Watzl, B. (2012). Critical review: Vegetables and fruit in the prevention of chronic diseases. *European Journal of Nutrition*, *51*(6), 637–663. https://doi.org/10.1007/s00394-012-0380-y

Borkowski, J., Szajdek, A., Borkowska, E. J., Ciska, E., & Zielinski, H. (2008). Content of selected bioactive components and antioxidant properties of broccoli (*Brassica oleracea* L.). *European Food Research and Technology*, *226*(3), 459–465. https://doi.org/10.1007/s00217-006-0557-9

Bosland, P. W. (1996). Capsicums: Innovative uses of an ancient crop. In J. Janick (Ed.), *Progress in New Crops* (pp. 479–487). ASHS Press.

Breitmaier, E. (2008). *Terpenes: Flavors, Fragrances, Pharmaca, Pheromones* (1st ed.). Wiley-VCH Verlag GmbH & Co. KGaA.

Brennan, P., Hsu, C. C., & Moullan, N. (2005). Effect of cruciferous vegetables on lung cancer in patients stratified by genetic status: A Mendelian randomisation approach. *The Lancet*, *366*(9496), 1558–1560. https://doi.org/10.1016/S0140-6736(05)67628-3

Brewster, J. L. (2008). *Onions and Other Vegetable Alliums* (2nd ed., Vol. 15). CAB International.

Cao, G., Sofic, E., & Prior, R. L. (1996). Antioxidant capacity of tea and common vegetables. *Journal of Agricultural and Food Chemistry*, *44*(11), 3426–3431. https://doi.org/10.1021/jf9602535

Cartea, M. E., Velasco, P., Obregón, S., Padilla, G., & de Haro, A. (2008). Seasonal variation in glucosinolate content in *Brassica oleracea* crops grown in northwestern Spain. *Phytochemistry*, *69*(2), 403–410. https://doi.org/10.1016/j.phytochem.2007.08.014

Chen, S., & Andreasson, E. (2001). Update on glucosinolate metabolism and transport. *Plant Physiology and Biochemistry*, *39*(9), 743–758.

Ciska, E., Martyniak-Przybyszewska, B., & Kozlowska, H. (2000). Content of glucosinolates in cruciferous vegetables grown at the same site for two years under different climatic conditions. *Journal of Agricultural and Food Chemistry*, *48*(7), 2862–2867. https://doi.org/10.1021/jf981373a

Clifford, M. N. (2000). Anthocyanins – Nature, occurrence and dietary burden. *Journal of the Science of Food and Agriculture*, *80*(7), 1063–1072.

Clinton, S. (1998). Lycopene: Chemistry, biology and implication for human health and disease. *Nutrition Reviews*, *56*(2), 35–51. https://doi.org/10.1111/j.1753-4887.1998.tb01691.x

Croteau, R., Kutchan, T. M., & Lewis, N. G. (2000). Natural products. In B. Buchanan, W. Gruissem, & R. Jones (Eds.), *Biochemistry and Molecular Biology of Plants* (pp. 1250–1318). American Society of Plant Physiologists.

Crozier, A., Burns, J., & Aziz, A. A. (2000). Antioxidant flavonols from fruits, vegetables and beverages: Measurements and bioavailability. *Biological Research*, *33*(2), 79–88. https://doi.org/10.4067/S0716-97602000000200007

Dai, J., & Mumper, R. (2010). Plant phenolics: Extraction, analysis and their antioxidant and anticancer properties. *Molecules*, *15*(10), 7313–7352.

Davis, L., Richards, P., & Hamilton, B. (2021). Anthocyanins and cognitive function: A clinical review. *Brain & Aging*, *12*(5), 220–235.

Desjardins, Y. (2008). Onions as a nutraceutical and functional food. *Chronica Horticulturae*, *48*(2), 8–14.

Dietary Guidelines Advisory Committee. (2005). *Report of the Dietary Guidelines Advisory Committee*. U.S. Department of Health and Human Services. http://www.health.gov/dietaryguidelines/dga2005/report/HTML/C_Methodology.htm

Dillard, C. J., & German, J. B. (2000). Phytochemicals: Nutraceuticals and human health. *Journal of the Science of Food and Agriculture*, *80*(12), 1744–1756.

Fahey, J. W., Zalcmann, A. T., & Talalay, P. (2001). The chemical diversity and distribution of glucosinolates and isothiocyanates among plants. *Phytochemistry*, *56*(1), 5–51. https://doi.org/10.1016/S0031-9422(00)00316-2

Foster, G., Collins, J., & Miller, S. (2021). Glucosinolates and gut microbiota interactions. *Journal of Functional Foods*, *30*(3), 410–428.

Friedman, M. (1997). Chemistry, biochemistry and dietary role of potato polyphenols: A review. *Journal of Agricultural and Food Chemistry*, *45*(5), 1523–1540. https://doi.org/10.1021/jf960900s

Garcia, M., Rodriguez, L., & Chen, Y. (2022). Lycopene and cardiovascular health: A systematic review of clinical studies. *Journal of Nutrition Research*, *45*(3), 233–247.

Gershenzon, J., & Dudareva, N. (2007). The function of terpene natural products in the natural world. *Nature Chemical Biology*, *3*(7), 408–414.

Gerster, H. (1997). The potential role of lycopene for human health. *Journal of the American College of Nutrition*, *16*(2), 109–126.

Giovannucci, E. (1999). Tomatoes, tomato-based products, lycopene and cancer: Review of the epidemiological literature. *Journal of the National Cancer Institute*, *91*(4), 317–331. https://doi.org/10.1093/jnci/91.4.317

Harbourne, N., Marete, E., Jacquire, J. C., & O'Riordan, D. (2013). Stability of phytochemicals as sources of anti-inflammatory nutraceuticals in beverages – A review. *Food Research International*, *50*(2), 480–486.

Hernandez, J., Wu, L., & Patel, A. (2022). Anthocyanins in sweet potatoes and skin aging. *Dermatology & Nutrition*, *19*(2), 134–150.

Howard, L. R., Talcott, S. T., Brenes, C. H., & Villalon, B. (2000). Changes in phytochemical and antioxidant activity of selected pepper cultivars (*Capsicum* species) as influenced by maturity. *Journal of Agricultural and Food Chemistry*, *48*(5), 1713–1720. https://doi.org/10.1021/jf990916t

Husen, A. (2021). *Traditional Herbal Therapy for the Human Immune System*. CRC Press. https://doi.org/10.1201/9781003137955

Husen, A. (2022). *Herbs, Shrubs and Trees of Potential Medicinal Benefits*. CRC Press. https://doi.org/10.1201/9781003205067

Husen, A. (2023a). *Exploring Poisonous Plants: Medicinal Values, Toxicity Response and Other Potential Uses*. CRC Press. https://doi.org/10.1201/b23017

Husen, A. (2023b). *Antidiabetic Medicinal Plants and Herbal Treatments*. CRC Press. https://doi.org/10.1201/b23347

Husen, A. (2024a). *Antimalarial Medicinal Plants*. Taylor & Francis Group. https://doi.org/10.1201/9781003378396

Husen, A. (2024b). *Medicinal Spices and Condiment Crops*. Taylor & Francis Group. https://doi.org/10.1201/9781003387046

Ip, C., & Lisk, D. J. (1994). Enrichment of selenium in allium vegetables for cancer prevention. *Carcinogenesis*, *15*(9), 1881–1885. https://doi.org/10.1093/carcin/15.9.1881

Kalra, N., Arora, A., & Shukla, Y. (2006). Involvement of multiple signaling pathways in diallyl sulfide mediated apoptosis in mouse skin tumors. *Asian Pacific Journal of Cancer Prevention*, *7*(4), 556–562.

Kardong, G., Verma, A. K., Upadhaya, S., & Borah, D. (2013). Phytochemical and cytotoxic properties of wild *Sarchoclamys pulcherrima* Goud from Assam, north eastern India. *International Journal of Pharmacy and Pharmaceutical Sciences*, *5*(4), 394–397.

Khan, M. I., Sharma, N., & Kumar, A. (2020). Health benefits of phytonutrients in vegetables. *Journal of Nutritional Science and Healthy Eating*, *7*(4), 34–46.

Kilian, S., Kritzinger, S., Rycroft, C., Gibson, G. R., & du Preez, J. (2002). The effects of the novel bifidogenic trisaccharide, neokestose, on the human colonic microbiota. *World Journal of Microbiology and Biotechnology*, *18*(7), 637–644. https://doi.org/10.1023/A:1016808015630

Kirsh, V. A., Peters, U., & Mayne, S. T. (2007). Prospective study of fruit and vegetable intake and risk of prostate cancer. *Journal of the National Cancer Institute*, *99*(15), 1200–1209. https://doi.org/10.1093/jnci/djm065

Krinsky, N. I., & Johnson, E. J. (2005). Carotenoid actions and their relation to health and disease. *Molecular Aspects of Medicine*, *26*(6), 459–516. https://doi.org/10.1016/j.mam.2005.10.001

Kurilich, A., Tsau, G. J., & Brown, A. (1999). Carotene, tocopherol and ascorbate contents in subspecies of *Brassica oleracea*. *Journal of Agricultural and Food Chemistry*, *47*(4), 1576–1581. https://doi.org/10.1021/jf9810158

Kushad, M. K., Brown, A. F., & Kurilich, A. C. (1999). Variation in glucosinolates in vegetable crops of *Brassica oleracea*. *Journal of Agricultural and Food Chemistry*, *47*(4), 1541–1548. https://doi.org/10.1021/jf980985s

Lee, T., & Kim, H. (2023). Allicin's antimicrobial effects against drug-resistant bacteria. *International Journal of Microbiology*, *29*(4), 312–325.

Lee, Y., Howard, R., & Villalon, B. (1995). Flavonoids and antioxidant activity of fresh pepper (*Capsicum annuum*) cultivars. *Journal of Food Science*, *60*(3), 473–476. https://doi.org/10.1111/j.1365-2621.1995.tb09806.x

Leonardi, C., Ambrosino, P., Esposito, F., & Fogliano, V. (2000). Antioxidant activity and carotenoid and tomatine contents in different typologies of fresh consumption tomatoes. *Journal of Agricultural and Food Chemistry*, *48*(10), 4723–4727. https://doi.org/10.1021/jf000225t

Lila, M. A. (2004). Anthocyanins and human health: An in vitro investigative approach. *Journal of Biomedicine and Biotechnology*, *2004*(5), 306–313. https://doi.org/10.1155/S111072430440401X

Liu, R. H. (2021). Phytonutrients in personalized nutrition: A growing frontier in health. *Journal of Nutritional Science Advances*, *15*(1), 12–19.

Maggi, L., Carmona, M., del Campo, C. P., Kanakis, C. D., Anastasaki, E., & Tarantilis, P. A. (2009). Worldwide market screening of saffron volatile composition. *Journal of the Science of Food and Agriculture*, *89*(11), 1950–1954.

Manach, C., Williamson, G., Morand, C., Scalbert, A., & Rémésy, C. (2005). Bioavailability and bioefficacy of polyphenols in humans: I. Review of 97 bioavailability studies. *American Journal of Clinical Nutrition*, *81*(Suppl.), 230S–242S.

Matsuzoe, N., Yamaguchi, M., Kawanobu, S., Watanabe, Y., Higashi, H., & Sakata, Y. (1999). Effect of dark treatment of the eggplant on fruit skin color and its anthocyanin components. *Journal of the Japanese Society for Horticultural Science*, *68*(1), 138–145. https://doi.org/10.2503/jjshs.68.138

Michikawa, T., Nishiwaki, Y., & Kikuchi, Y. (2009). Serum levels of retinol and other antioxidants for hearing impairment among Japanese older adults. *Journals of Gerontology Series A: Biological Sciences and Medical Sciences*, *64*(8), 910–915. https://doi.org/10.1093/gerona/glp038

Miean, K. H., & Mohamed, S. (2001). Flavonoid (myricetin, quercetin, kaempferol, luteolin and apigenin) content of edible tropical plants. *Journal of Agricultural and Food Chemistry*, *49*(1), 106–112. https://doi.org/10.1021/jf000892m

Miller, D., Green, P., & Carter, J. (2021). Lutein and cognitive function: Evidence from a double-blind study. *Brain & Nutrition*, *18*(2), 112–128.

Nilsson, J., Olsson, K., & Engqvist, G. (2006). Variation in the content of glucosinolates, hydroxycinnamic acids, carotenoids, total antioxidant capacity and low-molecular-weight carbohydrates in *Brassica* vegetables. *Journal of the Science of Food and Agriculture*, *86*(4), 528–538. https://doi.org/10.1002/jsfa.2355

Noda, Y., Kneyuki, T., & Igarashi, K. (2000). Antioxidant activity of nasunin, an anthocyanin in eggplant peels. *Toxicology*, *148*(2–3), 119–123. https://doi.org/10.1016/S0300-483X(00)00202-X

Ogidi, O. I. (2024). Recent advances in anticancer activity and bioinformatics approach from potential plants. In *Computational Approaches in Biomaterials, Bioinformatics and Biomedical Engineering Applications*. CRC Press, Taylor and Francis Group.

Ogidi, O. I., & Adigwe, P. C. (2025). Toxin concerns and safe consumption of edible mushroom. In S. C. Izah, M. C. Ogwu, & M. Akram (Eds.), *Bioactive Compounds in Edible Mushrooms*. Reference Series in Phytochemistry. Cham: Springer. https://doi.org/10.1007/978-3-031-52642-8_39-1

Ogidi, O. I., & Ajoko, I. (2024). Neuroprotective potentials of phytochemicals. In M. S. Rajput, T. Sarachana, P. D. Sarkar, & M. Raval (Eds.), *NeuroPhytomedicine*. CRC Press, Routledge Taylor and Francis Group.

Ogidi, O. I., Ajoko, I. T., & Chukwudi, P. (2024c). Natural antioxidant compounds for skin care. In *Eco-Friendly Skin Solutions for Natural Cosmeceuticals*. IGI Global. https://doi.org/10.4018/979-8-3693-5473-5.ch0058

Ogidi, O. I., Barwant, M. M., & Okara, E. (2024d). Biomarker discovery: Protein phosphorylation state. In *Spatially Variable Genes in Cancer: Development, Progression, and Treatment Response*. IGI Global. https://doi.org/10.4018/979-8-3693-7728-4.ch005

Ogidi, O. I., Deghinmotei, A. U., & Omolora, S. O. (2024b). Anti-diabetic and lipid profile effect of ethanol extract of *Ocimum tenuiflorum* (Holy Basil) in albino rats. *Pharmaceutical Sciences & Analytical Research Journal*, *6*(3), 180070.

Ogidi, O. I., & Emaikwu, N. G. (2024). Plant phenolic compound isolation and its bioinformatics approaches of molecular mechanisms in antimicrobial activities and resistance. In *Computational Approaches in Biomaterials, Bioinformatics and Biomedical Engineering Applications*. CRC Press, Taylor and Francis Group.

Ogidi, O. I., & Felagha, I. (2024). Gut microbiome and diet interaction. In *Nutrition Controversies and Advances in Autoimmune Disease*. IGI Global.

Ogidi, O. I., Ogoun, T. R., & Tobia, P. S. (2025a). Antimalarial activities and lipid profile of ethanolic extract of Ficus sur Forssk (described by Peter Forsskål) in Plasmodium berghei infected albino rats. *Journal of Ethnopharmacology*, *348*, 119901. https://doi.org/10.1016/j.jep.2025.119901

Ogidi, O. I., Ogregade, I., & Joshua, S. (2024a). Investigation of renal protective effects of *Bryophyllum pinnatum* (Lam) extract in acetaminophen induced toxicity in albino rats. *Medical and Pharmaceutical Journal*, *3*(2), 82–93. https://doi.org/10.55940/medphar202485

Ogidi, O. I., & Oguoma, L. M. O. (2025). Allergies and sensitivities associated with edible mushroom. In S. C. Izah, M. C. Ogwu, & M. Akram (Eds.), *Bioactive Compounds in Edible Mushrooms*. Reference Series in Phytochemistry. Cham: Springer. https://doi.org/10.1007/978-3-031-52642-8_38-1

Ogidi, O. I., & Omu, O. (2025). In vivo study of antioxidant activities of ethanolic extract of pistia stratiotes (L) medicinal plant. *World Journal Medical & Pharmaceutical Research*, *1*(4), 01–09. https://doi-doi.org/101555/wjmpr.3250

Ogidi, O. I., Orlu, H. A., & Poripo, B. E. (2025b). Evaluation of hepatoprotective activities of *Bryophyllum pinnatum* leaf extract in paracetamol induced toxicity in Wistar rats. *Trends in Pharmaceutical Sciences*, *11*(1). https://doi.org/10.30476/tips.2025.105055.1272

Oguoma, L. M. O., Ogidi, O. I., Igwe, O. O., & Enenebeaku, U. E. (2025). The evaluation of phytochemical and antimicrobial properties of *Cnidoscolus aconitifolius* leaf extracts on selected organisms. *Journal of Advanced Pharmacy Research*, *9*(2), 91–98. https://doi.org/10.21608/aprh.2025.352034.1304

Ogwu, M. C., Ogidi, O. I., Izah, S. C., & Odubo, T. C. (2025). Therapeutic and preventive effects of natural compounds derived from microbes. In K. G. Ramawat & J. M. Mérillon (Eds.), *Natural Products*. Berlin, Heidelberg: Springer. https://doi.org/10.1007/978-3-642-36202-6_287-1

Okara, E., Ogidi, O. I., & Bunu, E. M. (2025). An *in vivo* study of the anti-inflammatory activity of ethanolic extract of *Pistia stratiotes* (L) medicinal plant. *Mediterranean Journal of Medical Research, 2*(1), 98–103. https://doi.org/10.5281/zenodo.15839127

Parker, M., White, S., & Langford, J. (2020). Indole-3-carbinol and estrogen metabolism. *Hormone & Cancer Research, 25*(1), 76–89.

Pascal, S. V., & Segal, R. (2006). Phytosterols – Biologically active compounds in food. *Journal of Agroalimentary Processes and Technologies, 12*(1), 149–158.

Pasquali, M. A., Gelain, D. P., & Zanotto-Filho, A. (2008). Retinol and retinoic acid modulate catalase activity in Sertoli cells by distinct and gene expression-independent mechanisms. *Toxicology in Vitro, 22*(5), 1177–1183. http://doi.org/10.1016/j.tiv.2008.03.007

Patel, S., Kumar, R., & Das, T. (2018). Curcumin and neuroprotection: A randomized clinical trial. *Neuroscience & Aging, 55*(1), 89–105.

Rahn, C. R., Bending, G. D., Lillywhite, R. D., & Turner, M. K. (1999). Chemical characterisation of vegetable and arable crop residue materials: A comparison of methods. *Journal of the Science of Food and Agriculture, 79*, 1715–1721. http://doi.org/10.1002/(SICI)1097-0010(199909)79:123.0.CO;2-V

Ramakrishna, V., & Jailkhani, R. (2008). Oxidative stress in non–insulin-dependent diabetes mellitus (NIDDM) patients. *Acta Diabetologica, 45*(1), 41–46. http://doi.org/10.1007/s00592-007-0018-3

Randle, W. M., & Lancaster, J. E. (2002). Sulphur compounds in Alliums in relation to flavour quality. In J. L. Brewster (Ed.), *Onions and other Vegetable Alliums* (pp. 329–356). CAB International.

Rao, A. V. (2006). *Tomatoes, Lycopene and Human Health: Preventing Chronic Diseases*. Caledonian Science Press.

Rao, A. V., & Rao, L. G. (2007). Carotenoids and human health. *Pharmacological Research, 55*(3), 207–216. http://doi.org/10.1016/j.phrs.2007.01.012

Robinson, A., & Harris, W. (2021). Beetroot nitrates and exercise performance. *Sports Science Journal, 33*(6), 442–458.

Rodriguez, A., Silva, I. D., Gaspar, J., Maia, R., Laires, A., & Rueff, J. (2006). Mutagenicity of kaempferol in V79 cells: The role of cytochromes P450. *Teratogenesis, Carcinogenesis, and Mutagenesis, 16*, 229–241.

Rose, P., Whiterman, M., Moore, P. K., & Zhu, Y. Z. (2005). Bioactive S-alk(en)yl cysteine sulfoxide metabolites in the genus *Allium*: The chemistry of potential therapeutic agents. *Natural Product Reports, 22*, 351–368. http://doi.org/10.1039/b417639c

Rubec, R., Svobodova, M., & Velisek, J. (1999). Gas chromatographic determination of S-alk(en)yl-L-cysteine sulfoxides. *Journal of Chromatography, 862*, 85–94. http://doi.org/10.1016/S0021-9673(99)00902-4

Sasaki, N., Sasamura, T., Ishikawa, H. O., Kanai, M., Ueda, R., Saigo, K., & Matsuno, K. (2007). Polarized exocytosis and transcytosis of Notch during its apical localization in *Drosophila* epithelial cells. *Genes to Cells, 12*(1), 89–103.

Scholz-Ahrens, K. E., Schaafsma, G., van den Heuvel, E. G. H. M., & Schrezenmeir, J. (2001). Effects of prebiotics on mineral metabolism. *American Journal of Clinical Nutrition, 73*, 459S–464S.

Scott, K. J., & Hart, D. J. (1995). Development and evaluation of an HPLC method for the analysis of carotenoids in food and the measurement of the carotenoid content of vegetables and fruits commonly consumed in the UK. *Food Chemistry, 54*, 101–111. http://doi.org/10.1016/0308-8146(95)92669-B

Seow, A., Yuan, J. M., Sun, C. L., Van Den Berg, D., Lee, H. P., & Yu, M. C. (2002). Dietary isothiocyanates, glutathione S-transferase polymorphisms, and colorectal cancer risk in the Singapore Chinese Health Study. *Carcinogenesis, 23*, 2055–2061. http://doi.org/10.1093/carcin/23.12.2055

Slavin, J. L. (2013). Fiber and prebiotics: Mechanisms and health benefits. *Nutrients, 5*(4), 1417–1435. https://doi.org/10.3390/nu5041417

Smith, J., & Brown, K. (2019). Beta-carotene and vision health in aging populations. *Journal of Ophthalmology, 12*(7), 356–370.

Srinivasan, K. (2005). Plant foods in the management of diabetes mellitus: Spices as beneficial antidiabetic food adjuncts. *International Journal of Food Sciences and Nutrition, 56*, 399–414. http://doi.org/10.1080/09637480500512872

Stewart, A. J., Bozonnet, S., Mullen, W., Jenkins, G. I., Lean, M. E., & Crozier, A. (2000). Occurrence of flavonols in tomatoes and tomato-based products. *Journal of Agricultural and Food Chemistry, 48*, 2663–2669. http://doi.org/10.1021/jf000070p

Tholl, D. (2006). Terpene synthases and the regulation, diversity, and biological roles of terpene metabolism. *Current Opinion in Plant Biology, 9*, 297–304.

Thomas, D. J., & Parkin, K. L. (1994). Quantification of alk(en)yl-L-cysteine sulfoxide and related amino acids in *Alliums* by high-performance liquid chromatography. *Journal of Agricultural and Food Chemistry, 42*, 1632–1638. http://doi.org/10.1021/jf00044a010

Thompson, L., Martin, R., & Evans, B. (2020). The role of quercetin in cardiovascular health. *Heart & Circulation, 27*(5), 198–210.

Traka, M. (2010). Broccoli consumption interferes with prostate cancer progression: Mechanisms of action. *Acta Horticulturae, 867*, 19–25.

Traka, M., & Mithen, R. (2009). Glucosinolates, isothiocyanates, and human health. *Phytochemistry Reviews, 8*(1), 269–282. https://doi.org/10.1007/s11101-008-9103-7

Tsuda, T. (2008). Regulation of adipocyte function by anthocyanins: Possibility of preventing metabolic syndrome. *Journal of Agricultural and Food Chemistry, 56*(3), 642–654.

Turner, F., Hall, B., & Kim, R. (2022). Kale and cardiovascular health: A review of clinical trials. *Nutritional Sciences Journal, 40*(3), 312–328.

Vallejo, F., Tomas-Barberan, F. A., Banavent-Garcia, A. G., & Garcia-Viguera, C. (2003). Total and individual glucosinolate contents in inflorescences of eight broccoli cultivars grown under various climatic and fertilization conditions. *Journal of the Science of Food and Agriculture, 83*(3), 307–313. https://doi.org/10.1002/jsfa.1320

Vallejo, F., Tomas-Barberan, F. A., & Garcia-Viguera, C. (2002). Potential bioactive compounds in health promotion from broccoli cultivars grown in Spain. *Journal of the Science of Food and Agriculture, 82*(11), 1293–1297. https://doi.org/10.1002/jsfa.1183

Van, A. M. A., Eggelte, T. A., & Boxtel, C. J. (1999). Artemisinin drugs in the treatment of malaria: From medicinal herb to registered medication. *Trends in Pharmacological Sciences, 20*, 199–205.

Verma, S. K., & Verma, D. (2008). Garlic – "The spice of life": Composition, cooking, chemistry and preparations. *Journal of Herbal Medicine and Toxicology, 2*(2), 21–28.

Viroli, G., Kalmpourtzidou, A., & Cena, H. (2023). Exploring benefits and barriers of plant-based diets: Health, environmental impact, food accessibility and acceptability. *Nutrients, 15*(22), 4723.

Wang, L. J., Giovannucci, E. L., Hunter, D., Neuberg, D., Su, L., & Christiani, D. C. (2004). Dietary intake of cruciferous vegetables, glutathione S-transferase (GST) polymorphisms, and lung cancer risk in a Caucasian population. *Cancer Causes & Control, 15*, 977–985. http://doi.org/10.1007/s10552-004-1093-1

Williams, J., Nelson, T., & Roberts, M. (2019). Capsaicin and metabolic health: Effects on fat oxidation. *Metabolism & Nutrition, 14*(4), 275–289.

Yasmeen, M., Barwant, M. M., Balwant Singh, B., & Ogidi, O. I. (2025). Present challenges and future perspective of the herbal drug industry. In A. Farid (Ed.), *Herbal Pharmacopeia: Nanotechnology and Advancing Drug Discovery* (1st ed.). CRC Press. https://doi.org/10.1201/9781003544418

Yin, M., Hwang, S., & Chan, K. (2002). Nonenzymatic antioxidant activity of four organosulfur compounds derived from garlic. *Journal of Agricultural and Food Chemistry, 50*, 6143–6147. http://doi.org/10.1021/jf0204203

Yoo, K. S., & Pike, L. M. (1998). Determination of flavor precursor compound S-alk(en)yl-L-cysteine sulfoxides by HPLC and their distribution in *Allium* species. *Scientia Horticulturae, 75*, 1–10. http://doi.org/10.1016/S0304-4238(98)00107-1

Zhang, W., Lin, H., & Zhao, Y. (2020). Sulforaphane and prostate cancer prevention. *Cancer Prevention Journal, 14*(3), 256–270.

5 Methods of Cooking Vegetables for Various Health Benefits

Md. Abdul Kader, Kazi N.S. Rafi, and Akbar Hossain

5.1 INTRODUCTION

Vegetables are nutrient-dense food that provides essential vitamins, minerals, dietary fibre, and phytochemicals, crucial for good human health. The Food and Agriculture Organization (FAO) and World Health Organization (WHO) recommend consuming at least 400 grams of fruits and vegetables daily to prevent chronic diseases such as CVD, cancer, and diabetes (WHO, 2003). However, the health benefits of vegetables depend not only on consumption quantity but also on preparation methods.

Cooking vegetables serves multiple purposes: improving digestibility, enhancing flavour, reducing anti-nutritional factors, and ensuring microbiological safety. Yet, cooking can also lead to nutrient losses through various mechanisms including thermal degradation, leaching into cooking water, unwanted browning action, and oxidation. Different cooking methods affect nutrients differently. Some techniques are better for preserving or even enhancing certain beneficial compounds while reducing other anti-nutrients (Palermo et al., 2014).

This chapter provides a comprehensive analysis of 18 vegetable cooking methods, examining their effects on nutrient retention, bioavailability of health-promoting compounds, and practical applications. Understanding these techniques empowers health professionals, nutritionists, and consumers to make informed decisions about vegetable preparation that maximize health benefits.

5.2 FACTORS IMPACTING NUTRIENT RETENTION DURING COOKING

Responsible factors for nutrient retention in vegetables during cooking (Figure 5.1) have been discussed in detail under the following subheadings:

5.2.1 Heat Sensitivity of Nutrients

Nutrients vary considerably in their heat stability. Water-soluble vitamins, particularly vitamin C and B-complex vitamins, are highly susceptible to thermal degradation. Vitamin C can lose 50–80% of its activity during prolonged cooking at high temperatures (Yuan et al., 2009). Conversely, fat-soluble vitamins (A, D, E, K) and carotenoids generally show better heat stability.

5.2.2 Water Solubility and Leaching

Water-soluble nutrients can leach into cooking water, leading to significant losses when the cooking liquid is discarded. Studies indicate that boiling can result in 20–50% losses of water-soluble vitamins depending on vegetable type, cutting size, and cooking duration (Jiménez-Monreal et al., 2009).

DOI: 10.1201/9781003606710-5

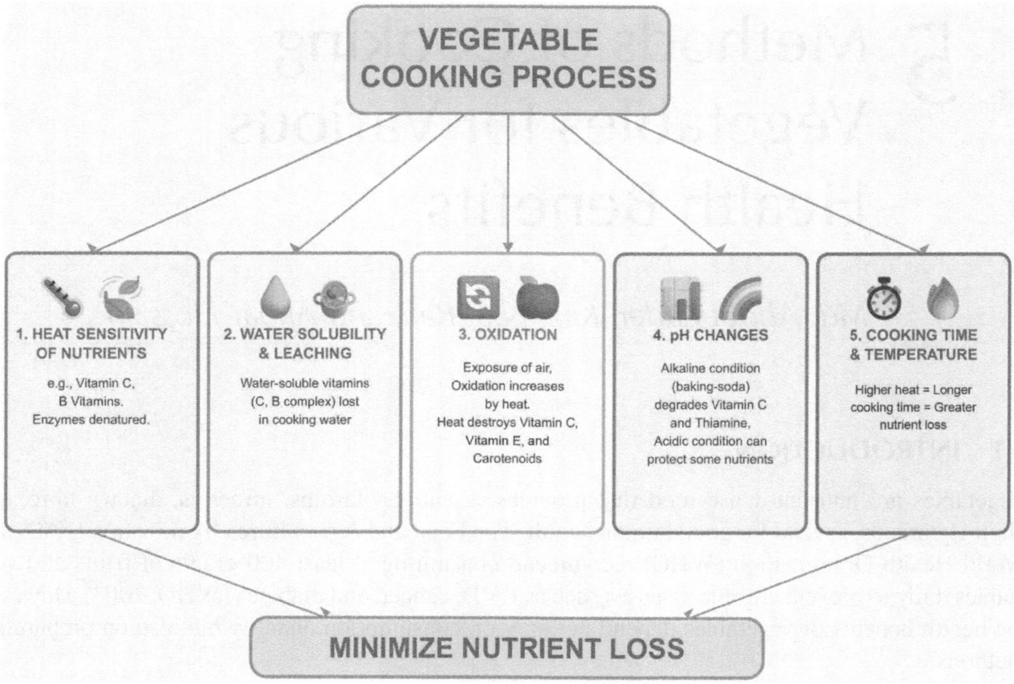

FIGURE 5.1 Responsible factors for nutrient retention in vegetables during cooking.

5.2.3 Oxidation

Exposure to oxygen during cooking can degrade sensitive nutrients and phytochemicals. Antioxidants like polyphenols, vitamin C, and vitamin E are particularly vulnerable to oxidative losses.

5.2.4 pH Changes

The pH of the cooking medium influences nutrient stability. Alkaline conditions can accelerate vitamin C degradation, while acidic environments may help preserve certain nutrients but affect colour and texture.

5.2.5 Cooking Time and Temperature

Extended cooking times and higher temperatures generally increase nutrient losses. However, optimal cooking can improve the bioavailability of certain compounds by breaking down cell walls and releasing bound nutrients.

5.3 THE 18 COOKING METHODS

General guidelines for cooking methods of vegetables (Table 5.1) are discussed in detail under the following subheadings:

5.3.1 Boiling

Boiling involves cooking vegetables in water or broth at 100°C at sea level. Boiling is one of the most commonly used cooking methods but can result in substantial losses of water-soluble

TABLE 5.1
General Guidelines for Cooking Methods of Vegetables

Cooking Method	Standard Temperature	Standard Pressure	Standard Time	Standard Amount of Water Usage	Reference
Air Frying	180°C–200°C (360°F–400°F)	N/A	10–20 minutes (depending on vegetable density)	Light coat of oil or oil spray	Téllez-Morales et al. (2024)
Slow Cooking	Low: 70°C–93°C (160°F–200°F); high: 93°C–116°C (200°F–240°F)	N/A	2–8 hours	Enough liquid (water/stock) to partially or completely cover the vegetables	Miglio et al. (2008)
Fermenting	Room temperature (18°C–24°C / 65°F–75°F)	N/A	Several days to several weeks	Enough brine (salt water) to fully submerge vegetables	Cheung et al. (2025)
Sous Vide	85°C–95°C (185°F–203°F)	N/A (vacuum condition)	30 minutes–2 hours	Enough water to submerge the vacuum-sealed bag	Baldwin (2012)
Poaching	71°C–82°C (160°F–180°F)	N/A	5–15 minutes	Enough liquid (water/stock) to submerge the vegetables	Miglio et al. (2008)
Braising	135°C–160°C (275°F–325°F) (Oven/ Stovetop)	N/A	30 minutes–1.5 hours	Enough liquid to cover approximately 31% of the vegetable height	Baldwin (2012)
Blanching	100°C (212°F) (boiling water)	N/A	30 seconds–3 minutes	Enough boiling water to fully submerge vegetables	Miglio et al. (2008)
Pressure Cooking	Above 100°C (temperature increases due to pressure)	High: 15 psi (standard); low: 6–8 psi	1–15 minutes (very rapid, depends on vegetable type)	1–2 cups minimum liquid required to build and maintain pressure	Fam et al. (2021)
Baking	175°C–200°C (350°F–400°F)	N/A	20–60 minutes (depending on thickness and type, e.g. potatoes take longer)	N/A (oil/fat is typically used)	Fellows (2022)
Grilling	Medium-high to high heat (175°C–230°C / 350°F–450°F)	N/A	4–15 minutes (over direct heat)	N/A (vegetables should be oiled)	Gerrard (2005)
Roasting	200°C–230°C (400°F–450°F)	N/A	20–45 minutes	N/A (oil/fat is typically used)	Gerrard (2005)
Sautéing	Medium-high heat (160°C–200°C / 325°F–400°F)	N/A	3–10 minutes	Small amount of fat (oil/butter)	Fellows (2022)
Stir Frying	High heat (200°C–240°C / 400°F–465°F)	N/A	3–8 minutes	Small amount of fat (oil)	Gerrard (2005)
Microwaving	N/A (uses electromagnetic radiation)	N/A	1–10 minutes (depends on wattage and amount)	1–3 tablespoons of water (often placed in the container and covered)	Fellows (2022)
Steaming	100°C (212°F) (steam from boiling water)	N/A	5–20 minutes	1–2 inches of water at the bottom of the pot (must not touch the steamer basket)	Yuan et al. (2009)
Boiling	100°C (212°F) (boiling water)	N/A	4–20 minutes (depending on density)	Enough water to fully submerge the vegetables	Yuan et al. (2009)
Juicing	N/A	N/A	N/A	N/A	Fellows (2022)
Raw consumption	N/A	N/A	N/A	N/A	Fellows (2022)

vitamins if optimal cooking conditions are not maintained. Studies show that boiling can reduce vitamin C content by 30–50% and folate by 40–60% (Munyaka et al., 2010). Minerals also leach into cooking water. However, boiling effectively reduces anti-nutritional factors like oxalates and phytates.

5.3.1.1 Health Benefits

Health benefits of boiling are as follows:

- Reduces anti-nutritional compounds
- Softens fibre for easier digestion
- Ensures food safety through pathogen destruction

5.3.1.2 Optimization Tips

Optimization tips of boiling are as follows:

- Use minimal water
- Reduce cooking time
- Utilize cooking water in soups or sauces
- Cut vegetables into larger pieces to minimize surface area

5.3.2 Steaming

Steaming cooks vegetables using hot vapor without direct water contact. Steaming is one of the best methods for nutrient retention. Studies indicate that steaming preserves 80–90% of vitamin C and retains more minerals compared to boiling (Yuan et al., 2009). Glucosinolates in cruciferous vegetables are better preserved with steaming than boiling.

5.3.2.1 Health Benefits

- Excellent retention of water-soluble vitamins
- Preservation of antioxidant capacity
- Maintains mineral content
- Retains natural colour and texture

5.3.2.2 Optimization Tips

- Avoid overcooking
- Use bamboo or stainless-steel steamers
- Steam until just tender-crisp
- Season after cooking to enhance flavour

5.3.3 Microwaving

Microwave cooking uses electromagnetic radiation to heat water molecules within a food. Microwaving with minimal water is highly effective for nutrient retention. Research demonstrates that microwaving can preserve 77–95% of vitamin C, which is higher than most conventional methods (Jiménez-Monreal et al., 2009). The short cooking time and minimal water use also reduce nutrient losses.

5.3.3.1 Health Benefits

- Superior vitamin retention
- Minimal leaching losses

- Time-efficient
- Energy-efficient

5.3.3.2 Optimization Tips

- Use microwave-safe containers with lids
- Add minimal water (one to two tablespoons)
- Arrange vegetables evenly for uniform cooking
- Let stand after cooking for heat distribution

5.3.4 Stir-Frying

Stir-frying involves cooking small pieces of vegetables quickly over high heat with minimal oil. The high heat and short cooking time help preserve heat-sensitive vitamins. Using oil enhances the absorption of fat-soluble vitamins (vitamin A, D, E and K) and carotenoids. Studies show stir-frying can increase lycopene bioavailability in tomatoes and beta-carotene in carrots (Turkmen et al., 2005).

5.3.4.1 Health Benefits

- Enhanced bioavailability of fat-soluble nutrients
- Good retention of vitamins
- Minimal water-soluble nutrient losses
- Improved antioxidant absorption

5.3.4.2 Optimization Tips

- Use healthy oils (olive, canola, avocado)
- Maintain high heat for quick cooking
- Keep vegetables crisp
- Avoid excessive oil

5.3.5 Sautéing

Sautéing cooks vegetables in a small amount of fat over medium-high heat. Similar to stir-frying, sautéing enhances fat-soluble nutrient absorption while preserving water-soluble vitamins due to short cooking times. The Maillard reaction during sautéing can create beneficial compounds with antioxidant properties.

5.3.5.1 Health Benefits

- Increased carotenoid bioavailability
- Good vitamin retention
- Enhanced flavour without excessive fat
- Preservation of texture

5.3.5.2 Optimization Tips

- Use moderate heat to prevent burning
- Choose heart-healthy fats
- Don't overcrowd the pan
- Add garlic or herbs for additional phytochemicals

5.3.6 Roasting

Roasting cooks vegetables in an oven using dry heat, typically at temperatures between 180°C and 230°C. Roasting can concentrate flavours and nutrients as water evaporates. While some heat-sensitive vitamins are lost, carotenoid content and antioxidant activity can increase. Research shows roasting enhances the bioavailability of lycopene and beta-carotene (Miglio et al., 2008).

5.3.6.1 Health Benefits

- Increased carotenoid bioavailability
- Enhanced antioxidant capacity in some vegetables
- Improved digestibility
- Natural sweetness development

5.3.6.2 Optimization Tips

- Cut vegetables uniformly
- Use minimal oil
- Roast at moderate temperatures (180–200°C)
- Avoid charring or burning

5.3.7 Grilling

Grilling applies direct heat from below, with less water during food preparation. This method is good for cooking vegetables, but it is comparatively better for animal food products like meat and fish. High-temperature grilling can result in losses of heat-sensitive vitamins but minimal leaching. Carotenoids and phenolic compounds may be retained or concentrated. However, excessive grilling can produce carcinogens and potentially harmful compounds like acrolein, acrylamide, and polycyclic aromatic hydrocarbons (PAHs).

5.3.7.1 Health Benefits

- No added fat required
- Enhanced flavour
- Retained minerals
- Good for low-calorie cooking

5.3.7.2 Optimization Tips

- Avoid direct flame contact
- Prevent charring
- Marinate vegetables to reduce harmful compound formation
- Use moderate heat and watch carefully

5.3.8 Baking

Baking cooks vegetables in an enclosed oven environment with dry heat. Baking at moderate temperatures (425°F or 220°C) preserves many nutrients while improving digestibility. Fat-soluble vitamins and minerals are well retained. However, prolonged baking can reduce vitamins C and B (Palermo et al., 2014).

5.3.8.1 Health Benefits

- Minimal nutrient leaching
- Enhanced digestibility of complex carbohydrates
- Retention of minerals
- Development of beneficial flavour compounds

5.3.8.2 Optimization Tips

- Use moderate temperatures (160–180°C)
- Cover with foil to retain moisture

- Minimize baking time
- Use the vegetable whole or in large pieces

5.3.9 Pressure Cooking

Pressure cooking uses steam under high pressure, increasing cooking temperature above 100°C and reducing cooking time. The shorter cooking time compensates for higher temperatures, resulting in good nutrient retention. Studies show pressure cooking preserves 90–95% of nutrients compared to conventional boiling (Jiménez-Monreal et al., 2009). It's particularly effective for legumes and root vegetables.

5.3.9.1 Health Benefits

- Excellent nutrient retention despite high heat
- Reduced cooking time preserves vitamins
- Effective anti-nutritional factor reduction
- Energy efficient

5.3.9.2 Optimization Tips

- Follow manufacturer's timing guidelines
- Use minimal water
- Quick-release pressure immediately after cooking
- Don't overfill the pressure cooker

5.3.10 Blanching

Blanching involves briefly immersing vegetables in boiling water followed by rapid cooling in ice water. Short blanching times (30 seconds) minimize nutrient losses while inactivating enzymes and prevent the enzymatic browning of vegetables which causes visual quality deterioration. Some vitamin C loss occurs (10–30%), but colour, texture, and other nutrients are well preserved (Gonçalves et al., 2009).

5.3.10.1 Health Benefits

- Enzyme inactivation preserves quality
- Minimal nutrient loss with proper technique
- Improved texture and colour
- Reduced anti-nutritional factors

5.3.10.2 Optimization Tips

- Use brief blanching times
- Immediately cool in ice water
- Maintain boiling temperature throughout
- Use for freezing preparation or salad preparation

5.3.11 Braising

Braising combines both moist and dry heat, typically browning vegetables first, then cooking slowly in liquid. The initial searing can enhance flavour compounds, while the gentle simmering allows nutrient retention. When cooking liquid is consumed, leached nutrients are retained in the dish.

5.3.11.1 Health Benefits

- Nutrients retained in cooking liquid
- Enhanced digestibility of tough vegetables

- Development of complex flavours
- Suitable for fibrous vegetables

5.3.11.2 Optimization Tips

- Brown quickly to minimize nutrient loss
- Use low heat for simmering
- Consume the cooking liquid
- Add acidic ingredients (tomatoes, wine) to enhance mineral absorption

5.3.12 Poaching

Poaching gently cooks vegetables in simmering liquid (85–95°C), below boiling point. The lower temperature compared to boiling reduces thermal degradation of sensitive nutrients (vitamin C). However, water-soluble nutrients still leach into the cooking liquid. Using flavourful poaching liquids (broth, wine) that are later consumed helps retain nutrients.

5.3.12.1 Health Benefits

- Less thermal damage than boiling
- Gentle cooking preserves texture
- Suitable for delicate vegetables
- Fat-free cooking method

5.3.12.2 Optimization Tips

- Use flavourful liquids that will be consumed
- Maintain temperature below boiling
- Cook just until tender
- Use poaching liquid in sauces or soups

5.3.13 Sous Vide

Sous vide involves vacuum sealing of vegetables in plastic bags and cooking in precisely temperature-controlled water (typically 80–90°C). The sealed environment prevents nutrient leaching and oxidation. Precise temperature control minimizes thermal degradation. Research indicates excellent retention of vitamins, minerals, and antioxidants (Renna et al., 2017).

5.3.13.1 Health Benefits

- Superior nutrient retention
- No leaching losses
- Excellent texture and colour preservation
- Enhanced bioavailability of certain compounds

5.3.13.2 Optimization Tips

- Use food-grade vacuum bags
- Set appropriate temperature for specific vegetables
- Include small amounts of healthy fats for fat-soluble vitamins
- Consider adding herbs for additional phytochemicals

5.3.14 Fermenting

Fermentation uses beneficial microorganisms to preserve vegetables and create new palatable test and antioxidant rich food. Fermentation can increase vitamin content (particularly B vitamins and

K2), improve mineral bioavailability, and create beneficial probiotics. Studies show that fermented vegetables have enhanced antioxidant activity (Septembre-Malaterre et al., 2018).

5.3.14.1 Health Benefits

- Increased B vitamin and vitamin K2 content
- Probiotic development supports gut health
- Enhanced mineral bioavailability
- Reduced anti-nutritional factors
- Improved digestibility

5.3.14.2 Optimization Tips

- Use fresh, high-quality vegetables
- Maintain proper salt concentration (2–3%)
- Ensure anaerobic conditions
- Control fermentation temperature (18–22°C)

5.3.15 Juicing

Juicing extracts liquid from vegetables, separating juice from fruit or vegetable pulp and fibre. Juicing concentrates water-soluble vitamins, minerals, and phytochemicals but removes most dietary fibre. This increases nutrient bioavailability but eliminates fibre's health benefits. Some oxidation may also occur during juicing (Gião et al., 2012).

5.3.15.1 Health Benefits

- Concentrated micronutrient content
- Enhanced nutrient absorption
- Hydration with nutrients
- Suitable for those with digestive issues

5.3.15.2 Optimization Tips

- Consume immediately to minimize oxidation
- Include pulp when possible
- Combine with whole vegetables for fibre
- Use cold-press or masticating juicers

5.3.16 Raw Consumption

This involves consuming fresh vegetables such as ripe and unripe papaya, cucumber, and carrot without any cooking or heat treatment. Raw vegetables retain 100% heat-sensitive and water-soluble nutrients like vitamins and enzymes. However, some nutrients have lower bioavailability due to intact cell walls. Certain vegetables contain anti-nutritional factors or require cooking for food safety (Fabbri & Crosby, 2016).

5.3.16.1 Health Benefits

- Maximum retention of heat-sensitive vitamins
- Preservation of natural enzymes
- High fibre content with intact structure
- Maximum antioxidant content in some vegetables

5.3.16.2 Optimization Tips

- Wash thoroughly to remove contaminants
- Combine with healthy fats for fat-soluble vitamins

- Chop or grate to increase nutrient release
- Choose appropriate vegetables (some require cooking)

5.3.17 Slow Cooking

Slow cooking uses low heat over extended periods (4–8 hours) in a covered pot. The extended cooking time can reduce heat-sensitive vitamins, but the low temperature and retained cooking liquid help preserve nutrients better than high-heat methods. Minerals and fat-soluble vitamins are well retained in this cooking method (Palermo et al., 2014).

5.3.17.1 Health Benefits

- Nutrients retained in cooking liquid
- Enhanced digestibility
- Breakdown of tough fibres
- Convenient for tough root vegetables

5.3.17.2 Optimization Tips

- Keep lid closed to prevent nutrient oxidation
- Use minimal liquid
- Consume cooking liquid
- Add delicate vegetables toward the end

5.3.18 Air Frying

Air frying circulates hot air around vegetables at high speed, creating a crispy exterior with minimal oil. Air frying uses significantly less oil than deep frying while achieving similar texture. Studies indicate that a good retention of vitamins and antioxidants is assured in this cooking method due to short cooking times and minimal oil use. Fat-soluble nutrient absorption is also enhanced because of the small amounts of oil application (Raychaudhuri et al., 2022).

5.3.18.1 Health Benefits

- Reduced fat content compared to deep frying
- Good nutrient retention
- Enhanced carotenoid bioavailability with minimal oil
- Lower calorie content

5.3.18.2 Optimization Tips

- Use light oil coating
- Avoid overcrowding basket
- Monitor cooking time closely
- Use moderate temperatures (160–180°C)

5.4 COMPARATIVE ANALYSIS OF COOKING METHODS

5.4.1 Vitamin C Retention

Ranking from highest to lowest retention:

1. Raw consumption (100%)
2. Microwaving (77–95%)
3. Steaming (80–90%)

4. Stir-frying (70–85%)
5. Pressure cooking (70–80%)
6. Boiling (40–60%)

5.4.2 Carotenoid Bioavailability

Cooking generally increases carotenoid bioavailability:

1. Sautéing/stir-frying with oil (highest)
2. Roasting with oil
3. Steaming
4. Boiling
5. Raw (lowest bioavailability despite highest content)

5.4.3 Glucosinolate Preservation (Cruciferous Vegetables)

1. Raw consumption
2. Steaming (3–5 minutes)
3. Microwaving
4. Stir-frying
5. Boiling (significant losses)

5.4.4 Mineral Retention

Methods that avoid leaching are best:

1. Roasting, baking, grilling (no water contact)
2. Steaming
3. Microwaving with minimal water
4. Stir-frying
5. Methods where cooking liquid is consumed

5.5 PRACTICAL RECOMMENDATIONS

5.5.1 Vegetable-Specific Guidelines

1. **Leafy greens (spinach, kale):** Steam or sauté quickly to preserve folate and vitamin C. Light cooking improves iron and calcium absorption by reducing oxalates.
2. **Cruciferous vegetables (broccoli, cauliflower):** Steam for 3–5 minutes to preserve glucosinolates and vitamin C. Avoid prolonged boiling which destroys beneficial compounds.
3. **Root vegetables (carrots, sweet potatoes):** Roast, steam, or microwave. Cooking with small amounts of fat enhances beta-carotene absorption. Pressure cooking is efficient for tough roots.
4. **Tomatoes:** Cook with oil (roast, sauté) to maximize lycopene bioavailability. Heating increases lycopene availability by two to three times.
5. **Alliums (Onions, Garlic):** Chop and let stand 10 minutes before cooking to allow beneficial compounds to form. Sauté gently to preserve OSCs.
6. **Legumes:** Pressure cook or boil to reduce anti-nutritional factors and improve protein digestibility. Soaking before cooking further reduces phytates.

5.5.2 General Principles for Nutrient Optimization

1. **Minimize water use:** Use steaming, microwaving, or stir-frying over boiling.
2. **Reduce cooking time:** Cook vegetables until just tender-crisp.
3. **Include healthy fats:** Add olive oil or other healthy fats to enhance fat-soluble vitamin absorption.
4. **Retain cooking liquid:** When boiling or braising, use the cooking liquid in the final dish.
5. **Cut strategically:** Larger pieces retain more nutrients than finely chopped vegetables.
6. **Cook immediately after cutting:** Minimize oxidation time between preparation and cooking.
7. **Use diverse methods:** Vary cooking techniques to obtain a range of nutrients and bioactive compounds.

5.5.3 Combining Methods

Combining cooking methods can optimize both nutrition and palatability:

- **Blanch then stir-fry:** Reduces cooking time while achieving desired texture
- **Steam then roast:** Ensures even cooking with enhanced flavour
- **Raw + cooked:** Consume both raw and cooked vegetables to maximize different nutrients

5.6 SPECIAL CONSIDERATIONS

5.6.1 Bioavailability versus Content

While raw vegetables contain maximum nutrients, cooking often increases bioavailability. The 'best' method depends on specific health goals:

- For vitamin C: Raw or lightly steamed
- For carotenoids: Cooked with oil
- For minerals: Methods that retain cooking liquid
- For overall antioxidant benefit: Moderate cooking with varied methods

5.6.2 Anti-Nutritional Factors

Some vegetables contain compounds that inhibit nutrient absorption:

- **Oxalates** (spinach, chard): Reduced by boiling or steaming
- **Phytates** (legumes): Reduced by soaking and cooking
- **Goitrogens** (cruciferous): Reduced by cooking
- **Lectins** (beans): Destroyed by adequate cooking

Proper cooking methods reduce these factors while preserving beneficial nutrients.

5.6.3 Food Safety

Cooking ensures microbiological safety by destroying pathogens. Minimum internal temperatures and proper handling are essential, particularly for:

- Vegetables intended for immunocompromised individuals.
- Root vegetables grown in soil.
- Leafy greens that may harbour bacteria.

5.7 CULTURAL AND DIETARY PATTERNS

5.7.1 Traditional Cooking Wisdom

Many traditional cooking practices have nutritional merit:

- **Mediterranean:** Sautéing with olive oil enhances nutrient absorption.
- **Asian:** Stir-frying preserves nutrients while adding beneficial spices.
- **Fermented vegetables:** Kimchi and sauerkraut provide probiotics and enhanced nutrients.
- **One-pot dishes:** Soups and stews retain nutrients in cooking liquid.

5.7.2 Modern Innovations

Contemporary cooking technologies offer advantages:

- **Sous vide:** Precise temperature control optimizes nutrient retention.
- **Air frying:** Reduces fat while maintaining palatability.
- **High-speed blending:** Breaks cell walls without heat for raw preparations.
- **Pressure cooking:** Reduces cooking time and energy while preserving nutrients.

5.8 ENVIRONMENTAL AND SUSTAINABILITY CONSIDERATIONS

Cooking method selection also impacts environmental sustainability.

5.8.1 Energy Efficiency

- Pressure cooking: Most energy-efficient
- Microwave: Very efficient for small quantities
- Slow cooking: Less efficient but useful for batch cooking
- Steaming: Efficient when cooking multiple layers

5.8.2 Resource Conservation

- Methods using minimal water reduce water consumption.
- Consuming cooking liquids eliminates waste.
- Batch cooking reduces overall energy use.
- Raw consumption requires no energy.

5.9 FUTURE DIRECTIONS AND RESEARCH NEEDS

5.9.1 Emerging Technologies

Novel cooking technologies are being developed:

- **Infrared heating:** May preserve nutrients better than conventional methods.
- **Ultrasound-assisted cooking:** Could reduce cooking time and energy use.
- **Pulsed electric field processing:** Nonthermal technology preserving nutrients.
- **High-pressure processing:** Preserves nutrients without heat.

5.9.2 Personalized Nutrition

Future research may identify optimal cooking methods for:

- Individual genetic variations impacting nutrient metabolism.
- Specific health conditions requiring particular nutrients.
- Age-related nutritional needs.
- Gut microbiome composition.

5.9.3 Research Gaps

Further investigation is needed regarding:

- Interactive effects of multiple nutrients and cooking methods.
- Long-term health outcomes of different cooking practices.
- Optimal combinations of raw and cooked vegetables.
- Effects on bioactive compound bioavailability beyond vitamins and minerals.

5.10 CONCLUSION

The chapter concludes that nutritional quality of vegetables is profoundly influenced by cooking methods. No single technique is universally superior; rather, optimal cooking depends on the specific vegetable, targeted nutrients, and health objectives. Steaming and microwaving generally preserve the most water-soluble vitamins, while cooking with oil enhances fat-soluble nutrient bioavailability. Methods that retain cooking liquids minimize mineral losses, and moderate heat reduces anti-nutritional factors while preserving beneficial compounds. Health-conscious consumers and food professionals should employ diverse cooking methods to maximize the nutritional benefits of vegetable consumption. Combining raw and cooked vegetables, using appropriate techniques for specific vegetables, and minimizing nutrient losses through careful preparation can significantly enhance the health-promoting properties of vegetable-rich diets. As research continues to reveal the complex interactions between cooking methods and nutrient bioavailability, evidence-based recommendations will become increasingly refined. Meanwhile, the principles outlined in this chapter provide practical guidance for optimizing vegetable preparation to support human health and disease prevention. The traditional wisdom of diverse cooking methods across cultures, combined with modern scientific understanding, offers a comprehensive approach to vegetable preparation that balances nutrition, safety, palatability, and sustainability. By thoughtfully selecting and applying appropriate cooking techniques, we can maximize the substantial health benefits that vegetables offer to human nutrition and well-being.

REFERENCES

Baldwin, D. (2012). Sous vide cooking: A review. *International Journal of Gastronomy and Food Science*, 1. https://doi.org/10.1016/j.ijgfs.2011.11.002

Cheung, M. M., Miller, L., Deutsch, J., Sherman, R., Katz, S. H., & Wise, P. M. (2025). Sensory properties and acceptability of fermented pearl millet, a climate-resistant and nutritious grain, among consumers in the United States – A pilot study. *Foods*, 14(5), 871. https://doi.org/10.3390/foods14050871

Fabbri, A. D., & Crosby, G. A. (2016). A review of the impact of preparation and cooking on the nutritional quality of vegetables and legumes. *International Journal of Gastronomy and Food Science*, 3, 2–11.

Fam, S., Khosravi, K., Massoud, R., & Massoud, A. (2021). High-pressure processing in food. *Biointerface Research in Applied Chemistry*, 11, 11553–11561. https://doi.org/10.33263/BRIAC114.1155311561

Fellows, P. J. (2022). *Food Processing Technology: Principles and Practice* (5th ed.). Academic Press.

Gerrard, J. (2005). The Maillard reaction: Chemistry, biochemistry and implications by Harry Nursten. *Australian Journal of Chemistry*, 58(10), 756–756. https://doi.org/10.1071/CH0505_BR

Gião, M. S., Pintado, M. E., Moura, A. P., Pereira, A. F., Malcata, F. X., & Azevedo-Silva, J. (2012). Effect of processing on the bioactivity and bioavailability of antioxidants in vegetables: A review. *Journal of Food Science*, 77(2), R11–R17.

Gonçalves, E. M., Pinheiro, J., Abreu, M., Brandão, T. R., & Silva, C. L. (2009). Modelling the kinetics of peroxidase inactivation, colour and texture changes of pumpkin (Cucurbita maxima L.) during blanching. *Journal of Food Engineering*, 92(1), 10–17.

Jiménez-Monreal, A. M., García-Diz, L., Martínez-Tomé, M., Mariscal, M., & Murcia, M. A. (2009). Influence of cooking methods on antioxidant activity of vegetables. *Journal of Food Science*, 74(3), H97–H103.

Miglio, C., Chiavaro, E., Visconti, A., Fogliano, V., & Pellegrini, N. (2008). Effects of different cooking methods on nutritional and physicochemical characteristics of selected vegetables. *Journal of Agricultural and Food Chemistry*, 56(1), 139–147. https://doi.org/10.1021/jf072304b

Munyaka, A. W., Makule, E. E., Oey, I., Van Loey, A., & Hendrickx, M. (2010). Thermal stability of L-ascorbic acid and ascorbic acid oxidase in broccoli (Brassica oleracea var. italica). *Journal of Food Science*, 75(4), C336–C340.

Palermo, M., Pellegrini, N., & Fogliano, V. (2014). The effect of cooking on the phytochemical content of vegetables. *Journal of the Science of Food and Agriculture*, 94(6), 1057–1070.

Raychaudhuri, U., Chakraborty, R., & Srivastava, P. P. (2022). Air frying: A healthier alternative to deep frying. *Journal of Food Science and Technology*, 59(12), 4463–4477.

Renna, M., Castellino, M., Leoni, B., Paradiso, V. M., & Santamaria, P. (2017). Microgreens production with low potassium content for patients with impaired kidney function. *Nutrients*, 9(5), 470.

Septembre-Malaterre, A., Remize, F., & Poucheret, P. (2018). Fruits and vegetables, as a source of nutritional compounds and phytochemicals: Changes in bioactive compounds during lactic fermentation. *Food Research International*, 104, 86–99.

Téllez-Morales, J., Rodríguez-Miranda, J., & Aguilar-Garay, R. (2024). Review of the influence of hot air frying on food quality. *Measurement: Food*, 14, 100153. https://doi.org/10.1016/j.meafoo.2024.100153

Turkmen, N., Sari, F., & Velioglu, Y. S. (2005). The effect of cooking methods on total phenolics and antioxidant activity of selected green vegetables. *Food Chemistry*, 93(4), 713–718.

World Health Organization. (2003). *Diet, Nutrition and the Prevention of Chronic Diseases* (WHO Technical Report Series No. 916). World Health Organization.

Yuan, G.-F., Sun, B., Yuan, J., & Wang, Q.-M. (2009). Effects of different cooking methods on health-promoting compounds of broccoli. *Journal of Zhejiang University. Science. B*, 10(8), 580–588. https://doi.org/10.1631/jzus.B0920051

6 Importance of Vegetables in the Diet and Their Classification

Anjali Gupta, Syeda Bushra Fatima, Mitesh Patel, and Malvi Surti

6.1 INTRODUCTION: OVERVIEW OF VEGETABLE CONSUMPTION AND DIETARY PATTERNS

Vegetables are known to be main source of vitamins, folates, amino acids, carotenoids, bioflavonoids, and dietary fibre (Kainat et al., 2022). The profound health benefits associated with sufficient vegetable intake are extensively documented, with regular consumption consistently linked to a decreased incidence of NCDs, including CVD, Type 2 Diabetes Mellitus (T2DM), and certain types of cancer, alongside a reduction in all-cause mortality (Wallace et al., 2020; Budreviciute et al., 2020). The Global Burden of Disease (GBD) study further underscores this critical relationship, identifying diets low in fruits and vegetables as a primary dietary risk factor contributing to the global burden of NCDs (Pan et al., 2023; Melaku et al., 2016). Despite this compelling evidence and clear public health messaging, actual intake levels globally often fall considerably short of recommended guidelines, highlighting a persistent and complex challenge rooted in multifaceted systemic, socioeconomic, cultural, and psychological factors (Cam, 2024). Dietary patterns, which encapsulate the overall composition, quantities, proportions, variety, and combination of foods and beverages typically consumed over a specific period, are pivotal in influencing nutrient intake, metabolic processes, and overall well-being (Black & Lozoff, 2008; Grundy et al., 2016). These patterns are not merely random collections of foods but are intrinsically shaped by cultural, social, and economic factors, reflecting the complex interplay between individuals and their food environments (Jayasinghe et al., 2025). Research consistently demonstrates that patterns characterized by high intakes of fruits, vegetables, whole grains, and lean proteins are associated with superior health outcomes when compared to diets dominated by processed foods and added sugars (Conrad et al., 2022; Mozaffarian, 2016). Understanding these broader patterns is crucial, as vegetable consumption is an integral part of achieving a healthy overall diet (Mason-D'Croz et al., 2019). From a nutritional and public health perspective, the term 'vegetable' generally refers to a plant cultivated for its edible part(s), encompassing a wide array of components such as stems, stalks, roots, tubers, bulbs, leaves, flowers, and even fruits (R. V Kumar, 2023). It is important to note that botanically, fruits are considered a subgroup of vegetables, a distinction that sometimes diverges from common culinary or dietary classifications. Typically, this definition excludes cereal grains, nuts, seeds, tea leaves, coffee beans, cacao beans, herbs, and spices (Monteiro et al., 2024). Vegetables are nutritionally dense, characterized by high fibre and water content, and serve as excellent sources of essential micronutrients including vitamins C and E, provitamin A, folate, potassium, magnesium, phosphorus, and calcium (Ebabhi & Adebayo, 2022).

Recommended daily intake levels for vegetables vary across different international and national guidelines. For instance the Dutch Health Council recommends 200 grams daily (Cámara et al., 2021), while the WHO generally advises at least 400 grams of fruits and vegetables combined per

 DOI: 10.1201/9781003606710-6

TABLE 6.1
Comparison of Daily Vegetable Intake Recommendations from Various Health Organizations and Studies for Adults, Children, and Adolescents

Guideline Name	Recommended Daily Vegetable Intake	Target Population	References
Dutch Health Council	200 grams daily	Adults	Kromhout et al. (2016)
WHO	At least 400 grams of F&V per day	Adults	Demydas (2010)
Chinese Dietary Guidelines	300 grams of vegetables daily	Adults	Yang et al. (2018)
GBD Study	< 280–320 grams of vegetables (low)	Adults	Kalmpourtzidou et al. (2020)
WHO	250 g/day F&V	Children aged 2–5 years	Wai and Myo (2022)
WHO	350 g/day F&V	Children aged 6–9 years	Adachi et al. (2025)
WHO	400 g/day F&V	Adolescents aged 10–17 years	Ramsay et al. (2014)

Abbreviation: F&V, Fruits and vegetables

day for adults (Guzman-Vilca et al., 2025). The Chinese Dietary Guidelines specifically advocate for 300 grams of vegetables daily, and the GBD study defines low vegetable intake as less than 280–320 grams per day (Zhang et al., 2025; Ding et al., 2024). The variability in defining 'vegetable' and the differing quantitative recommendations across various guidelines present a challenge for both researchers and the public, as is shown in Table 6.1 (Harris et al., 2022). Higher consumption of fruits and vegetables is associated with reduced cardiovascular mortality, with studies showing a significant inverse association. Furthermore, adopting vegetarian dietary patterns has been associated with a 22% lower cardiovascular mortality and a 28% lower incidence of CVD (Y. Wang et al., 2023).

T2DM: Adequate fruit and vegetable intake is associated with a reduced risk of T2DM (Nguyen et al., 2022). Plant-based diets, in particular, have demonstrated efficacy in managing and potentially reversing T2DM, partly due to their high fibre content, which helps slow glucose absorption and prevent significant blood sugar spikes.

Cancer: Vegetable consumption is linked to a reduced risk of certain types of cancer (Watling et al., 2022). Consuming more fruits and vegetables has been linked to a decreased risk of colon cancer and breast cancer. Specific benefits have been observed for leafy vegetables, fruiting vegetables, and raw tomatoes regarding breast cancer risk. However, the consistency of evidence for a reduced risk of cancer mortality across all studies, specifically from fruit and vegetable consumption, can be less uniform (Tsai et al., 2023).

All-Cause Mortality: An increased intake of fruits and vegetables is substantially linked to a decreased risk of death from all cause underscoring their fundamental role in promoting longevity and overall health.

6.2 HEALTH BENEFITS OF REGULAR VEGETABLE INTAKE

The health benefits of regular vegetable intake are cardiovascular health, weight management and obesity prevention, diabetes control, and cancer risk reduction.

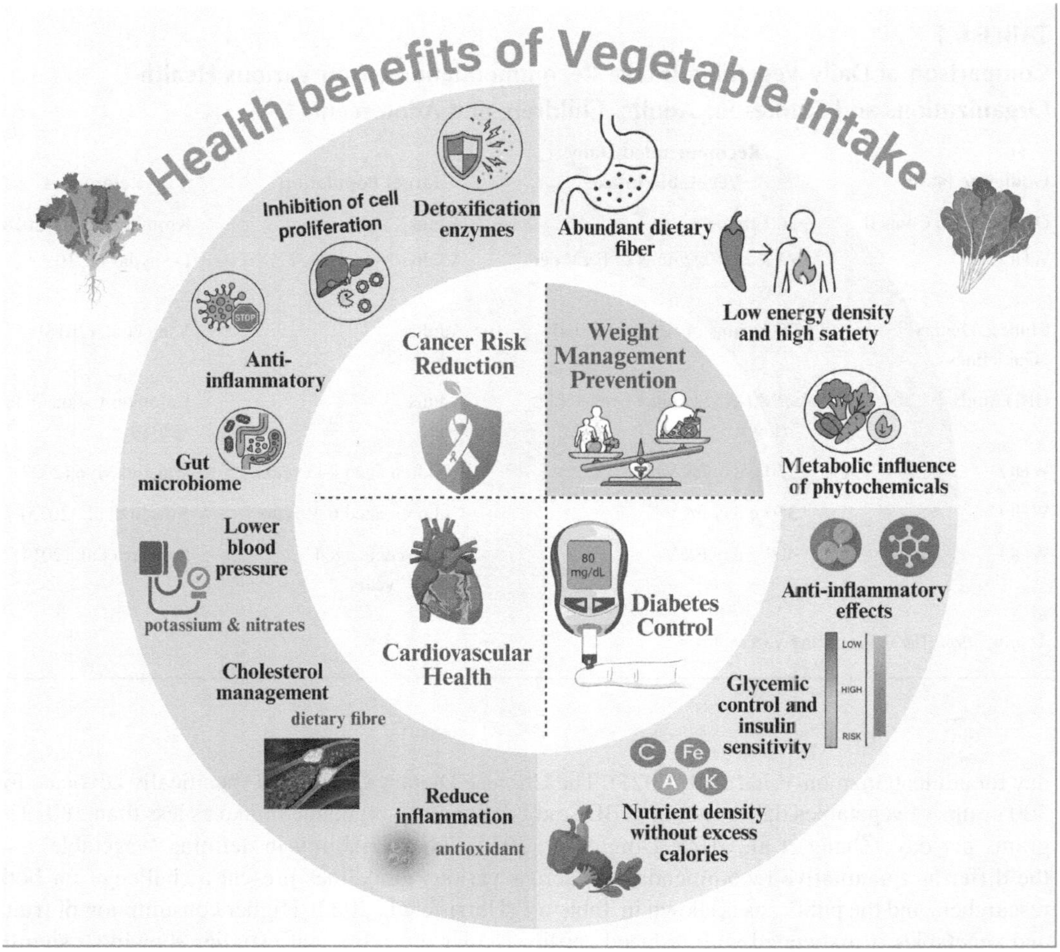

FIGURE 6.1 The diverse health benefits of consuming vegetables, organized into four categories: cancer risk reduction, weight management and prevention, cardiovascular health, and diabetes control.

The profound impact of dietary choices on human health is unequivocally established, and among the diverse components of a wholesome diet, the regular consumption of vegetables stands as a cornerstone of well-being (Borrego-Ruiz et al., 2024; Samuel et al., 2024). The efficacy of vegetables in these domains stems from a complex and synergistic interplay of their inherent nutritional richness, encompassing a vast array of vitamins, minerals, dietary fibre, potent antioxidants, and a diverse spectrum of phytochemicals, as shown in Figure 6.1.

6.2.1 Cardiovascular Health: Cultivating a Resilient Heart

The protective effects of vegetables on cardiovascular health are among the most extensively researched and compelling. Regular consumption significantly contributes to the prevention and management of various cardiovascular ailments through several critical mechanisms.

Blood Pressure Regulation: Vegetables are naturally rich in potassium, a vital electrolyte that counteracts the hypertensive effects of sodium. By promoting vasodilation (the widening of blood vessels) and maintaining proper fluid balance, potassium helps to lower and stabilize blood pressure, thereby reducing the strain on the heart and arteries. Studies consistently demonstrate an inverse relationship between potassium intake from vegetables and the incidence of hypertension (Kim et al., 2024; Caminiti et al., 2024). Furthermore, the presence of naturally occurring nitrates in leafy green vegetables like spinach, arugula, and beetroot, when consumed, is converted into nitric oxide (NO) in the body. Nitric oxide is a potent vasodilator, directly contributing to relaxed blood vessels, improved blood flow, and, consequently, reduced blood pressure (Kurhaluk & Tkaczenko, 2025).

Cholesterol Management: The dietary fibre, particularly soluble fibre found in vegetables such as oats, lentils, and many root vegetables, plays a crucial role in managing cholesterol levels. Soluble fibre binds to cholesterol and bile acids in the digestive tract, preventing their reabsorption and promoting their excretion. This mechanism necessitates the liver to draw more cholesterol from the bloodstream to produce new bile acids, leading to a reduction in LDL or 'bad' cholesterol levels. Lower LDL cholesterol is directly associated with a decreased risk of atherosclerosis, the hardening and narrowing of arteries, a primary precursor to heart attacks and strokes (Suresh et al., 2024).

Combating Oxidative Stress and Inflammation: Vegetables are veritable powerhouses of antioxidants, including vitamins C and E, beta-carotene, lycopene, and a vast array of flavonoids and polyphenols. These compounds actively neutralize free radicals, which are unstable molecules that cause oxidative damage to cells, including the endothelial lining of blood vessels. Oxidative stress is a key driver of endothelial dysfunction, inflammation, and the progression of atherosclerotic plaque formation (Akbari et al., 2022; Sharma et al., 2024). By mitigating oxidative damage, these antioxidants preserve arterial elasticity and function, reducing the risk of cardiovascular events. Moreover, many phytochemicals in vegetables exhibit potent anti-inflammatory properties, dampening the chronic low-grade inflammation that underlies many CVDs.

Improved Endothelial Function: The synergistic action of antioxidants, fibre, potassium, and nitrates contributes to enhanced endothelial function. A healthy endothelium is crucial for regulating vascular tone, blood clotting, and inflammatory responses. Regular vegetable intake supports the integrity and functionality of this critical layer, thereby contributing to overall cardiovascular resilience (Perera et al., 2024).

6.2.2 Weight Management and Obesity Prevention: A Satiating Solution

In the global fight against obesity, vegetables emerge as an indispensable dietary component due to their unique nutritional profile that promotes satiety and limits caloric intake.

Low Energy Density and High Satiety: Vegetables are characterized by their low caloric density and high-water content. This means that a large volume of vegetables can be consumed for a relatively small number of calories. This high volume, coupled with their water content, fills the stomach, sending signals of fullness to the brain, thereby promoting satiety and reducing overall energy intake without compromising nutrient delivery (Ahire et al., 2023). This 'eat more, weigh less' principle is a fundamental aspect of successful weight management.

Abundant Dietary Fibre: The high dietary fibre content in vegetables is paramount for weight control. Fibre adds bulk to the diet, slows down gastric emptying, and prolongs the feeling of fullness. This extended satiety helps to reduce appetite, minimizing snacking between meals, and ultimately contributes to a lower overall calorie consumption (Lynrah et al., 2024; Suárez et al., 2025). Furthermore, fibre aids in maintaining a healthy digestive system and regulates blood glucose levels, preventing rapid spikes and subsequent crashes that can trigger cravings and overeating.

Metabolic Influence of Phytochemicals: Beyond their fibre and low-calorie attributes, specific phytochemicals in vegetables may exert direct effects on fat metabolism and appetite regulation. For instance certain compounds found in chili peppers (capsaicin) can transiently increase thermogenesis, leading to a slight increase in calorie expenditure, and may also play a role in appetite suppression (Cecchini et al., 2022). Flavonoids, abundant in various vegetables and fruits, have been associated with improved glucose metabolism and insulin sensitivity, which indirectly supports weight management by reducing fat storage and controlling cravings (Hofer et al., 2021).

6.2.3 Diabetes Control: Stabilizing Blood Glucose for Better Health

For the individuals who have diabetes or are susceptible to type 2 diabetes, consistent vegetable intake is a critical dietary intervention due to its profound impact on glucose metabolism.

Glycaemic Control and Insulin Sensitivity: Most non-starchy vegetables possess a low glycaemic index (GI), which indicates that they lead to a gradual and moderate rise in blood glucose levels after consumption, in contrast to high-GI foods that lead to rapid spikes. The high dietary fibre further contributes to this by slowing down the absorption of carbohydrates, ensuring a more stable and sustained release of glucose into the bloodstream (Al-Adwi et al., 2023; Basak & Laskar, 2024). This helps to prevent post-meal hyperglycaemia and reduces the demand on the pancreas to produce insulin, thereby improving insulin sensitivity over time (O'Hearn et al., 2023).

Anti-Inflammatory Effects: The development of chronic low-grade inflammation significantly influences the progression of insulin resistance and type 2 diabetes. The vast array of antioxidants and anti-inflammatory compounds found in vegetables and pigmental compounds also contribute to overall antioxidant ability depending on amount and the compound (Polak-Szczybyło, 2023).

Nutrient Density without Excess Calories: Vegetables provide essential vitamins, minerals, and other micronutrients without adding excessive calories or rapidly digestible carbohydrates. This nutrient-dense profile is crucial for overall metabolic health in individuals managing diabetes, ensuring adequate nutrient intake while adhering to carbohydrate and calorie targets (Sharif et al., 2021).

6.2.4 Cancer Risk Reduction: Nature's Chemopreventive Arsenal

The compelling evidence linking regular vegetable consumption to a reduced risk of various cancers highlights their role as potent chemopreventive agents. The mechanisms are complex and involve a sophisticated interplay of various bioactive compounds.

Antioxidant Defence: As discussed previously, the abundant *antioxidants* in vegetables (e.g. vitamins C and E, beta-carotene, lycopene, selenium, and numerous polyphenols) are essential for suppressing free radicals. By preventing oxidative damage to DNA and cellular structures, they help to inhibit the initiation phase of carcinogenesis (Mukherjee et al., 2025; Alsulami, 2024). Oxidative stress is a well-recognized contributor to cellular mutations that can lead to cancer.

Modulation of Detoxification Enzymes: A remarkable feature of certain vegetables, particularly cruciferous vegetables (e.g. broccoli, cauliflower, Brussels sprouts, kale), is their content of glucosinolates. When these vegetables are chewed and digested, glucosinolates are hydrolysed into biologically active compounds, notably isothiocyanates (e.g. SFN, indole-3-carbinol). These compounds have been extensively studied for their ability to enhance the activity of phase I and phase II detoxification enzymes in the liver. These enzymes are crucial for metabolizing and eliminating carcinogens and other toxic compounds from the body, thereby reducing their opportunity to inflict cellular damage (S. Sharma et al., 2025; Melim et al., 2022).

Inhibition of Cell Proliferation and Induction of Apoptosis: Numerous phytochemicals in vegetables have demonstrated the ability to inhibit the uncontrolled proliferation of cancer cells and induce apoptosis (programmed cell death) in abnormal or pre-cancerous cells. For instance quercetin (found in onions, peppers), lycopene (in tomatoes), and resveratrol (in grapes and some vegetables) have shown anti-proliferative and pro-apoptotic effects in various cancer cell lines and *in vivo* models (Roszkowski, 2023; Awasthi & Srivastava, 2024). This ability to selectively target and eliminate potentially cancerous cells is a critical aspect of their chemopreventive action.

Anti-inflammatory and Immune Modulation: Chronic inflammation is a recognized risk factor for many cancers. The widespread anti-inflammatory properties of vegetable components help to suppress inflammatory pathways that can promote tumour growth and progression (Cuevas-Cianca et al., 2023). Furthermore, certain phytochemicals may modulate immune responses, enhancing the body's natural defences against emerging cancer cells. Fibre and Gut Microbiome Health: The dietary fibre in vegetables promotes a healthy gut microbiome. A balanced and diverse gut microbiota produces beneficial SCFAs like butyrate, which can have anticancer effects, particularly in the colon. A healthy gut environment also helps to reduce inflammation and promote regular bowel movements, thereby decreasing the exposure of the colon lining to potential carcinogens (Nakkarach et al., 2021). Epidemiological studies consistently demonstrate an inverse association between high vegetable intake and the risk of various cancers, including colorectal, lung, breast, prostate, and gastric cancers (Ubago-Guisado et al., 2021; Guo et al., 2024). While research continues to unravel the intricate molecular mechanisms, the collective evidence strongly positions vegetables as an indispensable component of a cancer-preventive diet.

In culmination, the scientific evidence overwhelmingly underscores the profound and multifaceted health benefits derived from the regular and diverse intake of vegetables (Chaudhari et al., 2024). From bolstering cardiovascular resilience through optimized blood pressure and cholesterol levels, facilitating effective weight management via enhanced satiety and low energy density, to providing robust control over blood glucose in the context of diabetes, and significantly mitigating cancer risk through a complex array of chemopreventive actions, vegetables are unequivocally an indispensable foundation for a healthy diet. The synergistic action of their rich nutrient profile – including fibre, vitamins, minerals, antioxidants, and a vast spectrum of phytochemicals – contributes powerfully to disease prevention and the cultivation of optimal well-being across the lifespan. Embracing a vibrant and varied array of vegetables as a central component of daily dietary patterns is not merely a recommendation; it is a fundamental and potent strategy for enhancing health outcomes and fostering longevity (Hu, 2024).

6.3 CLASSIFICATION OF VEGETABLES

Vegetables, forming the cornerstone of healthy diets worldwide, are incredibly diverse. Beyond their culinary applications, understanding their classification is crucial for agricultural practices, nutritional science, and public health initiatives. This chapter delves into a comprehensive classification of vegetables, exploring distinctions based on botanical families, edible plant parts, nutrient composition, and their significant functional and medicinal roles.

6.3.1 Classification Based on Botanical Families

Botanical classification groups vegetables based on their taxonomic relationships, offering insights into their evolutionary history, shared characteristics, and susceptibility to similar pests and diseases. This understanding is vital for crop rotation, breeding programmes, and sustainable agriculture (Chiquito-Contreras et al., 2024). Moreover, shared botanical ancestry often correlates with similar phytochemical profiles, influencing the nutritional and functional properties of the vegetables within a given family.

Alliaceae (Onion Family): This family is characterized by plants producing bulbs with pungent flavours due to sulphur compounds. Onion (*Allium cepa*), garlic (*Allium sativum*), leek (*Allium ampeloprasum*), and chives (*Allium schoenoprasum*) are examples (Bhat, 2021). It is often used for flavouring and recognized for its potential cardiovascular benefits. Beyond their flavour, Alliaceae vegetables are highly valued for their health-promoting properties. For instance recent reviews highlight the role of *Allium sativum* in ameliorating CVDs, often attributed to its OSCs, contributing to improved lipid profiles and blood pressure regulation (Tang et al., 2025).

Brassicaceae (Mustard/Cabbage Family): This family, also known as the Cruciferae family (referring to their cross-shaped flowers), includes Brassicaceae vegetables that are a diverse group celebrated for their crisp textures and often pungent, slightly bitter flavours (Wells, 2024).

Cucurbitaceae (Gourd Family): This family primarily consists of sprawling or climbing plants that produce large, often fleshy fruits that are botanically fruits but culinarily treated as vegetables. They are often characterized by their trailing vines and distinctively ribbed or smooth fruits (Verma et al., 2024), vining or sprawling growth habit, large fruits (often pepos), and prominent male and female flowers. It includes cucumber (*Cucumis sativus*), pumpkin (*Cucurbita maxima*, *Cucurbita moschata*, *Cucurbita pepo*), and many other cucurbit vegetables (Levi, 2023).

Fabaceae (Legume/Pea Family): Also called Leguminosae, this is one of the largest plant families, distinguished by their characteristic seed pods. Many members are staple food crops globally. It forms pod-bearing fruits and nitrogen-fixing root nodules (symbiotic relationship with *Rhizobium* bacteria). It can fix atmospheric nitrogen into the soil, making it invaluable for sustainable agriculture and crop rotation, and reducing the need for synthetic nitrogen fertilizers (Abusaief, 2021). Examples are peas (*Pisum sativum*), green beans (*Phaseolus vulgaris*), lima beans (*Phaseolus lunatus*), lentils (*Lens culinaris*), chickpeas (*Cicer arietinum*), and soybeans (*Glycine max*).

Solanaceae (Nightshade Family): This family is incredibly diverse, encompassing not only many globally significant food crops but also several toxic plants (Motti, 2021). The edible members are often recognized by their typically fleshy fruits or tubers. Examples are tomato (*Solanum lycopersicum*), potato (*Solanum tuberosum*), eggplant (*Solanum melongena*), bell pepper (*Capsicum annuum*), and chili pepper (*Capsicum annuum*).

Asteraceae (Daisy/Sunflower Family: This is one of the largest plant families, often characterized by their composite flower heads (like a daisy or sunflower). In terms of vegetables, it primarily includes leafy greens and some flower-head vegetables. Composite flower heads often contain bitter compounds (Bala et al., 2023). It includes (*Lactuca sativa*), artichoke (*Cynara scolymus*), chicory (*Cichorium intybus*), and endive (*Cichorium endivia*). Chenopodiaceae (Goosefoot Family): Traditionally recognized as a distinct family, most genera formerly in Chenopodiaceae, including many important vegetables, have been reclassified into the Amaranthaceae family due to phylogenetic studies. However, for practical purposes in vegetable classification, it's still common to discuss them as a functional group or acknowledge their traditional grouping. Examples are spinach (*Spinacia oleracea*), beetroot (*Beta vulgaris*), Swiss chard (*Beta vulgaris* var. *cicla*), and quinoa (*Chenopodium quinoa*).

6.3.2 Classification Based on Edible Plant Parts

This classification is perhaps the most intuitive from a culinary perspective, categorizing vegetables by the specific part of the plant consumed. Many vegetables can fall into multiple categories if different parts are edible (e.g. beetroot leaves and roots).

Root Vegetables: These are underground storage organs that absorb nutrients and water from the soil and often good sources of carbohydrates (especially starchy roots like potato and sweet potato), fibre, and various vitamins and minerals depending on the type (e.g. beta-carotene in carrots) (Pathak et al., 2021). Stem Vegetables: The edible stems, stalks, or shoots of plants. Generally low in calories and good sources of fibre, vitamins, and minerals. Asparagus, for instance, is known for its folate content and the presence of asparagusic acid, a sulphur-containing compound with potential antioxidant properties (Olas, 2024). Examples are asparagus (*Asparagus officinalis*) and kohlrabi (*Brassica oleracea* var. *gongylodes*). Leafy Vegetables: The edible leaves of plants, often rich in vitamins and minerals. Excellent sources of vitamins K, A, and C, folate, iron, and dietary fibre. They lower the chance of developing chronic diseases and are essential for preventing micronutrient deficits.

Flower Vegetables: The edible flowers or flower buds. Provide vitamins C and K, fibre, and potent antioxidants and phytochemicals (e.g. glucosinolates in broccoli and cauliflower). Examples are broccoli, cauliflower, artichoke, and broccoflower (Faisal et al., 2022). Fruit Vegetables: Botanically, these are fruits (containing seeds) but are commonly used in savoury culinary applications. Vary widely, but are often good sources of vitamins (especially C and A), antioxidants (like lycopene in tomatoes), and fibre. Their high-water content also contributes significantly to hydration.

Nutritional Aspects: Seed/Pod Vegetables: The edible seeds or pods containing seeds. Rich in protein, fibre, and various B vitamins. Leguminous seeds (beans, peas, lentils) are particularly valued for their plant-based protein content and are central to sustainable dietary patterns, contributing to both human health and environmental sustainability (M. Kumar et al., 2022).

Bulb Vegetables: Underground modified leaf shoots with fleshy scales or layers. Known for their distinctive flavours and sulphur compounds, which contribute to their unique health benefits, including potential cardiovascular protection and antimicrobial properties (Chauhan et al., 2021).

6.3.3 Classification Based on Nutrient Composition

This classification focuses on the primary nutritional contributions of vegetables, guiding dietary recommendations and promoting a balanced intake of essential nutrients.

Dark Green Leafy Vegetables (DGLVs): These are rich in Vitamin K, Vitamin A (as beta-carotene), Vitamin C, folate, iron, calcium, and dietary fibre. Examples include spinach, kale, collard greens, Swiss chard, and mustard greens, which are crucial for bone health, vision, immune function, red blood cell formation, and digestive health. Their high antioxidant content helps combat oxidative stress and inflammation (Sarkar et al., 2023).

Red and Orange Vegetables: Orange and yellow hues are mostly caused by carotenoids (e.g. beta-carotene, lycopene), vitamin C, and potassium. They include carrots, sweet potatoes, tomatoes, red bell peppers, and pumpkin. The role of dietary carotenoids in health is a well-established area of research, emphasizing their absorption and conversion into active forms in the body (Sravya, 2022).

Starchy Vegetables: These are primarily carbohydrates (starch), providing energy. Also contain fibre, potassium, and some B vitamins. Potatoes, sweet potatoes, corn, peas, cassava, yams are examples. They provide sustained energy and contribute to dietary fibre intake. While being higher in calories than non-starchy vegetables, they are nutrient-dense when consumed in appropriate portions and prepared healthily. Research continues to refine the understanding of their glycaemic index and overall health impact within balanced dietary patterns (Bai et al., 2021).

Beans, Peas, and Lentils (Legumes): These are excellent sources of plant-based protein, dietary fibre, complex carbohydrates, folate, iron, potassium, and magnesium. Their protein content makes them valuable in vegetarian and vegan diets, significantly contributing to protein recommendations. They include kidney beans, black beans, chickpeas (garbanzo beans), lentils, and peas.

6.3.4 Classification Based on Functional and Medicinal Roles

Beyond providing basic nutrition, many vegetables contain bioactive compounds that exert specific physiological effects, contributing to disease prevention and overall well-being. This area is a rapidly expanding field of research. Those vegetables are classified as antioxidant-rich vegetables, which combat oxidative stress by neutralizing free radicals, thereby reducing cellular damage and inflammation. Chronic illnesses like cancer, heart disease, and neurological disorders are less likely to occur as a result. Examples include dark green leafy vegetables (spinach, kale), red and orange vegetables (carrots, tomatoes, bell peppers), cruciferous vegetables (broccoli, cauliflower), garlic, and onions. Numerous studies from 2015 onwards emphasize the role of vegetable-derived antioxidants in mitigating oxidative stress and improving health outcomes.

Anti-inflammatory vegetables, which help reduce chronic inflammation, are a key factor in many NCDs. Broccoli, turmeric (often considered a spice, but its rhizome is a vegetable), ginger (a rhizome vegetable), spinach, and kale are examples. The bioactive compounds are glucosinolates and isothiocyanates (cruciferous vegetables), curcumin (turmeric), gingerols (ginger), and flavonoids (Yu et al., 2024). Cardioprotective vegetables contribute to heart health by lowering blood pressure, reducing cholesterol levels, and improving vascular function. The examples are garlic, onions, leafy greens (spinach, kale), tomatoes, asparagus, and beets (Velissaridou et al., 2024). Anticancer Vegetables: The vegetables in this group may inhibit cancer cell growth, promote apoptosis (programmed cell death), and reduce carcinogen activation. Examples are cruciferous vegetables (broccoli, cabbage), tomatoes, garlic, onions, and spinach. Extensive research, including epidemiological studies and in-vitro/in-vivo experiments, continues to support the chemopreventive properties of specific vegetables. For instance recent reviews highlight the potential of SFN from cruciferous vegetables in various cancer types, detailing its molecular targets and mechanisms (Mitra et al., 2022; Yuan et al., 2022). Digestive health-promoting vegetables, which improve gut motility, support a healthy gut microbiome and prevent constipation. It includes all high-fibre vegetables, including leafy greens, root vegetables, and legumes. The bioactive compounds are dietary fibre (soluble and insoluble) andprebiotics (e.g. inulin in onions, garlic, chicory).

The classification of vegetables, whether by botanical family, edible part, nutrient composition, or functional roles, provides a comprehensive framework for understanding their immense value. This multifaceted approach highlights their agricultural significance, nutritional prowess, and their increasingly recognized roles as functional foods and sources of medicinal compounds. Encouraging diverse vegetable consumption, guided by these classifications, remains a cornerstone of promoting global health and sustainable food systems.

6.4 CULINARY AND DIETARY APPLICATIONS

Raw vs. cooked consumption, fermented and pickled vegetables, and regional and seasonal preferences are the different applications in the form of dietary preference and need.

The journey of vegetables from their health benefits to classification culminates in their culinary and dietary applications. This stage is where their diverse forms, textures, flavours, and nutritional profiles are transformed and integrated into human diets. The methods of preparation and consumption vary widely across cultures and over time, significantly influencing nutrient bioavailability, sensory experiences, and even the functional health benefits derived from vegetables, as detailed in Figure 6.2.

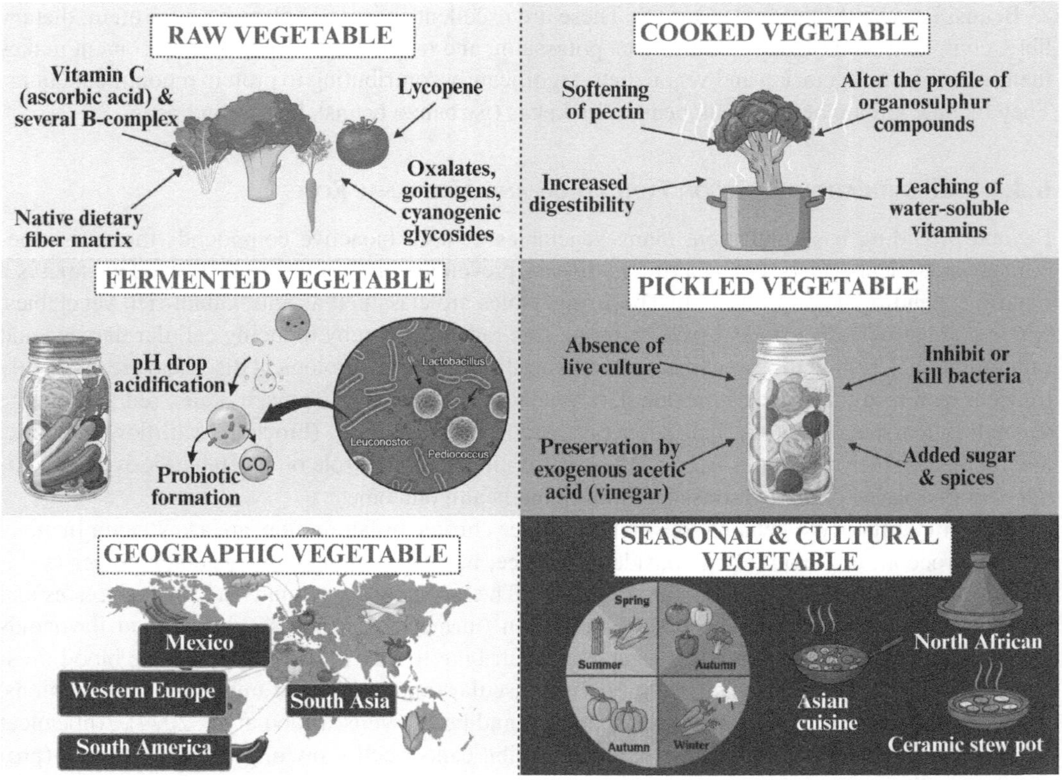

FIGURE 6.2 Various ways vegetables are prepared and consumed, highlighting the effects on their nutritional content and properties, showing raw vegetables, cooked vegetables, fermented vegetables, pickled vegetables, geographic, seasonal and cultural vegetables, illustrating how consumption patterns vary by region and season.

6.4.1 Raw versus Cooked Consumption: Optimizing Nutrient Bioavailability and Sensory Appeal

The consumption of vegetables in their raw state is often championed for the purported preservation of heat-labile nutrients and intrinsic enzymatic activity. Vitamin C (ascorbic acid) and several B-complex vitamins, particularly folate, are highly susceptible to thermal degradation and leaching into cooking water due to their water-soluble nature (Malik et al., 2023). Consequently, raw consumption often ensures a maximal intake of these micronutrients. Beyond micronutrient preservation, raw vegetables maintain their native dietary fibre matrix, contributing to optimal gastrointestinal motility, satiety, and modulation of glycaemic response (Arora, 2023; Janety, 2021). However, the advantages of raw consumption are counterbalanced by several scientific considerations. Many essential fat-soluble vitamins (e.g. provitamin A carotenoids like beta-carotene in carrots) and lipophilic phytochemicals (e.g. lycopene in tomatoes) are sequestered within robust plant cell walls. Thermal processing, by disrupting these cellular matrices, enhances the liberation and subsequent bioavailability of these compounds upon ingestion. For instance the bioaccessibility of lycopene is higher in cooked tomato as compared to raw tomato because of thermal breakdown of the cell wall structure (C. Wang et al., 2023).

Moreover, certain raw vegetables contain naturally occurring anti-nutritional factors or toxic compounds. Examples include oxalates, goitrogens, and cyanogenic glycosides (e.g. in some root vegetables). Cooking methods, especially boiling, can bring the levels of these harmful substances down to safe levels, which makes the vegetable safer to eat and more nutritious overall (Salim et al., 2023; Singh & Arora, 2023). Furthermore, the complex carbohydrate structures and tough fibres in some raw vegetables can pose digestive challenges for sensitive individuals; thermal softening can significantly improve palatability and assimilation.

6.4.1.1 The Transformative Effects of Cooked Vegetable Consumption

Cooking, encompassing a myriad of techniques such as steaming, boiling, roasting, stir-frying, and grilling, fundamentally alters the physical, chemical, and organoleptic properties of vegetables. A primary advantage of cooking is the enhancement of digestibility and palatability (Dihdashti, 2023). Heat induces the gelatinization of starches and the softening of pectin and other cell wall polysaccharides, rendering tough fibrous structures more manageable for mastication and subsequent enzymatic breakdown in the human digestive tract. This increased digestibility can lead to more efficient nutrient extraction (Jacob, 2021). While concerns regarding nutrient loss during cooking are valid, the extent of loss is highly dependent on the cooking method employed. Water-based methods like boiling can lead to significant leaching of water-soluble vitamins and minerals into the cooking medium. In contrast, methods that minimize contact with water and reduce cooking duration, such as steaming, microwaving, and stir-frying, are generally more effective at retaining these heat-sensitive nutrients (Siddique et al., 2024). The optimization of cooking parameters is therefore crucial for nutrient preservation.

Crucially, thermal processing can enhance the bioavailability of specific nutrients and phytochemicals. Beyond carotenoids, the heating of Allium vegetables (e.g. garlic, onions) can alter the profile of OSCs, with some becoming more bioavailable (Gul et al., 2025). Similarly, the cooking of cruciferous vegetables can influence the formation and stability of isothiocyanates from glucosinolates, compounds noted for their chemoprotective properties (Melim et al., 2022). From a culinary perspective, cooking elicits complex flavour and aroma transformations. Processes like caramelization (sugar browning in root vegetables) and Maillard reactions (non-enzymatic browning reactions involving amino acids and reducing sugars) generate new volatile compounds that contribute rich, savoury, and often sweet notes, significantly enhancing the overall palatability and desirability of cooked vegetables (Kattel & Antonious, 2025). Moreover, cooking serves a critical role in food safety by destroying pathogenic microorganisms, thereby reducing the risk of foodborne illnesses.

Therefore, to consume vegetables raw or cooked represents a nuanced interplay of nutritional goals, digestive tolerance, and culinary preferences. An optimal dietary approach integrates both raw and appropriately cooked vegetables. This synergistic strategy ensures a broad spectrum of nutrient intake, maximizes the bioavailability of diverse phytochemicals, mitigates potential anti-nutritional factors, and delivers a wide array of sensory experiences essential for long-term dietary adherence and overall health.

6.4.2 Fermented and Pickled Vegetables: Ancient Preservation Meets Modern Health Benefits

Fermentation and pickling are ancient food preservation techniques that have been practiced globally for millennia. Beyond extending shelf life, these processes profoundly transform vegetables, enhancing their flavour profiles, improving digestibility, and often augmenting their nutritional and functional properties. Understanding these processes from a scientific perspective reveals their intricate role in food security, cultural heritage, and emerging public health benefits.

6.4.2.1 Fermentation: Microbial Alchemy for Enhanced Food Functionality

Fermentation, particularly lactic acid fermentation (LAF), involves the controlled proliferation of beneficial microorganisms, predominantly lactic acid bacteria (LAB) such as *Lactobacillus*, *Leuconostoc*, and *Pediococcus* species (Sionek et al., 2023). These microorganisms metabolize fermentable carbohydrates (sugars) present in vegetables into lactic acid, acetic acid, carbon dioxide, and various other metabolites. The resultant accumulation of organic acids significantly lowers the pH of the substrate, creating an acidic environment (typically pH < 4.5) that effectively inhibits the growth of spoilage microorganisms and pathogenic bacteria, thereby extending the shelf life of the vegetable product (Okoye et al., 2025). Some examples of fermented food include Kimchi: A cornerstone of Korean gastronomy, primarily produced from Napa cabbage (*Brassica rapa subsp. pekinensis*) and other vegetables through LAF, seasoned with a complex blend of spices including gochugaru (chili powder), garlic, and ginger. Its intricate microbial ecology contributes to its distinct flavour profile and reported health benefits (Kwon et al., 2023). Sauerkraut: A traditional Central and Eastern European staple derived from fermented shredded cabbage (*Brassica oleracea var. capitata*). The sequential dominance of LAB species like *Leuconostoc mesenteroides*, *Lactobacillus plantarum*, and *Lactobacillus brevis* dictates its characteristic sour flavour development (Fijan et al., 2024). Fermented Pickles: Cucumbers and other vegetables can undergo natural LAF in brine, yielding sour pickles characterized by their tangy flavour and crisp texture, distinct from vinegar-pickled counterparts.

The health implications of consuming fermented vegetables are increasingly substantiated by scientific research. Fermented vegetables are a natural reservoir of probiotics, defined as 'live microorganisms that, when administered in adequate amounts, confer a health benefit on the host' (Parichat & Pongsak, 2023). These beneficial microbes contribute to the modulation and diversification of the gut microbiome, which is intimately linked to numerous physiological functions, including improved digestion, enhanced immune system regulation, synthesis of certain vitamins (e.g. Vitamin K2, some B vitamins), and even neurotransmitter production influencing mood and cognitive function via the gut-brain axis (Kirmizigul & Sengun, 2024; Sevim et al., 2023). Beyond probiotic effects, fermentation can increase nutrient bioavailability. Microbial enzymatic activities can break down complex plant components, making certain vitamins (e.g. B vitamins, Vitamin K) and minerals more accessible for absorption (Samtiya et al., 2021). Fermentative microorganisms can degrade anti-nutritional compounds such as phytates, oxalates, and certain tannins, which would otherwise hinder nutrient absorption in raw vegetables (Borrego-Ruiz et al., 2024).

6.4.2.2 Pickling: Acid-Mediated Preservation and Sensory Enhancement

While often colloquially interchanged with fermentation, 'pickling' broadly refers to the preservation of food in an acidic medium. This typically involves the use of exogenous acidic solutions, most commonly vinegar (acetic acid), rather than relying solely on endogenous microbial acid production. The low pH directly inhibits the proliferation of spoilage and pathogenic microorganisms. Pickling brines frequently include salt, sugar, and various spices that contribute to flavour and additional preservation (Li et al., 2025). Examples of vinegar-pickled vegetables include gherkins, pickled onions, and mixed vegetable pickles. Unlike traditionally fermented products, vinegar-pickled vegetables generally do not contain a significant population of live probiotic cultures, as the high initial acidity of vinegar might inhibit or kill the beneficial bacteria. Therefore, their primary health contributions are typically through their retained dietary fibre, vitamins, and minerals (similar to raw or lightly cooked vegetables), along with their distinct acidic flavour profile, which can stimulate appetite and complement meals. They serve as a convenient and stable method for preserving seasonal vegetable surpluses.

6.4.3 Regional and Seasonal Preferences: The Eco Dialect of Vegetable Consumption

The consumption patterns of vegetables are inextricably linked to a complex interplay of environmental, geographical, cultural, and temporal determinants. These interdependent factors collectively shape the availability, selection, and culinary integration of vegetable species within localized and regional dietary frameworks. Understanding this 'Eco dialect' of vegetable consumption is crucial for appreciating dietary diversity, promoting sustainable food systems, and addressing global nutritional security.

6.4.3.1 The Determinative Influence of Climate and Geographic Context on Agrobiodiversity

The success and prevalence of specific vegetable cultivars are directly governed by ambient climatic parameters, edaphic (soil) characteristics, and topographical features. Tropical and Subtropical Agro-Ecological Zones: These regions, characterized by warm temperatures, high humidity, or distinct arid seasons, naturally foster the cultivation of crops adapted to these conditions. Examples include various species of Cucurbitaceae (e.g. bitter gourd *Momordica charantia*, snake gourd *Trichosanthes cucumerina*), diverse root and tuber crops (e.g. cassava *Manihot esculenta*, taro *Colocasia esculenta*, yam *Dioscorea* spp.), and an expansive array of indigenous leafy greens (e.g. amaranth *Amaranthus* spp., drumstick leaves *Moringa oleifera*). The extensive reliance on plantains (*Musa paradisiaca*) in West African and Caribbean cuisines and various forms of cassava in sub-Saharan African and Latin American diets exemplify profound climatic adaptation and localized food system evolution (Roy et al., 2022; Otieno, 2024).

The rich culinary traditions spanning Asian cuisines are deeply rooted in their regional agrobiodiversity, incorporating vegetables like bok choy (*Brassica rapa subsp. chinensis*), daikon radish (*Raphanus sativus* var. *longipinnatus*), bamboo shoots (*Bambusa vulgaris*), and a diverse spectrum of legumes. The regional variation in species and cultivars within this vast continent underscores the powerful influence of localized ecological niches. Climate change is increasingly impacting these regional patterns, necessitating adaptation in agricultural practices and potentially shifting traditional consumption patterns as certain crops become less viable or new species are introduced.

6.4.3.2 Cultural Heritage and Culinary Embodiment of Dietary Patterns

Cultural heritage serves as a profound architect of vegetable preferences and of preparation techniques. Traditional culinary systems are often the culmination of centuries of adaptation to locally available flora, leading to the development of distinctive flavour matrices, sophisticated cooking methodologies, and ingrained food pairings that become defining hallmarks of a specific culture's diet (Rajan, 2023). In Indian cuisine, the historical and philosophical emphasis on plant-based diets in numerous communities has fostered an unparalleled diversity and depth in vegetable utilization. From the distinct bitterness of *karela* (bitter gourd) to the mucilaginous texture of *bhindi* (okra) and the earthy notes of various *saag* (leafy greens), vegetables are not merely accompaniments but often central components, prepared with intricate spice blends and highly localized culinary nuances. This diversity is a testament to the subcontinent's varied agro-climatic zones and rich culinary legacy (Nandal et al., 2025; Dwivedi et al., 2025).

Mesoamerican culinary traditions, as profoundly exemplified by Mexican cuisine, are historically intertwined with the symbiotic cultivation of the 'three sisters' – maize, beans, and squash – which form a nutritional bedrock, augmented by a vast array of chilies and tomatoes. This tripartite agricultural wisdom reflects ancient indigenous knowledge systems (Cárdenas-Castro et al., 2019; Gómez-Corona et al., 2024). European culinary traditions showcase region-specific vegetable specialities and preparation philosophies: French cuisine often elevates the subtle, inherent flavours of leeks, carrots, and potatoes through precise classical techniques; Italian cuisine relies heavily on tomatoes, basil, artichokes, and numerous regional greens. The pervasive integration of soybeans (processed into forms such as tofu, tempeh, miso, and soy sauce) into East Asian diets exemplifies a deeply ingrained cultural preference that has profoundly influenced the nutritional landscape, agricultural practices, and culinary identity of the region (Hernandez-Rodriguez, 2021).

6.4.3.3 The Imperative Role of Seasonality and Its Ecological and Economic Dimensions

Historically, and before the widespread advent of globalized food supply chains and controlled-environment agriculture (CEA), vegetable consumption was almost exclusively dictated by seasonal availability. The contemporary resurgence of seasonal eating reflects a growing appreciation for freshness, food quality, environmental sustainability, and cultural authenticity (Fusté-Forné, 2022). Vegetables harvested at their peak physiological maturity during their natural growing season generally exhibit superior taste, texture, aroma, and visual appeal, owing to optimal environmental conditions for metabolite synthesis (Gruda et al., 2025). In-season produce, allowed to ripen naturally on the plant, often accumulates higher concentrations of vitamins, minerals, and secondary metabolites (e.g. antioxidants, polyphenols, glucosinolates) compared to those harvested prematurely for extended transport or cultivated out of season under less ideal conditions (Krishna et al., 2024). Sourcing vegetables locally and seasonally significantly minimizes the energy expenditure associated with long-distance transportation, energy-intensive cold storage, and artificial ripening processes. This contributes to a lower carbon footprint and enhanced overall food system sustainability. Seasonal abundance often translates to more competitive pricing for consumers due to higher supply, while simultaneously supporting local agricultural economies and fostering direct producer-consumer relationships (Kanthaswamy et al., 2025).

6.5 CHALLENGES AND FUTURE PERSPECTIVES

6.5.1 Challenges and Future Perspectives in Vegetable Consumption

Despite the established health benefits and diverse culinary applications of vegetables, global consumption patterns often fall short of dietary recommendations. This discrepancy is increasingly exacerbated by complex socio-economic, environmental, and behavioural factors. Addressing these challenges requires multifaceted interventions spanning policy, education, and innovative food system approaches, as detailed in Figure 6.3.

6.5.2 Urbanization and Shifting Dietary Habits: A Global Paradigm Shift

The unprecedented rate of global urbanization presents significant challenges to the maintenance and promotion of healthy dietary patterns, particularly vegetable consumption. By 2050, it is projected that nearly 70% of the world's population will reside in urban areas, profoundly reshaping food systems (Bruin & Holleman, 2023). Urbanization often leads to a shift from traditional, often

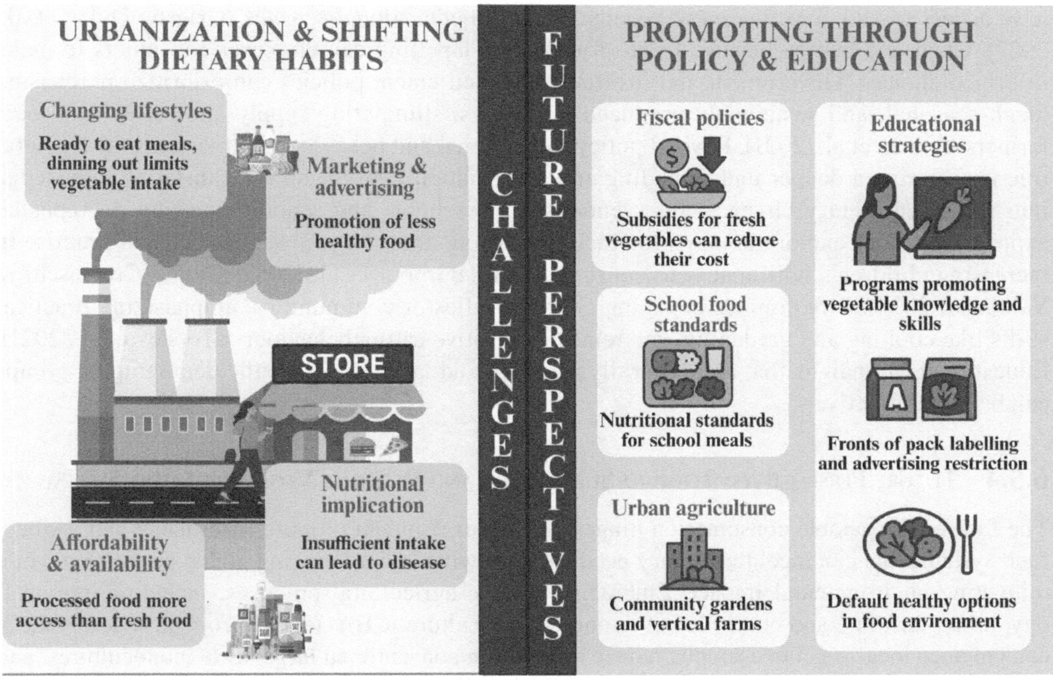

FIGURE 6.3 Challenges to vegetable consumption, such as urbanization and unhealthy dietary habits, alongside future solutions, proposing policy and educational strategies, including fiscal policies and urban agriculture, to promote healthier eating.

plant-rich diets towards more 'Westernized' dietary patterns characterized by increased consumption of processed foods; animal-sourced products; and foods high in fats, sugars, and salt, often at the expense of vegetables and fruits (Ecklu-Mensah et al., 2022).

Marketing and Advertising: Urban settings are saturated with aggressive marketing and advertising of less healthy food options, disproportionately influencing consumer choices, particularly among younger demographics.

Supply Chain Dynamics: Urbanization extends food supply chains, often prioritizing shelf-stable, less perishable goods over fresh, locally sourced vegetables, potentially impacting quality, cost, and access. This also contributes to higher carbon footprints due to increased transportation (Backholer et al., 2021).

6.5.3 Promoting Vegetable Consumption through Policy and Education: A Multifaceted Approach

Addressing the decline in vegetable consumption amidst global challenges requires synergistic efforts encompassing public health policy, targeted educational interventions, and innovative food system redesign. Effective policy frameworks are crucial for creating supportive food environments that facilitate healthy dietary choices. Fiscal Policies: Implementing subsidies for fresh vegetables can reduce their cost, making them more affordable and accessible, particularly for low-income households. Taxes on unhealthy, ultra-processed foods can disincentivize their consumption, with revenues potentially reinvested into promoting healthy options (Organization, 2022). Policies to regulate or restrict the marketing and advertising of unhealthy foods, especially to children, can counteract their pervasive influence (Organization, 2023). School Food Standards: Mandatory nutritional standards for school meals and vending machines can ensure that children have access to and consume more vegetables within educational settings (Grigsby-Duffy et al., 2022). Clear and comprehensive front-of-pack food labelling can empower consumers to make informed choices. Government and institutional procurement policies can prioritize purchasing local, seasonal, and sustainably produced vegetables, stimulating supply and supporting local farmers (Soares et al., 2021). Beyond policy, educational and behavioural strategies play a pivotal role in fostering a deeper understanding and appreciation for vegetables, translating knowledge into sustained dietary changes. Interventions in preschools and schools focusing on repeated exposure, sensory exploration, and interactive learning about vegetables have shown promise in increasing children's willingness to consume them (Barbour et al., 2022; Gaitán-Cremaschi & Valbuena, 2024). Programmes engaging entire families or communities, emphasizing practical skills like cooking and gardening, can reinforce positive eating behaviours (Tørslev et al., 2021). Educational campaigns that are culturally sensitive and tailored to specific demographic groups can be more effective.

6.5.4 Future Perspectives: Toward Sustainable and Resilient Vegetable Food Systems

The future of vegetable consumption hinges on the development of more sustainable and resilient food systems that can meet the dietary needs of a growing and urbanizing global population while mitigating environmental impact. Embracing diverse agricultural practices, including agroecology, urban farming, and controlled environment agriculture (CEA) (e.g. hydroponics, aeroponics), can enhance local vegetable supply, reduce reliance on conventional large-scale monocultures, and improve resilience to climate shocks (Snapp et al., 2021). Advances in food science and technology can improve nutrient retention in processed vegetables, develop novel plant-based foods, and optimize post-harvest handling to reduce waste. Digital platforms can also facilitate direct farmer-to-consumer links, promoting seasonal and local produce. Effective interventions require collaboration across government sectors (agriculture, health, education, urban planning), private industry, civil society organizations, and research institutions to develop coherent and impactful strategies (Marín-González et al., 2022). Empowering consumers with knowledge, skills, and access to healthy options, alongside fostering a renewed cultural appreciation for diverse vegetables, will be paramount. This includes promoting culinary education that celebrates traditional and innovative ways of preparing vegetables (Hart, 2024).

Therefore, ensuring adequate vegetable consumption in the future necessitates a holistic approach that acknowledges the intricate relationship between urbanization, changing dietary habits, and the broader food system. By strategically implementing evidence-based policies, promoting comprehensive education, and fostering innovation, societies can collectively work towards healthier and more sustainable vegetable consumption patterns globally.

6.6 CONCLUSION

The exploration of vegetable consumption, from raw and cooked forms to fermented and pickled preparations, and its interplay with regional and seasonal preferences, underscores the profound influence of culinary and dietary applications on human health and sustainable food systems. Our journey through these diverse modalities reveals that while each method offers unique benefits – be it the preserved heat-sensitive nutrients of raw produce, the enhanced bioavailability of carotenoids in cooked vegetables, or the probiotic and flavour complexities of fermented foods – a holistic approach is paramount. The pervasive impact of urbanization and evolving dietary habits presents significant challenges, necessitating concerted efforts through evidence-based policy interventions and comprehensive nutritional education to foster healthier choices. Ultimately, promoting vegetable consumption globally requires a synergistic strategy that champions both traditional wisdom and innovative food technologies, paving the way for more resilient and equitable food futures.

REFERENCES

Abusaief, H. (2021). A taxonomic study of twelve wild forage species of Fabaceae. *Heliyon*. Retrieved August 7, 2025, from https://www.cell.com/heliyon/fulltext/S2405-8440(21)00182-1

Adachi, R., Oono, F., Matsumoto, M., Yuan, X., Murakami, K., Sasaki, S., & Takimoto, H. (2025). Seasonal variation in the intake of food groups and nutrients in Japan: A systematic review and meta-analysis. *Journal of Epidemiology*, *35*(2), 53–62. https://doi.org/10.2188/jea.JE20240139

Ahire, E. D., Pathan, A. S., Pandagale, P. M., Khairnar, S. J., Surana, K. R., Kshirsagar, S. J., Talele, S. G., & Talele, G. S. (2023). Role of dietary in the prevention and management of obesity. In *The Metabolic Syndrome: Dietary Supplements and Food Ingredients* (pp. 139–161). Apple Academic Press. https://doi.org/10.1201/9781003329732-7

Akbari, B., Baghaei-Yazdi, N., Bahmaie, M., & Mahdavi Abhari, F. (2022). The role of plant-derived natural antioxidants in reduction of oxidative stress. *BioFactors*, *48*(3), 611–633. https://doi.org/10.1002/biof.1831

Al-Adwi, M. E., Al-Haswsa, Z. M., Alhmmadi, K. M., Eissa, Y. A., Hamdan, A., Bawadi, H., & Tayyem, R. F. (2023). Effects of different diets on glycemic control among patients with type 2 diabetes: A literature review. *Nutrition and Health*, *29*(2), 215–221. https://doi.org/10.1177/02601060221112805

Alsulami, F. (2024). Role of natural antioxidants in cancer. In S. Shaheed (Ed.), *Nutrition and Dietary Interventions in Cancer* (pp. 95–117). Springer. https://doi.org/10.1007/978-3-031-55622-7_4

Arora, S. K. (2023). Vitamins and minerals: A review on processing losses and strategies to control it. *Modern Concepts & Developments in Agronomy*, *12*(2). https://doi.org/10.31031/mcda.2023.12.000783

Awasthi, A., & Srivastava, Y. (2024). Effect of plant and vegetable-based secondary metabolites on cancer treatment. In *Plant Metabolites and Vegetables as Nutraceuticals* (pp. 145–199). Apple Academic Press. https://doi.org/10.1201/9781032680125-9

Backholer, K., Gupta, A., Zorbas, C., Bennett, R., Huse, O., Chung, A., Isaacs, A., Golds, G., Kelly, B., & Peeters, A. (2021). Differential exposure to, and potential impact of, unhealthy advertising to children by socio-economic and ethnic groups: A systematic review of the evidence. *Obesity Reviews*, *22*(3). https://doi.org/10.1111/obr.13144

Bai, Y., Li, X., Ji, H., Wang, Y., Zheng, D., Wang, Y., & Jin, Z. (2021). A review of the design and architecture of starch-based dietary foods. *Current Research in Food Science*, *4*, 1–12. Retrieved August 7, 2025, from https://www.sciencedirect.com/science/article/pii/S2095809920303805

Bala, M., Rehana, S., & Singh, M. P. (2023). Self-incompatibility: A targeted, unexplored pre-fertilization barrier in flower crops of Asteraceae. *Journal of Plant Research*, *136*(5), 587–612. https://doi.org/10.1007/s10265-023-01480-6

Barbour, L., Lindberg, R., Woods, J., Charlton, K., & Brimblecombe, J. (2022). Local urban government policies to facilitate healthy and environmentally sustainable diet-related practices: A scoping review. *Public Health Nutrition*, *25*(2), 471–487. https://doi.org/10.1017/S1368980021004432

Basak, M., & Laskar, M. A. (2024). Pathophysiology, lifestyle intervention and complications of type-2 diabetes: A review. *Journal of Applied Pharmaceutical Research*, *12*(3), 1–10. https://doi.org/10.69857/joapr.v12i3.482

Bhat, R. (2021). Bioactive compounds of Allium species. In *Reference Series in Phytochemistry* (pp. 277–295). Springer Science and Business Media B.V. https://doi.org/10.1007/978-3-030-57415-4_17

Black, M. M., & Lozoff, B. (2008). Nutrition and Diet. In *Encyclopedia of Infant and Early Childhood Development, Three-Volume Set* (Vol. 2, pp. V2-449–V2-459). https://doi.org/10.1016/B978-012370877-9.00113-4

Borrego-Ruiz, A., González-Domenech, C. M., & Borrego, J. J. (2024). The role of fermented vegetables as a sustainable and health-promoting nutritional resource. *Applied Sciences (Switzerland)*, *14*(23). https://doi.org/10.3390/app142310853

Bruin, S. De, & Holleman, C. (2023). *Urbanization Is Transforming Agrifood Systems Across the Rural–Urban Continuum Creating Challenges and Opportunities to Access Affordable Healthy Diets*. https://books.google.com/books?hl=en&lr=&id=kKH5EAAAQBAJ&oi=fnd&pg=PR5&dq=114.%09De+Bruin,+S.,+%26+Holleman,+C.+(2023).+Urbanization+is+transforming+agrifood+systems+across+the+rural–urban+continuum+creating+challenges+and+opportunities+to+access+afford ab

Budreviciute, A., Damiati, S., Sabir, D. K., Onder, K., Schuller-Goetzburg, P., Plakys, G., Katileviciute, A., Khoja, S., & Kodzius, R. (2020). Management and prevention strategies for non-communicable diseases (NCDs) and their risk factors. *Frontiers in Public Health*, *8*, 574111. Frontiers Media S.A. https://doi.org/10.3389/fpubh.2020.574111

Cam, G. (2024). *Exploring Young People's Understandings of Health and Healthy Lifestyles: A Qualitative Investigation.* https://openaccess.wgtn.ac.nz/articles/thesis/Exploring_Young_People_s_Understandings_of_Health_and_Healthy_Lifestyles_A_Qualitative_Investigation/25623657

Cámara, M., Giner, R. M., González-Fandos, E., López-García, E., Mañes, J., Portillo, M. P., Rafecas, M., Domínguez, L., & Martínez, J. A. (2021). Food-based dietary guidelines around the world: A comparative analysis to update aesan scientific committee dietary recommendations. *Nutrients, 13*(9). https://doi.org/10.3390/nu13093131

Caminiti, R., Carresi, C., Mollace, R., Macrì, R., Scarano, F., Oppedisano, F., Maiuolo, J., Serra, M., Ruga, S., Nucera, S., Tavernese, A., Gliozzi, M., Musolino, V., Palma, E., Muscoli, C., Rubattu, S., Volterrani, M., Federici, M., Volpe, M., & Mollace, V. (2024). The potential effect of natural antioxidants on endothelial dysfunction associated with arterial hypertension. *Frontiers in Cardiovascular Medicine, 11.* https://doi.org/10.3389/fcvm.2024.1345218

Cárdenas-Castro, A. P., Perales-Vázquez, G. del C., De la Rosa, L. A., Zamora-Gasga, V. M., Ruiz-Valdiviezo, V. M., Alvarez-Parrilla, E., & Sáyago-Ayerdi, S. G. (2019). Sauces: An undiscovered healthy complement in Mexican cuisine. *International Journal of Gastronomy and Food Science, 17*, 100154. https://doi.org/10.1016/j.ijgfs.2019.100154

Cecchini, A. L., Biscetti, F., Rando, M. M., Nardella, E., Pecorini, G., Eraso, L. H., Dimuzio, P. J., Gasbarrini, A., Massetti, M., & Flex, A. (2022). Dietary risk factors and eating behaviors in peripheral arterial disease (PAD). *International Journal of Molecular Sciences, 23*(18). https://doi.org/10.3390/ijms231810814

Chaudhari, V. M., Singh, O. B., Gouthami, N. S., Thakur, N., Singh, R., Singh, S., Thapa, U., & Nagar, B. L. (2024). Unlocking the nutritional power of vegetables: A guide to vibrant health. *European Journal of Nutrition & Food Safety, 16*(8), 247–261. https://doi.org/10.9734/ejnfs/2024/v16i81512

Chauhan, A., Sharma, D., Kumar, R., & Shiwani, K. (2021). Methods of propagation in vegetable crops. In *Recent Trends in Propagation of Forest and Horticultural Crops* (Issue October, pp. 270–281). https://www.researchgate.net/profile/Kumari-Shiwani/publication/355393673_Methods_of_Propagation_in_Vegetable_Crops/links/616e63feb90c51266269fc48/Methods-of-Propagation-in-Vegetable-Crops.pdf

Chiquito-Contreras, C. J., Meza-Menchaca, T., Guzmán-López, O., Vásquez, E. C., & Ricaño-Rodríguez, J. (2024). Molecular insights into plant–microbe interactions: A comprehensive review of key mechanisms. *Frontiers in Bioscience – Elite, 16*(1). https://doi.org/10.31083/j.fbe1601009

Conrad, Z., Kowalski, C., Dustin, D., Johnson, L. A. K., McDowell, A., Salesses, M., Nance, J., & Belury, M. A. (2022). Quality of popular diet patterns in the United States: Evaluating the effect of substitutions for foods high in added sugar, sodium, saturated fat, and refined grains. *Current Developments in Nutrition, 6*(9). https://doi.org/10.1093/cdn/nzac119

Cuevas-Cianca, S. I., Romero-Castillo, C., Gálvez-Romero, J. L., Juárez, Z. N., & Hernández, L. R. (2023). Antioxidant and anti-inflammatory compounds from edible plants with anti-cancer activity and their potential use as drugs. *Molecules, 28*(3). https://doi.org/10.3390/molecules28031488

Demydas, T. (2010). Consumer segmentation based on the level and structure of fruit and vegetable intake: An empirical evidence for US adults from the National Health and Nutrition Examination Survey (NHANES) 2005–2006. *Public Health Nutrition, 14*(6), 1088–1095. https://doi.org/10.1017/S1368980010003745

Dihdashti, M. (2023). *Gastronomy Foundation: Exploring the Culinary Arts.* https://papers.ssrn.com/sol3/papers.cfm?abstract_id=5207204

Ding, Q., Ma, X., Zhang, Z., Lu, P., & Liu, M. (2024). Pooled and global burdens and trends of five common cancers attributable to diet in 204 countries and territories from 1990 to 2019: An analysis of the Global Burden of Disease Study. *European Journal of Cancer Prevention, 33*(6), 485–492. https://doi.org/10.1097/CEJ.0000000000000884

Dwivedi, S., Ortiz, R., & Kole, C. (2025). *Next Generation Food Crops for Human Health.* https://books.google.com/books?hl=en&lr=&id=-CR2EQAAQBAJ&oi=fnd&pg=PP1&dq=106.%09Dwivedi,+S.+L.,+Ortiz,+R.,+%26+Kole,+C.+(Eds.).+(2025).+Next+Generation+Food+Crops+for+Human+Health.+CRC+Press.&ots=XcJ1Valm1b&sig=PRaIuClPhJFzAyHjymjNt0eMYDo

Ebabhi, A., & Adebayo, R. (2022). Nutritional values of vegetables. In *Vegetable Crops – Health Benefits and Cultivation.* https://doi.org/10.5772/intechopen.101090

Ecklu-Mensah, G., Gilbert, J., & Devkota, S. (2022). Dietary selection pressures and their impact on the gut microbiome. *CMGH, 13*(1), 7–18. https://doi.org/10.1016/j.jcmgh.2021.07.009

Faisal, Z., Saeed, F., Afzaal, M., Akram, N., Shah, Y. A., Islam, F., & Ateeq, H. (2022). Phytochemical profile and food applications of edible flowers: A comprehensive treatise. *Journal of Food Processing and Preservation, 46*(11), e17061. https://doi.org/10.1111/JFPP.17061

Fijan, S., Fijan, P., Wei, L., & Marco, M. L. (2024). Health benefits of kimchi, sauerkraut, and other fermented foods of the genus brassica. *Applied Microbiology (Switzerland), 4*(3), 1165–1176. https://doi.org/10.3390/applmicrobiol4030079

Fusté-Forné, F. (2022). Seasonality in food tourism: Wild foods in peripheral areas. *Tourism Geographies, 24*(4–5), 578–598. https://doi.org/10.1080/14616688.2018.1558453

Gaitán-Cremaschi, D., & Valbuena, D. (2024). Examining purchasing strategies in public food procurement: Integrating sustainability, nutrition, and health in Spanish school meals and social care centres. *Food Policy, 109*, 102478. Retrieved August 7, 2025, from https://www.sciencedirect.com/science/article/pii/S0306919224001532

Gómez-Corona, C., Casotti, L. M., Pino, F. J., & Deliza, R. (2024). Foods, dishes and flavours from Latin American tradition: A case study of Mexico, Brazil and Argentina. In *Sensory Evaluation and Consumer Acceptance of New Food Products* (pp. 317–350). https://doi.org/10.1039/bk9781839166655-00317

Grigsby-Duffy, L., Brooks, R., Boelsen-Robinson, T., Blake, M. R., Backholer, K., Palermo, C., & Peeters, A. (2022). The impact of primary school nutrition policy on the school food environment: A systematic review. *Health Promotion International, 37*(5). https://doi.org/10.1093/heapro/daac084

Gruda, N. S., Samuolienė, G., Dong, J., & Li, X. (2025). Environmental conditions and nutritional quality of vegetables in protected cultivation. *Comprehensive Reviews in Food Science and Food Safety, 24*(2), e70139. https://doi.org/10.1111/1541-4337.70139

Grundy, M. M. L., Edwards, C. H., Mackie, A. R., Gidley, M. J., Butterworth, P. J., & Ellis, P. R. (2016). Re-evaluation of the mechanisms of dietary fibre and implications for macronutrient bioaccessibility, digestion and postprandial metabolism. *British Journal of Nutrition, 116*(5), 816–833. https://doi.org/10.1017/S0007114516002610

Gul, F., Ullah, Z., Iqbal, J., Abbasi, B. A., Ijaz, S., Uddin, S., Yaseen, T., Ali, H., Jamil, M., Kanwal, S., & Mahmood, T. (2025). Bioactive phytochemicals from garlic (*Allium sativum* L.) by-products. In *Bioactive Phytochemicals in By-Products from Bulb, Flower and Fruit Vegetables* (pp. 23–83). Springer. https://doi.org/10.1007/978-3-031-77399-0_2

Guo, C., Liu, Y., Fu, H., Zhang, X., & Li, M. (2024). Effect of cruciferous vegetable intake on cancer: An umbrella review of meta-analysis. *Journal of Food Science, 89*(9), 5230–5244). https://doi.org/10.1111/1750-3841.17300

Guzman-Vilca, W. C., Carrillo-Larco, R. M., & Tarazona-Meza, C. (2025). Consumption of fruits and vegetables among Peruvian adults: Analysis of a national health survey 2017–2018. *PLoS Global Public Health, 5*(3). https://doi.org/10.1371/journal.pgph.0004222

Harris, J., Tan, W., Raneri, J. E., Schreinemachers, P., & Herforth, A. (2022). Vegetables for healthy diets in low- and middle-income countries: A scoping review of the food systems literature. *Food and Nutrition Bulletin, 43*(2), 232–248. https://doi.org/10.1177/03795721211068652

Hart, A. O. (n.d.). Institutional framework for sustainable community engagement for development in Nigeria. *Journal of Humanities and Social Policy, 10*(2), 14–37. https://doi.org/10.56201/jhsp.v10.no2.2024.pg14.37

Hernandez-Rodriguez, R. (2021). Food cultures of Mexico. *Food Cultures of Mexico*. https://doi.org/10.5040/9798400652561

Hofer, S. J., Davinelli, S., Bergmann, M., Scapagnini, G., & Madeo, F. (2021). Caloric restriction mimetics in nutrition and clinical trials. *Frontiers in Nutrition, 8*, 717343. https://doi.org/10.3389/fnut.2021.717343

Hu, F. B. (2024). Diet strategies for promoting healthy aging and longevity: An epidemiological perspective. *Journal of Internal Medicine, 295*(4), 508–531. https://doi.org/10.1111/joim.13728

Jacob, J. (2021). Food cultures of Japan: Recipes, customs, and issues. *Food Cultures of Japan: Recipes, Customs, and Issues*. https://www.torrossa.com/gs/resourceProxy?an=5527187&publisher=FZ0661

Janety, M. (2021). *Effect of Cooking on Folates and Ascorbic Acid Levels in Selected African Indigenous Vegetables from Githurai Market, Nairobi County, Kenya* [Master's thesis, Kenyatta University]. IR Library. https://ir-library.ku.ac.ke/bitstreams/8550fba7-ff10-498d-8645-1ac1f14447ec/download

Jayasinghe, S., Byrne, N. M., & Hills, A. P. (2025). Cultural influences on dietary choices. *Progress in Cardiovascular Diseases*. https://doi.org/10.1016/j.pcad.2025.02.003

Kainat, S., Arshad, M. S., Khalid, W., Zubair Khalid, M., Koraqi, H., Afzal, M. F., Noreen, S., Aziz, Z., & Al-Farga, A. (2022). Sustainable novel extraction of bioactive compounds from fruits and vegetables waste for functional foods: A review. *International Journal of Food Properties, 25*(1), 2457–2476. https://doi.org/10.1080/10942912.2022.2144884

Kalmpourtzidou, A., Eilander, A., & Talsma, E. F. (2020). Global vegetable intake and supply compared to recommendations: A systematic review. *Nutrients, 12*(6). https://doi.org/10.3390/nu12061558

Kanthaswamy, H., Das, A., Arunachalam, S. U. (2025). *Strengthening Livestock Systems Through Sorghum-Based Fodder Value Chain Innovations in Southern India*. Elsevier. Retrieved August 7, 2025, from https://www.sciencedirect.com/science/article/pii/S0308521X25000137

Kattel, S., & Antonious, G. F. (2025). Glucosinolates in cruciferous vegetables: Genetic and environmental regulation, metabolic pathways, and cancer-preventive mechanisms. *International Journal of Plant Biology, 16*(2), 58. https://doi.org/10.3390/ijpb16020058

Kim, B. S., Yu, M. Y., & Shin, J. (2024). Effect of low sodium and high potassium diet on lowering blood pressure and cardiovascular events. *Clinical Hypertension*, *30*(1). https://doi.org/10.1186/s40885-023-00259-0

Kirmizigul, A., & Sengun, I. Y. (2024). Traditional non-dairy fermented products: A candidate for probiotics. *Food Reviews International*, *40*(5), 1217–1237. https://doi.org/10.1080/87559129.2023.2212040

Krishna, H., Hebbar, S., Kumar, P., Sharma, S., Kumar, R., Tiwari, S. K., Maurya, S., Srivastava, K., Pal, G., Bahadur, A., & Behera, T. K. (2024). Navigating challenges and prospects in off-season vegetable production. *Vegetable Science*, *51*, 97–105. https://doi.org/10.61180/vegsci.2024.v51.spl.09

Kromhout, D., Spaaij, C. J. K., De Goede, J., Weggemans, R. M., Brug, J., Geleijnse, J. M., Van Goudoever, J. B., Hoes, A. W., Hopman, M. T. E., Iestra, J. A., Mensink, R. P., Pijl, H., Romijn, J. A., Schols, A. M. W. J., Seidell, J. C., Veer, P. V. T., Visser, M., & Zwietering, M. H. (2016). The 2015 Dutch food-based dietary guidelines. *European Journal of Clinical Nutrition*, *70*(8), 869–878. https://doi.org/10.1038/ejcn.2016.52

Kumar, M., Chandran, D., Tomar, M., Bhuyan, D. J., Grasso, S., Sá, A. G. A. M., Carciofi, B. A. M., & Radha, H. (2022). Valorization potential of tomato (*Solanum lycopersicum* L.) seed: Nutraceutical quality, food properties, safety aspects, and application as a health-promoting ingredient. *Horticulturae, 8*(3), 265. https://doi.org/10.3390/horticulturae8030265

Kumar, R. V. (2023). *Vegetable Science & Technology*. https://books.google.com/books?hl=en&lr=&id=4w3wEAAAQBAJ&oi=fnd&pg=PA1&dq=13.%09Kumar,+R.+V.+(2023).+Vegetable+Science+%26+Technology.+Academic+Guru+Publishing+House.&ots=pWikCWIQwo&sig=UjALBneAs6BfrcvFUc2UT_fKpbA

Kurhaluk, N., & Tkaczenko, H. (2025). L-arginine and nitric oxide in vascular regulation – Experimental findings in the context of blood donation. *Nutrients*, *17*(4). https://doi.org/10.3390/nu17040665

Kwon, D. Y., Soon-Hee, K., Chung, K. R., Daily, J., & Park, S. (2023). Science and philosophy of Korea traditional foods (K-food). *Journal of Ethnic Foods*, *10*(1), 26. https://doi.org/10.1186/s42779-023-00194-3

Levi, J. (2023). Plants as staple foods. In *A Cultural History of Plants: In the Seventeenth and Eighteenth Centuries* (Vol. 4, pp. 25–50). Bloomsbury Academic. https://doi.org/10.5040/9781474207010.ch-1

Li, W., Ghamry, M., Xu, X., Zhang, L., & Li, J. (2025). Organic acids: Key products of fermented foods and sources of anti-inflammatory potential. *Food Reviews International*. https://doi.org/10.1080/87559129.2025.2529546

Lynrah, L. V., Devi, C. B., Choudhuri, S., C.S., V., Niharika, M., Singh, R., Pandey, S. K., & Singh, B. V. (2024). The role of food science in combating obesity and related metabolic disorders. *European Journal of Nutrition & Food Safety*, *16*(8), 202–228. https://doi.org/10.9734/ejnfs/2024/v16i81508

Malik, D., Narayanasamy, N., Pratyusha, V. A., Thakur, J., & Sinha, N. (2023). Water-soluble vitamins. In *Textbook of Nutritional Biochemistry* (pp. 291–389). Springer. https://doi.org/10.1007/978-981-19-4150-4_10

Marín-González, F., Moganadas, S. R., Paredes-Chacín, A. J., Yeo, S. F., & Subramaniam, S. (2022). Sustainable local development: Consolidated framework for cross-sectoral cooperation via a systematic approach. *Sustainability (Switzerland)*, *14*(11). https://doi.org/10.3390/su14116601

Mason-D'Croz, D., Bogard, J. R., Sulser, T. B., Cenacchi, N., Dunston, S., Herrero, M., & Wiebe, K. (2019). Gaps between fruit and vegetable production, demand, and recommended consumption at global and national levels: An integrated modelling study. *The Lancet Planetary Health*, *3*(7), e318–e329. https://doi.org/10.1016/S2542-5196(19)30095-6

Melaku, Y. A., Temesgen, A. M., Deribew, A., Tessema, G. A., Deribe, K., Sahle, B. W., Abera, S. F., Bekele, T., Lemma, F., Amare, A. T., Seid, O., Endris, K., Hiruye, A., Worku, A., Adams, R., Taylor, A. W., Gill, T. K., Shi, Z., Afshin, A., & Forouzanfar, M. H. (2016). The impact of dietary risk factors on the burden of non-communicable diseases in ethiopia: Findings from the global burden of disease study 2013. *International Journal of Behavioral Nutrition and Physical Activity*, *13*(1), 17. https://doi.org/10.1186/s12966-016-0447-x

Melim, C., Lauro, M. R., Pires, I. M., Oliveira, P. J., & Cabral, C. (2022). The role of glucosinolates from cruciferous vegetables (Brassicaceae) in gastrointestinal cancers: From prevention to therapeutics. *Pharmaceutics*, *14*(1). https://doi.org/10.3390/pharmaceutics14010190

Mitra, S., Emran, T. Bin, Chandran, D., Zidan, B. M. R. M., Das, R., Mamada, S. S., Masyita, A., Salampe, M., Nainu, F., Khandaker, M. U., Idris, A. M., & Simal-Gandara, J. (2022). Cruciferous vegetables as a treasure of functional foods bioactive compounds: Targeting p53 family in gastrointestinal tract and associated cancers. *Frontiers in Nutrition*, *9*. https://doi.org/10.3389/fnut.2022.951935

Monteiro, J. S., Nakano, E. Y., Zandonadi, R. P., Botelho, R. B. A., & Araújo, W. M. C. (2024). How do Brazilian consumers understand food groups in the food-based dietary guidelines? *Foods*, *13*(2). https://doi.org/10.3390/foods13020338

Motti, R. (2021). The Solanaceae family: Botanical features and diversity. In *The Wild Solanums Genomes* (pp. 1–9). Springer. https://doi.org/10.1007/978-3-030-30343-3_1

Mozaffarian, D. (2016). Dietary and policy priorities for cardiovascular disease, diabetes, and obesity. *Circulation, 133*(2), 187–225. https://doi.org/10.1161/CIRCULATIONAHA.115.018585

Mukherjee, B., Al Hoque, A., Hota, S. H., Gope, S., Ray, M., Barman, M., Bhattacharya, S., Chakraborty, S., & Das, L. (2025). Antioxidants and their physiological role in free radical scavenging. In *Dietary Supplements and Nutraceuticals* (pp. 1–44). Springer Nature Singapore. https://doi.org/10.1007/978-981-97-9936-7_12-1

Nakkarach, A., Foo, H. L., Song, A. A. L., Mutalib, N. E. A., Nitisinprasert, S., & Withayagiat, U. (2021). Anti-cancer and anti-inflammatory effects elicited by short chain fatty acids produced by Escherichia coli isolated from healthy human gut microbiota. *Microbial Cell Factories, 20*(1), 36. https://doi.org/10.1186/s12934-020-01477-z

Nandal, K., Harshit, N., Nandal, R., & Kaur, M. (2025). Understanding the dietary nutrition and its health implications in historical food practices. *Advances in Research, 26*(3), 588–599. https://doi.org/10.9734/air/2025/v26i31371

Nguyen, H. D., Oh, H., & Kim, M. S. (2022). Higher intakes of nutrients are linked with a lower risk of cardiovascular diseases, type 2 diabetes mellitus, arthritis, and depression among Korean adults. *Nutrition Research, 100*, 19–32. https://doi.org/10.1016/j.nutres.2021.11.003

O'Hearn, M., Lara-Castor, L., Cudhea, F., Miller, V., Reedy, J., Shi, P., Zhang, J., Wong, J. B., Economos, C. D., Micha, R., Mozaffarian, D., Bas, M., Ali, J. H., Abumweis, S., Krishnan, A., Misra, P., Hwalla, N. C., Janakiram, C., Liputo, N. I., . . . Hakeem, R. (2023). Incident type 2 diabetes attributable to suboptimal diet in 184 countries. *Nature Medicine, 29*(4), 982–995. https://doi.org/10.1038/s41591-023-02278-8

Okoye, C., Ezenwanne, B., & Olasoji, O. (2025). *Mechanisms of Preservation by Lactic Acid Bacteria in Food Fermentation*. https://www.intechopen.com/online-first/1221812

Olas, B. (2024). A review of the pro-health activity of *Asparagus officinalis* L. and its components. *Foods, 13*(2). https://doi.org/10.3390/foods13020288

Otieno, M. (2024). *Sustainable Agroecological Practices in Sub-Saharan Africa in the Face of Climate Change*. https://books.google.com/books?hl=en&lr=&id=Bi0zEQAAQBAJ&oi=fnd&pg=PR8&dq=103.%09Otieno,+M.+(2024).+Sustainable+Agroecological+Practices+in+Sub-Saharan+Africa+in+the+Face+of+Climate+Change.+Springer+Nature.&ots=wzeRs0iyrX&sig=WzFfCyNeay-W3NP7LksGjRITtZI

Pan, S., Lin, Z., Yao, T., Guo, X., Xu, T., Sheng, X., Song, X., Chen, Z., Wei, W., Yan, Y., & Hu, Y. (2023). Corrigendum: Global burden of non-communicable chronic diseases associated with a diet low in fruits from 1990 to 2019. *Frontiers in Nutrition, 10*, 1202763. https://doi.org/10.3389/fnut.2023.1285805

Parichat, P., & Pongsak, R. (2023). Probiotics: Sources, selection and health benefits. *Research Journal of Biotechnology, 18*(5), 102–113. https://doi.org/10.25303/1805rjbt1020113

Pathak, M., Barik, S., & Das, S. K. (2021). Impact of climate change on root crops production. In *Advances in Research on Vegetable Production Under a Changing Climate* (Vol. 1, pp. 125–148). Springer. https://doi.org/10.1007/978-3-030-63497-1_6

Perera, G. C., Sathyaruban, S., & Rathnayake, R. (2024). Herbal additives as pigmentation sources in aquafeeds: Current usage and future potential in Asian ornamental fish farming. In *Sustainable Feed Ingredients and Additives for Aquaculture Farming* (pp. 341–364). Springer. https://doi.org/10.1007/978-981-97-4279-0_15

Polak-Szczybyło, E. (2023). Low-grade inflammation and role of anti-inflammatory diet in childhood obesity. *International Journal of Environmental Research and Public Health, 20*(3). https://doi.org/10.3390/ijerph20031682

Rajan, A. (2023). Gastronomic evolution: A review of traditional and contemporary Food Culture. *International Journal for Multidimensional Research Perspectives*. https://www.researchgate.net/profile/Chandigarh-Philosophers/publication/375925343_Gastronomic_evolution_A_review_of_traditional_and_contemporary_Food_Culture/links/6563ef52b1398a779db7fda0/Gastronomic-evolution-A-review-of-traditional-and-contemporary-Fo

Ramsay, S. A., Eskelsen, A. K., Branen, L. J., Armstrong Shultz, J., & Plumb, J. (2014). Nutrient intake and consumption of fruit and vegetables in young children. *Infant, Child, and Adolescent Nutrition, 6*(6), 332–344. https://doi.org/10.1177/1941406414549622

Roszkowski, S. (2023). Application of polyphenols and flavonoids in oncological therapy. *Molecules, 28*(10). https://doi.org/10.3390/molecules28104080

Roy, T., Kalambukattu, J. G., Biswas, S. S., & Kumar, S. (2022). Agro-climatic variability in climate change scenario: Adaptive approach and sustainability. In *Springer Climate* (pp. 313–348). Springer Science and Business Media B.V. https://doi.org/10.1007/978-3-031-15501-7_12

Salim, R., Nehvi, I. B., Mir, R. A., Tyagi, A., Ali, S., & Bhat, O. M. (2023). A review on anti-nutritional factors: Unraveling the natural gateways to human health. *Frontiers in Nutrition, 10*. https://doi.org/10.3389/fnut.2023.1215873

Samtiya, M., Aluko, R. E., Puniya, A. K., & Dhewa, T. (2021). Enhancing micronutrients bioavailability through fermentation of plant-based foods: A concise review. *Fermentation, 7*(2). https://doi.org/10.3390/fermentation7020063

Samuel, P. O., Edo, G. I., Emakpor, O. L., Oloni, G. O., Ezekiel, G. O., Essaghah, A. E. A., Agoh, E., & Agbo, J. J. (2024). Lifestyle modifications for preventing and managing cardiovascular diseases. *Sport Sciences for Health, 20*(1), pp. 23–36. https://doi.org/10.1007/s11332-023-01118-z

Sarkar, T., Salauddin, M., Roy, S., Chakraborty, R., Rebezov, M., Shariati, M. A., Thiruvengadam, M., & Rengasamy, K. R. R. (2023). Underutilized green leafy vegetables: Frontier in fortified food development and nutrition. *Critical Reviews in Food Science and Nutrition, 63*(33), 11679–11733. https://doi.org/10.1080/10408398.2022.2095555

Sevim, Y., Öztürk, H. N. O., & Ergün, M. (2023). Knowledge and consumption frequency of probiotics and fermented foods in elite volleyball players-A pilot study. *Journal of Research in Pharmacy, 27*(2), 519–528. https://doi.org/10.29228/jrp.334

Sharif, S., Van der Graaf, Y., Cramer, M. J., Kapelle, L. J., de Borst, G. J., Visseren, F. L. J., Westerink, J., van Petersen, R., Dinther, B. G. F., Algra, A., van der Graaf, Y., Grobbee, D. E., Rutten, G. E. H. M., Visseren, F. L. J., Kappelle, L. J., Leiner, T., & Nathoe, H. M. (2021). Low-grade inflammation as a risk factor for cardiovascular events and all-cause mortality in patients with type 2 diabetes. *Cardiovascular Diabetology, 20*(1). https://doi.org/10.1186/s12933-021-01409-0

Sharma, M., Bithel, N., Sharma, K. K., & Sharma, M. (2024). Potential of vegetables and plant metabolites in healthcare. In *Plant Metabolites and Vegetables as Nutraceuticals* (pp. 3–31). Apple Academic Press. https://doi.org/10.1201/9781032680125-2

Sharma, S., Rani, H., Kaur, G., Kumar, S., Sheikh, S., & Samota, M. K. (2025). Comprehensive overview of glucosinolates in crucifers: Occurrence, roles, metabolism, and transport mechanisms – a review. *Phytochemistry Reviews, 24*(3), 2417–2444. https://doi.org/10.1007/s11101-024-10021-5

Siddique, S. S., Gupta, S., Gupta, S., & Kumar, D. (2024). Impact of cooking and processing on mushroom nutrient retention. In *Mushroom Magic: Biochemistry and Nutritional Value of Fungi* (pp. 174–190). CRC Press. https://doi.org/10.1201/9781003570257-11

Singh, J., & Arora, S. K. (2023). Antinutritional factors in plant based foods. *International Journal of Agricultural Sciences, 19*(1), 366–375. https://doi.org/10.15740/has/ijas/19.1/366-375

Sionek, B., Szydłowska, A., Küçükgöz, K., & Kołożyn-Krajewska, D. (2023). Traditional and new microorganisms in lactic acid fermentation of food. In *Fermentation, 9*(12). https://doi.org/10.3390/fermentation9121019

Snapp, S., Kebede, Y., Wollenberg, E., Dittmer, K. M., Brickman, S., Egler, S., Egler, C., & Shelton, S. (2021). *Agroecology & Climate Change Rapid Evidence Review: Performance of Agroecological Approaches in Low-and Middle-Income Countries* (pp. 1–58). HAL Open Science. https://hal.science/hal-03278936/

Soares, P., Martinelli, S. S., Davó-Blanes, M. C., Fabri, R. K., Clemente-Gómez, V., & Cavalli, S. B. (2021). Government policy for the procurement of food from local family farming in brazilian public institutions. *Foods, 10*(7). https://doi.org/10.3390/foods10071604

Sravya, S. (2022). *Development of Value Added Products Using Green Leafy Vegetable Powders for the Prevention of Anemia*. https://krishikosh.egranth.ac.in/bitstreams/3fd6d8ae-bf52-4451-a346-9458797dc332/download

Suárez, R., Bautista-Valarezo, E., Matos, A., Calderón, P., Fascì-Spurio, F., Castano-Jimenez, J., Zambrano-Villacres, R., Sarno, G., & Frias-Toral, E. (2025). Obesity and nutritional strategies: Advancing prevention and management through evidence-based approaches. *Food and Agricultural Immunology, 36*(1). https://doi.org/10.1080/09540105.2025.2491597

Suresh, A., Shobna, Salaria, M., Morya, S., Khalid, W., Afzal, F. A., Khan, A. A., Safdar, S., Khalid, M. Z., & Mukonzo Kasongo, E. L. (2024). Dietary fiber: An unmatched food component for sustainable health. *Food and Agricultural Immunology, 35*(1), 2384420. https://doi.org/10.1080/09540105.2024.2384420

Tang, Y., Lv, D., Tao, Y., & Wang, J. (2025). The therapeutic effects of natural organosulfur compounds on atherosclerosis and their potential mechanisms: A comprehensive review. *Frontiers in Cardiovascular Medicine, 12*. https://doi.org/10.3389/fcvm.2025.1599154

Tørslev, M. K., Bjarup Thøgersen, D., Høstgaard Bonde, A., Bloch, P., & Varming, A. (2021). Supporting positive parenting and promoting healthy living through family cooking classes. *International Journal of Environmental Research and Public Health, 18*(9). https://doi.org/10.3390/ijerph18094709

Tsai, H. H., Yu, J. C., Hsu, H. M., Chu, C. H., Chang, T. M., Hong, Z. J., Feng, A. C., Fu, C. Y., Hsu, K. F., Dai, M. S., & Liao, G. S. (2023). The risk of breast cancer between western and mediterranean dietary patterns. *Nutrients*, *15*(9). https://doi.org/10.3390/nu15092057

Ubago-Guisado, E., Rodríguez-Barranco, M., Ching-López, A., Petrova, D., Molina-Montes, E., Amiano, P., Barricarte-Gurrea, A., Chirlaque, M. D., Agudo, A., & Sánchez, M. J. (2021). Evidence update on the relationship between diet and the most common cancers from the european prospective investigation into cancer and nutrition (Epic) study: A systematic review. *Nutrients*, *13*(10). https://doi.org/10.3390/nu13103582

Velissaridou, A., Panoutsopoulou, E., Prokopiou, V., & Tsoupras, A. (2024). Cardio-protective-promoting properties of functional foods inducing HDL-cholesterol levels and functionality. *Nutraceuticals*, *4*(4), 469–502. https://doi.org/10.3390/nutraceuticals4040028

Verma, S., Anand, M. K., Ram, M. M., Pathak, S., & Atal, M. K. (2024). *Cultivation of Underutilized Vegetable Crops*. https://www.researchgate.net/profile/Stuti-Pathak/publication/389439984_CULTIVATION_OF_UNDERUTILIZED_VEGETABLE_CROPS_VOLUME-1_Cultivation_of_Underutilized_Vegetable_Crops_Volume-1/links/67c29945207c0c20fa9bef3e/CULTIVATION-OF-UNDERUTILIZED-VEGETABLE-CROPS

Wai, H., & Myo, L. (2022). *Food Environment, Food Acquisition Behavior, and Fruit and Vegetable Consumption among Burmese Immigrants and Refugees: A Socio-Ecological Study* (pp. 1–14). https://doi.org/10.54014/GPP2-TTM9

Wallace, T. C., Bailey, R. L., Blumberg, J. B., Burton-Freeman, B., Chen, C. Y. O., Crowe-White, K. M., Drewnowski, A., Hooshmand, S., Johnson, E., Lewis, R., Murray, R., Shapses, S. A., & Wang, D. D. (2020). Fruits, vegetables, and health: A comprehensive narrative, umbrella review of the science and recommendations for enhanced public policy to improve intake. *Critical Reviews in Food Science and Nutrition*, *60*(13), 2174–2211. https://doi.org/10.1080/10408398.2019.1632258

Wang, C., Li, M., Duan, X., Abu-Izneid, T., Rauf, A., Khan, Z., Mitra, S., Emran, T. Bin, Aljohani, A. S. M., Alhumaydhi, F. A., Thiruvengadam, M., & Suleria, H. A. R. (2023). Phytochemical and nutritional profiling of tomatoes; impact of processing on bioavailability – a comprehensive review. *Food Reviews International*, *39*(8), 5986–6010. https://doi.org/10.1080/87559129.2022.2097692

Wang, Y., Liu, B., Han, H., Hu, Y., Zhu, L., Rimm, E. B., Hu, F. B., & Sun, Q. (2023). Associations between plant-based dietary patterns and risks of type 2 diabetes, cardiovascular disease, cancer, and mortality – a systematic review and meta-analysis. *Nutrition Journal*, *22*(1). https://doi.org/10.1186/s12937-023-00877-2

Watling, C. Z., Schmidt, J. A., Dunneram, Y., Tong, T. Y. N., Kelly, R. K., Knuppel, A., Travis, R. C., Key, T. J., & Perez-Cornago, A. (2022). Risk of cancer in regular and low meat-eaters, fish-eaters, and vegetarians: A prospective analysis of UK Biobank participants. *BMC Medicine*, *20*(1). https://doi.org/10.1186/s12916-022-02256-w

Wells, R. (2024). Breeding, genetics and models. In *Vegetable Brassicas and Related Crucifers* (2nd ed., pp. 42–89). https://doi.org/10.1079/9781789249170.0002

World Health Organization (2022). *Fiscal Policies to Promote Healthy Diets: Policy Brief*. https://books.google.com/books?hl=en&lr=&id=pHkOEQAAQBAJ&oi=fnd&pg=PA3&dq=117.%09World+Health+Organization.+(2022).+Fiscal+policies+to+promote+healthy+diets:+policy+brief.+World+Health+Organization.&ots=k2Q4oQddVK&sig=7kWo6oaIXdCaPn8MimGxR9uEuV8

World Health Organization (2023). *Policies to Protect Children from the Harmful Impact of Food Marketing: WHO Guideline*. https://books.google.com/books?hl=en&lr=&id=7aIOEQAAQBAJ&oi=fnd&pg=PR2&dq=118.%09World+Health+Organization.+(2023).+Policies+to+protect+children+from+the+harmful+impact+of+food+marketing:+WHO+guideline.+World+Health+Organization.&ots=ioyG2kA8mX&sig=lfAWvm

Yang, Y. X., Wang, X. L., Leong, P. M., Zhang, H. M., Yang, X. G., Kong, L. Z., Zhai, F. Y., Cheng, Y. Y., Guo, J. S., & Su, Y. X. (2018). New Chinese dietary guidelines: Healthy eating patterns and food-based dietary recommendations. *Asia Pacific Journal of Clinical Nutrition*, *27*(4), 908–913. https://doi.org/10.6133/apjcn.072018.03

Yu, X., Pu, H., & Voss, M. (2024). Overview of anti-inflammatory diets and their promising effects on non-communicable diseases. *British Journal of Nutrition*, *132*(7), 898–918. https://doi.org/10.1017/S0007114524001405

Yuan, M., Zhang, G., Bai, W., Han, X., Li, C., & Bian, S. (2022). The role of bioactive compounds in natural products extracted from plants in cancer treatment and their mechanisms related to anticancer effects. *Oxidative Medicine and Cellular Longevity*, *2022*, 1429869. https://doi.org/10.1155/2022/1429869

Zhang, Y., Du, W., Xiang, X., Ji, M., & Wang, X. (2025). Popular vegetarian diets in China: A sustainable dietary option for the globe? *Frontiers in Sustainable Food Systems*, *9*. https://doi.org/10.3389/fsufs.2025.1578127

7 Food Safety Standard for Phyto-Nutraceuticals

Komal Janiyani, Mohd Adnan, Mitesh Patel, and Malvi Surti

7.1 INTRODUCTION

Phyto-nutraceuticals are plant-derived bioactive compounds that provide both nutritional and therapeutic benefits. The term 'nutraceutical', combining 'nutrition' and 'pharmaceutical', describes substances that offer health-promoting effects beyond supplying basic nutrients such as proteins or carbohydrates (Borodina et al., 2020). These compounds include vitamins, minerals, and a variety of plant chemicals like polyphenols, flavonoids, and carotenoids. Many of them help prevent diseases by supporting normal cell functions and reducing the risks of cancer, CVDs, inflammation, and neurodegenerative disorders (Ahmad et al., 2016; Asiwe et al., 2021; Castro-Castaneda et al., 2022; Husen, 2021, 2022, 2023a, 2023b, 2024a, 2024b; Husen & Pant, 2024a, 2024b; Pant et al., 2025; Pant & Husen, 2025). Their use in healthcare is increasing because they can complement conventional treatments, improving patient outcomes and lowering side effects (Castro-Castaneda et al., 2022). For example berries rich in antioxidants have been linked to brain protection and slower age-related memory decline in long-term population studies (Malve & Bhalerao, 2023; Singh, 2018). Many of these natural products also support immune health and cardiovascular function (AlAli et al., 2021).

The global market for phyto-nutraceuticals is expanding quickly due to increasing health awareness, preference for natural remedies, and a focus on preventive healthcare. The Asia-Pacific region leads this trend, supported by its large population and long history of plant-based medicine (Jadhav et al., 2023). These products are obtained from various plant sources such as herbs, fruits, vegetables, and algae, with their popularity further strengthened by scientific studies confirming their benefits (Elkhalifa et al., 2021; Puri et al., 2022; Sajad et al., 2024). However, as consumption increases, there is a critical need for strong food safety standards to ensure that these products are safe, pure, and effective. Food safety issues in phyto-nutraceuticals encompass several areas, including contamination, adulteration, variations in bioactive content, and misleading health claims. Contamination with pesticides, heavy metals, mycotoxins, and harmful microorganisms during cultivation or processing poses major risks to product safety and effectiveness (Dima et al., 2020; Santini et al., 2018). Such contaminants can reduce bioavailability and cause unpredictable health effects (Preciado-Rangel et al., 2019; Santini & Novellino, 2017). Adulteration, such as mixing with synthetic drugs or using undeclared plant species, further threatens consumer safety, especially where there is no clear definition or consistent regulation for nutraceuticals (Kinth & Suman, 2019; Santini et al., 2018). The number of bioactive compounds can also vary due to differences in soil quality, climate, harvesting, and storage, making it difficult to maintain consistent product quality (Ceccanti et al., 2020; Hewavitharana et al., 2013). Such variations can affect the reliability of claimed health benefits (Ting et al., 2015). Additionally, many products on the market make health claims without sufficient scientific evidence, and regulatory agencies often struggle to verify these claims promptly (Baruah et al., 2018; Gonçalves et al., 2018).

Regulatory systems for phyto-nutraceuticals differ widely between countries. Organizations such as the U.S. Food and Drug Administration (FDA), the European Food Safety Authority (EFSA), and the Food Safety and Standards Authority of India (FSSAI) follow different rules for safety

DOI: 10.1201/9781003606710-7

checks, labelling, and health claim approval (Bansal et al., 2020; Blaze, 2021; Danielski, 2023). For example the FDA requires less evidence than pharmaceuticals, whereas EFSA demands strong scientific proof for claims made on food and supplements (Bansal et al., 2020). These differences can confuse consumers, as a product considered safe in one country may not meet the same standards in another (Blaze, 2021). Developing international guidelines that require clinical evidence and strict adherence could strengthen consumer trust, support innovation, and enhance the overall quality and reputation of the global nutraceutical market (Arora et al., 2019; Santini et al., 2018; Visioli, 2022). As the use of phyto-nutraceuticals becomes more common, it is essential to set clear food safety standards. These standards help ensure that products are safe, pure, and effective. They protect consumers and support trust in the use of natural plant-based health products. Setting strong guidelines will also help health professionals and consumers make better choices when using phyto-nutraceuticals for nutrition and therapy.

7.2 CLASSIFICATION OF PHYTO-NUTRACEUTICALS

The classification of phyto-nutraceuticals encompasses a diverse range of categories and methodological approaches, facilitating a comprehensive understanding of their roles in health promotion and disease management. This multidimensional framework incorporates classification based on food sources, mechanisms of action, and chemical composition (Rehmani, 2024). Derived predominantly from plant sources, phyto-nutraceuticals play pivotal roles in supporting health, preventing the onset of diseases, and enhancing the therapeutic efficacy of conventional pharmacological interventions.

From an origin-based perspective, phyto-nutraceuticals are broadly categorized into dietary supplements, functional foods, and herbal medicines (Bhardwaj et al., 2022). Each category is characterized by distinct properties that contribute to its specific health-promoting functions. For example flavonoids, one of the most extensively studied phytochemical groups, exhibit potent antioxidative and anti-inflammatory properties, contributing to a wide array of physiological benefits (Mabena et al., 2023). Other classification systems emphasize their mechanism of action, grouping them into categories such as anticancer agents, lipid-regulating agents, or anti-inflammatory compounds (AlAli et al., 2021). Such targeted classification enables the identification of phyto-nutraceuticals with potential efficacy against specific pathological conditions. In recent years, considerable attention has been directed toward the bioavailability of phyto-nutraceuticals, recognizing it as a critical determinant of their therapeutic potential. As noted by Dima et al. (2020), understanding the interactions of these bioactive compounds within the gastrointestinal tract is fundamental to improving their absorption and systemic availability (Dima et al., 2020). Innovative delivery systems, including phytosomes and nanocarriers, have emerged as promising strategies to enhance the solubility, stability, and bioaccessibility of plant-derived bioactives (Kattyar et al., 2022).

Furthermore, therapeutic classification of phytochemicals within nutraceuticals highlights their relevance in the management of chronic diseases such as diabetes (Mabena et al., 2023), cardiovascular disorders, and obesity (Cencic & Chingwaru, 2010). These applications underscore the potential of phyto-nutraceuticals in dietary interventions and integrative therapeutic regimens. The continuous advancement of scientific knowledge in this domain, coupled with evolving regulatory and safety frameworks, further refines their classification and supports their evidence-based incorporation into healthcare systems (Table 7.1) (Chopra et al., 2022).

7.3 REGULATORY FRAMEWORKS AND GUIDELINES FOR PHYTO-NUTRACEUTICALS

Nutraceuticals and dietary supplements are gaining increasing scientific recognition for their potential to mitigate disease risk and enhance overall quality of life, with the global nutraceutical market projected to experience a twofold expansion. In response to this growing demand, regulatory

TABLE 7.1
Classification of Phyto-Nutraceuticals with Corresponding Plant Sources and Documented Therapeutic Applications

Phyto-nutraceutical	Sources	Disease/Treatment Effects	References
		Carotenes	
α-Carotene	Carrots, sweet potatoes, winter squash, pumpkins, maize, tangerine	Antimetastatic agent, provitamin A for immune enhancement, prevention of cataracts, and macular degeneration	Liu et al. (2015)
β-Carotene	Carrots, sweet potatoes, winter squash, dark leafy greens, red/orange/yellow fruits, and vegetables	Provitamin A, antimetastatic, prevention of cataracts, and macular	Avraham et al. (2019), Cooperstone and Schwartz (2024), Mehrad et al. (2018), Ravanfar et al. (2018)
		Terpenes and Terpenoids	
Triterpenoid	Soybeans, beans, legumes, maize, alfalfa	Anticancer, antidiabetic, anti-inflammatory, and antioxidant-related disease prevention	Grishko and Galaiko (2016), Salvador et al. (2017), Siddiqui et al. (2024), J. Xu et al. (2018)
Diterpenes	Mustard, bugleweeds, common skullcap, germanders	Antioxidant-related disease prevention, antimicrobial, and anti-inflammatory	de Oliveira et al. (2018), Faiella et al. (2014), Grishko and Galaiko (2016)
Monoterpenes	Citrus oils, cherries, spearmint, dill, garlic, celery, maize, rosemary, ginger, basil, caraway, mints	Antimicrobial, bioactive for disease prevention	Mewalal et al. (2017), Zebec et al. (2016), Zhang et al. (2017)
Steroids	Almonds, cashews, peanuts, sesame seeds, sunflower seeds, whole wheat, maize, soybeans, avocado, rice bran, wheat germ, corn oils, fennel, basil, buckwheat	Neuroprotective, immunomodulatory, bone and muscle health, reproductive health support	Chrbolka et al. (2017)
		Phenolic Compounds	
Natural monophenols	Parsley, celery leaf, rosemary, sage, oregano, thyme, pepperwort, wild bergamot	Antimicrobial, antioxidant-related disease prevention	Gutiérrez-Larraínzar et al. (2012)
Flavonoids	Red/blue/purple pigments, tea, strawberries, gooseberries, cranberries, grapefruit, apples, peas, brassicas, chives, spinach, endive, leek, tomatoes	Antioxidant, antimicrobial, and chronic disease prevention	Ouédraogo et al. (2018), K. Xu et al. (2024)
Isoflavonoids	Soy, alfalfa sprouts, red clover, chickpeas, peanuts, kudzu, legumes	Antimycobacterial, oestrogenic (menopause health), wound healing, antimicrobial	Araya-Cloutier et al. (2017), Coronado-Aceves et al. (2017)
Aurones	Lemons, oranges, grapefruit	Active against hepatitis C, neurodegenerative diseases, antifungal, anticancer, antiviral, antimalarial, anti-inflammatory	Boucherle et al. (2017), Liew et al. (2015), Liew et al. (2017)

TABLE 7.1 *(Continued)*
Classification of Phyto-Nutraceuticals with Corresponding Plant Sources and Documented Therapeutic Applications

Phyto-nutraceutical	Sources	Disease/Treatment Effects	References
Chalconoids	Licorice root, citrus peel, apple peel	Anticancer, antimalarial, antimicrobial, anti-inflammatory, anti-protozoal, antioxidant-related disease prevention	Mirzaei et al. (2017), Mostofi et al. (2015), P. Singh et al. (2014)
Lignans	Apricots, strawberries, broccoli, kale, oats, wheat	Anticancer, neuroprotective, cognition enhancement, anti-inflammatory	Jiang et al. (2017)
Stilbenoids	Grapes, peanuts, Japanese knotweed root	Anticancer, antioxidant-related disease prevention, anti-inflammatory	Basheer et al. (2017), Cao et al. (2016), H.-B. Hu et al. (2018)
Curcuminoids	Turmeric, mustard	Anticancer, antidiabetic, antimicrobial, and pain relief	Bahramsoltani et al. (2017), Ullah et al. (2025), Verma et al. (2017)
Tannins	Tea, berries, horse chestnut, cranberry juice, peanut skin	Wound healing, antimicrobial	Adamczyk et al. (2017), Barrett et al. (2018), Sekowski et al. (2018)
Aromatic acids	Peppermint, licorice, peanut, wheat, vanilla beans, cloves, soy	Anticancer, antibacterial	Cvijetić et al. (2018), Garza et al. (2024)
Glucosinolates	Broccoli, cabbage, kale, cauliflower, turnip, mustard greens	Anticancer, heart protection (low doses)	Biegańska-Marecik et al. (2017), Radošević et al. (2017)
Betalains	Beets, chard, Amaranthus tricolor	Anticancer, anti-lipidemic, antimicrobial	Gengatharan et al. (2015), Kumar et al. (2015)

authorities across various countries are establishing standards and guidelines to promote the safety and efficacy of these products. Despite these efforts, regulatory frameworks for nutraceuticals remain diverse, lacking harmonization, and differ significantly among nations. Santini and coworkers (Santini et al., 2018) have highlighted the urgent need for unified international regulations, advocating for a coordinated global approach in developing new or amended frameworks. Notably, many countries, including India, Europe, Japan, and Russia, align their food additive standards with the Codex Alimentarius, developed by the FAO of the United Nations and the WHO (Codex Alimentarius Commission, 1995).

7.3.1 Food and Drug Administration (FDA), the United States

In the United States, dietary supplements are regulated under the Dietary Supplement Health and Education Act of 1994 (DSHEA) as a category of food that may contain vitamins, minerals, herbs, or other substances, including their concentrates, metabolites, constituents, extracts, or combinations. Under DSHEA, supplements must not contain components posing a 'significant or unreasonable risk' when used as directed. Health claims are approved under the Nutrition Labeling and Education Act (NLEA) of 1990, requiring substantiation by publicly available scientific evidence and expert consensus. For new dietary ingredients introduced after 15 October 1994, manufacturers must provide the FDA with safety evidence from historical use or scientific literature before marketing; however,

DSHEA does not require clinical trials for safety assessment. The FDA may restrict or remove products shown to present health risks, but supplements are generally deemed unsafe only after harm is reported, unlike drugs, which undergo pre-market safety evaluation (Bansal et al., 2020; Scarbrough, 2004). Registration typically involves declaring the botanical name, specifying whether ingredients are pre- or post-1994, and providing relevant safety data (Jain et al., 2018; Kessler et al., 2001).

7.3.2 Food Safety and Standards Authority of India (FSSAI)

In India, the regulation of functional foods and nutraceuticals is governed by the Food Safety and Standards (FSS) Act of 2006, implemented by the FSSAI. Provisions specifically addressing functional foods and nutraceuticals were incorporated into the Act in 2016. Under the FSS Act, nutraceuticals are defined as products derived from food sources that provide physiological benefits, whereas health supplements are concentrated sources of specific nutrients intended to complement the normal diet (Bansal et al., 2020).

According to the FSS Act, all functional foods and nutraceuticals must meet the quality criteria specified in the Indian Pharmacopoeia. Colours and additives used must comply with standards stipulated in the Act, and nutrient quantities must align with the Recommended Dietary Allowances (RDA) set by the Indian Council of Medical Research (ICMR). In the absence of ICMR specifications, Codex Alimentarius international standards are applied (Bansal et al., 2020). For products bearing health claims, the principal components include the nutritional ingredients and their associated health benefits. Manufacturers may seek approval for nutraceuticals based on documented usage history (minimum of 15 years in India or 30 years if originating abroad). Novel ingredients, those without a history of use in India, require prior approval, supported by robust scientific evidence, which may include controlled clinical trials, nutra-epidemiological data, and published literature. FSSAI also stipulates specific labelling requirements such as declaring the organic status of products along with other mandatory information (Bansal et al., 2020). Registration is mandatory for all nutraceuticals and dietary supplements entering the Indian market. For public safety, applications for new products must be accompanied by essential documents, including the food category list, raw material sources, food safety management system plans, and pesticide residue reports for water, among others (Bansal et al., 2020; Jain et al., 2018).

7.3.3 European Food Safety Authority (EFSA)

The EFSA, established in 2002, regulates food supplements in the EU under Directive 2002/46/EC, defining functional foods as concentrated sources of nutrients or other substances with nutritional or physiological effects in a dose form. While overarching legislation (Regulation 178/2002) applies, the classification of botanicals varies across Member States; some, like Italy, Belgium, and France, maintain national positive or negative lists (Gulati et al., 2014). Marketing must comply with EU safety, labelling, composition, additive, and contaminant rules, with the mutual recognition principle (Articles 34–36, TFEU) enabling cross-border acceptance despite differing national regulations. Key unharmonized areas include maximum vitamin/mineral levels and the regulation of non-vitamin/mineral ingredients unless classified as novel foods. Novel botanicals require pre-market safety assessments per Scientific Committee on Food (SCF, 2001) criteria, supported by in-vitro, in-vivo, or clinical data (Getman, 2011; Gulati et al., 2014). Traditional Herbal Medicinal Products (THMPs), regulated under Directive 2004/24/EC, follow simplified registration based on traditional use (≥30 years, including ≥15 in the EU), requiring proof of quality and safety but not efficacy (Gulati et al., 2014; Peschel, 2014).

7.3.4 Other Regional Guidelines

In China, health foods, regulated by the State Food and Drug Administration (SFDA, 2003), are defined as foods with specific health functions or supplying vitamins/minerals for targeted populations, excluding therapeutic intent or harm potential (Bansal et al., 2020). Under the Interim Administrative Measure of Health Food Registration (Article 21), claims (maximum two per product) require validation by government-approved laboratories, with comprehensive data and safety assessments; approved claims are monitored for three years before potential inclusion in the official list (C. Hu, 2014). In Russia, Biologically Active Food Supplements (BAFS) are regulated by Rospotrebnadzor under federal sanitary laws, defined as natural or nature-identical bioactive substances and probiotics for dietary enrichment, with therapeutic claims prohibited; regulations, harmonized with the EU and US, impose stricter bans on certain substances, microorganisms, and child use, with registration involving sensory and sanitary evaluations per Codex Alimentarius principles (Bagchi, 2014; Bansal et al., 2020; Sukhanov et al., 2011; Tutelyan & Sukhanov, 2008). In Canada, Natural Health Products (NHPs) under the Food and Drugs Act and Natural Health Products Regulations (NHPRs, 2004) include plant-, animal-, microbial-, or marine-derived substances such as vitamins, minerals, herbal remedies, probiotics, and amino acids, classified into three risk tiers from low (Class 1) to high (Class 3), with evidence ranging from literature to phase III/IV clinical trials; license holders must report serious adverse reactions within 15 days and maintain safety summaries (Bansal et al., 2020; Harrison & Nestmann, 2014). In Australia, complementary medicines under the Therapeutic Goods Act (TGA, 1989) include approved active ingredients such as amino acids, essential oils, herbal materials, vitamins, and minerals, with regulation based on risk: high-risk products require full evaluation and ARTG registration, while low-risk products are listed with simplified evidence; all proprietary active ingredients must meet GMP standards, supported by post-marketing surveillance (Bansal et al., 2020; Ghosh, 2014).

7.4 SAFETY ASSESSMENT PROTOCOLS FOR PHYTO-NUTRACEUTICALS

Safety assessment protocols for phyto-nutraceuticals are essential to ensuring both their efficacy and safety in consumer applications, given their complex bioactive profiles and potential health implications. Globally, evaluation strategies are broadly categorized into toxicological testing, clinical assessment, and compliance with established regulatory frameworks. Toxicological evaluations, encompassing both in-vitro and in-vivo studies, remain a cornerstone of safety determination, with particular emphasis on assessing cytotoxicity and genotoxicity. For instance Badolati et al. (2020) demonstrated the application of the Ames test in evaluating the mutagenic potential of nutraceuticals derived from agri-food waste biomasses, aligning with EFSA) requirements. Such assessments are further strengthened by methodologies that integrate structure, activity relationship analysis, and advanced bioinformatics tools, enhancing the reliability of phytochemical safety profiles (Ceniti et al., 2024).

Beyond standard toxicity testing, comprehensive safety evaluation must also consider bioactivity and the potential for adverse interactions with pharmaceuticals or pre-existing health conditions. For example Marx et al. (2020) reported that certain nutraceuticals, such as resveratrol, may exacerbate symptoms in individuals with multiple sclerosis, underscoring the need for rigorous clinical evaluation, especially in vulnerable populations (Marx et al., 2020). International frameworks such as Codex Alimentarius provide harmonized safety criteria for the incorporation of phyto-nutraceuticals into functional foods and dietary supplements, thereby minimizing risk and fostering consumer trust (Santini et al., 2018). The establishment of standardized protocols that integrate toxicological data with robust clinical evidence is consistent with recommendations by Ahmad (2024), who emphasizes the necessity of regulatory alignment for effective risk management (S. J. A. J. o. P. R. Ahmad & Development, 2024).

7.4.1 Toxicological Testing

Toxicological evaluation constitutes a fundamental component of safety assessment protocols for phyto-nutraceuticals, aimed at identifying potential adverse effects associated with their consumption. Standard testing includes acute and chronic toxicity studies, typically involving the administration of varying dosages to animal models to monitor toxicological endpoints. For example curcumin, a widely investigated phyto-nutraceutical, has consistently demonstrated a favourable safety profile, with no evidence of toxic, genotoxic, or teratogenic effects in multiple studies (Prakash et al., 2014). Such findings provide critical preclinical evidence supporting its progression into clinical trials. Globally, standardized protocols, such as those outlined in OECD (Organisation for Economic Co-operation and Development) guidelines, are routinely implemented to establish safety thresholds, enabling the identification and mitigation of risks before market authorization (Teng et al., 2022).

7.4.2 Genotoxicity and Carcinogenicity Studies

Assessment of genotoxicity and carcinogenicity is essential to ensure that phyto-nutraceuticals do not compromise genomic stability or promote oncogenesis. Genotoxicity testing assesses the potential of compounds to induce DNA damage, which may lead to mutations or neoplastic transformation, using assays such as the in-vivo comet assay to detect strand breaks in treated animal models. Investigations into plant-derived compounds, including curcumin, have consistently shown an absence of genotoxic activity, reinforcing their suitability for human consumption (O'Keefe, 2016). Carcinogenicity studies, conducted after genotoxicity evaluation, investigate whether chronic exposure may induce tumour formation. Regulatory authorities mandate comprehensive, long-term investigations to confirm that such products do not trigger malignant cellular changes, thereby safeguarding public health (Teng et al., 2022).

7.4.3 Allergenicity and Immunogenicity

Allergenicity and immunogenicity testing are particularly relevant for phyto-nutraceuticals derived from botanicals with known allergenic potential. These evaluations determine the capacity of a product to elicit immune-mediated hypersensitivity reactions, employing both in-vitro assays with human cell lines and in-vivo animal models. Characterization of allergenic proteins within product formulations is critical, as even trace amounts may provoke adverse responses in susceptible individuals. The outcomes underscore the necessity for accurate allergen labelling and consumer education, enabling risk mitigation through targeted formulation and premarket assessment (Sheik Abdulazeez & Thiruvengadam, 2013).

7.4.4 ADME Studies (Absorption, Distribution, Metabolism, and Excretion)

ADME profiling is integral to elucidating the pharmacokinetics of phyto-nutraceuticals, providing insight into their absorption, systemic distribution, metabolic transformation, and excretion. Such studies inform bioavailability assessments and optimal dosing strategies. For instance bioactive constituents of curcumin undergo complex metabolic pathways that influence both efficacy and safety. Analytical modeling, coupled with in-vitro simulations of metabolic processes, facilitates the prediction of in-vivo behaviour and supports evidence-based dosing recommendations. Comprehensive ADME data are therefore essential for defining safety margins and guiding the integration of phyto-nutraceuticals into nutritional and therapeutic applications (O'Keefe, 2016).

7.5 QUALITY CONTROL AND STANDARDIZATION FOR PHYTO-NUTRACEUTICALS

Quality control and standardization are indispensable for ensuring the safety, efficacy, and reproducibility of phyto-nutraceuticals. A robust quality assurance framework integrates GMPs, with the verification of identity, purity, and potency; the application of advanced analytical methodologies such as chromatographic and spectroscopic techniques; and comprehensive contaminant testing.

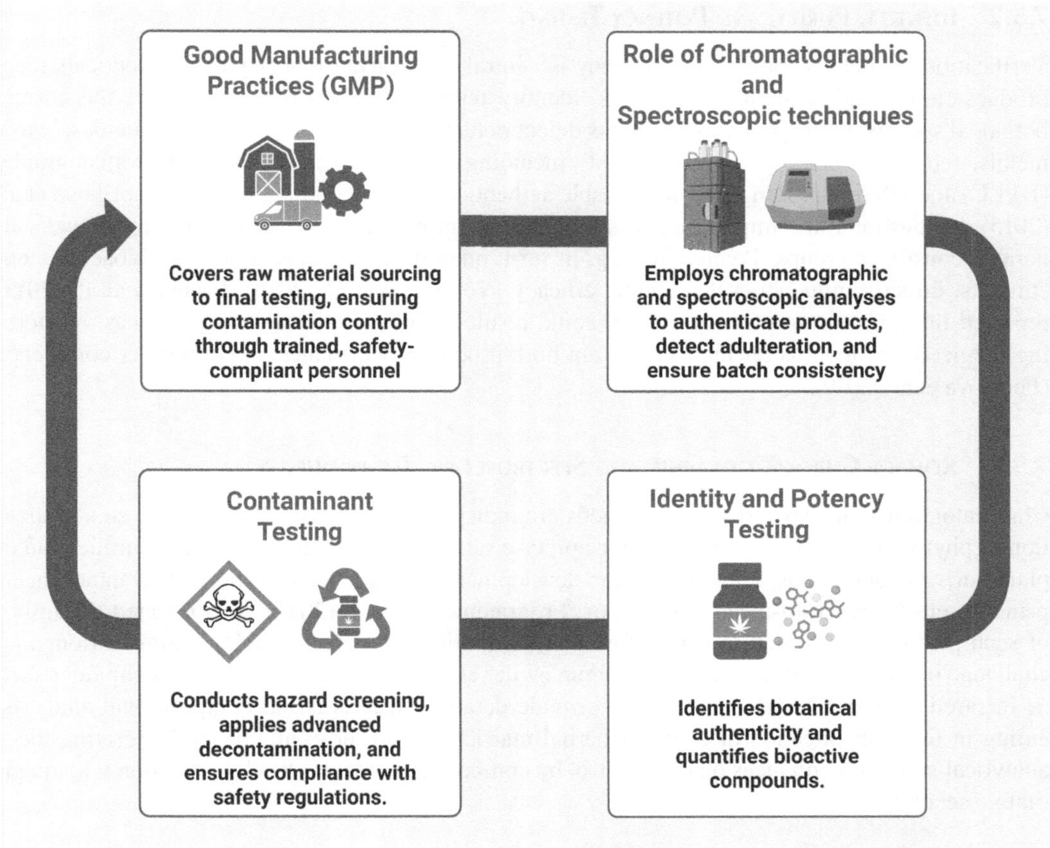

FIGURE 7.1 Framework for quality control and standardization of phyto-nutraceuticals, integrating GMP; advanced analytical authentication; contaminant screening; and bioactive quantification to ensure safety, consistency, and efficacy.

Collectively, these measures form the basis for regulatory compliance, consumer trust, and sustained therapeutic value (Figure 7.1).

7.5.1 Good Manufacturing Practices (GMP)

GMP forms the cornerstone of a controlled production environment, encompassing all stages from raw material sourcing to manufacturing, packaging, and final product testing. These practices ensure that phyto-nutraceuticals are produced consistently in compliance with predefined quality standards, minimizing variability and contamination risks. Novanda and Hidayat (2022) emphasized that GMP implementation is a vital element of quality control, conferring both regulatory compliance and competitive advantage in the nutraceutical sector (Novanda & Hidayat, 2022). Effective GMP adherence relies not only on infrastructure but also on the competence of personnel; training programmes on hygiene, food safety, and contamination prevention are critical. Campos and Barreto (2018) highlighted that targeted training in these domains significantly reduces the risk of foodborne illness, underscoring the role of workforce education in safeguarding public health. Thus, GMP is not merely a regulatory requirement but an essential pillar supporting the safety and reliability of phyto-nutraceuticals (Campos et al., 2018).

7.5.2 Identity, Purity, and Potency Testing

Verification of identity, purity, and potency is central to ensuring that phyto-nutraceuticals meet product claims and regulatory standards. Identity testing confirms the presence of the correct botanical species, while purity assessments detect potential contaminants such as pesticides, heavy metals, and mycotoxins. Advanced methods, including High-Performance Liquid Chromatography (HPLC) and DNA barcoding, provide reliable authentication of herbal ingredients. Veprikova et al. (2015) demonstrated the importance of stringent contaminant testing, particularly, to protect vulnerable consumer groups. Potency testing, in turn, quantifies the concentration of bioactive constituents, directly influencing therapeutic efficacy (Veprikova et al., 2015). Fenclova et al. (2019) reported that inadequate potency control could result in inconsistent clinical outcomes, reinforcing the need for rigorous testing to maintain both product performance and consumer confidence (Fenclova et al., 2019).

7.5.3 Role of Chromatographic and Spectroscopic Techniques

Chromatographic and spectroscopic methods are indispensable analytical tools in the standardization of phyto-nutraceuticals. HPLC enables precise separation, identification, and quantification of plant-derived compounds, allowing for the development of characteristic chromatographic fingerprints to ensure batch-to-batch consistency. Frommenwiler et al. (2022) demonstrated the utility of such profiling in validating herbal formulation quality and detecting adulteration, an ongoing challenge in the nutraceutical market (Frommenwiler et al., 2022). Complementary techniques such as Infrared (IR) and NMR spectroscopy provide detailed structural and compositional analyses, aiding in the authentication of complex herbal matrices (Ranftler et al., 2022). Integrating these analytical strategies enhances quality control by combining molecular-level verification with quantitative accuracy.

7.5.4 Contaminant Testing (Pesticides, Heavy Metals, Mycotoxins)

Comprehensive contaminant testing safeguards consumer health by detecting hazardous substances, including pesticides, heavy metals, and mycotoxins, which may be introduced during cultivation, processing, or storage. Numerous studies have revealed that certain commercial herbal products exceed established safety limits. For example Abdullah et al. (2025) identified mycotoxin contamination in herbal supplements at levels of public health concern, highlighting the need for stricter surveillance measures (Abdullah et al., 2025). Technological advances, such as liquid chromatography coupled with tandem MS, enable simultaneous detection of multiple mycotoxins with high sensitivity, an approach particularly valuable for assessing products intended for immunocompromised populations. Furthermore, Gammoh et al. (2023) proposed novel decontamination strategies to reduce fungal toxin loads, offering promising interventions for risk mitigation. Rigorous contaminant screening thus forms a non-negotiable element of quality control, ensuring that phyto-nutraceuticals meet both safety and regulatory benchmarks (Gammoh et al., 2023).

7.6 RISK–BENEFIT ANALYSIS OF PHYTO-NUTRACEUTICALS

A comprehensive risk–benefit analysis is fundamental for evaluating the overall health impact and clinical relevance of phyto-nutraceuticals. This assessment involves weighing potential therapeutic benefits, such as disease prevention, symptom management, and overall health maintenance, against the risks of adverse effects, which may range from mild discomfort to severe health outcomes. For example Derosa et al. (2020) reported that a supplement containing quercetin, rutin, bromelain, and L-carnosine demonstrated significant hypouricemic activity without notable adverse events, supporting its favourable safety profile. However, not all phyto-nutraceuticals exhibit such safety; some are associated with substantial adverse reactions, particularly when safety evaluations are insufficient (Derosa et al., 2020). Li and Wertheimer (2023) noted an increase in FDA-reported adverse events for certain dietary supplements, underscoring the urgent need for rigorous pre-market evaluation and post-market monitoring to ensure consumer safety (Li & Wertheimer, 2023).

The establishment of safety threshold levels plays a pivotal role in optimizing the risk-benefit balance of phyto-nutraceuticals. These thresholds define acceptable daily intakes or exposure limits that prevent significant adverse effects while preserving therapeutic efficacy. Toxicological assessments and controlled clinical trials are commonly employed to determine these limits, which can vary by compound, formulation, and target population. For instance the FDA's report of over a thousand adverse events linked to Garcinia cambogia illustrates the consequences of inadequate safety evaluation and the necessity for robust regulatory oversight (Li & Wertheimer, 2023). Such cases emphasize the need for risk-benefit evaluation to be integrated into product development and regulatory decision-making, ensuring both efficacy and safety.

7.6.1 Safety Threshold Levels

Safety threshold levels represent the highest exposure to a phyto-nutraceutical that can be tolerated without producing clinically significant adverse effects in the general population. Determining these thresholds requires systematic toxicological evaluations, pharmacokinetic and pharmacodynamic studies, and clinical trials. Natale et al. (2022) demonstrated the utility of defining reference values for nutraceuticals intended for hypercholesterolemia management, showing that adherence to recommended intake ranges enhanced cardiovascular outcomes while minimizing risks (Natale et al., 2022). Variations in safety thresholds can occur across demographic and clinical subgroups. Vulnerable populations, such as the elderly, pregnant individuals, or those with preexisting metabolic disorders, may require lower thresholds due to altered metabolism or increased susceptibility to toxicity. Consequently, safety thresholds should be periodically revised, considering emerging evidence to account for changes in product composition, usage patterns, and population health profiles. Integrating such thresholds into public health guidelines not only strengthens consumer protection but also enables healthcare practitioners to provide evidence-based recommendations for phyto-nutraceutical use.

7.6.2 Risk Management Strategies

Effective risk management for phyto-nutraceuticals requires a multitiered approach encompassing pre-market evaluation, post-market surveillance, and consumer education. Pre-market assessment involves comprehensive toxicological studies and human clinical trials to establish safety and efficacy before market entry. Regulatory authorities such as the U.S. Food and Drug Administration (FDA) require manufacturers to submit safety data; however, adverse events still occur, particularly for products with insufficient testing, such as Garcinia cambogia (Li & Wertheimer, 2023). Strengthening pre-market requirements could significantly reduce the incidence of safety-related withdrawals. Post-market surveillance is equally critical for detecting and addressing adverse events that arise after widespread consumer use. For example Oleñik (2014) reported that supplementation with omega-3 fatty acids and antioxidants was generally well tolerated, with gastrointestinal discomfort being the most frequent side effect, which supports their continued safe use with appropriate labelling (Oleñik, 2014). Furthermore, transparent adverse event reporting systems facilitate timely regulatory interventions when necessary.

Several high-profile cases underscore the importance of stringent oversight. Hydroxycut, a popular weight-loss supplement, was recalled by the FDA in 2009 after reports of severe hepatotoxicity and fatalities, highlighting the need for stricter evaluation of products claiming rapid weight-loss benefits (Li & Wertheimer, 2023). Similarly, Choi and Shin (2024) documented multiple serious adverse events, including hepatobiliary drug reactions and anaphylaxis, associated with herbal medicines in South Korea between 2012 and 2021, emphasizing the necessity of robust pharmacovigilance systems (Choi & Shin, 2024). Ultimately, effective risk management demands coordinated efforts among manufacturers, regulatory agencies, healthcare providers, and consumers. By combining stringent pre-market evaluation, continuous post-market surveillance, and targeted public education, the safety and credibility of phyto-nutraceuticals can be maximized while minimizing risks to public health.

7.7 CHALLENGES AND LIMITATIONS

The exploration and utilization of phyto-nutraceuticals face several critical challenges that limit their global acceptance, safety assurance, and therapeutic efficacy. One major obstacle is the lack of harmonization in regulatory frameworks across different countries. Herbal products are variably classified as foods, dietary supplements, or medicines depending on the jurisdiction, creating discrepancies in quality control, safety evaluation, and labelling requirements (Ichim, 2019; Raclariu et al., 2018). Such inconsistencies not only hinder international trade but also complicate compliance for manufacturers operating in multiple markets, thereby impacting product reliability and consumer safety (Ekor, 2014). Moreover, the absence of unified standards facilitates fraudulent practices, including adulteration and mislabelling, which compromise product authenticity and potentially endanger public health (Urumarudappa et al., 2019).

The authentication of herbal sources is another significant challenge for ensuring the safety and efficacy of phyto-nutraceuticals. Research indicates that nearly half of the commercial herbal products tested globally show evidence of adulteration or mislabelling, raising concerns about their therapeutic potential (M. C. Ichim et al., 2020; Seethapathy et al., 2019). Techniques such as DNA barcoding offer robust tools for ingredient verification, yet their application remains inconsistent across the industry (Niu et al., 2013; Urumarudappa et al., 2022). The complexity of botanical mixtures and the presence of contaminants further necessitate the development and enforcement of standardized analytical protocols to ensure the authenticity and purity of raw materials (Riswanto et al., 2022; Zafar et al., 2019). In addition, stability and shelf-life issues present another barrier to the optimal use of phyto-nutraceuticals. Bioactive compounds are susceptible to degradation under adverse storage conditions such as high humidity, elevated temperatures, and light exposure, which can reduce potency and therapeutic value over time (Chandrakar et al., 2024). Although advanced delivery systems, including phyto-phospholipid complexes, have been developed to enhance bioavailability, they may still face stability limitations (Chandrakar et al., 2024). The scarcity of comprehensive stability data, particularly in diverse environmental conditions, contributes to uncertainty regarding the long-term efficacy and safety of these products (Divella et al., 2020).

Looking forward, the future prospects of phyto-nutraceuticals are closely tied to advancements in regulatory science, technological integration, and the establishment of global quality standards. The development of harmonized regulatory frameworks could address inconsistencies in product classification, quality control, and market surveillance, thereby promoting fair trade practices and improving consumer confidence (Hastig et al., 2020; H. Zhou et al., 2023). Comprehensive regulatory pathways should encompass both pre-market evaluation and post-market monitoring to ensure the ongoing quality, safety, and efficacy of products (X. Zhou et al., 2024; X. Zhou & Xu, 2022). Incorporating scientifically rigorous approaches, including targeted pharmacovigilance programmes for herbal products, would further enhance safety monitoring (Qiu et al., 2023). Emerging technologies such as Artificial Intelligence (AI) and blockchain hold substantial potential for transforming safety monitoring and supply chain transparency. AI can process and analyse large datasets to detect adverse events, predict trends, and support regulatory decision-making (Ahmed & MacCarthy, 2021). Blockchain technology offers immutable, transparent supply chain records, enabling complete traceability from raw material sourcing to final product delivery (Hastig et al., 2020; Qiu et al., 2023; X. Zhou et al., 2024). This integration not only enhances safety oversight but

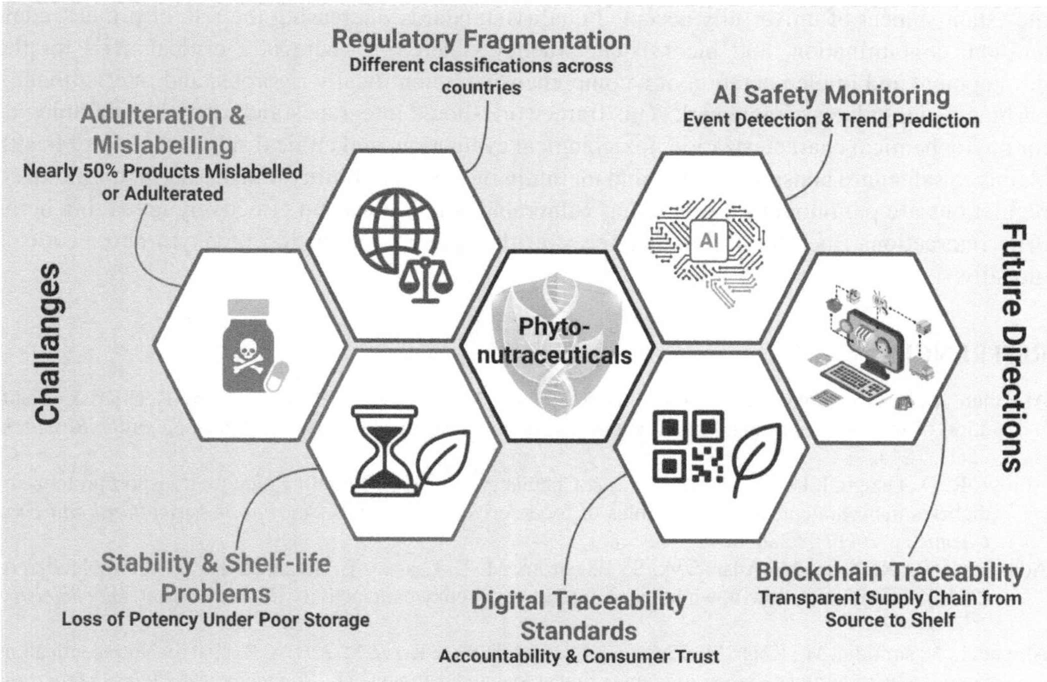

FIGURE 7.2 Challenges and future directions for phyto-nutraceuticals, highlighting regulatory gaps, adulteration, stability issues, and the role of AI, blockchain, and digital traceability in improving safety and authenticity.

also facilitates compliance with international regulations, potentially streamlining market access for manufacturers (Bahn et al., 2021; Hew et al., 2020).

Equally important is the establishment of global standards and robust digital traceability frameworks. Standardized criteria for cultivation, processing, testing, and labelling would help reduce variability in product quality and reinforce consumer trust (Suksasilp et al., 2021; X. Zhou & Xu, 2022). Digital traceability systems, by documenting the complete product journey, can ensure compliance with safety requirements while holding producers and suppliers accountable (Abioye et al., 2022; Grote et al., 2020). The combined implementation of stringent standards and advanced traceability systems could significantly enhance the integrity, safety, and market viability of phyto-nutraceuticals (Bahn et al., 2021; Sarkar, 2024) (Figure 7.2). While the sector's growth potential remains considerable, realizing its full benefits requires overcoming regulatory disparities, ensuring product authenticity, improving stability, and leveraging technological innovations to safeguard public health. Addressing these challenges through collaborative global efforts will be essential to shaping a safer, more reliable, and sustainable future for the phyto-nutraceutical industry.

7.8 CONCLUSION

Despite the rapid expansion of the global phyto-nutraceutical market and accumulating evidence supporting their therapeutic efficacy, regulatory frameworks remain markedly heterogeneous across different countries. This lack of harmonization engenders significant challenges in ensuring product safety, efficacy, and quality control. Comparative analyses reveal substantial variability among regulatory authorities such as the FDA, FSSAI, and EFSA with respect to safety assessment protocols, labelling requirements, and validation of health claims. Such regulatory discordance compromises the establishment of universally accepted quality standards, increasing the risk of product adulteration, contamination, and inconsistent potency. Therefore, there is a critical need for the development and implementation of a comprehensive, scientifically rigorous, and internationally harmonized regulatory framework. This framework should integrate standardized methodologies for phytochemical characterization, toxicological evaluation, and clinical substantiation of health claims to safeguard consumer health and maintain industry credibility. Furthermore, harmonized regulations are paramount for protecting vulnerable populations and effectively managing herb-drug interactions, thereby optimizing the safe therapeutic application of phyto-nutraceuticals globally.

REFERENCES

Abdullah, R., Kamarozaman, N. S., Ab Dullah, S. S., Aziz, M. Y., & Aziza, H. B. A. (2025). Health risks evaluation of mycotoxins in plant-based supplements marketed in Malaysia. *Journal of Scientific Research, 15*(1), 1244.

Abioye, R. O., Okagu, I. U., Udenigwe, C. C., & Chemistry, F. (2022). Targeting glucose transport proteins for diabetes management: Regulatory roles of food-derived compounds. *Journal of Agricultural and Food Chemistry, 70*(17), 5284–5290.

Adamczyk, B., Karonen, M., Adamczyk, S., Engström, M. T., Laakso, T., Saranpää, P., . . . & Smolander, A. (2017). Tannins can slow-down but also speed-up soil enzymatic activity in boreal forest. *Biochemistry, 107*, 60–67.

Ahmad, I., Nasiruddin, M., Khan, M., Zafaryab, M., Mehdi, S., & Rizvi, M. J. B. A. B. (2016). Nutraceuticals as chemopreventive and therapeutic agents in gut associated pathogenesis. *Biomedicine & Pharmacotherapy, 5*(298), 2161–1009.1000298.

Ahmad, S. (2024). India's growing demand of nutraceuticals. *Asian Journal of Pharmaceutical Research and Development, 12*(1), 126–132.

Ahmed, W. A., & MacCarthy, B. L. (2021). Blockchain-enabled supply chain traceability in the textile and apparel supply chain: A case study of the fiber producer, Lenzing. *Sustainability, 13*(19), 10496.

AlAli, M., Alqubaisy, M., Aljaafari, M. N., AlAli, A. O., Baqais, L., Molouki, A., . . . & Lim, S.-H. E. J. M. (2021). Nutraceuticals: Transformation of conventional foods into health promoters/disease preventers and safety considerations. *Emerging Microbes & Infections, 26*(9), 2540.

Araya-Cloutier, C., den Besten, H. M., Aisyah, S., Gruppen, H., & Vincken, J.-P. (2017). The position of prenylation of isoflavonoids and stilbenoids from legumes (Fabaceae) modulates the antimicrobial activity against Gram positive pathogens. *Food Chemistry, 226*, 193–201.

Arora, M., Arora, M., Bansal, P., & Baldi, A. (2019). Need and recommendations for universal guidelines on regulatory status and quality control/safety of probiotic products. *Asian Clinical Research & Clinical Trials Affairs, 6*(3), 231–249.

Asiwe, E., Igwe, C., Iheanacho, K., Onyeocha, I., & Onwuliri, V. (2021). Bioactive composition and acute oral toxicity studies on persea americana seed ethyl acetate fraction. *Asian Journal of Research in Biotechnology, 8*(4), 10–17.

Avraham, Y., Berry, E. M., Donskoy, M., Ahmad, W. A., Vorobiev, L., Albeck, A., & Mankuta, D. (2019). Beta-carotene as a novel therapy for the treatment of "Autistic like behavior" in animal models of Autism. *Biomedicine & Pharmacotherapy Research, 364*, 469–479.

Badolati, N., Masselli, R., Maisto, M., Di Minno, A., Tenore, G. C., Stornaiuolo, M., & Novellino, E. (2020). Genotoxicity assessment of three nutraceuticals containing natural antioxidants extracted from agri-food waste biomasses. *Foods, 9*(10), 1461.

Bagchi, D. (2014). *Nutraceutical and Functional Food Regulations in the United States and Around the World*. Elsevier.

Bahn, R. A., Yehya, A. A. K., & Zurayk, R. (2021). Digitalization for sustainable agri-food systems: Potential, status, and risks for the MENA region. *Sustainability, 13*(6), 3223.

Bahramsoltani, R., Rahimi, R., & Farzaei, M. H. (2017). Pharmacokinetic interactions of curcuminoids with conventional drugs: A review. *Journal of Ethnopharmacology, 209*, 1–12.

Bansal, R., & Dhiman, A. (2020). Nutraceuticals: A comparative analysis of regulatory framework in different countries of the world. *Expert Review of Metabolic & Disease Targets, 20*(10), 1654–1663.

Barrett, A. H., Farhadi, N. F., & Smith, T. J. (2018). Slowing starch digestion and inhibiting digestive enzyme activity using plant flavanols/tannins – a review of efficacy and mechanisms. *LWT – Food Science and Technology, 87*, 394–399.

Baruah, L., Malik, P. K., Kolte, A. P., Dhali, A., & Bhatta, R. (2018). Methane mitigation potential of phyto-sources from Northeast India and their effect on rumen fermentation characteristics and protozoa in vitro. *Veterinary World*, *11*(6), 809.

Basheer, L., Schultz, K., Guttman, Y., & Kerem, Z. (2017). In silico and in vitro inhibition of cytochrome P450 3A by synthetic stilbenoids. *Food Chemistry*, *237*, 895–903.

Bhardwaj, S., Kunj, P., Sharma, S. J. G. J. o. P., & Biomaterials. (2022). Nutraceuticals: An inspiring therapy for lifestyle diseases. *1*(4), 3–13.

Biegańska-Marecik, R., Radziejewska-Kubzdela, E., & Marecik, R. (2017). Characterization of phenolics, glucosinolates and antioxidant activity of beverages based on apple juice with addition of frozen and freeze-dried curly kale leaves (Brassica oleracea L. var. acephala L.). *Food Chemistry*, *230*, 271–280.

Blaze, J. J. (2021). A comparison of current regulatory frameworks for nutraceuticals in Australia, Canada, Japan, and the United States. *Issues in Pharmaceuticals*, *12*(2), https://doi.org/10.24926/iip.v24912i24922.23694.

Borodina, I., Kenny, L. C., McCarthy, C. M., Paramasivan, K., Pretorius, E., Roberts, T. J., . . . & Kell, D. B. (2020). The biology of ergothioneine, an antioxidant nutraceutical. *Nature Reviews Research*, *33*(2), 190–217.

Boucherle, B., Peuchmaur, M., Boumendjel, A., & Haudecoeur, R. J. P. (2017). Occurrences, biosynthesis and properties of aurones as high-end evolutionary products. *Phytochemistry*, *142*, 92–111.

Campos, T. C. C., & Barreto, M. F. (2018). Effectiveness of Good Manufacturing Practice training for food manipulators. *Journal of Nutrition & Food Sciences*, *8*(6), 1–5.

Cao, H., Jia, X., Shi, J., Xiao, J., & Chen, X. (2016). Non-covalent interaction between dietary stilbenoids and human serum albumin: Structure–affinity relationship, and its influence on the stability, free radical scavenging activity and cell uptake of stilbenoids. *Food Chemistry*, *202*, 383–388.

Castro-Castaneda, C. R., Altamirano-Lamarque, F., Ortega-Macías, A. G., Santa Cruz-Pavlovich, F. J., Gonzalez-De la Rosa, A., Armendariz-Borunda, J., . . . & Navarro-Partida, J. (2022). Nutraceuticals: A promising therapeutic approach in ophthalmology. *Nutrients*, *14*(23), 5014.

Ceccanti, C., Landi, M., Incrocci, L., Pardossi, A., Venturi, F., Taglieri, I., . . . & Guidi, L. (2020). Comparison of three domestications and wild-harvested plants for nutraceutical properties and sensory profiles in five wild edible herbs: Is domestication possible? *Foods*, *9*(8), 1065.

Cencic, A., & Chingwaru, W. (2010). The role of functional foods, nutraceuticals, and food supplements in intestinal health. *Nutrients*, *2*(6), 611–625.

Ceniti, C., Di Vito, A., Ambrosio, R. L., Anastasio, A., Bria, J., Britti, D., & Chiarella, E. (2024). Food safety assessment and nutraceutical outcomes of dairy by-products: Ovine milk whey as wound repair enhancer on injured human primary gingival fibroblasts. *Foods*, *13*(5), 683.

Chandrakar, R., Vyas, A., Kumar, N., Sahu, U., & Jain, V. (2024). Phyto-phospholipid complex vesicles: A revolutionary approach for enhancing bioavailability and optimizing therapeutic potential in herbal medicine. *Journal of Research in Unani*, *37*(2), 80–95.

Choi, Y., & Shin, H.-K. (2024). Adverse events associated with herbal medicine products reported in the Korea Adverse Event Reporting System from 2012 to 2021. *Frontiers in Pharmacology*, *15*, 1378208.

Chopra, A. S., Lordan, R., Horbańczuk, O. K., Atanasov, A. G., Chopra, I., Horbańczuk, J. O., . . . & Banach, M. (2022). The current use and evolving landscape of nutraceuticals. *Progress in Research*, *175*, 106001.

Chrbolka, P., Palúch, Z., Hill, M., & Alušík, Š. (2017). Circulating steroids negatively correlate with tinnitus. *Steroids*, *123*, 37–42.

Codex Alimentarius Commission (1995). General Standard for Food Additives (GSFA). CODEX STAN 192-1995. FAO/WHO Codex Alimentarius. https://www.fao.org/fao-who-codexalimentarius/codex-texts/list-standards/en.

Cooperstone, J., & Schwartz, S. (2024). Insights into the health benefits of carotenoids. In *Handbook on Natural Pigments in Food and Beverages* (pp. 555–575). Elsevier.

Coronado-Aceves, E. W., Gigliarelli, G., Garibay-Escobar, A., Zepeda, R. E. R., Curini, M., Cervantes, J. L., . . . & Marcotullio, M. C. (2017). New isoflavonoids from the extract of *Rhynchosia precatoria* (Humb. & Bonpl. ex Willd.) DC. and their antimycobacterial activity. *Journal of Ethnopharmacology*, *206*, 92–100.

Cvijetić, I. N., Verbić, T. Ž., de Resende, P. E., Stapleton, P., Gibbons, S., Juranić, I. O., . . . & Zloh, M. (2018). Design, synthesis and biological evaluation of novel aryldiketo acids with enhanced antibacterial activity against multidrug resistant bacterial strains. *European Journal of Medicinal Chemistry*, *143*, 1474–1488.

Danielski, R. (2023). Challenges hindering the commercialization of nutraceuticals derived from agri-food by-products. *Journal of Food Bioactives*, *24*.

de Oliveira, N. A., Cornelio-Santiago, H. P., Fukumasu, H., & de Oliveira, A. L. (2018). Green coffee extracts rich in diterpenes–Process optimization of pressurized liquid extraction using ethanol as solvent. *Journal of Food Engineering*, *224*, 148–155.

Derosa, G., D'Angelo, A., & Maffioli, P. (2020). Evaluation of the effects of a supplement composed by quercetin, rutin, bromelain and L-carnosine in patients with borderline uricemia. *Journal of Functional Nutrition Research*, *8*(10), 550–555.

Dima, C., Assadpour, E., Dima, S., & Jafari, S. M. (2020). Bioavailability of nutraceuticals: Role of the food matrix, processing conditions, the gastrointestinal tract, and nanodelivery systems. *Current Research in Food Science*, *19*(3), 954–994.

Divella, R., Daniele, A., Savino, E., & Paradiso, A. (2020). Anticancer effects of nutraceuticals in the Mediterranean diet: An epigenetic diet model. *Clinical Genetics & Proteomics*, *17*(4), 335–350.

Ekor, M. (2014). The growing use of herbal medicines: Issues relating to adverse reactions and challenges in monitoring safety. *Frontiers in Pharmacology*, *4*, 177.

Elkhalifa, A. E. O., Al-Shammari, E., Adnan, M., Alcantara, J. C., Mehmood, K., Eltoum, N. E., . . . & Ashraf, S. A. (2021). Development and characterization of novel biopolymer derived from Abelmoschus esculentus L. extract and its antidiabetic potential. *Materials*, *26*(12), 3609.

Faiella, L., Dal Piaz, F., Bader, A., & Braca, A. (2014). Diterpenes and phenolic compounds from *Sideritis pullulans*. *Phytochemistry*, *106*, 164–170.

Fenclova, M., Novakova, A., Viktorova, J., Jonatova, P., Dzuman, Z., Ruml, T., . . . & Stranska-Zachariasova, M. (2019). Poor chemical and microbiological quality of the commercial milk thistle-based dietary supplements may account for their reported unsatisfactory and non-reproducible clinical outcomes. *Scientific Reports*, *9*(1), 11118.

Frommenwiler, D. A., Reich, E., Sharaf, M. H., Cañigueral, S., & Etheridge, C. J. (2022). Investigation of market herbal products regulated under different categories: How can HPTLC help to detect quality problems? *Frontiers in Pharmacology*, *13*, 925298.

Gammoh, S., Alu'datt, M. H., Alhamad, M. N., Tranchant, C. C., Rababah, T., Kanakri, K., . . . & Tan, T. C. (2023). Determination of mycotoxins in nuts, cereals, legumes, and coffee beans and effectiveness of a selenium-based decontamination treatment. *Journal of Food Science*, *43*(6), e13087.

Garza, M. C., Pérez-Calahorra, S., Rodrigo-Carbó, C., Sánchez-Calavera, M. A., Jarauta, E., Mateo-Gallego, R., . . . & Lamiquiz-Moneo, I. (2024). Effect of aromatic herbs and spices present in the Mediterranean diet on the glycemic profile in type 2 diabetes subjects: A systematic review and meta-analysis. *Nutrients*, *16*(6), 756.

Gengatharan, A., Dykes, G. A., & Choo, W. S. (2015). Betalains: Natural plant pigments with potential application in functional foods. *LWT – Food Science and Technology*, *64*(2), 645–649.

Getman, S. (2011). EU regulations on food supplements, health foods, herbal medicines. *USA: United States Commercial Service, United States of America Department of Commerce*.

Ghosh, D. (2014). Complementary medicine regulation in Australia. In *Nutraceutical and Functional Food Regulations in the United States and Around the World* (pp. 301–308). Elsevier.

Gonçalves, R. F., Martins, J. T., Duarte, C. M., Vicente, A. A., & Pinheiro, A. C. (2018). Advances in nutraceutical delivery systems: From formulation design for bioavailability enhancement to efficacy and safety evaluation. *Trends in Food Science & Technology*, *78*, 270–291.

Grishko, V. V., & Galaiko, N. V. (2016). Structural diversity, natural sources and pharmacological potential of naturally occurring A-seco-triterpenoids. *Studies in Natural Products Chemistry*, *51*, 51–149.

Grote, G., & Pfrombeck, J. J. (2020). Uncertainty in aging and lifespan research: Covid-19 as catalyst for addressing the elephant in the room. *Work, Aging and Retirement*, *6*(4), 246–250.

Gulati, O. P., Ottaway, P. B., & Coppens, P. (2014). Botanical nutraceuticals,(food supplements, fortified and functional foods) in the European Union with main focus on nutrition and health claims regulation. In *Nutraceutical and Functional Food Regulations in the United States and Around the World* (pp. 221–256). Elsevier.

Gutiérrez-Larraínzar, M., Rúa, J., Caro, I., de Castro, C., de Arriaga, D., García-Armesto, M. R., & del Valle, P. (2012). Evaluation of antimicrobial and antioxidant activities of natural phenolic compounds against foodborne pathogens and spoilage bacteria. *Food Control*, *26*(2), 555–563.

Harrison, J. R., & Nestmann, E. R. (2014). Current Canadian regulatory initiatives and policies for natural health products (dietary supplements). In *Nutraceutical and Functional Food Regulations in the United States and Around the World* (pp. 183–200). Elsevier.

Hastig, G. M., Sodhi, M. S. J. P., & Management, O. (2020). Blockchain for supply chain traceability: Business requirements and critical success factors. *Production Planning & Control*, *29*(4), 935–954.

Hew, J.-J., Wong, L.-W., Tan, G. W.-H., Ooi, K.-B., & Lin, B. (2020). The blockchain-based Halal traceability systems: A hype or reality? *Science of Computer Programming*, *25*(6), 863–879.

Hewavitharana, A. K., Tan, Z. W., Shimada, R., Shaw, P. N., & Flanagan, B. M. (2013). Between fruit variability of the bioactive compounds, β-carotene and mangiferin, in mango (M angifera indica). *Nutrition & Dietetics*, *70*(2), 158–163.

Hu, C. (2014). Historical change of raw materials and claims of health food regulations in China. In *Nutraceutical and Functional Food Regulations in the United States and Around the World* (pp. 363–388). Elsevier.

Hu, H.-B., Liang, H.-P., Li, H.-M., Yuan, R.-N., Sun, J., Zhang, L.-L., . . . & Wu, Y. (2018). Isolation, modification and cytotoxic evaluation of stilbenoids from Acanthopanax leucorrhizus. *Fitoterapia*, *124*, 167–176.

Husen, A. (2021). *Traditional Herbal Therapy for the Human Immune System*. CRC Press. https://doi.org/10.1201/9781003137955

Husen, A. (2022). *Herbs, Shrubs and Trees of Potential Medicinal Benefits*. CRC Press. https://doi.org/10.1201/9781003205067

Husen, A. (2023a). *Exploring Poisonous Plants: Medicinal Values, Toxicity Response and Other Potential Uses*. CRC Press. https://doi.org/10.1201/b23017

Husen, A. (2023b). *Antidiabetic Medicinal Plants and Herbal Treatments*. CRC Press. https://doi.org/10.1201/b23347

Husen, A. (2024a). *Antimalarial Medicinal Plants*. Taylor & Francis Group. https://doi.org/10.1201/9781003378396

Husen, A. (2024b). *Medicinal Spices and Condiment Crops*. Taylor & Francis Group. https://doi.org/10.1201/9781003387046

Husen, H., & Pant, M. (2024a). *Exploring Medicinal Orchids*. Taylor & Francis Group. https://doi.org/10.1201/b23375

Husen, H., & Pant, M. (2024b). *Tissue Culture Techniques and Medicinal Plants: Enhancing Propagation and Production*. Taylor & Francis Group. https://doi.org/10.1201/b23374

Ichim, M. C. (2019). The DNA-based authentication of commercial herbal products reveals their globally widespread adulteration. *Frontiers in Pharmacology*, *10*, 1227.

Ichim, M. C., Häser, A., & Nick, P. (2020). Microscopic authentication of commercial herbal products in the globalized market: Potential and limitations. *Frontiers in Pharmacology*, *11*, 876.

Jadhav, H. B., Sablani, S., Gogate, P., Annapure, U., Casanova, F., & Nayik, G. A. (2023). Factors governing consumers buying behavior concerning nutraceutical product. *Foods*, *11*(9), 4988–5003.

Jain, P. N., Rathod, M. H., Jain, C. V., & Vijayendraswamy, S. (2018). Current regulatory requirements for registration of nutraceuticals in India and USA. *International Journal of Drug Regulatory Affairs*, *6*(2), 22–29.

Jiang, Z.-H., Chai, L., Liu, Y.-P., Qiu, H.-C., Wang, T.-T., Feng, X.-Y., . . . & Fu, Y.-H. (2017). Bioactive lignans from the stems of Mappianthus iodoides. *Phytochemistry Letters*, *22*, 194–198.

Kattyar, S. L., Patil, P. S., Patil, S. V., & Kadam, S. S. (2022). Phytosomes and recent research on phytosomal drugs. *Asian Journal of Pharmaceutical Analysis*, *12*(1), 61–69.

Kessler, R. C., Davis, R. B., Foster, D. F., Van Rompay, M. I., Walters, E. E., Wilkey, S. A., . . . Eisenberg, D. M. (2001). Long-term trends in the use of complementary and alternative medical therapies in the United States. *Annals of Internal Medicine*, *135*(4), 262–268.

Kinth, P., & Suman, Y. (2019). Ambiguous definition of nutraceuticals: Challenges & possible solutions. *International Journal of Science and Research Reviews*. https://doi.org/10.36106/ijsrr

Kumar, S. S., Manoj, P., Giridhar, P., Shrivastava, R., & Bharadwaj, M. (2015). Fruit extracts of Basella rubra that are rich in bioactives and betalains exhibit antioxidant activity and cytotoxicity against human cervical carcinoma cells. *Journal of Food and Feed*, *15*, 509–515.

Li, W., & Wertheimer, A. (2023). Narrative review: The FDA's perfunctory approach of dietary supplement regulations giving rise to copious reports of adverse events. *Innovations in Pharmacy*, *14*(1). https://doi.org/10.24926/iip.v14i1.24989.

Liew, K.-F., Chan, K.-L., & Lee, C.-Y. (2015). Blood–brain barrier permeable anticholinesterase aurones: Synthesis, structure–activity relationship, and drug-like properties. *European Journal of Medicinal Chemistry*, *94*, 195–210.

Liew, K.-F., Hanapi, N. A., Chan, K.-L., Yusof, S. R., & Lee, C.-Y. (2017). Assessment of the blood-brain barrier permeability of potential neuroprotective aurones in parallel artificial membrane permeability assay and porcine brain endothelial cell models. *Journal of Pharmaceutical Sciences*, *106*(2), 502–510.

Liu, Y.-Z., Yang, C.-M., Chen, J.-Y., Liao, J.-W., & Hu, M.-L. (2015). Alpha-carotene inhibits metastasis in Lewis lung carcinoma in vitro, and suppresses lung metastasis and tumor growth in combination with taxol in tumor xenografted C57BL/6 mice. *The Journal of Nutritional Biochemistry*, *26*(6), 607–615.

Mabena, P., Fasemore, T. M., & Nkomozepi, P. (2023). Impact of nutraceuticals on type 1 and type 2 diabetes mellitus-induced micro-and macrovasculopathies. *Advances in Science*, *14*(1), 64.

Malve, H., & Bhalerao, P. J. (2023). Past, present, and likely future of nutraceuticals in India: Evolving role of pharmaceutical physicians. *Journal of Pharmaceutical Sciences and Research*, *15*(2), 68–74.

Marx, W., Hockey, M., McGuinness, A. J., Lane, M., Christodoulou, J., Van Der Mei, I., . . . & Disorders, R. (2020). The effect of emerging nutraceutical interventions for clinical and biological outcomes in multiple sclerosis: A systematic review. *Multiple Sclerosis and Related Disorders*, *37*, 101486.

Mehrad, B., Ravanfar, R., Licker, J., Regenstein, J. M., & Abbaspourrad, A. (2018). Enhancing the physicochemical stability of β-carotene solid lipid nanoparticle (SLNP) using whey protein isolate. *Food Research International*, *105*, 962–969.

Mewalal, R., Rai, D. K., Kainer, D., Chen, F., Külheim, C., Peter, G. F., & Tuskan, G. A. (2017). Plant-derived terpenes: A feedstock for specialty biofuels. *Trends in Biotechnology*, *35*(3), 227–240.

Mirzaei, H., Shokrzadeh, M., Modanloo, M., Ziar, A., Riazi, G. H., & Emami, S. (2017). New indole-based chalconoids as tubulin-targeting antiproliferative agents. *Bioorganic Chemistry*, *75*, 86–98.

Mostofi, M., Ziarani, G. M., Mahdavi, M., Moradi, A., Nadri, H., Emami, S., . . . & Shafiee, A. (2015). Synthesis and structure-activity relationship study of benzofuran-based chalconoids bearing benzylpyridinium moiety as potent acetylcholinesterase inhibitors. *European Journal of Medicinal Chemistry*, *103*, 361–369.

Natale, F., Molinari, R., Covino, S., Piccinocchi, G., Salvetti, A., Monda, E., . . . & Disease, D. (2022). Effectiveness in the short-term of a novel nutraceutical for the management of hypercholesterolemia: An observational multicenter primary care experience. *Nutrients*, *12*(11), 627–638.

Niu, L., Mantri, N., Wohlmuth, H., Li, C., Xue, C. C., & Pang, E. J. (2013). First report of fingerprinting dried herbal products using a subtractive diversity array. *Journal of Chromatography & Medicine*, *10*(1), 17–27.

Novanda, A. D., & Hidayat, A. W. (2022). Study of implementation of Good Manufacturing Practices (GMP) in CV "XYZ". *Journal of Nursing Science: Proceedings*, 51–57.

O'Keefe, S. J. (2016). Diet, microorganisms and their metabolites, and colon cancer. *Nature Reviews Gastroenterology & Hepatology*, *13*(12), 691–706.

Oleñik, A. (2014). Effectiveness and tolerability of dietary supplementation with a combination of omega-3 polyunsaturated fatty acids and antioxidants in the treatment of dry eye symptoms: Results of a prospective study. *Clinical Ophthalmology*, *8*, 169–176.

Ouédraogo, J. C. W., Dicko, C., Kini, F. B., Bonzi-Coulibaly, Y. L., & Dey, E. S. (2018). Enhanced extraction of flavonoids from Odontonema strictum leaves with antioxidant activity using supercritical carbon dioxide fluid combined with ethanol. *The Journal of Supercritical Fluids*, *131*, 56–71.

Pant, M., & Husen, A. (2025). *Exploring Medicinal Bamboos*. Taylor & Francis Group. https://doi.org/10.1201/9781003485933

Pant, M., Sohrab, S. S., & Husen, A. (2025). *Genetic Improvement and Conservation Practice of Medicinal Plants*. Taylor & Francis Group. https://doi.org/10.1201/9781003482659

Peschel, W. J. (2014). The use of community herbal monographs to facilitate registrations and authorisations of herbal medicinal products in the European Union 2004–2012. *Journal of Ethnopharmacology*, *158*, 471–486.

Prakash, A., Khan, S., Kumar, D., Telang, A. G., & Malik, J. K. (2014). Concurrent administration of curcumin mitigates arsenic–and chlorpyrifos–induced apoptosis in rat thymocytes. *Asian-Australasian Journal of Veterinary Sciences*, *2*(7), 407–413.

Preciado-Rangel, P., Reyes-Pérez, J. J., Ramírez-Rodríguez, S. C., Salas-Pérez, L., Fortis-Hernández, M., Murillo-Amador, B., & Troyo-Diéguez, E. (2019). Foliar aspersion of salicylic acid improves phenolic and flavonoid compounds, and also the fruit yield in cucumber (Cucumis sativus L.). *Plants*, *8*(2), 44.

Puri, V., Nagpal, M., Singh, I., Singh, M., Dhingra, G. A., Huanbutta, K., . . . Sangnim, T. (2022). A comprehensive review on nutraceuticals: Therapy support and formulation challenges. *Nutrients*, *14*(21), 4637.

Qiu, S., Feng, X., Zhu, M., & Luo, Q. (2023). Research on key technologies of supply chain traceability based on blockchain. In *Applied Mathematics, Modeling and Computer Simulation* (pp. 127–133). IOS Press.

Raclariu, A. C., Heinrich, M., Ichim, M. C., & de Boer, H. J. (2018). Benefits and limitations of DNA barcoding and metabarcoding in herbal product authentication. *Phytochemical Analysis*, *29*(2), 123–128.

Radošević, K., Srček, V. G., Bubalo, M. C., Brnčić, S. R., Takács, K., & Redovniković, I. R. (2017). Assessment of glucosinolates, antioxidative and antiproliferative activity of broccoli and collard extracts. *Journal of Food Composition and Analysis*, *61*, 59–66.

Ranftler, C., Röhrich, A., Sparer, A., Tschegg, C., & Nagl, D. (2022). Purified clinoptilolite-tuff as an efficient sorbent for gluten derived from food. *International Journal of Molecular Sciences*, *23*(9), 5143.

Ravanfar, R., Comunian, T. A., Dando, R., & Abbaspourrad, A. J. F. C. (2018). Optimization of microcapsules shell structure to preserve labile compounds: A comparison between microfluidics and conventional homogenization method. *Food Chemistry*, *241*, 460–467.

Rehmani, A. (2024). Nutraceuticals: Revolutionizing healthcare in 21st century. *International Journal of Health Care Management and Practice*, *12*(3), 45–58.

Riswanto, F. D. O., Windarsih, A., Lukitaningsih, E., Rafi, M., Fadzilah, N. A., & Rohman, A. (2022). Metabolite fingerprinting based on 1H-NMR spectroscopy and liquid chromatography for the authentication of herbal products. *Medicines*, *27*(4), 1198.

Sajad, M., Shabir, S., Singh, S. K., Bhardwaj, R., Alsanie, W. F., Alamri, A. S., . . . & Singh, M. P. (2024). Role of nutraceutical against exposure to pesticide residues: Power of bioactive compounds. *Frontiers in Nutrition*, *11*, 1342881.

Salvador, J. A., Leal, A. S., Valdeira, A. S., Goncalves, B. M., Alho, D. P., Figueiredo, S. A., . . . & Mendes, V. I. (2017). Oleanane-, ursane-, and quinone methide friedelane-type triterpenoid derivatives: Recent advances in cancer treatment. *European Journal of Medicinal Chemistry*, *142*, 95–130.

Santini, A., Cammarata, S. M., Capone, G., Ianaro, A., Tenore, G. C., Pani, L., & Novellino, E. (2018). Nutraceuticals: Opening the debate for a regulatory framework. *British Journal of Clinical Pharmacology*, *84*(4), 659–672.

Santini, A., & Novellino, E. J. F. (2017). *To Nutraceuticals and Back: Rethinking a Concept* (Vol. 6, p. 74). MDPI.

Sarkar, S. (2024). The future of digital drug traceability in the global supply chain. *World Journal of Supply Chain*, *4*, 896.

Scarbrough, B. (2004). Dietary supplements: A review of United States regulation with emphasis on the Dietary Supplement Health and Education Act of 1994 and subsequent activity. *Regulatory Toxicology and Pharmacology*, *40*(3), 234–246.

Seethapathy, G. S., Raclariu-Manolica, A.-C., Anmarkrud, J. A., Wangensteen, H., & de Boer, H. J. (2019). DNA metabarcoding authentication of ayurvedic herbal products on the European market raises concerns of quality and fidelity. *Frontiers in Pharmacology*, *10*, 68.

Sekowski, S., Bitiucki, M., Ionov, M., Zdeb, M., Abdulladjanova, N., Rakhimov, R., . . . & Zamaraeva, M. (2018). Influence of valoneoyl groups on the interactions between Euphorbia tannins and human serum albumin. *Journal of Luminescence*, *194*, 170–178.

Sheik Abdulazeez, S., & Thiruvengadam, D. (2013). Effect of lycopene on oxidative stress induced during D-galactosamine/lipopolysaccharide-sensitized liver injury in rats. *Pharmacological Biology*, *51*(12), 1592–1599.

Siddiqui, T., Khan, M. U., Sharma, V., & Gupta, K. (2024). Terpenoids in essential oils: Chemistry, classification, and potential impact on human health and industry. *Phytochemistry: Plants*, *4*(2), 100549.

Singh, P., Anand, A., & Kumar, V. (2014). Recent developments in biological activities of chalcones: A mini review. *European Journal of Medicinal Chemistry*, *85*, 758–777.

Singh, R. J. (2018). Current Alzheimer's management with berries fruits therapy. *Journal of Pharmaceutical and Healthcare*, *2*.

Sukhanov, B., Kerimova, M., Elizarova, E., & Chigireva, É. (2011). Sanitary and epidemiological examination of foods under carrying sanitary and epidemiological surveillance on the customer border and at in the territory of the Customers Union. *Voprosy Pitaniia (Questions of Nutrition)*, *80*(4), 25–31.

Suksasilp, C., Griffiths, S., Sebastian, C. L., & Norbury, C. J. (2021). Reliability and validity of a temporal distancing emotion regulation task in adolescence. *Emotion*, *21*(4), 830.

Teng, K., Loganathan, R., Siew, E. L., Lau, H. L. N., & Rajab, N. F. (2022). Toxicological assessment of refined palm-pressed mesocarp fibre oil. *Journal of Oil Palm Research*, *34*(3), 562–570.

Ting, Y., Zhao, Q., Xia, C., & Huang, Q. (2015). Using in vitro and in vivo models to evaluate the oral bioavailability of nutraceuticals. *Journal of Agricultural and Food Chemistry*, *63*(5), 1332–1338.

Tutelyan, V., & Sukhanov, B. (2008). Food supplements: Modern approaches to quality and safety. *Voprosy Pitaniia (Questions of Nutrition)*, *4*.

Ullah, S., Khan, T., Khan, T., Ali, M., Jan, A. K., Shinwari, Z. K., . . . & Waqas, M. (2025). Phytochemicals for cancer treatment: An update on plant-derived anti-cancer compounds and their mechanisms of action. *Current Topics in Medicinal Chemistry*, *25*(8), 871–890.

Urumarudappa, S. K. J., Tungphatthong, C., Jaipaew, J., Pornputtapong, N., Pakdeesattayapong, D., Vimolmangkang, S., & Sukrong, S. (2022). Development of a DNA barcode library of plants in the Thai Herbal Pharmacopoeia and Monographs for authentication of herbal products. *Scientific Reports*, *12*(1), 9624.

Urumarudappa, S. K. J., Tungphatthong, C., & Sukrong, S. (2019). Mitigating the impact of admixtures in Thai herbal products. *Frontiers in Pharmacology*, *10*, 1205.

Veprikova, Z., Zachariasova, M., Dzuman, Z., Zachariasova, A., Fenclova, M., Slavikova, P., . . . & Chemistry, F. (2015). Mycotoxins in plant-based dietary supplements: Hidden health risk for consumers. *Food Chemistry*, *63*(29), 6633–6643.

Verma, S., Mundkinajeddu, D., Agarwal, A., Chatterjee, S. S., & Kumar, V. J. (2017). Effects of turmeric curcuminoids and metformin against central sensitivity to pain in mice. *Journal of Traditional and Complementary Medicine*, *7*(2), 145–151.

Visioli, F. (2022). Science and claims of the arena of food bioactives: Comparison of drugs, nutrients, supplements, and nutraceuticals. *Frontiers in Food Function*, *13*(24), 12470–12474.

Xu, J., Cao, J., Yue, J., Zhang, X., & Zhao, Y. (2018). New triterpenoids from acorns of *Quercus liaotungensis* and their inhibitory activity against α-glucosidase, α-amylase and protein-tyrosine phosphatase 1B. *Journal of Functional Foods*, *41*, 232–239.

Xu, K., Qu, L., Li, H., Ren, X., Yan, N., & Fu, X. (2024). Functional properties of dietary quercetin in cardiovascular health and disease. *Journal of Functional Foods*, *5*(5), 1951–1967.

Zafar, M., Ahmad, M., Sultana, S., Lubna, Anjum, F., Ozdemir, F. A. (2019). Light microscopy and scanning electron microscopy: Implications for authentication of misidentified herbal drugs. *Microscopy Research and Technique*, *82*(10), 1779–1786.

Zebec, Z., Wilkes, J., Jervis, A. J., Scrutton, N. S., Takano, E., & Breitling, R. (2016). Towards synthesis of monoterpenes and derivatives using synthetic biology. *Current Opinion in Chemical Biology*, *34*, 37–43.

Zhang, L., Xiao, W.-H., Wang, Y., Yao, M.-D., Jiang, G.-Z., Zeng, B.-X., . . . & Yuan, Y.-J. (2017). Chassis and key enzymes engineering for monoterpenes production. *Bioresource Technology*, *35*(8), 1022–1031.

Zhou, H., Wang, Q., Li, L., Teo, T. S., & Yang, S. (2023). Supply chain digitalization and performance improvement: A moderated mediation model. *Supply Chain Management: An International Journal*, *28*(6), 993–1008.

Zhou, X., Lu, H., & Kumar Mangla, S. (2024). The impact of digital traceability on sustainability performance: Investigating the roles of sustainability-oriented innovation and supply chain learning. *Supply Chain Management: An International Journal*, *29*(3), 497–522.

Zhou, X., & Xu, Z. (2022). Traceability in food supply chains: A systematic literature review and future research directions. *International Food Research Journal*, *25*(2), 173–196.

8 Nutritional Spectrum of Vegetables

Arpan Modi, Megha Barot, and Azamal Husen

8.1 INTRODUCTION

The production of vegetables has remarkably increased over a period of time and across the globe. This is due to the technological advancement which started from labour-saving techniques to the use of hybrids and low-cost fertilizers to enhance the productivity (Welbaum, 2015). Vegetables have been consumed worldwide since ancient times. Ancient literature of India and Israel revealed that vegetables have had both the nutritional and spiritual significance. Many plant species belonging to different families are mentioned in Yajurveda, and these species are being cultivated and consumed even today (Tripathi et al., 2024; Włodarczyk, 2020). Cultivation of specific vegetables is also connected to ethnicity. Several ethnic vegetables are underutilized, and their use have been limited to specific region or people. These include African baobab, amaranth, spider plant, sea kale, bitter melon, Indian poke, purslane, yellow cresses, parsnip, water spinach, prickly pear, black nightshade, and water leaf. Ethnic vegetables are rich in proteins, micronutrients, vitamins, and antioxidants, and many of them have been found to be superior to regularly used vegetables such as tomato or cabbage (Jha, 2024).

Vegetables are classified on the basis of the part being used. However, based on the nutritional status of the edible part of the plant, the same can be done with the emphasis on health benefits they impart due to key ingredients. It is well known that any food item is made up of organic and inorganic constituents. Organic constituents are metabolites (either primary or secondary) synthesized by living organisms. All the essential minerals form the inorganic part of the food. Therefore, in this context, nutritional classification of the vegetables can be done with the fact that the vegetable is rich in primary/secondary metabolites and/or micronutrients (Vicente et al., 2014). Table 8.1 represents different vegetables categorized by nutrients they carry and the health benefits they impart. As water is also essential part of the food, it is also included into the categorization.

8.2 BIOFORTIFICATION

A plant-based diet has been shown to lower the risk of long-term illnesses and conditions like cancer, diabetes, obesity, heart problems, and other conditions that affect people's health (Nayak et al., 2021). Agronomists, breeders, and agricultural biotechnologists strive to increase the nutritional value of plant-based functional meals as they are now an essential component of healthy lifestyles. To maintain nutritional benefits, agronomists create good agricultural practices (GAP). After using strict selection techniques, breeders create new hybrids which are believed to have improved qualities. The same goal is being served by agricultural biotechnologists, albeit in a different method. But before putting their variety on the market, they have to adhere to stringent regulations, unlike breeders (Sudha et al., 2018). Although vegetables are rich sources of nutrients like carbohydrates, proteins, minerals, phytochemicals, and vitamins, their complex matrix of biomolecules and post-harvest to cooking practices allow these nutrients to be absorbed less or inadequate availability. Some group of people consume specific vegetables only, which may not provide all the nutrients. Therefore, it is essential to enrich the vegetables with essential nutrients. There

DOI: 10.1201/9781003606710-8

TABLE 8.1
List of Vegetables According to the Metabolites or Inorganic Nutrients They Contain and Their Use in Prevention and/or Cure during Various Health Conditions

Category	Sub-Category	Vegetables	Health Benefits	Key reference
Carbohydrate rich	Soluble in water (digestible)	Beets, sweet potatoes (rich in soluble sugars) Onion, garlic, leeks (rich in inulins) • Carrots, fava beans, tomatoes (rich in pectin)	• Quick energy • Beneficial modifications of gut microbiota composition and functions and improved nutritional behaviour • Improve starch digestibility by increasing the activity of amylase; impart anti-hyperglycaemic, anti-hyperlipidaemic, anti-obesity, immunomodulatory, gastro-protective, and anti-tumour activities	Arazi and Eghbali (2021), Mohanraj and Sivasanakr (2024) Hiel et al. (2019) Valladares and Vio (2025)
	Insoluble in water (rich in cellulose, hemicellulose, gums, mucilage, non-carbohydrate components of lignin, resistant starch)	• Bitter gourd, beetroot, fenugreek leaves, lady finger, raw spinach, eggplant (dietary fibre-rich) • Broad beans, white potatoes, sweet potatoes, sweetcorn, yam (rich in resistant starch)	• Reduce appetite, lower variances in blood sugar, lower total and LDL cholesterols, add bulk to stool, balance intestinal pH, stimulate intestinal fermentation leading to the production of short-chain fatty acids. • Body weight management; prevention of obesity, diabetes, and colon cancer.	Dhingra et al. (2012) Birt et al. (2013)
Protein rich	-	Soybean, garden pea	Modulate gut microbiota, cholesterol-lowering effects.	Busnelli et al. (2018)
Lipids and fatty acids rich	• Saturated fatty acids (C16:0 and C18:0), MUFA (C16:1 and C18:1), PUFA (C18:2 and C18:3) • Glyco-glycerolipids	• Red cabbage, soya sprouts, parsley, French beans, eggplant, green pepper; kale, artichoke, radish sprout, chickpea; Shiitake, champignons, oyster mushroom, fennel, tarragon, green lettuce, zucchini, broccoli • Spinach	• Reduce the risk of chronic diseases like cardiovascular and cancer. • Inhibit the cancer cell growth	Vidrih et al. (2009) Maeda et al. (2010)
Vitamin rich	Vitamin C	Embelica officinalis Garten	Help in lowering the triglyceride and LDL content, as well as increasing HDL	Kavita et al. (2016)
Secondary metabolites rich	Secondary metabolites-rich plant extract	Leaves of African potato	Showed caspase-3 activity and mitochondrial membrane depolarization rate comparable with commercially available anticancer drug, Doxorubicin, in the treatment of breast and skin cancer.	Kasumbwe and Mohanlall (2025)
Mineral rich	Magnesium	Non-starchy vegetables	Lowers the risk of colorectal cancer.	Wark et al. (2012)

are strategies to increase the metabolite or mineral content of the vegetables. These strategies include agronomic techniques (ATs), conventional breeding techniques (CBTs), and new breeding technologies (NBTs). Such enrichment is called as biofortification (Buturi et al., 2021). In addition to this, some post-harvest management (which are known to preserve the nutritional status of the fruit and vegetables) practices can also be used to serve the same objective (Singh et al., 2024). In a direct manner, the term biofortification refers to the enhancement of macro or micro-elements. However, fortification of such elements also leads to the qualitative and quantitative improvement of other metabolites associated with them. Biofortification of vegetables using ATs starts with good agronomical practices where the required mineral supplement is administered to the plant and reproductive organs via foliar spray of specific minerals. This mineral supplement may be the direct requirement, or such supplement is required for proper functioning of particular enzyme engaged in the synthesis of other nutrients. Some of the micronutrients are absorbed by the plant in specific form; otherwise their bioaccumulation in vegetative parts will be hindered. This is due to the physiology of the root architecture in which specific transporter proteins are key players. These proteins can only interact with specific form of the minerals, and therefore such minerals need to be supplied in their ionic forms such as $^{+}$, $^{2+}$, or $^{3+}$ (White & Broadley, 2009). Unlike horticultural crops, vegetables are sown with the intensification (high-density plantation) which requires the fertigation in higher amount. Foliar sprays are also preferred sometimes for the plants whose root systems are poorly developed, or the minerals present in the soil do not comply with the physicochemical or biological properties of the soil. One such example is the application of iodine on *Capsicum annum* L. plants under field conditions which showed an enhanced iodine uptake and the pepper contained significantly higher amounts of iodine than recommended by WHO. It was also observed that fortification of iodine enhanced the soluble sugar content and ascorbic acid content of the fruit, suggesting the nutritional enrichment (Li et al., 2017). In green leafy vegetables like spinach, biofortification of selenium and iodine showed significantly higher values of ascorbic acid, total phenol, and pigments like carotenoids and chlorophylls, as well as other essential nutrients in the leaves. The result indicated that the biofortification of these minor elements at agronomic practice stage can significantly boost the nutritional value of spinach (Eslamiparvar et al., 2025). Equivalent results were observed when a foliar spray of combined nitrogen and zinc was applied on broccoli plants. The results showed an increased accumulation of Ca, Fe, Mg, Zn, glucosinolates, and total phenols in leaves (Rivera-Martin et al., 2021). The use of microbes as biofertilizers is also a part of the GAP nowadays. The classical example of such case is the application of arbuscular mycorrhizal fungi in a variety of field crops (Thakur et al., 2022). Zinc-solubilizing bacteria had been applied in tomato plants to increase the bioavailability of zinc that could result in increased plant growth and vegetation (Karnwal, 2021). Similarly, Fe solubilizing plant growth promoting rhizobacteria were applied on potato plants, and the result showed increased starch, vitamin C, and relative water content of the tuber (Mushtaq et al., 2022).

One of the most rapid and reliable sources of nutrition from plant are microgreens. They are seedlings, grown under controlled environment with high nutritional value. Besides these, they have several advantages such as they can be farmed without land, freshness, customized nutritional profile, and year-round availability (Seth et al., 2025). Particularly for microgreens, seed priming is the preferred way for the biofortification. By this, seedling quality in terms of phytochemicals increases significantly. Moreover, reduction in phytic acid was also observed in zinc-fortified seedlings of peas and sunflowers (Poudel et al., 2023). Application of different carbon nanomaterials as a seed priming treatment on a variety of crops including vegetables showed enhanced germination rate and dry weight of the seedlings (Fonseca et al., 2024). Apart from the carbon nanomaterials-based seed priming, the seed treated with magnetic field also showed enhanced germination rate and biochemical properties in garlic seeds. The treatment is cheaper and chemical-free and is a promising alternative for the sustainable solution for agronomical practices for biofortification (Ali et al., 2025). Agronomical practices are treatment and effect base approaches for the biofortification of nutritional components in vegetables (Figure 8.1). However, the genetic basis of the plant can also be modulated

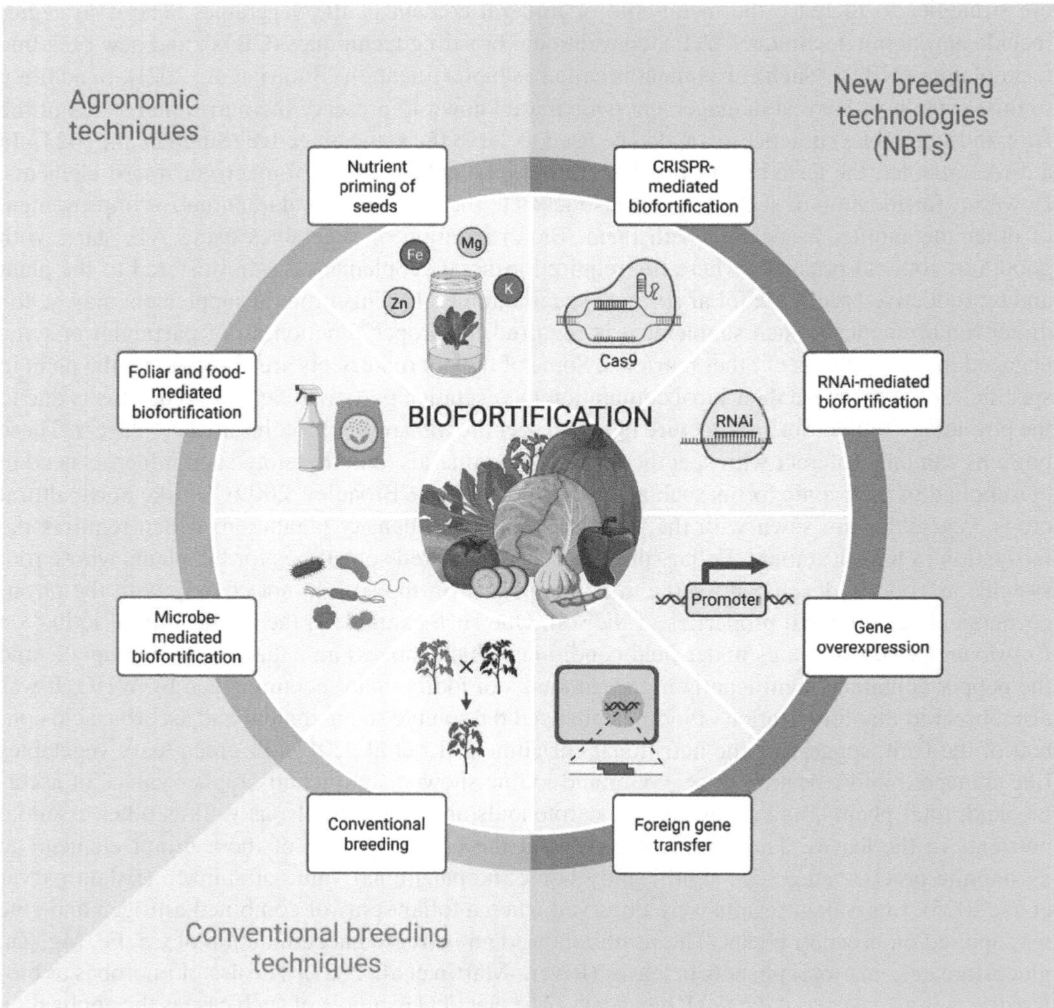

FIGURE 8.1 Conventional and modern techniques to enhance the nutritional spectrum of the vegetables.

to enhance the metabolite production. The strategy starts with the selection of superior genotypes with higher production of the desired compound and strategically inserting the responsible gene(s) into the existing genotypes with other important characteristics. The development of new variety can take years of breeding cycles and extensive screening programmes. Genomic selection in vegetable crops is done before commencing the actual breeding program. Such selection helps in the primary screening. This includes molecular, biochemical, and phenotypic characterization of the plant (Lee et al., 2024). This practice has been implemented in major vegetable crops like tomato (Khan et al., 2017), potato (Mahgoub et al., 2015), chilli (Litoriya et al., 2009), cauliflower (Rakshita et al., 2021), cabbage (Parkash et al., 2023), eggplant (Kaur et al., 2020). garlic (Pasupula et al., 2024), and onion (Bal et al., 2021). Molecular marker-assisted screening has a significant impact on such programmes (Weiss & Gruda, 2025). Ilodibia et al. (2015) developed F1 hybrid from the interspecific hybridization of *Capsicum annuum* L. and *Capsicum frutescens* L. The F1 hybrid showed increased dry weight, carbohydrates, fibre, lipid, protein, and ash content in leaves, seeds, and fruits as compared to their parental lines, although secondary metabolites are equally important in chilli which contribute towards pungency. *Pusa Jwala* is the chilli variety developed by IARI with high pungency (more than 1% of capsaicin content). For the richness of secondary metabolite in leafy vegetables like spinach, *Pusa Bharti* variety has been developed which has high iron and vitamin A content (Nehra & Verma, 2025). Conventional and advanced breeding programmes have already been developed in the important vegetable crops like and cauliflower (Natalini et al., 2021). Recent studies also showed that changing breeding strategies from conventional (natural cycle of plants) to speed breeding (induction of flowering in controlled environments) helps to reduce time for the release of superior genotype (Potts et al., 2023). This is done by modulating the physical parameters such as temperature, light, photoperiodic regime, and humidity, required for the transformation from vegetative to flowering stage (Acharya et al., 2023).

Compared to CBTs, the NBTs are costly, skill-intensive, regimented, but fast. Besides these, in conventional breeding, the new line or genotype is developed from two parents, and gametic cells are involved in the development. Compared to that, NBTs rely mostly on the source of the gene and plant tissue culture operations, and gametic cells are rarely involved; instead, somatic cells are used in order to maintain the fidelity of the superior genotype taken for the experimental purposes (Rai et al., 2022; Singh et al., 2021). Genetic engineering and genome editing using CRISPR-Cas9 are two leading approaches widely used under NBTs. By overexpressing important biosynthetic and transporter genes that increase the accumulation of vitamins, antioxidants, and minerals, genetic engineering has been successfully used to improve the nutritional quality of vegetables. For instance overexpressing phytoene synthase (PSY) or lycopene β-cyclase in tomatoes and carrots increases the content of β-carotene (pro-vitamin A), while increasing GDP-mannose pyrophosphorylase raises the levels of vitamin C in leafy vegetables, and introducing strong promoters to drive ferritin expression increases iron storage and bioavailability (Dutta & Chatterjee, 2024). On the other hand, with genome editing using Clustered Regularly Interspaced Short Palindromic Repeats (CRISPR) – Cas9 (CRISPR Associated Protein 9), a new mutation is created in the target gene within the same organism, and the mutation is beneficial. Jinek et al. (2012) synthesized a new molecule, called the single-guide RNA, which combines key features of the two bacterial RNAs and directs Cas9 to cut a specific site in DNA. For their work on CRISPR-Cas9 genome editing, Emmanuelle Charpentier and Jennifer A. Doudna were awarded the Noble Prize in 2020. The CRISPR-Cas9 genome editing tool consists of a plasmid construct with a short guide RNA (20 nucleotides) complementary to the target sequence (in this case, the sequence codes for the active site), followed by a scaffold sequence to which the Cas9 protein – which is known to cause double strand breaks in DNA – is attached. The constitutive promoter drives both the guide RNA sequences and the Cas9 protein. Thus, the process begins with the construction of a guide RNA sequence (from accessible or isolated gene sequences), followed by the preparation and transfer of the plasmid containing the aforementioned machinery into *Agrobacterium*. Genome editing techniques produce mutation in the target gene, and, therefore, undesired genes are targeted in vegetables which indirectly improves the nutritional quality. Fresh fruits and vegetables often experience enzymatic browning because the enzyme polyphenol oxidase (PPO) is released. The food becomes unhealthy due to the product. Additionally, the enzyme is released when the product is being stored. Because pH, temperature, and oxygen have a significant impact on enzyme activity, they are treated with specific chemical or physical agents that negatively affect the food's nutritional value and organoleptic qualities. In this regard, Maioli et al. (2020) effectively edited three genes (*SmelPPO4*, *SmelPPO5*, and *SmelPPO6*) by targeting the *SmelPPO1-10* genes in eggplant. Up to the second generation, these lines displayed decreased PPO activity.

8.3 TECHNOLOGICAL ADVANCES FOR THE PRESERVATION OF NUTRIENTS

As vegetables are harvested and stored before being transported to the market, it is also essential to maintain the freshness and nutritional quality of the fruit. Post-harvest management includes sorting, storage, packaging, transport, and market outreach. At every stage, there are chances of loss of crops due to several factors. These losses are bruising, decay of the fruit, physical damage, disease infestation, high or low temperatures stress, over-ripening, rodents, microbial growth, contamination, and enzymatic browning (Verma et al., 2025). Harvesting the fruits and vegetables at proper maturity stage is the first important aspect of preservation of nutritional quality contained within (Hailu & Derbew, 2015). Exotic fruits and vegetables, which are asynchronously matured on the plant, may be treated with abscisic acid before harvest. Such practices not only induce the synchronous ripening but also help to preserve the nutritional value of the produce (Lama et al., 2019; Soto et al., 2013). Apart from such techniques, various technologies have been developed to preserve the nutritional value of fruits and vegetables. Advancement in post-harvest management allowed farmers to sell their produce across the globe and hold their commodities till the market touches the desired price. The decay in the vegetables or any other food items during post-harvest attributes towards the changes in physical parameters of the environment which include temperature, humidity, carbon dioxide, ethylene, and other volatile substances. All these parameters can be monitored using biosensors. This technology is known as IoT (Internet of Things). The sensors placed at relevant locations in the storage facility send the data in real time to farmers. By combining IoT with predictive analytics, farmers can act proactively to prevent losses, improve efficiency, and reduce food waste (Nayak & Mishra, 2025). Treatment is equally important as monitoring during post-harvest management. To reduce the post-harvest waste of fruits and vegetables, simple and advanced methodologies have been developed. These include washes with chlorinated water; sorting and grading; pre-cooling such as hydro cooling, forced air cooling, room cooling, evaporative forced air cooling, and ice packaging; commodity coating with natural compounds such as chitosan, alginate, and pectin; chemical and biocidal treatments; ozone treatment; ultrasound treatment; pulse electric field treatment; high-pressure processing; and cold plasma techniques (Rajapakshe et al., 2025).

8.4 CONCLUSION AND FUTURE PROSPECTS

People today are increasingly getting conscious of the significance of nutrition in food, and eating a balanced diet has become a regular aspect of modern life. Nowadays, food is viewed as a source of health as well as a means of satisfying hunger. Due to their high nutritious content – from simple minerals to complex compounds – and ability to fend against chronic illnesses, vegetables are essential. Their diverse range, which includes leafy greens, roots, and tubers, yields a wealth of beneficial natural compounds. Dietary fibre, which is found in most veggies, aids in blood sugar regulation, promotes digestion, and keeps you feeling full – all of which make them perfect for managing weight. To boost vegetables' nutritional value, researchers are also focusing on biofortification and better farming techniques. These methods guarantee that vegetables maintain their inherent health advantages while also offering higher vitamin and mineral content to satisfy contemporary dietary requirements. The health benefits of various vegetable families vary. This nutritional diversity has led to advancements in farming methods which increase quality, enable year-round output, and increase farmers' profits. In this regard, the nutritional spectrum of vegetables not only covers the nutritional values but also emphasizes on enrichment and management strategies. As it covers different fields like agronomy, breeding, and biotechnology, the collaborative work among the institutions will bring the significant outcomes in a most sustainable way covering both the ecology and economy.

REFERENCES

Acharya, S., Manisha, Chatterjee, S. and Nayak, P., 2023. Speed breeding in vegetable crops. In Biradar, N., Shah, R. A. and Ahmad, A. (eds.), *Recent Advances in Agricultural Science and Technology* (pp. 330–336). ICAR-IGFRI, SRRS & NADC and country - India.

Ali, K.A., Junaid, M., El-Meligy, M.A., Dar, M.A., Altarawneh, R., Ul-Haq, Z. and Mahmoud, H.A., 2025. Magnetic field treatment as a catalyst for enhanced germination, growth, and biochemical properties of garlic seeds. *Scientia Horticulturae*, *350*, p.114356.

Arazi, H. and Eghbali, E., 2021. Possible effects of beetroot supplementation on physical performance through metabolic, neuroendocrine, and antioxidant mechanisms: A narrative review of the literature. *Frontiers in Nutrition*, *8*, p.660150.

Bal, S., Maity, T.K. and Sharangi, A.B., 2021. Morphological and biochemical characterization of onion (Allium cepa L.) germplasm by principal component analysis. *Journal of Pharmacognosy and Phytochemistry*, *10*, pp.121–124.

Birt, D.F., Boylston, T., Hendrich, S., Jane, J.L., Hollis, J., Li, L., McClelland, J., Moore, S., Phillips, G.J., Rowling, M. and Schalinske, K., 2013. Resistant starch: Promise for improving human health. *Advances in Nutrition*, *4*(6), pp.587–601.

Busnelli, M., Manzini, S., Sirtori, C.R., Chiesa, G. and Parolini, C., 2018. Effects of vegetable proteins on hypercholesterolemia and gut microbiota modulation. *Nutrients*, *10*(9), p.1249.

Buturi, C.V., Mauro, R.P., Fogliano, V., Leonardi, C. and Giuffrida, F., 2021. Mineral biofortification of vegetables as a tool to improve human diet. *Foods*, *10*(2), p.223.

Dhingra, D., Michael, M., Rajput, H. and Patil, R.T., 2012. Dietary fibre in foods: A review. *Journal of Food Science and Technology*, *49*(3), pp.255–266.

Dutta, A. and Chatterjee, A., 2024. Biotechnological approach to improve nutritional quality of plants. *International Journal of Agriculture and Nutrition*, *6*(2), pp.71–74.

Eslamiparvar, A., Hosseinifarahi, M., Amiri, S. and Radi, M., 2025. Combined bio fortification of spinach plant through foliar spraying with iodine and selenium elements. *Scientific Reports*, *15*(1), p.6722.

Fonseca, J.D.D.S., Wojciechowska, E., Kulesza, J. and Barros, B.S., 2024. Carbon nanomaterials in seed priming: Current possibilities. *ACS Omega*, *9*(45), pp.44891–44906.

Hailu, G. and Derbew, B., 2015. Extent, causes and reduction strategies of postharvest losses of fresh fruits and vegetables–A review. *Journal of Biology, Agriculture and Healthcare*, *5*(5), pp.49–64.

Hiel, S., Bindels, L.B., Pachikian, B.D., Kalala, G., Broers, V., Zamariola, G., Chang, B.P., Kambashi, B., Rodriguez, J., Cani, P.D. and Neyrinck, A.M., 2019. Effects of a diet based on inulin-rich vegetables on gut health and nutritional behavior in healthy humans. *The American Journal of Clinical Nutrition*, *109*(6), pp.1683–1695.

Ilodibia, C.V., Ugwoke, C.E., Egboka, T.P., Akachukwu, E.E., Chukwuma, U.M. and Aziagba, B.O., 2015. Breeding pepper for enhanced food nutrients. *Asian Journal of Crop Science*, *7*(3), pp.214–218.

Jha, A.K., 2024. Ethnic vegetables as nutraceuticals: A promising source of bioactive compounds for disease prevention and management. *Journal of Applied Biochemistry and Microbiology*, *8*(1).

Jinek, M., Chylinski, K., Fonfara, I., Hauer, M., Doudna, J.A. and Charpentier, E., 2012. A programmable dual-RNA–guided DNA endonuclease in adaptive bacterial immunity. *Science*, *337*(6096), pp.816–821.

Karnwal, A., 2021. Pseudomonas spp., a zinc-solubilizing vermicompost bacteria with plant growth-promoting activity moderates zinc biofortification in tomato. *International Journal of Vegetable Science*, *27*(4), pp.398–412.

Kasumbwe, K. and Mohanlall, V., 2025. In vitro cytotoxicity and phytochemical analysis of the crude extract from Hypoxis hemerocallidea and its apoptosis-inducing effect on breast and skin can. *Brazilian Journal of Biology*, *85*, p.e298009.

Kaur, A., Kaur, N., Bassi, G., Kaur, N. and Dhatt, A.S., 2020. Morphological and molecular markers based assessment of genetic diversity in eggplant. *Indian Journal of Horticulture*, *77*(1), pp.116–124.

Kavita, M.B., Mallika, K.J. and Poornima, B., 2016. A clinical study on effect of Amalaki (Indian gooseberry) as food supplement in dyslipidemia. *International Journal of Research in Ayurveda and Pharmacy*, *7*(4), pp.59–64.

Khan, M.A., Butt, S.J., Nadeem, F., Yousaf, B. and Javed, H.U., 2017. Morphological and physico-biochemical characterization of various tomato cultivars in a simplified soilless media. *Annals of Agricultural Sciences*, *62*(2), pp.139–143.

Lama, K., Modi, A., Peer, R., Izhaki, Y. and Flaishman, M.A., 2019. On-tree ABA application synchronizes fruit ripening and maintains keeping quality of figs (Ficus carica L.). *Scientia Horticulturae*, *253*, pp.405–411.

Lee, A.M.J., Foong, M.Y.M., Song, B.K. and Chew, F.T., 2024. Genomic selection for crop improvement in fruits and vegetables: A systematic scoping review. *Molecular Breeding*, *44*(9), p.60.

Li, R., Li, D.W., Liu, H.P., Hong, C.L., Song, M.Y., Dai, Z.X., Liu, J.W., Zhou, J. and Weng, H.X., 2017. Enhancing iodine content and fruit quality of pepper (Capsicum annuum L.) through biofortification. *Scientia Horticulturae*, *214*, pp.165–173.

Litoriya, N., Modi, A.R. and Talati, J.G., 2009. Varietal identification of chilli (Capsicum annuum L.) using randomly amplified polymorphic DNA markers. *Indian Journal of Agricultural Biochemistry*, *22*(2), pp.83–86.

Maeda, N., Yoshida, H. and Mizushina, Y., 2010. Spinach and health: Anticancer effect. In *Bioactive Foods in Promoting Health* (pp. 393–405). Academic Press.

Mahgoub, H.A.M., Eisa, G.S.A. and Youssef, M.A.H., 2015. Molecular, biochemical and anatomical analysis of some potato (Solanum tuberosum L.) cultivars growing in Egypt. *Journal of Genetic Engineering and Biotechnology*, *13*(1), pp.39–49.

Maioli, A., Gianoglio, S., Moglia, A., Acquadro, A., Valentino, D., Milani, A.M., Prohens, J., Orzaez, D., Granell, A., Lanteri, S. and Comino, C., 2020. Simultaneous CRISPR/Cas9 editing of three PPO genes reduces fruit flesh browning in Solanum melongena L. *Frontiers in Plant Science*, *11*, p.607161.

Mohanraj, R. and Sivasankar, S., 2014. Sweet potato (Ipomoea batatas [L.] Lam)-A valuable medicinal food: A review. *Journal of Medicinal Food*, *17*(7), pp.733–741.

Mushtaq, Z., Asghar, H.N., Zahir, Z.A. and Maqsood, M., 2022. The interactive approach of rhizobacteria and l-tryptophan on growth, physiology, tuber characteristics, and iron concentration of potato (Solanum tuberosum L.). *Journal of Plant Growth Regulation*, *41*(3), pp.1359–1366.

Natalini, A., Acciarri, N. and Cardi, T., 2021. Breeding for nutritional and organoleptic quality in vegetable crops: The case of tomato and cauliflower. *Agriculture*, *11*(7), p.606.

Nayak, A., Mishra, N., 2025. Innovations in post-harvest technology and smart storage systems. *National Journal of Plant Sciences and Smart Horticulture*, pp.31–37.

Nayak, S.N., Aravind, B., Malavalli, S.S., Sukanth, B.S., Poornima, R., Bharati, P., Hefferon, K., Kole, C. and Puppala, N., 2021. Omics technologies to enhance plant based functional foods: An overview. *Frontiers in Genetics*, 12, p.742095.

Nehra, M.R. and Verma, M., 2025. Chemical and molecular advances in vegetable breeding: From genomics to nutritional enhancement. *Chemical Science Review and Letters*, *14*(55).

Parkash, C., Kumar, S., Thakur, N., Singh, S. and Sharma, B.B., 2023. Cabbage: Breeding and genomics. *Vegetable Science*, pp.231–243.

Pasupula, K., Verma, P., Zimik, M., Kaur, C., Vasudev, S. and Khar, A., 2024. Morphological, biochemical and molecular characterization of short-day tropical Indian garlic (Allium sativum L.). *Heliyon*, *10*(18).

Potts, J., Jangra, S., Michael, V.N. and Wu, X., 2023. Speed breeding for crop improvement and food security. *Crops*, *3*(4), pp.276–291.

Poudel, P., Di Gioia, F., Lambert, J.D. and Connolly, E.L., 2023. Zinc biofortification through seed nutri-priming using alternative zinc sources and concentration levels in pea and sunflower microgreens. *Frontiers in Plant Science*, *14*, p.1177844.

Rai, A.C., Kumar, A., Modi, A. and Singh, M., eds., 2022. *Advances in Plant Tissue Culture: Current Developments and Future Trends*. Academic Press.

Rajapakshe, P., Rathnasinghe, N., Guruge, K., Nilmini, R., Jayasinghe, R., Karunaratne, V., Wijesena, R. and Priyadarshana, G., 2025. Strategies to minimize post-harvest waste of fruits and vegetables: Current solutions and future perspectives. *Journal of Future Foods*, *6*(3), May 2026, pp.400–412.

Rakshita, K.N., Singh, S., Verma, V.K., Sharma, B.B., Saini, N., Iquebal, M.A., Sharma, A., Dey, S.S. and Behera, T.K., 2021. Agro-morphological and molecular diversity in different maturity groups of Indian cauliflower (Brassica oleracea var. botrytis L.). *PLoS ONE*, *16*(12), p.e0260246.

Rivera-Martin, A., Reynolds-Marzal, D., Martin, A., Velazquez, R. and Poblaciones, M.J., 2021. Combined foliar zinc and nitrogen application in broccoli (Brassica oleracea var. italica L.): Effects on growth, nutrient bioaccumulation, and bioactive compounds. *Agronomy*, *11*(3), p.548.

Seth, T., Mishra, G.P., Chattopadhyay, A., Deb Roy, P., Devi, M., Sahu, A., Sarangi, S.K., Mhatre, C.S., Lyngdoh, Y.A., Chandra, V., Dikshit, H.K., 2025. Microgreens: Functional food for nutrition and dietary diversification. *Plants*, *14*(4), p.526.

Singh, S., Sharma, S., Panwar, S., Krishna, H., Singh, S.K., Bahadur, A. and Behera, T.K., 2024. Post-harvest management and value addition in vegetable crops. *Vegetable Science*, *51*, pp.34–42.

Singh, S., Yadav, S. and Singh, A., 2021. Recent approaches for breeding vegetable crops for quality, rich in nutrients and nutraceuticals. *Journal of Pharmacognosy and Phytochemistry*, *10*(1), pp.2061–2071.

Soto, A., Ruiz, K.B., Ravaglia, D., Costa, G. and Torrigiani, P., 2013. ABA may promote or delay peach fruit ripening through modulation of ripening-and hormone-related gene expression depending on the developmental stage. *Plant Physiology and Biochemistry*, *64*, pp.11–24.

Sudha Rani, M., Satish, Y., Rani, C., Prasad NVVSD, B.S., Sri Lakshmi, B. and Ratna Kumari, S., 2018. History, status and impact of genetically modified crops in India. *International Journal of Chemical Studies*, *6*(5), pp.2735–2739.

Thakur, V., Sharma, A., Sharma, P., Kumar, P. and Shilpa, 2022. Biofortification of vegetable crops for vitamins, mineral and other quality traits. *The Journal of Horticultural Science and Biotechnology*, *97*(4), pp.417–428.

Tripathi, A.N., Dubey, R.K. and Singh, J., 2024. Historical and spiritual perspectives of vegetables in India. *Indian Horticulture*, *68*(2), pp.27–29.

Valladares, L. and Vio, F., 2025. Pectin and its beneficial effect on health: New contributions in research and the need to increase fruits and vegetables consumption – a review. *International Journal of Molecular Sciences*, *26*(14), p.6852.

Verma, S., Kumar, S., Kumar, V., Yadav, S., Singh, P., Saurabh, A., Kumar, A. and Jain, M., 2025. Recent advancements and innovations in post-harvest handling, storage, and technology for vegetables: A review. *Archives of Current Research International*, *25*(2), pp.161–180.

Vicente, A., Ortiz, C., Sozzi, G., Manganaris, G. and Crisosto, C., 2014. Nutritional properties of fruits and vegetables. *Postharvest Handling a Systems Approach*, p.60e122.

Vidrih, R., Filip, S. and Hribar, J., 2009. Content of higher fatty acids in green vegetables. *Czech Journal of Food Sciences*, *27*(1), pp. S125–S129.

Wark, P.A., Lau, R., Norat, T. and Kampman, E., 2012. Magnesium intake and colorectal tumor risk: A case-control study and meta-analysis. *The American Journal of Clinical Nutrition*, *96*(3), pp.622–631.

Weiss, J. and Gruda, N.S., 2025. Enhancing nutritional quality in vegetables through breeding and cultivar choice in protected cultivation. *Scientia Horticulturae*, *339*, p.113914.

Welbaum, G.E., 2015. Vegetable history, nomenclature, and classification. In *Vegetable Production and Practices* (pp. 1–15). Wallingford, UK: CABI.

White, P.J. and Broadley, M.R., 2009. Biofortification of crops with seven mineral elements often lacking in human diets–iron, zinc, copper, calcium, magnesium, selenium and iodine. *New Phytologist*, *182*(1), pp.49–84.

Włodarczyk, Z., 2020. The seven plant species—A basis of nutrition of ancient Israel. *Biomed. J. Sci. Tech. Res*, 25, pp.19369–19373.

9 Clinical Studies on the Role of Vegetable-Based Nutrients in Disease Prevention and Treatment

Prity Chakravarty, Pankaj Barman, and Rituraj Konwar

9.1 INTRODUCTION

Non-communicable diseases (NCDs) such as CVDs, T2DM, cancer, and other metabolic syndromes are among the leading causes of morbidity and mortality worldwide. A substantial body of epidemiological, experimental, and clinical trial evidence suggests that vegetable-based nutrients encompassing vitamins, minerals, phytochemicals, dietary fibre, and other bioactive compounds found in vegetables and plant foods play a key role in preventing, slowing, or even reversing certain disease processes (Ramya & Patel, 2019; Liu, 2013).

Vegetables provide multiple classes of bioactive compounds, including antioxidants (vitamin C and vitamin E, carotenoids), flavonoids, phenolic acids, phytoestrogens, sulphur-containing compounds (e.g. those in cruciferous vegetables, onions, garlic), and dietary fibre (Liu, 2013). These agents can exert beneficial effects via several mechanisms by reducing oxidative stress, lowering inflammation, improving lipid profile and endothelial function, modulating glucose metabolism, enhancing immune responses, and altering gene expression (Figure 9.1) (Yu et al., 2024).

While much of the evidence to date has come from observational cohort studies and meta-analysis, there is a growing number of clinical trials which directly examine the effect of increasing vegetable intake or supplementing with specific vegetable-based compounds on the human population. These trials provide more rigorous evidence about dose, timing, and measurable outcomes. Examples include randomized clinical trials of fruit and vegetable-based nutraceuticals improving cognitive functions in healthy adults, interventions increasing cruciferous or leafy green vegetable intake in cancer survivors, and trials using vegetable juice in combination with dietary advice to improve cardiovascular risk factors (Cartmel et al., 2005).

However, despite promising findings, clinical evidence is still inconclusive. In many cases, sample sizes are small, durations are relatively short, and outcomes are surrogate or intermediate (biomarker) rather than 'hard' clinical endpoints (disease incidence, mortality). Moreover, there are methodological challenges: heterogeneity of intervention (type of vegetable and extract, form, whole food vs extract vs juice); variance in adherence; bioavailability issues; as well as confounding in observational studies. Some systematic reviews caution that while increasing fruit and vegetable consumption reliably lowers risk for CVD and stroke, the evidence is less consistent for cancer or type 2 diabetes when confounding factors are considered.

As interest in lifestyle and nutritional intervention grows, especially in light of rising healthcare costs and the need for low-cost and accessible preventive strategies, it is essential to critically examine clinical cases in which vegetable-based nutrients have been used for prevention or treatment. Such case studies shed light on what works, for whom, under what conditions, and what gaps remain to achieve translation into public health or clinical practice.

 DOI: 10.1201/9781003606710-9

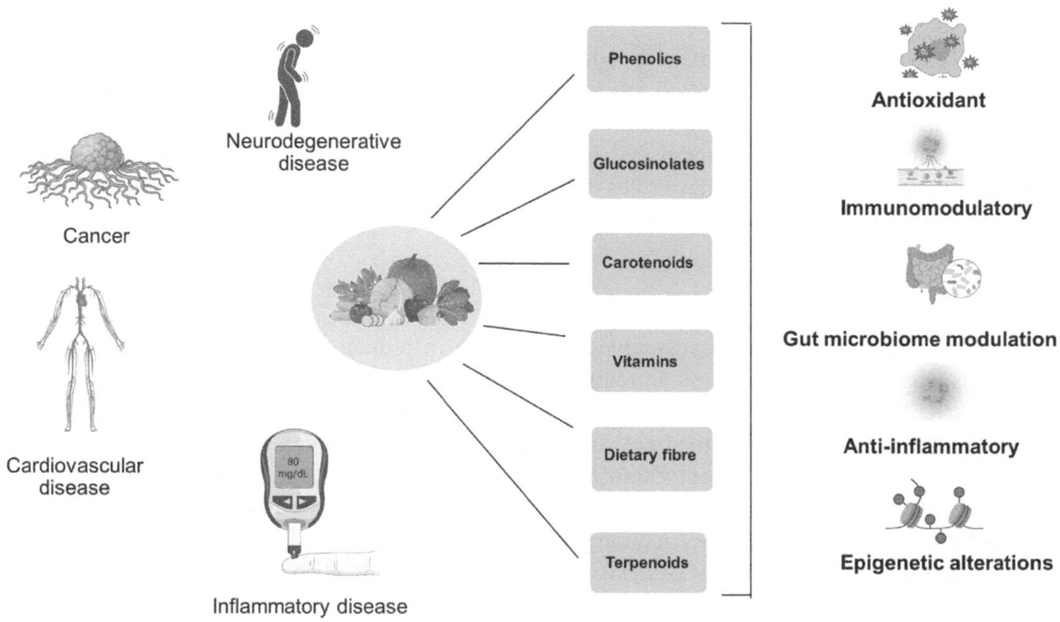

FIGURE 9.1 Vegetable-based nutrients and their pharmacological prospects in human health.

In this chapter, we present selected clinical studies and trials which illustrate the role of vegetable-derived nutrients in disease prevention and treatment. We analyse their designs; interventions (type of vegetable or extract, doses, duration); outcomes (biomarkers, clinical endpoints); and limitations, to draw lessons for future research and for possible implementation in clinical and public health settings.

9.2 PHYTOCHEMICAL COMPONENTS IN VEGETABLES

Phytochemical components from vegetables have certain protective effects on the human body, although some of them are present in trace amounts. When consumed in adequate quantities, vegetables can protect against different stress conditions. Some of the important phytochemicals and component of vegetable having known roles in protecting against various diseases are discussed below.

9.2.1 Dietary Fibre

Dietary fibre includes the macromolecules usually found in plant cell wall components, lignins and polysaccharides (Khanum et al., 2000). Twenty to thirty grams of dietary fibre is recommended for a healthy adult per day (Nishida et al., 2007). Vegetables are a good source of fibre, both soluble and insoluble. These dietary fibres are involved in functions like absorption of nutrients, protection of the gut lining, and maintaining microbiome ecology and physiology of the gut (Dai et al., 2017). The soluble dietary fibres, including pectin and mucilage, are mainly involved in nutrient absorption in the small intestine and are directly associated with lowering inflammatory diseases, cancer, and gastrointestinal diseases (Sonnenberg et al., 2014). The insoluble dietary fibres like cellulose, hemicellulose, and lignin are mostly found in citrus fruits, cereals, and some vegetables; bind to water in the intestines and add bulk to faeces, which is related to alleviation of constipation; and they also facilitate the removal of toxins from the colon (Cui et al., 2019).

9.2.2 Phenolic Compounds

Phenolics are an important group of secondary metabolites, also referred to as polyphenols, characterized by the presence of a hydroxyl group containing aromatic rings in their molecular structure (Croft et al., 1998). Phenolic compounds, like flavonoids (flavanols, anthocyanins, flavones, etc.); phenols (hydroxybenzoic acids and hydroxycinnamic acid); lignans; tannins; and silbenes, are most abundant in vegetables and also have roles in antioxidant, anticancer, anti-inflammatory, and neuroprotective functions (Hu et al., 2022; Khoddami et al., 2013).

9.2.3 Vitamins

Vitamins are micronutrients in vegetables which help in reducing the risk of certain diseases, weight management, digestion, and an improved immune system. They are classified as fat-soluble (vitamins K, E, and D), which are stored in the body, and water-soluble vitamins (C and B complex), which are excreted out of the body and need regular uptake. Vitamin A found in carrots and leafy vegetables is vital for vision; vitamin C found in bell peppers and some other vegetables is involved in boosting immune system, iron absorption, and collagen formation; vitamin K in leafy vegetables like spinach contributes in bone growth and blood clotting; likewise, vitamin B complex helps in DNA synthesis, function of nervous system, and blood cell production (Poiroux-Gonord et al., 2010).

9.2.4 Terpenoids

Terpenoids are a class of plant secondary metabolites having repeating $(C_5H_8)_n$ units, which are responsible for the characteristic fragrances of the plants (Mabou et al., 2021). These compounds have potential functions against various human diseases; for example the widely known antimalarial drug artemisinin is a terpenoid. Other than that, these terpenoids also have antioxidant properties and have applications in the perfume and cosmetics industry as well (Tetali, 2019).

9.2.5 Carotenoids

Beta-carotene, lycopene, zeaxanthin, and lutein are major carotenoids which contribute to characteristic pigments in fruits and vegetables. They are known antioxidants and are involved in protecting the body from free radical attacks and hence from diseases related to the eyes, CVDs, and certain cancers (Ramya & Patel, 2019)

9.2.6 Glucosinolates

Glucosinolates are sulphur-containing compounds present mostly in vegetables of the Brassicaceae family, like broccoli, cabbage, cauliflower, and kale (Horbowicz, 2003). The pungent aroma of these vegetables is a contribution of these compounds. In plants, glucosinolates are mainly involved in the defence system of the plants, and as a dietary source, they are known to have anticancer, anti-inflammatory, and antimicrobial properties (Possenti et al., 2016).

9.3 MECHANISM OF ACTION OF VEGETABLE-BASED NUTRIENTS

Vegetables are an important dietary component from the nutrient point of view as they possess diverse range of compounds. Along with nutritional components, metabolites, vegetables provide bioactive phytomolecules which have diverse and potent roles in various complications and are protective against various clinical complications. These nutrients and phytomolecules possess diverse pharmacological and health-promoting activities, such as anti-inflammatory, antioxidant,

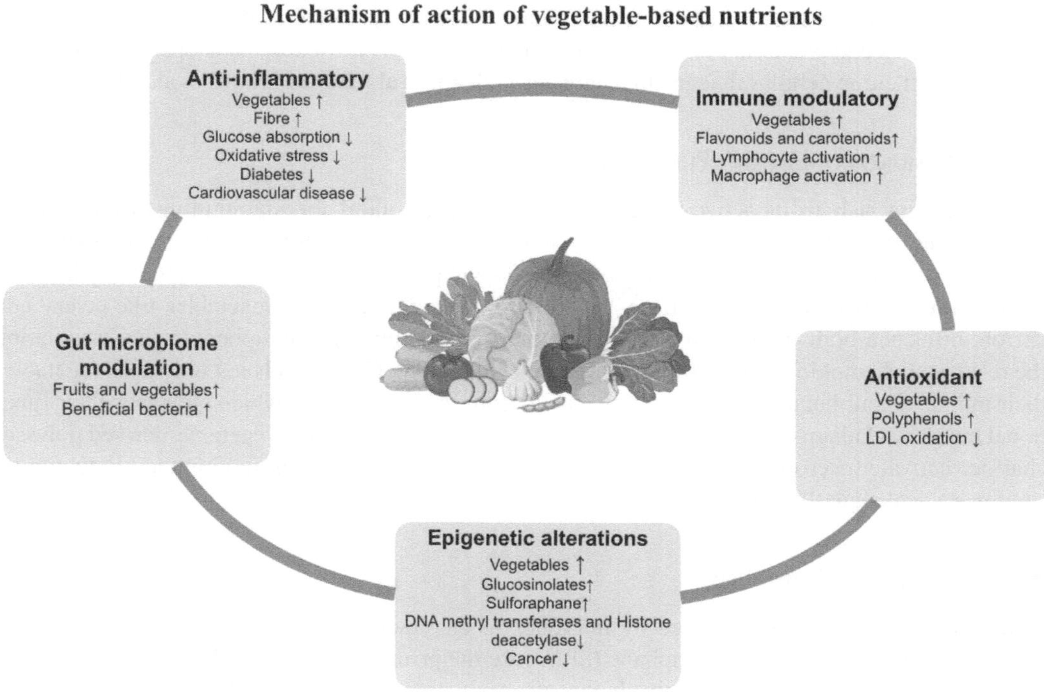

FIGURE 9.2 Vegetable-derived components and their mechanism of action through different biological events in various clinical conditions.

immune-modulatory, epigenetic, and gut microbial modulatory activity, with a clear mechanism in such disease conditions (Figure 9.2). In this section, we have discussed the mechanism of such vegetables and vegetable-derived phytomolecules with different pharmacological activities.

9.3.1 Anti-Inflammatory Effects

The presence of dietary fibre, vitamins, minerals, and other important phytochemicals makes vegetables very crucial for a balanced diet. These compounds, specifically, fibre and carotenoids, alleviate diabetes as fibre reduces the glucose absorption in the intestine, and carotenoids reduce oxidative stress in the cells, which hinders glucose uptake by the cells (Dias et al., 2017; Ramya & Patel, 2019). In a study conducted in the Netherlands by Veer et al., an estimated 16% reduction in cardiovascular-related deaths was observed with increased consumption of vegetables and fruits (Veer et al., 2000). Dietary fibre was also seen to reduce the risk of hypertension and cholesterol levels when consumed regularly (Mazur et al., 2024). A study in asthmatic children and adolescents of Iran showed up to 50% reduction in wheezing with regular intake of vegetables compared to a group who did not take or occasionally took vegetables (Rostampour et al., 2025). However, these studies are still in their preliminary state, limited to a small cohort of samples.

9.3.2 Antioxidant Properties

The major compounds present in vegetables that contribute to their antioxidant properties are carotenoids and polyphenols (Lako et al., 2007). Cruciferous vegetables such as broccoli contain glucosinolates like sulforaphane (antioxidants), which have significant growth inhibitory effects on A-549 lung cancer cell line (Farag & Motaal, 2010). Polyphenols in lettuce have been demonstrated to reduce LDL oxidation, thereby promoting cardiovascular protection in albino rat (Ayo et al., 2024), vitamins and phenolic compounds in peppers can act as pro-oxidants at specific concentrations, strengthening cellular defence mechanisms (Jideani et al., 2021; Howard et al., 2000).

9.3.3 Immunomodulatory Functions

Vegetables are rich in bioactive molecules, which are essential for maintaining overall health and immune function. Vitamin C, which is abundant in peppers, chillies, and green vegetables, improves immune cell function by boosting lymphocyte proliferation and activation in conjunction with vitamin A from carrots (Mitra et al., 2022). Flavonoids found in vegetables like celery and carrots influence both innate and adaptive immunity by stimulating lymphocytes and regulating the release of chemokines and cytokines (Cherng et al., 2008). Carotenoids not only give vegetables their unique colour but also protect immune cells from oxidative stress, which enhances their functional integrity (Jadaun et al., 2022). Furthermore, it has been shown that vegetable-derived polysaccharides activate macrophages, boosting their phagocytic activity and promoting the production of immunological signalling molecules (Gasmi et al., 2023).

9.3.4 Epigenetic Alterations

Glucosinolates, which are degraded by microbes or enzymes into bioactive isothiocyanates like indole-3-carbinol (I3C) and sulforaphane (SFN), are the primary mechanism by which cruciferous vegetables of the Brassicaceae family show cancer preventive effects. SFN lowers nuclear level of DNA methyltransferase 1 (DNMT1), which leads to DNA hypomethylation. It also links chromatin remodelling to cell cycle arrest and reduction of telomerase reverse transcriptase production in prostate cancer models by increasing acetylation of histone H3 (e.g. H3K18ac) and dimethylation of H3K4 (Abbas et al., 2016; Pan et al., 2025). In hepatocellular carcinoma cells, I3C stabilizes p53 and upregulates miR-34a, which targets LDHA to limit aerobic glycolysis, linking metabolic reprogramming with non-coding RNA regulation (Qi et al., 2022). Additionally, glucosinolates are broken down into other anti-carcinogenic compounds like SFN and I3C. These compounds are seen to modulate DNA methyltransferases (DNMTs) and Histone Deacetylases (HDACs), which are often overexpressed in cancers (Royston, 2015). Another polyphenol, curcumin, produced from vegetables, has been demonstrated to modify histone methylation (e.g. lowering H3K27me3 via Enhancer if zeste homolog 2 (EZH2) suppression) in cancer cells and to affect histone acetylation by inhibiting p300 and other Histone acetyltransferases (HATs) (Kondo et al., 2008; Zhao et al., 2018). Additionally, curcumin uses promoter hypomethylation to restore tumour suppressor gene expression while downregulating DNMT expression (Rajendran et al., 2022, Smail et al., 2025). Curcumin, lycopenes, and isothiocyanates are some of the well studied vegetable-derived compounds which are found to be effective against prostate cancer. Their mode of action involves DNA and histone modification, as well as alterations in miRNA (Singh et al., 2024).

9.3.5 Gut Microbiome Modulation

Nutrients derived from vegetables are essential for supporting human physiology, normal metabolic activity, and they also influence gut bacteria composition. In a randomized clinical study, Lakshmanan et al. (2022) showed that supplementing with concentrated phytonutrients derived from fruits and vegetables changes the Short chain fatty acid (SCFA) profiles and microbial community structure, including decreases in *Ruminococcus* spp. in the human gut. Research on lacto-fermented vegetables (Guse et al., 2023; Baron et al., 2024) shows that regular consumption of these products significantly alters the faecal metabolome, especially raising the concentrations of acetate, butyrate, and valerate, and causes no significant changes in diversity of the gut microbiota in healthy Human adults. Further research demonstrates that inulin, a fermentable fibre found in many vegetables, promotes microbiota-dependent activation of immune cells which produce IL-22 and aids in the remodelling of the epithelial barrier, demonstrating direct connections between diet, microbiota, and immunity (Corrêa et al.'s (2023). Furthermore, plant-associated bacteria from raw vegetables can temporarily integrate into the human gut microbiome, highlighting the possible role of the edible plant microbiome in microbial transmission (Wicaksono et al. (2023). All of these studies shows that the gut microbial ecology, SCFA production, and host metabolic and immunological pathways are significantly impacted by vegetable-based nutrition.

9.4 CLINICAL EVIDENCE: PREVENTION AND TREATMENT OF CHRONIC DISEASES

Studies using *In vitro* and *in vivo* animal model give insights into effects of vegetables or vegetable-based nutrients for various disease prevention and medication. Several clinical trials, randomized clinical trials, and case studies (Tables 9.1 and 9.2) have been carried out to validate the nutritional benefits and mechanisms in normal healthy or clinical conditions. Reported clinical trials, or case studies, validated the benefits in clinical conditions like cancer, metabolic disorders, neurodegenerative disorders, inflammation, and CVDs (Figure 9.3). Details of the clinical trials using vegetables and vegetable-based nutrients are discussed below.

9.4.1 Cancer Prevention and Adjunct Treatment

In a study, vegetable and fruit juice reported to reduce concurrent chemotherapy-induced oral mucositis in stage III and IV head and neck carcinoma patients (Chang et al., 2022). Curcumin and piperine, natural compounds extracted, showed anti-proliferative effect against thyroid cancer cells TPC1 (Esposito et al., 2019). Polyphenols from fruits and vegetables, including apigenin, hesperetin, caffeic acid, resveratrol, quercetin, etc., show promising results against different cancers both *in vitro* and *in vivo* (Murakami et al., 2008; Chen et al., 2007; Zarebczan et al., 2011; Prasad et al.,

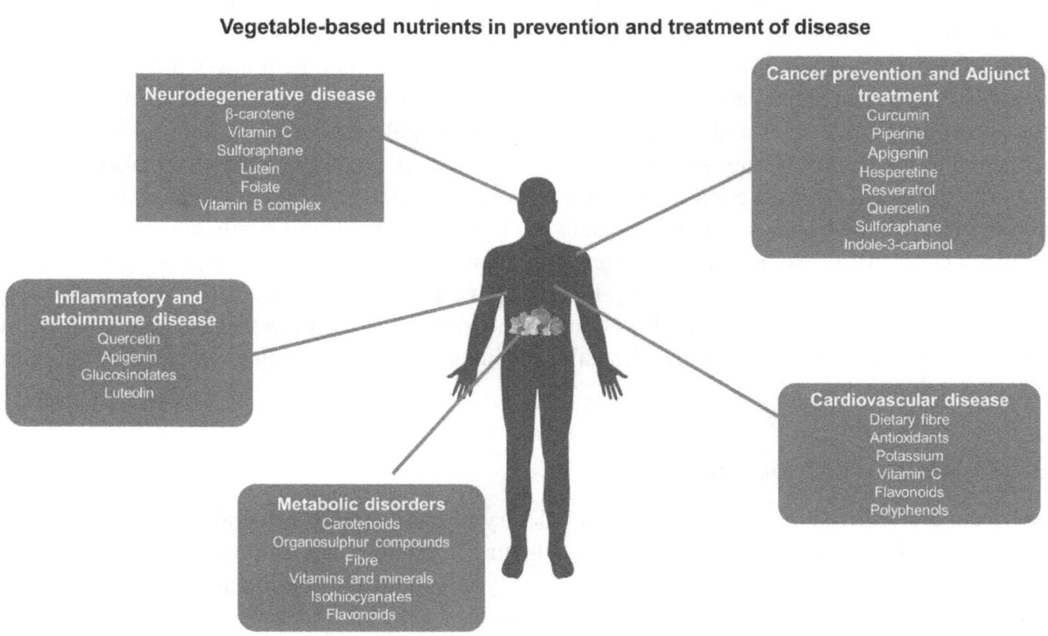

FIGURE 9.3 Different vegetable-derived phytochemicals in physiological disorders.

2011; Asensi et al., 2002). An epidemiological study conducted across continents showed an inverse proportional relationship between the intake of allium vegetables and occurrence of gastric cancer, which is significantly relevent in Asian Population (Dalmartello et al., 2022). Compounds like SFN and indole-3-carbinol, which are found in cruciferous vegetables like broccoli, cauliflower, and Brussels sprouts, have been shown to alter the metabolism of carcinogens, cause cancer cells to undergo apoptosis, and stop tumour growth (Ademosun et al., 2024). Additionally, these substances seem to protect healthy cells from related toxicities while boosting the effectiveness of chemotherapeutic agents. SFN, for instance, has been demonstrated to decrease cardiotoxic effects while simultaneously making tumour cells more sensitive to chemotherapeutic medications like doxorubicin (Jeffery & Araya, 2009). Carotenoids and flavonoids like lycopene, quercetin, and apigenin, which have anti-inflammatory, anti-metastatic, and antioxidant qualities, are found in red and orange vegetables like tomatoes and carrots. By altering important signalling pathways involved in the regulation of cell cycle and apoptosis, these phytochemicals may also work as a support of primary cancer treatment. These findings are supported by epidemiological research, which shows that diets rich in vegetables are associated with lower risk of cancer mortality and recurrence (Liu et al., 2013). Vegetables by themselves cannot cure cancer, but, when included in diet as part of an adjunct therapeutic approach, they may improve treatment results, reduce side effects, and improve quality of life in general (Key et al., 2011). Some brief details of the clinical trials in different types of cancer patients are tabulated in Table 9.1.

9.4.2 Cardiovascular Disease (CVD)

Cardiovascuar Disease (CVD) refers to the group of diseases of vital organs including cardiac muscle (Ischemic heart disease (IHD), stroke and congestive heart failure (CHF)). It is one of the leading cause of death worldwide (Gaziano et al., 2006). Because of high levels of dietary fibre, antioxidants, potassium, and other bioactive phytochemicals, vegetables are recommended for the prevention and cure of CVDs. Lower blood pressure, better lipid profiles, and less arterial stiffness have all been achieved with a regular consumption of a wide variety of vegetables, especially leafy greens, cruciferous vegetables, and root vegetables (Matsumoto et al., 2016). Vegetables high in nitrate, like beetroot and spinach, increase nitric oxide availability and endothelial function, which causes vasodilation and better vascular health (Lidder & Webb, 2013). Vegetables contain antioxidants such as vitamin C, polyphenols, and flavonoids which help lower inflammation and oxidative stress, two major causes of atherosclerosis (Khan et al., 2014). Furthermore, by preventing the absorption of bile acids and inducing the excretion of cholesterol, dietary fibre from vegetables is known to reduce serum cholesterol levels (Bennekum et al., 2005). Certain vegetable types, particularly green leafy and cruciferous vegetables, seem to provide better protection; for instance consuming leafy greens or cruciferous vegetables per day has shown significantly lower risk of CVDs (Pollock et al., 2016). Higher vegetable consumption is associated with lower systolic and diastolic blood pressure, as well as reduced LDL cholesterol and BMI (Yang et al., 2018),. Vegetable-based diets have also been associated with lower rates of heart failure, stroke, and coronary artery disease (He et al., 2007). Some details of the clinical trials involving cardiovascular complications are represented in Table 9.1.

9.4.3 Metabolic Disorders

Nutrients derived from vegetables, especially dietary fibres, vitamins, minerals, and bioactive phytochemicals like polyphenols, OSCs (Organosulphur compounds), and carotenoids are essential for managing and preventing metabolic diseases like type 2 diabetes, obesity, and insulin resistance. This is supported by epidemiological and clinical evidence; higher consumption of vegetables high in phytochemicals is associated with a lower chance of developing metabolic syndrome and its components, such as abdominal obesity and hypertension. Meta-analyses involving vegetable-based

TABLE 9.1
Some Clinical Trials Using Vegetables and Vegetable-Based Compounds in Cancer, Cardiovascular Disease, Inflammatory Disease, and Neurodegenerative Disorders

S No.	Trail Number (NCT)	Condition	Sample size	Dose	Duration	Mode of Action	Reference
				Cancer			
1	NCT00535977	Prostate cancer	20	400 g broccoli sprout extract/week	12 months	Regulation of EGF, TGFβ1, and insulin signalling in GSTM1 positive patients	Traka et al. (2008)
2	NCT01950143	Prostate cancer	Male (49)	300 mL glucoraphanin-rich broccoli soup/week	12 months	Attenuation of oncogenic pathways	Traka et al. (2019)
3	NCT01265953	Prostate cancer	Male (98)	200 μmol daily	4–8 weeks	Downregulation of ARLNC1, AMACR	Zhang et al. (2020)
4	NCT00843167	Breast cancer	Female (54)	180 mg daily glucoraphanin	8 weeks	Downregulation of Ki-67 and HDAC	Atwell et al. (2015)
5	NCT00433797	Prostate cancer	Male (79)	30 mg lycopene-rich tomato/day	3 weeks	Reduced PSA	Paur et al. (2017)
6	NCT01753908	Breast cancer	Female (57)	Indole-3-carbinol tablets	4 weeks	Favourable changes in oestrogen metabolism (increase in 2-hydroxylation)	Saito et al. (2025)
				Cardiovascular Disease			
7	NCT01405898	Hypertension	68	250 mL/day beetroot juice	4 weeks	Reduction in blood pressure	Kapil et al. (2015)
8	NCT01534910	Cardiovascular disease	55	2,400 mg AGE/day	12 months	Decrease-adverse cardiovascular events	Matsumoto et al. (2016)
				Inflammatory disease			
9	NCT03763240.	Diabetes	74	150 μmol/day	12 weeks	Reduced blood sugar levels by 0.3 mmol/l	Dwibedi et al. (2025)
10	NCT01766570	Diabetes	116	1.84 g strawberries and cranberries	6 weeks	Improve insulin sensitivity and prevent an increase in compensatory insulin secretion without impacting plasma lipids, CRP, pro-inflammatory cytokines, and antioxidant capacity	Paquette et al. (2017)
11	NCT04584372	Oxidative stress and inflammation related to HYPERTENSION	15	800 mg beetroot juice/day	4 weeks	Decreased oxLDL/NOx ratio, increase in GSH/GSSG ratio, decrease in monocytes (CD14+ CD16+)	Fejes et al. (2024)
12	NCT02817074	Inflammatory disease	604	Mediterranean diet	3 years	Non-significant	Barnes et al. (2023)
				Neurodegenerative disorders			
13	NCT04084457	Cognition in healthy older participants	61	26g freeze-dried wild blueberry powder/day	12 weeks	Improved vascular and cognitive function	Wood et al. (2023)

dietary patterns have demonstrated significant improvements in insulin sensitivity, lipid profiles, and inflammatory markers in individuals with metabolic syndrome (Zhang et al., 2018; Shin et al., 2014). Cruciferous vegetables containing glucosinolates and their metabolites, isothiocyanate-like SFN, activate Nrf2 pathways, enhancing antioxidant properties, reducing oxidative stress, and improving glucose and fatty acid metabolism in both animal and human trials. SFN, in particular, has shown promise as a therapeutic agent in obesity and type 2 diabetes by modulating IRS-1, attenuating inflammation, and regulating glucose homeostasis (Masoumvand et al., 2025). Furthermore, through AMPK, PI3K/Akt, and other signalling pathways, polyphenols, including flavonoids and carotenoids found in leafy greens, berries, and legumes, further improve insulin sensitivity (Ansari et al., 2024). These findings, which are supported by clinical data, highlight the fact that including vegetables high in fibre and phytochemicals provides a dietary approach which reduces metabolic dysregulation and enhances cardiometabolic health. Additionally, some clinical trials in metabolic disorders are described briefly in Table 9.1.

9.4.4 Inflammatory and Autoimmune Diseases

Bioactive compounds obtained from vegetables are known to reduce inflammation by regulating immune function in autoimmune and inflammatory diseases. Brussels sprouts, broccoli, cabbage, kale, and other cruciferous vegetables are rich in glucosinolates, which bind to the aryl hydrocarbon receptor (AhR) through metabolites including SFN and indole-3-carbinol. In clinical studies, AhR activation decreased proinflammatory cytokines such as TNF-α, IL-6, and IL-1β. It support regulatory T-cell responses and promote intestinal barrier integrity (Schanz et al., 2020, He et al., 2022). Women who consume enough cruciferous foods have demonstrated reductions of approximately 12–25% in these inflammatory markers (Jiang et al., 2014; Tilg, 2015). Meta-analyses of trials showed that higher vegetable consumption significantly lowers circulating CRP and TNF-α levels and enhances γδ-T cell populations, which indicates improved immune function (Neale et al., 2016). Further, plant polyphenols such as quercetin, apigenin, luteolin, and resveratrol present in various vegetables and legumes inhibit NF-κB, COX-2, and iNOS pathways, thus suppressing inflammatory gene expression and promoting antioxidant activity (Karunaweera et al., 2015; Dai et al., 2010). These phytochemicals also help modulate gut microbiota and support SCFAs synthesis like butyrate, which reinforces mucosal immunity, reduces gut permeability, and helps regulate systemic inflammation associated with autoimmune conditions like rheumatoid arthritis and IBD (Yu et al., 2024). Some details of the clinical trials in various inflammatory conditions are represented in Table 9.1.

9.4.5 Neuro-Degenerative Conditions

Vegetable-based nutrient, particularly antioxidants, vitamins, carotenoids, polyphenols, dietary fibre, and SFN are known for reducing neurodegeneration related to Alzheimer's disease, Parkinson's disease, and other cognitive disorders. Higher vegetable intake is directly associated with a lower risk of dementia and Alzheimer's disease. Observational studies and long-term cohorts, such as the 24-year Hisayama study (Cannas et al., 2024; Kimura et al., 2022), support this finding. Antioxidants such as β-carotene and vitamin C, abundant in carrots, leafy greens, broccoli, and sweet potatoes, suppress neuronal oxidative stress by activating NRF2 and enhancing glutathione and superoxide dismutase signalling, preserving neuronal integrity. Polyphenols and carotenoids, common in colourful vegetables, scavenge reactive species, modulate neuroinflammatory pathways, and reduce amyloidogenesis or neuronal death in experimental neurodegenerative model (APP/PS1 mice) (Huang et al., 2025). Specific vegetables, such as cruciferous greens rich in SFN, also induce detoxifying enzymes and neuroprotective signalling, demonstrated in experimental models of Parkinson's disease (Nrf2-knockdown mice) to protect dopaminergic neurons and alleviate neuroinflammation (Jaźwa et al., 2011). Furthermore, leafy greens containing lutein, folate, and B vitamins help support cognitive flexibility, reduce homocysteine levels, and protect against age-related cognitive decline. (Morris et al., 2017; Smith et al., 2010) Together, these epidemiological, mechanistic, and clinical observations suggest that regular consumption of vegetables, particularly dark leafy greens, cruciferous, and carotenoid-rich vegetables, can serve as a reasonably effective dietary plan to slow down cognitive ageing and reduce neurodegenerative risk. Some other clinical trials are discussed in Table 9.1.

9.5 FINDINGS FROM RANDOMIZED CLINICAL TRIALS (RCTS)

Randomized clinical trials (RCTs) are the gold standard in clinical research, which offer better evidence on the safety and efficacy of potential nutrient supplements and potential drug candidates. Recently, vegetable-based nutrients are gaining more attention due to their potential in modulating disease risks, improving metabolic health, and encouraging overall well-being. RCTs provide controlled conditions to determine the effect of vegetables or vegetable-derived compounds like flavonoids, carotenoids, dietary fibre, and glucosinolates. Vegetable-based compounds have showed their efficacy both *in vitro* and *in vivo*, for example, in colorectal cancer (CRC). *Brassica rapa*-derived glucosinolates were found to be cytotoxic towards colorectal cancer cell lines (Kim et al., 2022). In another study involving the Db/Db mouse model (genetic model for type 2 diabetes and obesity), SFN from cruciferous vegetables was found to be effective in protecting against diabetic cardiomyopathy (Wang et al., 2025).

In this section, we have discussed a series of randomized clinical trials that explore the role of vegetables and vegetable-derived nutrients in different clinical study. In an eight-year-long study conducted in females with breast cancer, subjects were fed cruciferous vegetable-rich diet, which resulted in reduced invasive events (Tang et al., 2025). In another study, two servings of vegetables per day increased plasma carotenoid levels in head and neck cancer patients in six months (Cartmel et al., 2005). Similarly, Indole-3-carbinol from cruciferous vegetables was found to increase chemopreventive biomarkers in the urine of breast cancer patients (Reed et al., 2005). In other studies, broccoli-sprout-derived SFN and glucoraphanin were found to regulate prostate cancer progression-specific gene expression in human males (Taraka et al., 2008; Zhang et al., 2020). Another study revealed that daily consumption of fruits and vegetables resulted in improved memory and motor response in humans. Anthocyanin, in particular, reduces the concentration of TNF-α, which is related to mild cognitive improvements (Rosario et al., 2021). Other than anticancer and neuroprotective functions, vegetable-based nutrients are also linked to inflammation-related disorders. For example tomato-based complex containing carotenoids was able to reduce systolic blood pressure in human subjects (Wolak et al., 2019). Broccoli sprouts powder reduced oxidative stress in type 2 diabetes patients and also improved insulin sensitivity after four weeks of consumption (Bahadoran et al., 2012). Vegetable-based nutrients, such as flavonoids, carotenoids, and glucosinolates, are involved in the modulation of numerous diseases, including cancer and neurodegenerative disorders. A few RCTs are represented in Table 9.2.

9.6 LIMITATION AND FUTURE DIRECTION

Several clinical trials have reported the effect of vegetablse and vegetable-derived nutrients on various pathological conditions resulting in significant improvements in disease conditions. Numerous vegetable-based phytomolecules have been reported to be effective in stratifying participants by specific factors or patients with the severity of clinical stages. Although some case studies and clinical trials showed significant improvement in disease prevention, cure, and biomarker expression, longitudinal studies are limited. Additionally, most of these studies have involved a limited number of participants, and only a few have reported favourable outcomes. More extensive studies with large sample sizes, longer follow-up, repeated doses, and post-trial observation will be helpful to validate these favourable observations. As the standardized clinical trial of vegetables as a curative medicine has several limitations, however, preventive clinical studies can be explored in long-term follow up setting with a large sample.

TABLE 9.2
Some Reported RCTs Using Vegetables and Vegetable-Derived Compounds in Cancer, Cardiovascular Disease, Inflammatory Disease, and Neurodegenerative Disorders

S No.	Condition	Sample size	Dose	Duration	Study Outcome	Reference
				Cancer		
1	Colon cancer	51	Aged Garlic Extract (AGE) 2.4 mL/day	12 months	Reduced the size and number of adenomas	Lee et al. (2021)
2	Breast cancer	3656	Cruciferous vegetable	8 years	Reduced invasive events	Tang et al. (2025)
3	Head and neck cancer	75	Vegetable (2 servings per day)	6 months	Rise in plasma carotenoids	Cartmel et al. (2005)
4	Breast cancer	17 Females	Cruciferous vegetables (indole-3-carbinol-rich)	8 weeks	Increase in urinary biomarker (2-hydroxyestrone/16-alpha-hydroxyestrone, which is related to chemoprevention).	Reed et al. (2005)
5	Prostate cancer	98 Males	Broccoli sprout (sulforaphane-rich)	53 months	Changes in gene expression related to prostate cancer progression	Zhang et al. (2020)
6	Prostate cancer	49 Males	Broccoli (glucoraphanin-rich)	12 months	Reduced progression of prostate cancer	Taraka et al. (2019)
7	Prostatic intraepithelial neoplasia	Males	Tomato extract capsule (lycopene-rich)	6 months	Possible modulation of prostate tissue	Gann et al. (2015)
				Cardiovascular disease		
8	Hypertension	459	DASH diet	8 weeks	Reduced systolic and diastolic blood pressure	Appel et al. (1997)
9	Cardiovascular disease	7447	Lyon diet and PREDIMED	4.8 years	Reduced cardiovascular events	Estruch et al. (2013)
10	Hypertension		Tomato nutrient complex (carotenoid-rich)	8 weeks	Reduction in mean systolic blood pressure	Wolak et al. (2019)
				Inflammatory disease		
11	Diabetes	22	500 mL beet juice	3 times in total	Augmented NO bioavailability and promoted insulin sensitivity	Beals et al. (2017)
12	Type II diabetes mellitus	1,517	7.6–8.3 g dietary fibre/day		Reduced glycosylated haemoglobin, fasting glucose, and insulin	Xie et al. (2021)
13	Ulcerative colitis	89	2 g curcumin/day	16 months	Maintained remission in patients with quiescent UC	Hanai et al. (2006)
14	Rheumatoid arthritis	70 Females	1 g garlic/day	8 weeks	improvements in oxidative stress, HAQ in women	Moosavian et al. (2020)
15	Rheumatoid arthritis and ulcerative colitis		250–1,500 mg curcumin/day	8–12 weeks	Decreases in CRP and ESR	Ebramhimzadeh et al. (2021)
16	Inflammation	93	Anthocyanin-rich (anthocyanin capsule)	12 weeks	Reduced platelet function in dyslipidaemic patients	Tian et al. (2021)

17	COVID-19	140	Curcumin and piperine capsules (curcumin and piperine)	3 months	Expedition for the recovery of COVID-19 patients and preventing secondary infections	Pawar et al. (2021)
			Neurodegenerative disorders			
18	Cognition	60	400 mg curcumin/day	4 weeks	Improvement in sustained attention and significant improvement in working memory	Cox et al. (2015)
19	Age-related cognitive disorders	-	-	-	Reduced risk of dementia and Alzheimer's disease	Fekete et al. (2025)
20	Cognitive functions	180	Fruit, vegetable, and berry juice concentrate	16 weeks	Daily consumption for over 4 months; improved memory and motor response	Carrillo et al. (2021)
21	Mild cognitive impairment	31	Fruit-derived anthocyanins (anthocyanin-rich)	8 weeks	Reduced concentration of TNF-α	Rosario et al. (2021)
			Metabolic disorders			
22	Hyperlipidaemia	74	Semi-dried PR-7 tomato (lycopene-rich)	12 weeks	Reduced LDL-C	Nishimura et al. (2019)
23	Type 2 diabetes	81	Broccoli sprouts powder	4 weeks	Reduced oxidative stress in type 2 diabetes patients, improved insulin sensitivity	Bahadoran et al. (2012)
24	Prediabetes	38	Freeze-dried bitter gourd powder (Momordica charantia L)	12 weeks	Extended bitter gourd consumption reduced fasting plasma glucose levels	Mes et al. (2025)
25	Type 2 diabetes	-	Anthocyanin	4 weeks	Anti-inflammatory potential	Nikbakht et al. (2021)

9.7 CONCLUSION

It is nothing new that vegetables are an important part of the diet, providing beneficial nutrients, helping in digestion, and promoting gut health. Beyond the dietary benefits, vegetables serve as source of important phytomolecules (phenolic acid, flavonoids, vitamins, fibres, etc.) with desired pharmacological activities such as anti-inflammatory, immunomodulatory, antioxidant, cardioprotective, anticancer, and neuroprotective activities. The mechanism of such pharmacological activities is unfolding in recent times. Several clinical trials have been conducted with vegetable, or specific phytomolecule-rich vegetables or phytomolecules extracted from vegetables, which reported promising results with both preventive and curative points of view in various diseases such as cancer, metabolic disorders, neurodegenerative, cardiac complications, and inflammatory diseases. Although there are limitations in conducted trials in terms of sample size, dosing, duration and follow-up, there is considerable indication of beneficial role of vegetables and vegetable-based nutrients in prevention and treatment of several diseases.

DECLARATIONS

The authors declare that they have no known competing financial interests or personal relationships which could have appeared to influence the work reported in this chapter.

ACKNOWLEDGEMENT

The authors gratefully acknowledge the valuable insights and constructive feedback provided by colleagues and reviewers which have strengthened the quality of this chapter. The authors are grateful to the Director, CSIR-NEIST, for the support for this research and to the Academy of Scientific and Innovative Research (AcSIR) for providing the scope for interdisciplinary research and innovative PhD studies. Prity Chakravarty (NTA-ref. no.: 241610209310) and Pankaj Barman (UGC-ref. no: 1008/ (CSIR-UGC NET JUNE 2019) sincerely acknowledge the University Grants Commission (UGC), New Delhi, for providing the Fellowship.

REFERENCES

Abbas, A., Hall, J. A., Patterson, W. L. III, Ho, E., Hsu, A., Al-Mulla, F., et al. (2016). Sulforaphane modulates telomerase activity via epigenetic regulation in prostate cancer cell lines. *Biochemistry and Cell Biology*, *94*(1), 71–81.

Ansari, P., Khan, J. T., Chowdhury, S., Reberio, A. D., Kumar, S., Seidel, V., . . . & Flatt, P. R. (2024). Plant-based diets and phytochemicals in the management of diabetes mellitus and prevention of its complications: A review. *Nutrients*, *16*(21), 3709.

Appel, L. J., Moore, T. J., Obarzanek, E., Vollmer, W. M., Svetkey, L. P., Sacks, F. M., . . . & Harsha, D. W. (1997). A clinical trial of the effects of dietary patterns on blood pressure. *New England Journal of Medicine*, *336*(16), 1117–1124.

Asensi, M., Medina, I., Ortega, A., Carretero, J., Baño, M. C., Obrador, E., & Estrela, J. M. (2002). Inhibition of cancer growth by resveratrol is related to its low bioavailability. *Free Radical Biology and Medicine*, *33*(3), 387–398.

Atwell, L. L., Zhang, Z., Mori, M., Farris, P. E., Vetto, J. T., Naik, A. M., . . . & Shannon, J. (2015). Sulforaphane bioavailability and chemopreventive activity in women scheduled for breast biopsy. *Cancer Prevention Research*, *8*(12), 1184–1191.

Ayo, V. I., Adondua, M. A., Morayo, A. E., Ekele, J., Amilo, D. A. V. I. D., Ochuele, D. A., . . . & Barde, C. E. (2023). Effect of Lactuca sativa supplemented diet on Poloxamer 407 induced hyperlipidemic albino rats (Rattus norvegicus). *Asian Journal of Natural Product Biochemistry*, 21, 67–78.

Bahadoran, Z., Tohidi, M., Nazeri, P., Mehran, M., Azizi, F., & Mirmiran, P. (2012). Effect of broccoli sprouts on insulin resistance in type 2 diabetic patients: A randomized double-blind clinical trial. *International Journal of Food Sciences and Nutrition*, *63*(7), 767–771.

Barnes, L. L., Dhana, K., Liu, X., Carey, V. J., Ventrelle, J., Johnson, K., . . . & Sacks, F. M. (2023). Trial of the MIND diet for prevention of cognitive decline in older persons. *New England Journal of Medicine, 389*(7), 602–611.

Baron, M., Zuo, B., Chai, J., Zhao, J., Jahan-Mihan, A., Ochrietor, J., & Arikawa, A. Y. (2024). The effects of fermented vegetables on the gut microbiota for prevention of cardiovascular disease. *Gut Microbiome, 5*, e6.

Beals, J. W., Binns, S. E., Davis, J. L., Giordano, G. R., Klochak, A. L., Paris, H. L., . . . & Bell, C. (2017). Concurrent beet juice and carbohydrate ingestion: Influence on glucose tolerance in obese and nonobese adults. *Journal of Nutrition and Metabolism, 2017*(1), 6436783.

Cannas, C., Lostia, G., Serra, P. A., Peana, A. T., & Migheli, R. (2024). Food and food waste antioxidants: Could they be a potent defence against Parkinson's disease?. *Antioxidants (Basel, Switzerland), 13*(6), 645. https://doi.org/10.3390/antiox13060645.

Carrillo, J. Á., Arcusa, R., Zafrilla, M. P., & Marhuenda, J. (2021). Effects of fruit and vegetable-based nutraceutical on cognitive function in a healthy population: Placebo-controlled, double-blind, and randomized clinical trial. *Antioxidants (Basel, Switzerland), 10*(1), 116. https://doi.org/10.3390/antiox10010116.

Cartmel, B., Bowen, D., Ross, D., Johnson, E., & Mayne, S. T. (2005). A randomized trial of an intervention to increase fruit and vegetable intake in curatively treated patients with early-stage head and neck cancer. *Cancer Epidemiology, Biomarkers & Prevention: A Publication of the American Association for Cancer Research, Cosponsored by the American Society of Preventive Oncology, 14*(12), 2848–2854. https://doi.org/10.1158/1055-9965.EPI-05-0191.

Chang, H. P., Huang, M. C., Lei, Y. P., Chuang, Y. J., Wang, C. W., & Sheen, L. Y. (2022). Phytochemical-rich vegetable and fruit juice alleviates oral mucositis during concurrent chemoradiotherapy in patients with locally advanced head and neck cancer. *Journal of Traditional and Complementary Medicine, 12*(5), 488–498.

Chen, D., Landis-Piwowar, K. R., Chen, M. S., & Dou, Q. P. (2007). Inhibition of proteasome activity by the dietary flavonoid apigenin is associated with growth inhibition in cultured breast cancer cells and xenografts. *Breast Cancer Research, 9*(6), R80.

Cherng, J. M., Chiang, W., & Chiang, L. C. (2008). Immunomodulatory activities of common vegetables and spices of Umbelliferae and its related coumarins and flavonoids. *Food Chemistry, 106*(3), 944–950.

Corrêa, R. O., Castro, P. R., Fachi, J. L., Nirello, V. D., El-Sahhar, S., Imada, S., . . . & Vinolo, M. A. R. (2023). Inulin diet uncovers complex diet-microbiota-immune cell interactions remodeling the gut epithelium. *Microbiome, 11*(1), 90.

Cox, K. H., Pipingas, A., & Scholey, A. B. (2015). Investigation of the effects of solid lipid curcumin on cognition and mood in a healthy older population. *Journal of Psychopharmacology, 29*(5), 642–651.

Cui, J., Lian, Y., Zhao, C., Du, H., Han, Y., Gao, W., . . . & Zheng, J. (2019). Dietary fibers from fruits and vegetables and their health benefits via modulation of gut microbiota. *Comprehensive Reviews in Food Science and Food Safety, 18*(5), 1514–1532.

Dalmartello, M., Turati, F., Zhang, Z. F., Lunet, N., Rota, M., Bonzi, R., . . . & Pelucchi, C. (2022). Allium vegetables intake and the risk of gastric cancer in the Stomach cancer Pooling (StoP) project. *British Journal of Cancer, 126*(12), 1755–1764.

da Silva Dias, J. C., & Imai, S. (2017). Vegetables consumption and its benefits on diabetes. *Journal of Nutritional Therapeutics, 6*(1), 1–10.

Dai, F. J., & Chau, C. F. (2017). Classification and regulatory perspectives of dietary fiber. *Journal of Food and Drug Analysis, 25*(1), 37–42.

Dai, J., & Mumper, R. J. (2010). Plant phenolics: Extraction, analysis and their antioxidant and anticancer properties. *Molecules, 15*(10), 7313–7352.

Dwibedi, C., Axelsson, A. S., Abrahamsson, B., Fahey, J. W., Asplund, O., Hansson, O., . . . & Rosengren, A. H. (2025). Effect of broccoli sprout extract and baseline gut microbiota on fasting blood glucose in prediabetes: A randomized, placebo-controlled trial. *Nature Microbiology, 10*(3), 681–693.

Ebrahimzadeh, A., Abbasi, F., Ebrahimzadeh, A., Jibril, A. T., & Milajerdi, A. (2021). Effects of curcumin supplementation on inflammatory biomarkers in patients with Rheumatoid Arthritis and Ulcerative colitis: A systematic review and meta-analysis. *Complementary Therapies in Medicine, 61*, 102773.

Esposito, T., Lucariello, A., Hay, E., Contieri, M., Tammaro, P., Varriale, B., . . . & Perna, A. (2019). Effects of curcumin and its adjuvant on TPC1 thyroid cell line. *Chemico-Biological Interactions, 305*, 112–118.

Estruch, R., Ros, E., Salas-Salvadó, J., Covas, M. I., Corella, D., Arós, F., . . . & Martínez-González, M. A. (2013). Primary prevention of cardiovascular disease with a Mediterranean diet. *New England Journal of Medicine*, *368*(14), 1279–1290.

Farag, M. A., & Motaal, A. A. A. (2010). Sulforaphane composition, cytotoxic and antioxidant activity of crucifer vegetables. *Journal of Advanced Research*, *1*(1), 65–70.

Fejes, R., Pilat, N., Lutnik, M., Weisshaar, S., Weijler, A. M., Krüger, K., . . . & Neubauer, O. (2024). Effects of increased nitrate intake from beetroot juice on blood markers of oxidative stress and inflammation in older adults with hypertension. *Free Radical Biology and Medicine*, *222*, 519–530.

Fekete, M., Varga, P., Ungvari, Z., Fekete, J. T., Buda, A., Szappanos, Á., . . . & Győrffy, B. (2025). The role of the Mediterranean diet in reducing the risk of cognitive impairement, dementia, and Alzheimer's disease: A meta-analysis. *Geroscience*, *47*(3), 3111–3130.

Gann, P. H., Deaton, R. J., Rueter, E. E., van Breemen, R. B., Nonn, L., Macias, V., . . . Ananthanarayanan, V. (2015). A phase II randomized trial of lycopene-rich tomato extract among men with high-grade prostatic intraepithelial neoplasia. *Nutrition and Cancer*, *67*(7), 1104–1112. https://doi.org/10.1080/01635581.2015.1075560.

Gasmi, A., Shanaida, M., Oleshchuk, O., Semenova, Y., Mujawdiya, P. K., Ivankiv, Y., . . . & Bjørklund, G. (2023). Natural ingredients to improve immunity. *Pharmaceuticals*, *16*(4), 528.

Gaziano, T., Reddy, K. S., Paccaud, F., Horton, S., & Chaturvedi, V. (2006). Cardiovascular disease. *Disease Control Priorities in Developing Countries.* 2nd edn. The International Bank for Reconstruction and Development / The World Bank, Washington (DC); Oxford University Press, New York.

Guse, K., Sharma, A., Weyenberg, E., Davison, S., Ma, Y., Choi, Y., . . . & Gomez, A. (2023). Regular consumption of lacto-fermented vegetables has greater effects on the gut metabolome compared with the microbiome. *Gut Microbiome*, *4*, e11.

Hanai, H., Iida, T., Takeuchi, K., Watanabe, F., Maruyama, Y., Andoh, A., . . . & Koide, Y. (2006). Curcumin maintenance therapy for ulcerative colitis: Randomized, multicenter, double-blind, placebo-controlled trial. *Clinical Gastroenterology and Hepatology*, *4*(12), 1502–1506.

He, C., Gao, M., Zhang, X., Lei, P., Yang, H., Qing, Y., & Zhang, L. (2022). The protective effect of sulforaphane on dextran sulfate sodium-induced colitis depends on gut microbial and Nrf2-related mechanism. *Frontiers in Nutrition*, *9*, 893344.

He, F. J., Nowson, C. A., Lucas, M., & MacGregor, G. A. (2007). Increased consumption of fruit and vegetables is related to a reduced risk of coronary heart disease: Meta-analysis of cohort studies. *Journal of Human Hypertension*, *21*(9), 717–728.

Horbowicz, M. (2003). The occurrence, role and contents of glucosinolates in Brassica vegetables. *Vegetable Crops Research Bulletin*, *58*, 23–40.

Howard, L. R., Talcott, S. T., Brenes, C. H., & Villalon, B. (2000). Changes in phytochemical and antioxidant activity of selected pepper cultivars (Capsicum species) as influenced by maturity. *Journal of agricultural and food chemistry*, *48*(5), 1713–1720.

Hu, W., Sarengaowa, Guan, Y., & Feng, K. (2022). Biosynthesis of phenolic compounds and antioxidant activity in fresh-cut fruits and vegetables. *Frontiers in Microbiology*, *13*, 906069.

Huang, L., Wang, Z., Wu, M., Hailili, G., Shen, T., Fei, L., . . . & Yuan, C. (2025). Long-term oral administration of dietary carotenoids mixture alleviates cognitive impairment and AD neuropathology in APP/PS1 mice. *Food Science and Human Wellness*. https://doi.org/10.26599/FSHW.2025.9250715

Jadaun, J. S., Chownk, M., Bose, S. K., Kumari, S., & Sangwan, N. S. (2022). Food and vegetables as source of phytoactives for immunomodulation. In *Plants and Phytomolecules for Immunomodulation: Recent Trends and Advances* (pp. 439–468). Singapore: Springer Nature Singapore.

Jazwa, A., Rojo, A.I., Innamorato, N.G., Hesse, M., Fernández-Ruiz, J. and Cuadrado, A., 2011. Pharmacological targeting of the transcription factor Nrf2 at the basal ganglia provides disease modifying therapy for experimental parkinsonism. *Antioxidants & redox signaling*, *14*(12), 2347–2360.

Jeffery, E. H., & Araya, M. (2009). Physiological effects of broccoli consumption. *Phytochemistry Reviews*, *8*(1), 283–298.

Jiang, Y., Wu, S. H., Shu, X. O., Xiang, Y. B., Ji, B. T., Milne, G. L., Cai, Q., Zhang, X., Gao, Y. T., Zheng, W., & Yang, G. (2014). Cruciferous vegetable intake is inversely correlated with circulating levels of proinflammatory markers in women. *Journal of the Academy of Nutrition and Dietetics*, *114*(5), 700–708.e2. https://doi.org/10.1016/j.jand.2013.12.019.

Jideani, A. I., Silungwe, H., Takalani, T., Omolola, A. O., Udeh, H. O., & Anyasi, T. A. (2021). Antioxidant-rich natural fruit and vegetable products and human health. *International Journal of Food Properties*, *24*(1), 41–67.

Kapil, V., Khambata, R. S., Robertson, A., Caulfield, M. J., & Ahluwalia, A. (2015). Dietary nitrate provides sustained blood pressure lowering in hypertensive patients: A randomized, phase 2, double-blind, placebo-controlled study. *Hypertension*, *65*(2), 320–327.

Karunaweera, N., Raju, R., Gyengesi, E., & Münch, G. (2015). Plant polyphenols as inhibitors of NF-κB induced cytokine production—a potential anti-inflammatory treatment for Alzheimer's disease?. *Frontiers in Molecular Neuroscience*, *8*, 24.

Key, T. J. (2011). Fruit and vegetables and cancer risk. *British Journal of Cancer*, *104*(1), 6–11.

Khan, F., Ray, S., Craigie, A. M., Kennedy, G., Hill, A., Barton, K. L., . . . & Belch, J. J. (2014). Lowering of oxidative stress improves endothelial function in healthy subjects with habitually low intake of fruit and vegetables: A randomized controlled trial of antioxidant-and polyphenol-rich blackcurrant juice. *Free Radical Biology and Medicine*, *72*, 232–237.

Khanum, F., Siddalinga Swamy, M., Sudarshana Krishna, K. R., Santhanam, K., & Viswanathan, K. R. (2000). Dietary fiber content of commonly fresh and cooked vegetables consumed in India. *Plant Foods for Human Nutrition*, *55*(3), 207–218.

Khoddami, A., Wilkes, M. A., & Roberts, T. H. (2013). Techniques for analysis of plant phenolic compounds. *Molecules*, *18*(2), 2328–2375.

Kim, J. S., Han, S., Kim, H., Won, S. Y., Park, H. W., Choi, H., . . . & Lee, S. G. (2022). Anticancer effects of high glucosinolate synthesis lines of Brassica rapa on colorectal cancer cells. *Antioxidants*, *11*(12), 2463.

Kimura, Y., Yoshida, D., Ohara, T., Hata, J., Honda, T., Hirakawa, Y., Shibata, M., Oishi, E., Sakata, S., Furuta, Y., Chen, S., Uchida, K., Nakao, T., Kitazono, T., & Ninomiya, T. (2022). Long-term association of vegetable and fruit intake with risk of dementia in Japanese older adults: The Hisayama study. *BMC geriatrics*, *22*(1), 257.

Kondo, Y., Shen, L., Cheng, A. *et al.* (2008). Gene silencing in cancer by histone H3 lysine 27 trimethylation independent of promoter DNA methylation. *Nat Genet*, *40*, 741–750. https://doi.org/10.1038/ng.159 and https://doi.org/10.1186/s12877-022-02939-2.

Lako, J., Trenerry, V. C., Wahlqvist, M., Wattanapenpaiboon, N., Sotheeswaran, S., & Premier, R. (2007). Phytochemical flavonols, carotenoids and the antioxidant properties of a wide selection of Fijian fruit, vegetables and other readily available foods. *Food Chemistry*, *101*(4), 1727–1741.

Lakshmanan, A. P., Mingione, A., Pivari, F., Dogliotti, E., Brasacchio, C., Murugesan, S., . . . & Terranegra, A. (2022). Modulation of gut microbiota: The effects of a fruits and vegetables supplement. *Frontiers in Nutrition*, *9*, 930883.

Lee, J., Zhao, N., Fu, Z., Choi, J., Lee, H. J., & Chung, M. (2021). Effects of garlic intake on cancer: A systematic review of randomized clinical trials and cohort studies. *Nutrition Research and Practice*, *15*(6), 773–788.

Lidder, S., & Webb, A. J. (2013). Vascular effects of dietary nitrate (as found in green leafy vegetables and beetroot) via the nitrate-nitrite-nitric oxide pathway. *British Journal of Clinical Pharmacology*, *75*(3), 677–696.

Liu, R. H. (2013). Health-promoting components of fruits and vegetables in the diet. *Advances in Nutrition*, *4*(3), 384S–392S.

Mabou, F. D., & Yossa, I. B. N. (2021). TERPENES: Structural classification and biological activities. *Journal of Pharmacy and Biological Sciences*, *16*(3), 25–40.

Masoumvand, M., Ramezani, E., Eshaghi Milasi, Y., Baradaran Rahimi, V., & Askari, V. R. (2025). New horizons for promising influences of sulforaphane in the management of metabolic syndrome: A mechanistic review. *Naunyn-Schmiedeberg's Archives of Pharmacology*, *398*(5), 4933–4946. https://doi.org/10.1007/s00210-024-03706-3.

Matsumoto, S., Nakanishi, R., Li, D., Alani, A., Rezaeian, P., Prabhu, S., . . . & Budoff, M. J. (2016). Aged garlic extract reduces low attenuation plaque in coronary arteries of patients with metabolic syndrome in a prospective randomized double-blind study. *The Journal of Nutrition*, *146*(2), 427S–432S.

Mazur, M., Przytuła, A., Szymańska, M., & Popiołek-Kalisz, J. (2024). Dietary strategies for cardiovascular disease risk factors prevention. *Current Problems in Cardiology*, 102746.

Mes, J. J., van den Belt, M., van der Haar, S., Oosterink, E., Luijendijk, T., Manusama, K., . . . & Esser, D. (2025). Bitter gourd (Momordica charantia L.) supplementation for twelve weeks improves biomarkers of glucose homeostasis in a prediabetic population. *Journal of Ethnopharmacology*, *347*, 119756.

Mitra, S., Paul, S., Roy, S., Sutradhar, H., Bin Emran, T., Nainu, F., . . . & Mubarak, M. S. (2022). Exploring the immune-boosting functions of vitamins and minerals as nutritional food bioactive compounds: A comprehensive review. *Molecules*, *27*(2), 555.

Moosavian, S. P., Paknahad, Z., & Habibagahi, Z. (2020). A randomized, double-blind, placebo-controlled clinical trial, evaluating the garlic supplement effects on some serum biomarkers of oxidative stress, and quality of life in women with rheumatoid arthritis. *International Journal of Clinical Practice*, *74*(7), e13498.

Morris, M. C., Wang, Y., Barnes, L. L., Bennett, D. A., Dawson-Hughes, B., & Booth, S. L. (2018). Nutrients and bioactives in green leafy vegetables and cognitive decline: Prospective study. *Neurology*, *90*(3), e214–e222.

Murakami, A., Ashida, H., & Terao, J. (2008). Multitargeted cancer prevention by quercetin. *Cancer Letters*, *269*(2), 315–325.

Neale, E. P., Batterham, M. J., & Tapsell, L. C. (2016). Consumption of a healthy dietary pattern results in significant reductions in C-reactive protein levels in adults: A meta-analysis. *Nutrition Research*, *36*(5), 391–401.

Nikbakht, E., Singh, I., Vider, J., Williams, L. T., Vugic, L., Gaiz, A., . . . & Colson, N. (2021). Potential of anthocyanin as an anti-inflammatory agent: A human clinical trial on type 2 diabetic, diabetic at-risk and healthy adults. *Inflammation Research*, *70*(3), 275–284.

Nishida, C., Uauy, R., Kumanyika, S., & Shetty, P. (2004). The joint WHO/FAO expert consultation on diet, nutrition and the prevention of chronic diseases: Process, product and policy implications. *Public Health Nutrition*, *7*(1a), 245–250.

Nishimura, M., Tominaga, N., Ishikawa-Takano, Y., Maeda-Yamamoto, M., & Nishihira, J. (2019). Effect of 12-week daily intake of the high-lycopene tomato (*Solanum lycopersicum*), a variety named "PR-7", on lipid metabolism: A randomized, double-blind, placebo-controlled, parallel-group study. *Nutrients*, *11*(5), 1177. https://doi.org/10.3390/nu11051177.

Pan, Yuxin, et al. (2025). Cancer-specific regulation of metabolic and epigenetic pathways by dietary phytochemicals. *Pharmaceutical Research*, 1–15.

Paquette, M., Larqué, A. S. M., Weisnagel, S. J., Desjardins, Y., Marois, J., Pilon, G., . . . & Jacques, H. (2017). Strawberry and cranberry polyphenols improve insulin sensitivity in insulin-resistant, non-diabetic adults: A parallel, double-blind, controlled and randomised clinical trial. *British Journal of Nutrition*, *117*(4), 519–531.

Paur, I., Lilleby, W., Bøhn, S. K., Hulander, E., Klein, W., Vlatkovic, L., . . . & Blomhoff, R. (2017). Tomato-based randomized controlled trial in prostate cancer patients: Effect on PSA. *Clinical Nutrition*, *36*(3), 672–679.

Pawar, K. S., Mastud, R. N., Pawar, S. K., Pawar, S. S., Bhoite, R. R., Bhoite, R. R., . . . & Deshpande, A. R. (2021). Oral curcumin with piperine as adjuvant therapy for the treatment of COVID-19: A randomized clinical trial. *Frontiers in Pharmacology*, *12*, 669362.

Poiroux-Gonord, F., Bidel, L. P., Fanciullino, A. L., Gautier, H., Lauri-Lopez, F., & Urban, L. (2010). Health benefits of vitamins and secondary metabolites of fruits and vegetables and prospects to increase their concentrations by agronomic approaches. *Journal of Agricultural and Food Chemistry*, *58*(23), 12065–12082.

Pollock, R. L. (2016). The effect of green leafy and cruciferous vegetable intake on the incidence of cardiovascular disease: A meta-analysis. *JRSM Cardiovascular Disease*, *5*, 2048004016661435.

Possenti, M., Baima, S., Raffo, A., Durazzo, A., Giusti, A. M., & Natella, F. (2016). Glucosinolates in food. In *Glucosinolates* (pp. 1–46). Cham: Springer.

Qi, Y., Zhang, C., Wu, D., Zhang, Y., Zhao, Y., & Li, W. (2022). Indole-3-carbinol stabilizes p53 to induce miR-34a, which targets LDHA to block aerobic glycolysis in liver cancer cells. *Pharmaceuticals*, *15*(10), 1257.

Rajendran, P., Abdelsalam, S. A., Renu, K., Veeraraghavan, V., Ben Ammar, R., & Ahmed, E. A. (2022). Polyphenols as potent epigenetics agents for cancer. *International Journal of Molecular Sciences*, *23*(19), 11712.

Rajendra Prasad, N., Karthikeyan, A., Karthikeyan, S., & Venkata Reddy, B. (2011). Inhibitory effect of caffeic acid on cancer cell proliferation by oxidative mechanism in human HT-1080 fibrosarcoma cell line. *Molecular and Cellular Biochemistry*, *349*(1), 11–19.

Ramya, V., & Patel, P. (2019). Health benefits of vegetables. *International Journal of Chemical Studies*, *7*(2), 82–87.

Reed, G. A., Peterson, K. S., Smith, H. J., Gray, J. C., Sullivan, D. K., Mayo, M. S., Crowell, J. A., & Hurwitz, A. (2005). A phase I study of indole-3-carbinol in women: Tolerability and effects. *Cancer Epidemiology, Biomarkers & Prevention: A Publication of the American Association for Cancer Research, Cosponsored by the American Society of Preventive Oncology*, *14*(8), 1953–1960. https://doi.org/10.1158/1055-9965.EPI-05-012.

Rosario, V. A., Fitzgerald, Z., Broyd, S., Paterson, A., Roodenrys, S., Thomas, S., . . . & Charlton, K. (2021). Food anthocyanins decrease concentrations of TNF-α in older adults with mild cognitive impairment: A randomized, controlled, double blind clinical trial. *Nutrition, Metabolism and Cardiovascular Diseases*, *31*(3), 950–960.

Rostampour, K., Sasanfar, B., Reshadfar, A., Emarati, A., Nafei, Z., Behniafard, N., . . . & Salehi-Abargouei, A. (2025). The association between fruit and vegetable intake and the odds of asthma among children and adolescents. *Journal of Health, Population and Nutrition*, *44*(1), 99.

Royston, K. J., & Tollefsbol, T. O. (2015). The epigenetic impact of cruciferous vegetables on cancer prevention. *Current Pharmacology Reports*, *1*, 46–51.Singh, V., Shirbhate, E., Kore, R., Mishra, A., Johariya, V., Veerasamy, R., . . . & Rajak, H. (2024). Dietary plant metabolites induced epigenetic modification as a novel strategy for the management of prostate cancer. *Mini Reviews in Medicinal Chemistry*, *24*(15), 1409–1426.

Saito, A., Ishikawa, S., Yang, K., Sawa, A., & Ishizuka, K. (2025). Sulforaphane as a potential therapeutic agent: A comprehensive analysis of clinical trials and mechanistic insights. *Journal of Nutritional Science*, *14*, e65.

Shin, J. Y., Kim, J. Y., Kang, H. T., Han, K. H., & Shim, J. Y. (2015). Effect of fruits and vegetables on metabolic syndrome: A systematic review and meta-analysis of randomized controlled trials. *International Journal of Food Sciences and Nutrition*, *66*(4), 416–425.

Smail, S. W., Bergsten, P., Taha, K. O., Yashooa, R. K., Hawezy, D. J., Abbas, M. A., & Shekha, M. S. (2025). Curcumin: Biochemistry, pharmacology, advanced drug delivery systems its epigenetic role in combating cancer. *Frontiers in Pharmacology*, *16*, 1695200.

Smith, A. D., Smith, S. M., De Jager, C. A., Whitbread, P., Johnston, C., Agacinski, G., . . . & Refsum, H. (2010). Homocysteine-lowering by B vitamins slows the rate of accelerated brain atrophy in mild cognitive impairment: a randomized controlled trial. *PloS one*, *5*(9), e12244.

Sonnenburg, E. D., & Sonnenburg, J. L. (2014). Starving our microbial self: The deleterious consequences of a diet deficient in microbiota-accessible carbohydrates. *Cell Metabolism*, *20*(5), 779–786.

Tang, L., Wang, Z., Hsiao, H. H., Kwan, M. L., Ergas, I. J., Roh, J. M., . . . & Kushi, L. H. (2025). Associations of cruciferous vegetable intake with breast cancer survival in a diverse population in the pathways study. *Cancer Epidemiology, Biomarkers & Prevention*, *34*(7), 1177–1187.

Tetali, S. D. (2019). Terpenes and isoprenoids: A wealth of compounds for global use. *Planta*, *249*(1), 1–8.

Thakur, A., & Sharma, R. (2018). Health promoting phytochemicals in vegetables: A mini review. *International Journal of Food and Fermentation Technology*, *8*(2), 107–117.

Tian, Z., Li, K., Fan, D., Zhao, Y., Gao, X., Ma, X., . . . & Yang, Y. (2021). Dose-dependent effects of anthocyanin supplementation on platelet function in subjects with dyslipidemia: A randomized clinical trial. *EBioMedicine*, *70*.

Tilg, H. (2015). Cruciferous vegetables: Prototypic anti-inflammatory food components. *Clinical Phytoscience*, *1*, 10. https://doi.org/10.1186/s40816-015-0011-2.

Traka, M. H., Gasper, A. V., Melchini, A., Bacon, J. R., Needs, P. W., Frost, V., et al. (2008). Broccoli consumption interacts with *GSTM1* to perturb oncogenic signalling pathways in the prostate. *PLoS ONE*, *3*(7), e2568. https://doi.org/10.1371/journal.pone.0002568.

Traka, M. H., Melchini, A., Coode-Bate, J., Al Kadhi, O., Saha, S., Defernez, M., . . . & Mithen, R. F. (2019). Transcriptional changes in prostate of men on active surveillance after a 12-mo glucoraphanin-rich broccoli intervention – results from the Effect of Sulforaphane on prostate CAncer PrEvention (ESCAPE) randomized controlled trial. *The American Journal of Clinical Nutrition*, *109*(4), 1133–1144.

van Bennekum, A. M., Nguyen, D. V., Schulthess, G., Hauser, H., & Phillips, M. C. (2005). Mechanisms of cholesterol-lowering effects of dietary insoluble fibres: Relationships with intestinal and hepatic cholesterol parameters. *British Journal of Nutrition*, *94*(3), 331–337.

Van't Veer, P., Jansen, M. C., Klerk, M., & Kok, F. J. (2000). Fruits and vegetables in the prevention of cancer and cardiovascular disease. *Public Health Nutrition*, *3*(1), 103–107.

Wang, P., Wang, Z., Jin, X., Zhang, M., Shen, M., & Li, D. (2025). Oral sulforaphane intervention protects against diabetic cardiomyopathy in db/db mice: Focus on cardiac lipotoxicity and substrate metabolism. *Antioxidants (Basel, Switzerland)*, *14*(5), 603.

Wicaksono, W. A., Cernava, T., Wassermann, B., Abdelfattah, A., Soto-Giron, M. J., Toledo, G. V., . . . & Berg, G. (2023). The edible plant microbiome: Evidence for the occurrence of fruit and vegetable bacteria in the human gut. *Gut Microbes*, *15*(2), 2258565.

Wolak, T., Sharoni, Y., Levy, J., Linnewiel-Hermoni, K., Stepensky, D., & Paran, E. (2019). Effect of tomato nutrient complex on blood pressure: A double blind, randomized dose response study. *Nutrients*, *11*(5), 950. https://doi.org/10.3390/nu11050950.

Wood, E., Hein, S., Mesnage, R., Fernandes, F., Abhayaratne, N., Xu, Y., . . . & Rodriguez-Mateos, A. (2023). Wild blueberry (poly) phenols can improve vascular function and cognitive performance in healthy older individuals: A double-blind randomized controlled trial. *The American Journal of Clinical Nutrition*, *117*(6), 1306–1319.

Xie, Y., Gou, L., Peng, M., Zheng, J. and Chen, L., 2021. Effects of soluble fiber supplementation on glycemic control in adults with type 2 diabetes mellitus: A systematic review and meta-analysis of randomized controlled trials. *Clinical Nutrition*, *40*(4), 1800–1810.

Yang, Y., Dong, B., Zou, Z., Wang, S., Dong, Y., Wang, Z., & Ma, J. (2018). Association between vegetable consumption and blood pressure, stratified by BMI, among Chinese adolescents aged 13–17 years: A national cross-sectional study. *Nutrients*, *10*(4), 451.

Yu, X., Pu, H., & Voss, M. (2024). Overview of anti-inflammatory diets and their promising effects on non-communicable diseases. *The British Journal of Nutrition*, *132*(7), 898–918. https://doi.org/10.1017/S0007114524001405

Zarebczan, B., Pinchot, S. N., Kunnimalaiyaan, M., & Chen, H. (2011). Hesperetin, a potential therapy for carcinoid cancer. *The American Journal of Surgery*, *201*(3), 329–333.

Zhang, Y., & Zhang, D. Z. (2018). Associations of vegetable and fruit consumption with metabolic syndrome. A meta-analysis of observational studies. *Public Health Nutrition*, *21*(9), 1693–1703.

Zhang, Z., Garzotto, M., Davis, E. W., Mori, M., Stoller, W. A., Farris, P. E., Wong, C. P., Beaver, L. M., Thomas, G. V., Williams, D. E., Dashwood, R. H., Hendrix, D. A., Ho, E., & Shannon, J. (2020). Sulforaphane bioavailability and chemopreventive activity in men presenting for biopsy of the prostate gland: A randomized controlled trial. *Nutrition and Cancer*, *72*(1), 74–87. https://doi.org/10.1080/01635581.2019.1619783.

Zhao, W., Zhou, X., Qi, G., & Guo, Y. (2018). Curcumin suppressed the prostate cancer by inhibiting JNK pathways via epigenetic regulation. *Journal of Biochemical and Molecular Toxicology*, *32*, e22049. https://doi.org/10.1002/jbt.22049.

10 Challenges and Future Perspectives about Nutraceuticals Derived from Vegetables

Aadya Gupta, Sparsh Saxena, Noopur Khare, Ayush Chaurasiya, Anshika Rai, Harvin Selin, and Tanvi Jain

10.1 INTRODUCTION

With the rapid advancement of modern science and technology, the distinction between food and medicine has become increasingly blurred. This convergence has given rise to the concept of *nutraceuticals*, a term that represents the intersection of nutrition and therapeutics – where foods and their bioactive components not only provide nourishment but also contribute to disease prevention and health promotion (Antony & Vijayan, 2021). In today's era, characterized by the prevalence of lifestyle-related disorders, malnutrition, and growing awareness of the role of diet in maintaining health, nutraceuticals have emerged as a cornerstone of preventive healthcare and functional nutrition (Qureshi et al., 2024). The term *nutraceutical* is a portmanteau derived from the words 'nutrition' and 'pharmaceutical'. It was first introduced in 1989 by Dr Stephen DeFelice, the founder of the Foundation for Innovation in Medicine (FIM), Cranford, New Jersey, to describe foods or food-derived substances which offer medical or health benefits (Daliri & Lee, 2015).

According to DeFelice (1995) a nutraceutical may be defined as 'a food, or fraction of a food, that has medical or health benefits, including prevention and/or treatment of disease'. This definition highlights that nutraceuticals go beyond the basic role of conventional foods, which primarily provide essential nutrients such as carbohydrates, proteins, vitamins, and minerals. Instead, nutraceuticals encompass biologically active compounds capable of modulating physiological processes, enhancing immune function, and offering therapeutic effects. These bioactive constituents such as polyphenols, flavonoids, carotenoids, omega-3 fatty acids, probiotics, and phytochemicals are known to influence molecular pathways linked to oxidative stress, inflammation, and metabolic regulation (Mokoena et al., 2016). Consequently, the regular inclusion of nutraceuticals in the diet can retard the ageing process; strengthen the body's defence mechanisms; and lower the risk of chronic diseases such as cardiovascular disorders, diabetes, cancer, and neurodegenerative conditions (Ristori et al., 2025). In essence, nutraceuticals embody the ancient wisdom of 'food as medicine', blending traditional nutritional principles with modern biomedical understanding to promote holistic health and longevity (Nagpal et al., 2025).

10.1.1 Significance of Nutraceuticals

Nutraceuticals are bioactive compounds derived from natural food sources such as plants, animals, or microorganisms which provide health benefits beyond basic nutrition. They offer physiological advantages and protection against chronic diseases such as cardiovascular disorders, diabetes, and

DOI: 10.1201/9781003606710-10

neurodegenerative conditions (Husen & Iqbal, 2021; Riar & Panesar, 2024). Unlike conventional pharmaceuticals, nutraceuticals are not stringent drugs but act as preventive agents that promote overall health and wellness. They can be consumed as dietary supplements, fortified foods, or medications and are generally safe when taken in the recommended amounts. Among the various natural sources of nutraceuticals, vegetables hold a particularly significant position due to their rich nutritional and bioactive composition (Jaglan et al., 2025). Vegetables are nutrient-dense rather than energy-dense, offering high levels of vitamins, minerals, dietary fibre, and phytochemicals while being low in calories. These bioactive compounds contribute to vital physiological functions, enhance immunity, and reduce the risk of chronic diseases, making vegetables essential for development, growth, and disease prevention. Their regular consumption supports cellular functions, combats oxidative stress, and maintains metabolic balance, thereby promoting long-term health and well-being. Thus, vegetables serve not only as vital dietary components but also as a potent source of natural nutraceuticals, playing a dual role in nourishment and disease prevention (Kussmann et al., 2023).

10.1.2 Source of Vitamins and Minerals

Vegetables serve as an excellent source of essential vitamins, minerals, dietary fibre, and bioactive compounds, all of which contribute significantly to overall health and disease prevention. They are abundant in vital micronutrients required for proper body functioning. For instance vitamin A found in carrots, spinach, and pumpkin helps maintain vision and supports immune health; vitamin C present in tomatoes, bell peppers, and broccoli acts as a potent antioxidant that promotes collagen synthesis and tissue repair; and vitamin K from green leafy vegetables is essential for blood clotting and bone formation (Garza-Juárez et al., 2023). Additionally, vegetables provide key minerals such as iron, calcium, magnesium, and potassium, which play important roles in oxygen transport, bone strength, nerve transmission, and electrolyte balance. Alongside micronutrients, vegetables are rich in both soluble and insoluble dietary fibres that aid digestion, promote regular bowel movement, regulate cholesterol and blood sugar levels, and enhance satiety, thereby supporting weight management. Furthermore, vegetables are an abundant source of antioxidants and phytochemicals, including carotenoids, flavonoids, polyphenols, and glucosinolates, which help neutralize harmful free radicals, reduce oxidative stress, and lower the risk of chronic diseases such as cardiovascular disorders, diabetes, and cancer. These bioactive compounds also exhibit anti-inflammatory and immune-modulating properties, further enhancing the body's defence mechanisms. Altogether, the diverse nutritional and therapeutic components of vegetables make them indispensable in maintaining health, preventing diseases, and improving overall quality of life.

10.1.3 Types of Vegetables, Nutraceuticals, and Their Benefits

Vegetables are broadly categorised into chlorophyll-rich (like spinach and lettuce), root (like beetroot), stem (like celery), flower (like broccoli and cauliflower), and fruit vegetables (like tomato and cucumber), each providing essential nutrients. They consists of certain nutraceuticals like organosulphur compounds (high in onions), carotenoids (high in carrots), and some polyphenols (spinach). Such elements increase and boost the immunity, helps in declining the inflammation that may lower the risk of chronic diseases like heart defects, cancer etc,. Continuous intake of these types of vegetables always supports the proper body functioning and overall health.

10.1.3.1 Carotenoids

Carotenoids are naturally occurring pigments responsible for the bright red, orange, and yellow colours in many vegetables. Prominent examples include β-carotene found in carrots, pumpkin, and sweet potatoes; lycopene in tomatoes; and lutein and zeaxanthin in spinach, kale, and corn. These compounds act as potent antioxidants which neutralize free radicals, thereby preventing oxidative damage to cells. β-Carotene serves as a precursor to vitamin A, essential for maintaining healthy vision, immune function, and skin health. Lycopene has been associated with a reduced risk of prostate cancer and CVDs, while lutein and zeaxanthin play a vital role in maintaining eye health and protecting against AMD.

10.1.3.2 Polyphenols (Flavonoids and Phenolic Acids)

Polyphenols are a diverse group of plant compounds with strong antioxidant and anti-inflammatory properties. Examples include quercetin found in onions, kale, and broccoli; kaempferol in spinach and beans; and chlorogenic acid in potatoes and eggplants. These compounds help scavenge reactive oxygen species (ROS), thereby reducing oxidative stress and inflammation. Polyphenols contribute to cardiovascular health by lowering cholesterol levels and blood pressure, improving endothelial function, and preventing platelet aggregation. Additionally, they support cognitive health by enhancing brain function and reducing the risk of neurodegenerative diseases such as Alzheimer's and Parkinson's disease. Some polyphenols also exhibit antimicrobial activity, helping to combat bacterial and viral infections (Kussmann et al., 2023).

10.1.3.3 Glucosinolates and Isothiocyanates

These sulphur-containing compounds are primarily found in cruciferous vegetables such as broccoli, cabbage, Brussels sprouts, cauliflower, kale, and mustard greens. Upon enzymatic hydrolysis, glucosinolates form isothiocyanates – bioactive molecules known for their detoxifying and anti-cancer properties (Prabhakar et al., 2025). They promote the elimination of harmful toxins from the body by inducing phase II detoxification enzymes and inhibiting the activation of carcinogens. Isothiocyanates also suppress tumour growth by inducing apoptosis (programmed cell death) and inhibiting angiogenesis (formation of new blood vessels in tumours). Regular consumption of cruciferous vegetables is linked to a lower risk of cancers, particularly of the breast, lung, and colon. Moreover, these compounds support liver function and strengthen immune regulation.

10.1.3.4 Organosulphur Compounds

Organosulphur compounds, such as allicin from garlic and diallyl sulphide from onions and leeks, are known for their broad spectrum of biological activities. These compounds exhibit strong antimicrobial, antiviral, and antifungal properties, making them natural agents against infections. They also contribute to cardiovascular health by lowering cholesterol levels, reducing blood pressure, and preventing platelet aggregation. OSCs enhance liver detoxification and protect against oxidative stress by boosting the activity of antioxidant enzymes like glutathione peroxidase. Their regular consumption supports heart health and may reduce the risk of certain cancers.

10.1.3.5 Dietary Fibre and Prebiotics

Vegetables are rich in both soluble and insoluble fibres such as inulin (found in onions, garlic, and asparagus); cellulose; and pectin (present in carrots, beets, and leafy greens). Dietary fibre improves digestion, prevents constipation, and supports healthy bowel function. Soluble fibre helps in lowering cholesterol levels by binding bile acids, while insoluble fibre aids in maintaining bowel regularity. Prebiotic fibres, such as inulin, serve as food for beneficial gut bacteria, promoting a healthy gut microbiome and enhancing immune function. Fibre also helps in regulating blood sugar levels and contributes to weight management by promoting a feeling of fullness and reducing overeating.

10.1.3.6 Vitamins and Minerals as Nutraceuticals

Vegetables are abundant in essential vitamins and minerals which function as vital nutraceuticals. Vitamin C, found in bell peppers, broccoli, and leafy greens, serves as a powerful antioxidant that strengthens the immune system, enhances collagen synthesis, and accelerates wound healing. Vitamin K, abundant in spinach and kale, is crucial for blood clotting and bone metabolism. Folate, present in spinach and lettuce, supports DNA synthesis and nervous system function, making it particularly important during pregnancy. Potassium, obtained from potatoes and spinach, helps regulate blood pressure, nerve signalling, and muscle contraction. Collectively, these micronutrients maintain physiological balance and prevent deficiency-related disorders.

10.1.3.7 Chlorophyll and Phytosterols

Chlorophyll, the green pigment found in leafy vegetables, possesses detoxifying, antioxidant, and wound-healing properties. It assists in cleansing the body by binding to and eliminating toxins and supports liver function. Phytosterols, plant-derived compounds structurally similar to cholesterol, are found in vegetables like broccoli and Brussels sprouts. They help lower LDL (bad) cholesterol levels by competing with dietary cholesterol absorption in the intestine, thereby supporting cardiovascular health. Additionally, phytosterols possess anti-inflammatory and anticancer properties, further enhancing their nutraceutical value.

10.2 CURRENT PROBLEMS IN THE DEVELOPMENT AND USE OF VEGETABLE-BASED NUTRACEUTICALS

10.2.1 Extraction and Processing Issues

Many vegetables have relatively small quantities of desirable compounds, such as carotenoids, flavonoids, glucosinolates, and polyphenols, and in addition, many can be unstable, and extracting them without impairing bioactivity is difficult. A key concern is that heat, air, and light will degrade these functional properties relatively quickly following harvest. While there are some technologies available (e.g. nanoencapsulation or supercritical fluid extraction) which allow for the extraction of any variety of compounds while maintaining functionality, these methods are expensive, not widely available in the market, and require very specific equipment (Jarouliya et al., 2024).

10.2.2 Stability and Shelf-Life

Vegetable compounds lose potency over time due to storage, transport, and cooking. For instance vitamin C and many polyphenols degrade fairly quickly. As the functional benefits of nutraceutical products are not guaranteed even with a developed product, with shelf-stable nutraceuticals, it is very difficult to ensure that the product is stable and effective in an ex-vivo model (outside a living organism) for an extended period (Shirodkar et al., 2025).

10.2.3 Bioavailability and Absorption

Vegetables have high bioactive contents; however, the human capacity to absorb (bioavailability) such nutrients can present issues. For example the bioactive from turmeric (curcumin) and carotenoids from carrots, spinach, and tomato are poorly bioavailable, and ultimately, only a very small amount gets into the bloodstream following ingestion. So far in the real world, bioactives perform poorly compared with research laboratory study results (Ingale, 2025).

10.2.4 Non-Standardization

There is no equivalent standardization between vegetable nutraceuticals and pharmaceutical drugs. The level of biotic will vary as a result of their soil type, growing season, harvesting method, and processing. This leads to variability in quality and dosage, making safety and effectiveness impossible to guarantee for consumers.

10.2.5 Evidence Gap

Numerous studies have drawn attention to the health benefits derived from vegetable-derived compounds, specifically antioxidants and anti-inflammatory properties. Most previous studies were testing compounds in vitro (test tubes) or on animals. Clinical testing on humans is not only costly but lengthy as well. Consequently, scientists have produced a knowledge gap of proven clinical evidence to convince healthcare professionals and governing authorities of common usage.

10.2.6 Regulatory Challenges

Every country views nutraceuticals differently – some countries deem nutraceuticals food supplements, others consider them a drug. The result of not having a universal regulatory framework is the fact that fraud and misleading health claims are rife, leading to poor quality control, poor safety, and corruption of the industry as a whole by false claims contained in products launched with high health claims. This leads to consumer mistrust of many (Shinde et al., 2014) nutraceuticals in the marketplace (Rane et al., 2024).

10.2.7 Cost and Accessibility

Increased extraction, encapsulation, and stabilization technologies have increased production costs to the point that many high-quality vegetable nutraceuticals are not affordable for most consumers, meaning that they are limited in population sampling (Galanakis, 2013). This gap of scientific advances and practical application thereof remains unbridged.

10.2.8 Consumer Awareness and Misuse

A lot of consumers do not actually understand the difference between a healthy diet, supplements, and nutraceuticals. Overconsumption, or misuse of supplements such as taking them at the same time with no direction, can lead to both toxicity and interactions with medications. Education and awareness campaigns are needed to prevent such instances from becoming more commonplace.

10.3 ISSUES IN STANDARDIZATION, EXTRACTION, AND FORMULATION

Vegetables are among the most vital components of a healthy diet, not only because they provide essential nutrients like vitamins, minerals, and fibre but also due to their rich content of bioactive compounds which possess therapeutic and preventive properties. These bioactive substances – commonly referred to as nutraceuticals – play a crucial role in disease prevention, immune enhancement, and overall physiological well-being. The major classes of nutraceuticals derived from vegetables include carotenoids, polyphenols, glucosinolates, OSCs, dietary fibre, vitamins, minerals, chlorophyll, and phytosterols. The basic issues in standardization, extraction and various formulations is also depicted in Figure 10.1.

10.3.1 Carotenoids

Carotenoids are naturally occurring pigments responsible for the bright red, orange, and yellow colours in many vegetables. Prominent examples include β-carotene found in carrots, pumpkin, and sweet potatoes; lycopene in tomatoes; and lutein and zeaxanthin in spinach, kale, and corn. These compounds act as potent antioxidants which neutralize free radicals, thereby preventing oxidative damage to cells. β-Carotene serves as a precursor to vitamin A, essential for maintaining healthy vision, immune function, and skin health. Lycopene has been associated with a reduced risk of prostate cancer and CVDs, while lutein and zeaxanthin play a vital role in maintaining eye health and protecting against AMD.

10.3.2 Polyphenols (Flavonoids and Phenolic Acids)

Polyphenols are a diverse group of plant compounds with strong antioxidant and anti-inflammatory properties. Examples include quercetin found in onions, kale, and broccoli; kaempferol in spinach

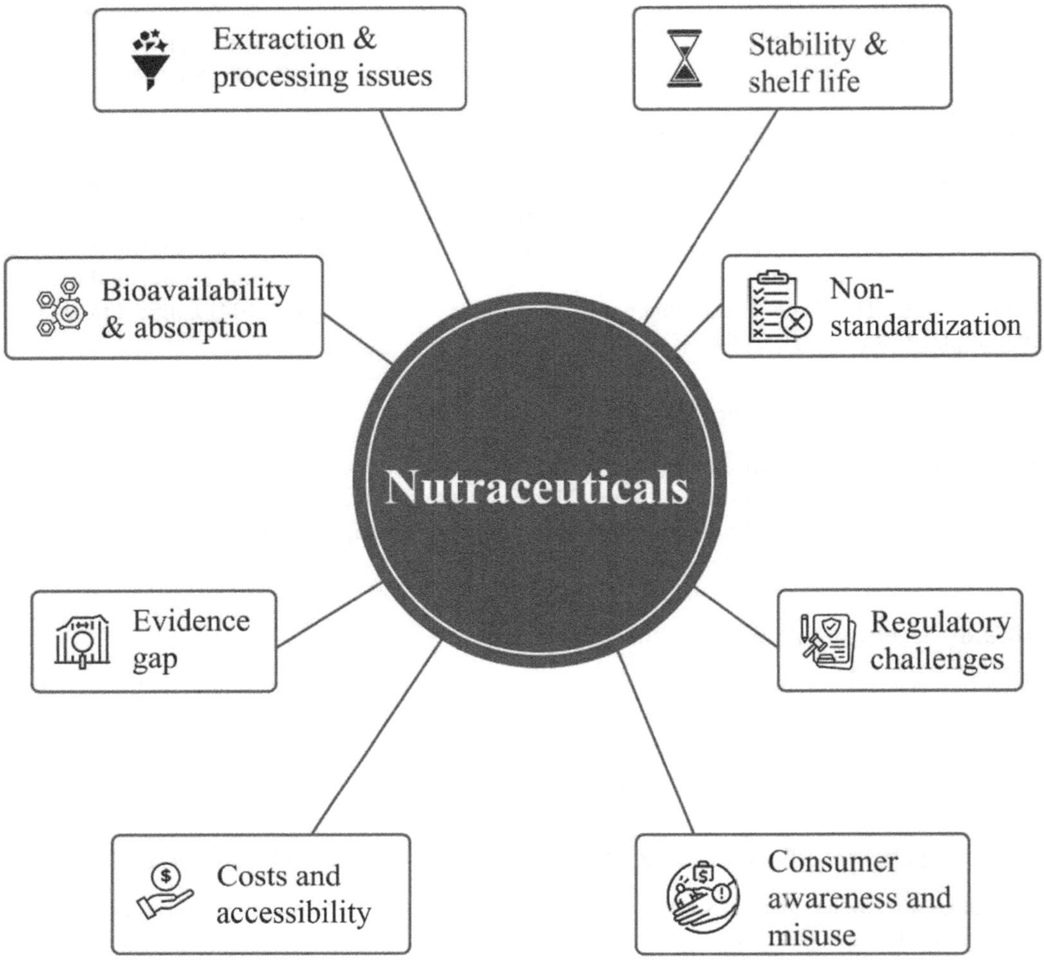

FIGURE 10.1 Issues in standardization, extraction, and formulation of nutraceuticals.

and beans; and chlorogenic acid in potatoes and eggplants. These compounds help scavenge ROS, thereby reducing oxidative stress and inflammation. Polyphenols contribute to cardiovascular health by lowering cholesterol levels and blood pressure, improving endothelial function, and preventing platelet aggregation. Additionally, they support cognitive health by enhancing brain function and reducing the risk of neurodegenerative diseases such as Alzheimer's and Parkinson's disease. Some polyphenols also exhibit antimicrobial activity, helping to combat bacterial and viral infections (Singh et al., 2024).

10.3.3 Glucosinolates and Isothiocyanates

These sulphur-containing compounds are primarily found in cruciferous vegetables such as broccoli, cabbage, Brussels sprouts, cauliflower, kale, and mustard greens. Upon enzymatic hydrolysis, glucosinolates form isothiocyanates – bioactive molecules known for their detoxifying and anticancer properties. They promote the elimination of harmful toxins from the body by inducing phase II detoxification enzymes and inhibiting the activation of carcinogens. Isothiocyanates also suppress tumour growth by inducing apoptosis (programmed cell death) and inhibiting angiogenesis (formation of new blood vessels in tumours). Regular consumption of cruciferous vegetables is linked to a lower risk of cancers, particularly of the breast, lung, and colon. Moreover, these compounds support liver function and strengthen immune regulation.

10.3.4 Organosulphur Compounds

Organosulphur compounds, such as allicin from garlic and diallyl sulphide from onions and leeks, are known for their broad spectrum of biological activities. These compounds exhibit strong antimicrobial, antiviral, and antifungal properties, making them natural agents against infections. They also contribute to cardiovascular health by lowering cholesterol levels, reducing blood pressure, and preventing platelet aggregation. OSCs enhance liver detoxification and protect against oxidative stress by boosting the activity of antioxidant enzymes like glutathione peroxidase. Their regular consumption supports heart health and may reduce the risk of certain cancers.

10.3.5 Dietary Fibre and Prebiotics

Vegetables are rich in both soluble and insoluble fibres such as inulin (found in onions, garlic, and asparagus); cellulose; and pectin (present in carrots, beets, and leafy greens). Dietary fibre improves digestion, prevents constipation, and supports healthy bowel function. Soluble fibre helps in lowering cholesterol levels by binding bile acids, while insoluble fibre aids in maintaining bowel regularity (Vu, 2025). Prebiotic fibres, such as inulin, serve as food for beneficial gut bacteria, promoting a healthy gut microbiome and enhancing immune function. Fibre also helps in regulating blood sugar levels and contributes to weight management by promoting a feeling of fullness and reducing overeating.

10.3.6 Vitamins and Minerals as Nutraceuticals

Vegetables are abundant in essential vitamins and minerals which function as vital nutraceuticals. Vitamin C, found in bell peppers, broccoli, and leafy greens, serves as a powerful antioxidant that strengthens the immune system, enhances collagen synthesis, and accelerates wound healing. Vitamin K, abundant in spinach and kale, is crucial for blood clotting and bone metabolism. Folate, present in spinach and lettuce, supports DNA synthesis and nervous system function, making it particularly important during pregnancy (Xiang et al., 2023). Potassium, obtained from potatoes and spinach, helps regulate blood pressure, nerve signalling, and muscle contraction. Collectively, these micronutrients maintain physiological balance and prevent deficiency-related disorders (Yadav et al., 2024).

10.3.7 Chlorophyll and Phytosterols

Chlorophyll, the green pigment found in leafy vegetables, possesses detoxifying, antioxidant, and wound-healing properties. It assists in cleansing the body by binding to and eliminating toxins and supports liver function. Phytosterols, plant-derived compounds structurally similar to cholesterol, are found in vegetables like broccoli and Brussels sprouts. They help lower LDL (bad) cholesterol levels by competing with dietary cholesterol absorption in the intestine, thereby supporting cardiovascular health. Additionally, phytosterols possess anti-inflammatory and anticancer properties, further enhancing their nutraceutical value.

10.4 LACK OF CONSUMER AWARENESS AS AN ISSUE

One of the most significant hurdles in the successful growth of marketable nutraceuticals from vegetables is a lack of general consumer awareness. Although there are scientific foundations for nutraceuticals having health-promoting and disease-preventing properties, most consumers do not know what nutraceuticals are, how they represent a different category than standard foods or supplements, or how they relate to the consumer (Radha et al., 2023).

10.4.1 Lack of Knowledge of Nutraceuticals

For the average consumer, the terminology surrounding nutraceuticals, bioactive compounds, phytochemicals, and functional foods often remains unclear, leading to considerable confusion. Most people tend to classify products simply as either 'food' or 'medicine', leaving nutraceuticals in an ambiguous middle ground that blurs understanding of their true purpose and benefits. This lack of clear distinction can make it difficult for consumers to appreciate the scientific basis and therapeutic potential of nutraceuticals (Kavitha Shree et al., 2025). Furthermore, many individuals assume that consuming vegetables alone is sufficient to obtain all possible health benefits, without realizing that the bioactive compounds responsible for these effects must be identified, extracted, and concentrated through proper processing to exert measurable therapeutic actions. As a result, there exists a significant gap between public perception and scientific reality regarding how nutraceuticals work, emphasizing the need for greater awareness, education, and clarity in communication about these functional food products (Augustin & Sanguansri, 2015).

10.4.2 Confusion from Marketing and Labels

The modern marketplace is flooded with a wide variety of health-oriented products, including vitamins, herbal supplements, fortified foods, and nutraceuticals – all of which often share similar health-focused terminology and promotional claims on their packaging. While this abundance reflects growing consumer interest in wellness, it also contributes to significant confusion and scepticism. Many consumers find it challenging to distinguish between genuine nutraceuticals backed by scientific evidence and products that rely on exaggerated or misleading health claims for marketing appeal. The overlap in terminology and lack of regulatory clarity further blur the lines between these categories, making it difficult to identify products with proven efficacy (Danielski, 2023). Additionally, instances of misrepresentation and false advertising have, at times, damaged the credibility of the nutraceutical industry as a whole, overshadowing legitimate scientific advancements and products with demonstrated benefits (El Sohaimy, 2012). As a result, ensuring transparency, accurate labelling, and evidence-based communication is essential to maintain consumer trust and uphold the integrity of authentic nutraceuticals.

10.5 SAFETY, DOSAGE, AND REGULATION CONCERNS

Ensuring the safety, efficacy, and credibility of nutraceuticals requires meticulous attention to ingredient identity, purity, formulation, and population-specific safety. The first and foremost step is identity and standardization of ingredients, where it is essential to verify the plant species, plant part used, and standardized bioactive content – for instance formulations specifying 'curcuminoids ≥ 95%' or 'sulforaphane yield per serving'. Non-standardized ingredients can lead to significant variability in efficacy and increased safety risks. Contaminants and adulteration represent another major concern; hence, each batch should be rigorously tested for heavy metals, pesticides, mycotoxins, residual solvents, and microbiological load. Adulteration – such as the intentional addition of unrelated substances to mimic activity – must be prevented through robust analytical testing and reliable sourcing. The U.S. FDA provides comprehensive guidance on supplement labelling and truthfulness, emphasizing that all health claims must be scientifically supported and labels must not mislead consumers (Nasri et al., 2014).

Additionally, allergenicity and idiosyncratic reactions should be carefully evaluated as some individuals may exhibit sensitivity to plant proteins, carriers, or excipients (e.g. soy lecithin).

Hence, proper labelling, allergen reporting, and documentation of user experiences are essential. Pharmacokinetic variations and interactions must also be taken into account, as many phytochemicals show low or inconsistent bioavailability. Bioavailability enhancers such as piperine or curcumin may alter drug metabolism pathways, particularly cytochrome P450 enzymes (CYPs) and P-glycoprotein (P-gp), potentially leading to drug–nutrient interactions. These effects should be thoroughly assessed when defining inclusion and exclusion criteria and while designing safety monitoring protocols (Shinde et al., 2014).

Awareness of population-specific risks is equally vital, especially for vulnerable groups such as pregnant or lactating women, children, or individuals with hepatic, renal, or thyroid dysfunctions. Excessive intake of certain compounds, like brassica-derived glucosinolates or oxalate-rich plants, can lead to adverse effects such as low blood pressure or kidney stones. Finally, the dosage form and stability of nutraceuticals significantly influence their long-term potency and safety. Factors such as encapsulant type, particle size, emulsification techniques, and packaging materials can alter bioavailability and shelf life. Therefore, it is essential to provide validated stability data consistent with label claims to ensure the product remains effective and safe throughout its intended use period.

10.6 FUTURE PROSPECTS AND INNOVATIONS IN NUTRACEUTICALS DERIVED FROM VEGETABLES

Vegetable-based nutraceuticals are gaining popularity worldwide because they could assist with promoting health and the prevention and management of long-term diseases such as cancer, diabetes, and CVDs (Daliri & Lee, 2015). Vegetables contain several biologically active compounds, including polyphenols, flavonoids, carotenoids, glucosinolates, vitamins, and minerals, which may exhibit antioxidant, anti-inflammatory, and immuno-activating effects. A possible determinant of the future successes of vegetable nutraceuticals is likely to be the advances in fields such as biotechnology, food science, and humanized medicine (Das et al., 2012).

10.6.1 Personalized Nutrition and Nutrigenomics

The most exciting development lies in the area of nutrigenomics, where dietary recommendations and nutraceutical development and recommendations are catered to the individual's genetic and metabolic profile. Some individuals with genetic variations may receive a greater benefit from SFN from broccoli or lycopene from tomatoes (Kussmann & Fay, 2008).

10.6.2 Advances in Delivery Systems

The challenges of low bioavailability and stability for vegetable bioactives are one of the most limiting factors. Curcumin, lycopene, and quercetin are examples of compounds which are often susceptible to degradation during storage or poorly absorbed in the gut. Future innovative developments, such as those which involve nanotechnology, liposomes, and microencapsulation, will greatly enhance the solubility, absorption, and controlled release of these bioactives. Researchers are leaning towards using these systems to protect sensitive bioactive compounds as well as to enhance their therapeutic benefit (Shao et al., 2020; Espro et al., 2021).

10.6.3 Sustainability and Green Extraction

With growing concerns about food waste and the environment, sustainability will need to take centre stage in the nutraceutical landscape. Edible vegetable by-products including peels, stalks, and seeds are now recognized as valuable species of bioactives and are effective sources of bioactive compounds. Environmentally friendly methods of extraction, such as supercritical CO_2 extraction, UAE, and MAE, will help meet the need for the most effective, safe, and sustainable production

(Balakrishnan et al., 2025). Furthermore, biofortification strategies through plant breeding and biotechnology are predicted to elevate the natural levels of bioactive compounds found in vegetables, enabling a continued supply of quality nutraceuticals.

10.6.4 Functional Foods and Consumer Incorporation

A key trend is the incorporation of vegetable-based nutraceuticals into functional foods and beverages (Bansal et al., 2023). Instead of consuming nutraceuticals in a capsule form, more consumers will prefer these nutraceuticals in soups, smoothies, baked goods, or customized foods made with a 3D printer.

10.6.5 Synergistic Formulations and Clinical Verification

Future research in nutraceuticals will implement synergistic formulations. Rather than test one bioactive compound in isolation, researchers will combine multiple bioactive compounds for improved efficacy. An example may include using lycopene from tomatoes with quercetin from onions, as this combination may be more effective in reducing oxidative stress collectively. Clinical verification will be paramount. To be accepted into the medical community, large-scale human trials and research based on evidence must still be conducted. While vegetable-based nutraceuticals may not yet be recommended by physicians as complementary interventions in managing lifestyle-related diseases, anecdotal accounts of success reinforce that vegetable nutraceuticals can be effective in many instances (Yang et al., 2024).

10.6.6 Biotechnology and the Development of Functional Foods

Recent advancements in biotechnology have transformed the development of functional foods, including those fortified with nutraceuticals based on plant sources. Biotechnology provides tools to enhance nutritional intake from vegetables, enhance the extraction and stability of bioactive compounds, and develop new delivery systems. One major development is biofortification, an approach of genetically improving crops so they contain greater concentrations of the specified bioactive and/or other essential nutrients. For example higher carotenoid levels can be engineered in carrots and therefore increase potential nutraceutical benefits, while the concentrations of lycopene and flavonoids can be developed to be higher in tomatoes. Another major area is utilizing plant tissue culture and metabolic engineering at scale, where the bioactives being produced are generally rare or unstable and can generate long-term, consistent supplies through intentional production regardless of seasonal factors (Meléndez-Martínez et al., 2021). Also, biotechnology will help with creating probiotic and prebiotic functional foods which target the microbiome and develop gut health-focused products.

The functional food sector has also expanded rapidly. It's moving beyond traditional dietary supplements and introducing nutraceuticals into more conventional uses. Functional foods not only deliver nutritional needs but also promise the prevention of disease and longevity in wider health (Çakmakçı et al., 2024). We see plant-based nutraceuticals entering beverages, snacks, soups, dairy replacements, and even microwave meals. There are new delivery systems (e.g. nanotechnology) which increase the solubility and uptake of poorly absorbed compounds such as curcumin or quercetin for maximum efficacy in functional foods. Smart food technologies such as 3D food printing and encapsulation systems allow for customized products that align with health needs – incorporating functional food elements into a more personalized approach. In all these areas of progress involving functional foods, biotechnology, and food science, the intersection is closing the gap between medicine and food. It is an exciting time for researchers, food developers, and health practitioners as consumers demand plant-based products which are natural and sustainable. The combination of functional, nutraceutical-enhanced, plant-based products driven by biotechnology will be a critical aspect of global health and wellness going forward (Shan et al., 2025).

10.7 NUTRITION AND NUTRIGENOMICS

Personalized nutrition is a new frontier for vegetable-based nutraceuticals, based on the science of nutrigenomics. Nutrigenomics studies the effects of metagenomic differences in individuals on how the body processes nutrients and bioactive compounds (Rane et al., 2024). By contrast to the more traditional 'one-size-fits-all' approach, this method can create personalized diet plans for a few or a nutraceutical product for any individual's genetic expression, lifestyle, health condition, and other co-morbidities or health status specific to the consumer. For instance consumers with certain genetic variations in detoxification pathways may experience additional benefits from high SFN cruciferous vegetables. Likewise, fermented products or tomato-based products may drive an improvement in health for those with a genetic predisposition to CVD or risk. While genomics and genetic sequencing offer the ability to select different vegetable bioactives for individual needs, metabolomics and gut microbiota research now help evaluate multiple levels of impact to better demonstrate the preventive and therapeutic benefits of these products.

In addition, individualized nutrition is likely to allow functional foods to double as precision tools for health management. With the bioinformatics from nutrigenomics and novel food technologies, it will be possible to deliver personalized foods or drinks containing the nutraceuticals suited to an individual's unique genetic code (Rao et al., 2024). One could imagine a smoothie with vegetable extracts specifically suited to someone who suffers from oxidative stress or is more vulnerable to inflammatory responses. This not only increases efficacy, but it also engenders trust because evidence-based recommendations will be based on specific data and not because population data suggest that leafy greens are a healthy food choice. As biotech, artificial intelligence, and big data continue to advance, nutrigenomic-based individualized nutrition is poised to play an increasingly prominent role in future health care, as nutraceuticals will more closely link food and medicine using vegetable-based nutraceuticals.

10.8 ROLE IN PREVENTIVE HEALTHCARE

Vegetable-derived nutraceuticals belong to the realm of preventive healthcare. Preventive healthcare focuses primarily on reducing disease, rather than on treating disease once it occurs. Vegetables are a rich source of bioactive compounds, including carotenoids, flavonoids, glucosinolates, polyphenols, vitamins, and minerals, which reduce disease progression and support health maintenance (Pai et al., 2022). For example lycopene found in tomatoes and β-carotene from carrots is a potent antioxidant that neutralizes free radicals and appears to reduce risk for cancer as well as cardiovascular disorders. Glucosinolates in cruciferous vegetables, such as broccoli and cabbage, promote data that indicates a reduction in risks for certain cancers by supporting our cell detoxification and excretion processes (Saiwal et al., 2019). Flavonoids in green leafy vegetables are shown to support vascular health and suppress inflammation. Therapeutic drugs typically suppress disease conditions, whereas nutraceuticals derived from vegetables, at the molecular and cellular levels, support normal body processes and enhance function and immune mechanisms (Puri et al., 2022). Therefore, nutraceuticals derived from vegetables offer a tremendous utility in disease management, including those resulting from lifestyle habits, such as obesity, type 2 diabetes, hypertension, and neurodegenerative diseases.

Moreover, vegetable nutraceuticals are well aligned with increasing attention to holistically and sustainably focused health initiatives. As the incidence of NCDs continues to escalate globally, health-care systems are taking a shift away from costly long-term treatment plans towards prioritizing prevention. Regular use of vegetable-based nutraceuticals in the form of dietary supplements or functional food provides a reasonable opportunity for lowering disease burden and healthcare costs. Understanding preventive care through nutraceuticals is not only effective (usually backed by the research of other studies) but also part of widespread public acceptance, particularly due to being food-sourced (natural plant sources consumed as part of traditional diets). It is believed that future research will provide even more solid connections between bioactives from certain vegetables and the prevention of chronic disease, as research stakeholders should progressively increase clinical studies demonstrating nutritional science and, by then, produce supporting and tailored research studies (Pandey et al., 2021).

10.9 SUSTAINABILITY IN PRODUCTION

Sustainability in the production of vegetable-based nutraceuticals is important to ensure productive innovations. The industry has to balance consumer demand with environmental responsibility (Escamilla-García, 2025). Most of the current extraction and processing methods use large quantities of chemical solvents and generate waste and ecological problems. To address this issue, researchers are searching for greener and sustainable technologies, including supercritical carbon dioxide extraction, UAE, and MAE. These methods have less environmental impact, along with higher yield and purity of bioactive substances (Jayan & Wani, 2025).

Another important measure is the use of by-products from agricultural processes. Peels, seeds, stems, and leaves have traditionally been disposed of as waste and have been considered bad for the environment, even with some potential for nutraceutical applications (Bhattacharyya et al., 2024). Lots of nutraceuticals come from peel by-products, tomato skins contain an abundance of lycopene, carrot pomace has some carotenoids, and broccoli stalks are rich in glucosinolates. The addition of agricultural by-products would also reduce the waste produced in a lifestyle and create value in a circular production scheme (Kaur, 2016).

The integrated approach of eco-extraction, waste valorization, and sustainable methods reduces the amount of waste and resources in producing nutraceuticals in both economies and environments, and as consumers continue to choose naturally derived nutraceuticals, from sustainably and ethically sourced vegetables, encouraging the uptake of regenerative technologies will be beneficial, not only to the sustainability agenda but also for enhancing the competitiveness of global trade of vegetable-derived nutraceuticals (Espro et al., 2021).

10.10 REGULATORY FRAMEWORK, MARKET POTENTIAL, CHALLENGES

Nutraceuticals occupy a unique position at the interface of conventional foods and pharmaceuticals, which makes their regulation more complex than standard medicines. Unlike pharmaceutical drugs which require strict and lengthy trials to establish safety and efficacy, nutraceuticals are generally governed by food-based regulatory frameworks, leading to significant variation in rules across countries and challenges for manufacturers, policymakers, and consumers (Santini & Novellino, 2018). In India, nutraceuticals fall under the FSSAI and are regulated through the Food Safety and Standards Regulations, 2016. Nutraceuticals are officially treated as food products rather than medicines (Kalra, 2014), and permissible claims are limited to functional benefits such as 'supports bone health' or 'aids digestion', while disease-curing claims are strictly prohibited (FSSAI, 2016). Manufacturers must ensure product safety by controlling heavy metals, pesticide residues, and microbial contamination (Ghosh & Playne, 2021), and all health claims must be backed by scientific evidence (Santini et al., 2018). Dosage limits for vitamins, minerals, and herbal extracts are also prescribed to prevent toxicity (FSSAI, 2016). Despite this framework, differentiating nutraceuticals from pharmaceuticals remains confusing, and enforcement of regulations is often inconsistent, with misleading labelling and exaggerated claims persisting in the Indian market (Rishi et al., 2020).

Internationally, regulatory approaches vary widely. In the United States, nutraceuticals are categorized as dietary supplements under the Dietary Supplement Health and Education Act (DSHEA) of 1994. Pre-market approval is not required, and manufacturers are responsible for ensuring safety. Claims are restricted to general health benefits such as 'supports heart health' or 'promotes immunity', and all products must include a disclaimer stating that the FDA has not evaluated the statements and that the product is not intended to diagnose, treat, cure, or prevent any disease

(FDA, 1994; Bailey et al., 2013; Dwyer et al., 2018). In contrast, the European Union follows a stricter approach, regulating nutraceuticals as functional or novel foods. Health claims require prior approval by the EFSA, and functional claims must be supported by robust clinical data. Any ingredient not widely consumed before May 15, 1997, must undergo a novel food approval process, which ensures consumer safety but can slow down innovation (Martirosyan & Singh, 2015; EFSA, 2006; Santini et al., 2018). In Canada, nutraceuticals are considered NHPs, requiring product licensing, scientific proof of safety, efficacy, and quality, as well as compliance with GMP (Health Canada, 2018; Dwyer et al., 2018). Japan operates the FOSHU system (Foods for Specified Health Uses), where products are reviewed by the Ministry of Health, Labour, and Welfare and can carry the FOSHU logo to signal verified safety and efficacy, exemplified by plant sterol-containing beverages approved for cholesterol-lowering effects (Shimizu, 2012; Santini et al., 2018).

The global nutraceutical market has experienced tremendous growth, driven by rising health awareness, ageing populations, and an increasing burden of lifestyle-related diseases. The global market was valued at approximately USD 400 billion in 2023 and is projected to reach USD 700 billion by 2030, with a CAGR of 7–9% (Fortune Business Insights, 2023). North America and Europe dominate the market due to high consumer awareness and purchasing power, while the Asia-Pacific region, particularly India and China, is experiencing the fastest growth owing to preventive healthcare adoption and local bioresource availability (Mordor Intelligence, 2023; Invest India, 2023). India's nutraceutical market, valued at USD 6 billion in 2023, is expected to triple to USD 18 billion by 2030, supported by high prevalence of lifestyle diseases, rich biodiversity for bioactive compounds, and government initiatives such as 'Make in India' (Rishi et al., 2020; FICCI, 2022; Invest India, 2023). The major market segments include functional beverages (e.g. probiotics, energy drinks); dietary supplements (proteins, vitamins, herbal extracts); and functional foods (fortified cereals, dairy products) (FICCI, 2022).

Despite growing consumer demand, nutraceuticals face scepticism from the medical community. Physicians often hesitate to recommend them due to limited clinical evidence, inconsistent product quality, potential drug–nutrient interactions, regulatory ambiguities, and aggressive marketing without robust data (Kalra, 2014; Dwyer et al., 2018; Rishi et al., 2020; Gupta et al., 2013; Santini & Novellino, 2018). Most studies are either in vitro or animal-based, with few large-scale human clinical trials. Bioactive content in plants can vary with soil, season, and processing techniques, leading to unpredictable results, while compounds like curcumin can interfere with conventional medications such as anticoagulants. To build confidence among healthcare providers, several measures are necessary: conducting large-scale human clinical trials to generate reliable data (Martirosyan & Singh, 2015); ensuring transparent labelling with precise dosage instructions and bioactive content (FSSAI, 2016)\; integrating nutraceutical education into medical training programmes (Rishi et al., 2020); and fostering collaboration between industry and medical professionals to provide safe, science-based guidance to patients (Dwyer et al., 2018).

10.11 CONCLUSION

Nutraceuticals made from vegetables are essential for preventing illness and bridging the gap between medicine and diet. In contemporary nutrition research, their capacity to lower the risk of chronic illnesses, boost immunity, and promote healthy ageing makes them essential. But there are obstacles in the industry, including inconsistent regulations, expensive production costs, a lack of clinical validation, and low physician adoption. An approach involving several stakeholders is necessary for nutraceuticals to reach their full potential. Globally standardized regulations must be put into effect by policymakers. Researchers must use human subjects in large-scale experiments. Transparency, affordability, and quality must be guaranteed by the industry. Safe and efficient usage has to be taught to patients and healthcare professionals. The field of nutraceuticals has a promising future. Innovations in biotechnology, functional foods, nutrigenomics, and sustainable manufacturing have made vegetable-based nutraceuticals a promising component of global health plans. It is anticipated that nutraceuticals will evolve from being an 'alternative option' to a mainstream healthcare component as the globe shifts from treating illnesses to preventing them, greatly enhancing global health and wellness.

REFERENCES

Antony, P., & Vijayan, R. (2021). Bioactive peptides as potential nutraceuticals for diabetes therapy: A comprehensive review. *International Journal of Molecular Sciences*, *22*(16), 9059.

Augustin, M. A., & Sanguansri, L. (2015). Challenges and solutions to incorporating nutraceuticals in foods. *Annual Review of Food Science and Technology*, *6*(1), 463–477.

Bailey, R. L., Gahche, J. J., Miller, P. E., Thomas, P. R., & Dwyer, J. T. (2013). Why US adults use dietary supplements. *JAMA Internal Medicine*, *173*(5), 355–361.

Balakrishnan, S., Chandrasekaran, I. R., Jegadeesan, R., Ramachandran, K., Govindasamy, A. K., & Kandasamy, G. (2025). Eco-innovation in action: Valorizing vegetable waste into high-value ingredients for sustainable applications. *Frontiers in Sustainable Food Systems*, *9*, 1507120.

Bansal, P., Maithani, M., Gupta, V., Kaur, G., & Bansal, R. (2023). Prospects of nutraceutical and functional food with herbs and spices. In *Herbs, spices and their roles in nutraceuticals and functional foods* (pp. 361–381). Academic Press.

Bhattacharyya, S., Roy, K., Bhattacharyya, R., Nazir, M., Saha, A., & Saha, D. (2024). Sustainable treatment of agro-wastes for the development of novel products especially bioenergy: Prospects and constraints. In *Agro-waste to microbe-assisted value-added product: Challenges and prospects: Recent developments in agro-waste valorization research* (pp. 229–245). Cham: Springer Nature Switzerland.

Çakmakçı, S., Polatoğlu, B., & Çakmakçı, R. (2024). Foods of the future: Challenges, opportunities, trends, and expectations. *Foods*, *13*(17), 2663.

Daliri, E. B. M., & Lee, B. H. (2015). Current trends and future perspectives on functional foods and nutraceuticals. In *Beneficial microorganisms in food and nutraceuticals* (pp. 221–244). Cham: Springer International Publishing.

Danielski, R. (2023). Challenges hindering the commercialization of nutraceuticals derived from agri-food by-products. *Journal of Food Bioactives*, *24*.

Das, L., Bhaumik, E., Raychaudhuri, U., & Chakraborty, R. (2012). Role of nutraceuticals in human health. *Journal of Food Science and Technology*, *49*, 173–183. https://doi.org/10.1007/s13197-011-0269-4.

DeFelice, S. L. (1995). The nutraceutical revolution: Its impact on food industry R&D. *Trends in Food Science & Technology*, *6*, 59–61.

Dwyer, J. T., Coates, P. M., & Smith, M. J. (2018). Dietary supplements: Regulatory challenges and research resources. *Nutrients*, *10*(1), 41. https://doi.org/10.3390/nu10010041

El Sohaimy, S. A. (2012). Functional foods and nutraceuticals: A modern approach to food science. *World Applied Sciences Journal*, *20*(5), 691–708.

Escamilla-García, E., Arevalo-Niño, K., García-Fajardo, J. A., Sánchez-Romo, D., Avilés-Arnaut, H., De la Garza, A. L., . . . & Pereyra-Alférez, B. (2025). Future of nutraceuticals as a safe and intelligent food. In *Food bioactives and nutraceuticals: Dietary medicine for human health and well-being* (pp. 35–58). Singapore: Springer Nature Singapore.

Espro, C., Paone, E., Mauriello, F., Gotti, R., Uliassi, E., Bolognesi, M. L., Rodríguez-Padrón, D., & Luque, R. (2021). Sustainable production of pharmaceutical, nutraceutical and bioactive compounds from biomass and waste. *Chemical Society Reviews*, *50*(20), 11191–11207.

European Food Safety Authority (EFSA). (2006). Regulation (EC) No. 1924/2006 on nutrition and health claims made on foods. *Official Journal of the European Union*.

FICCI (Federation of Indian Chambers of Commerce & Industry). (2022). Nutraceuticals: Paving the way for preventive healthcare. *FICCI Reports*.

Food and Drug Administration (FDA). (1994). Dietary Supplement Health and Education Act (DSHEA). *U.S. Food and Drug Administration*.

Food Safety and Standards Authority of India (FSSAI). (2016). *Food Safety and Standards* (Health Supplements, Nutraceuticals, Food for Special Dietary Use, Food for Special Medical Purpose, Functional Food and Novel Food) Regulations.

Fortune Business Insights. (2023). *Nutraceuticals market size, share & trends, 2023–2030*.

Galanakis, C. M. (2013). Emerging technologies for the production of nutraceuticals from agricultural by-products: A viewpoint of opportunities and challenges. *Food and Bioproducts Processing*, *91*(4), 575–579.

Garza-Juárez, A., Pérez-Carrillo, E., Arredondo-Espinoza, E. U., Islas, J. F., Benítez-Chao, D. F., & Escamilla-García, E. (2023). Nutraceuticals and their contribution to preventing noncommunicable diseases. *Foods, 12*(17), 3262.

Ghosh, D., & Playne, M. J. (2021). Functional foods and nutraceuticals: Opportunities and challenges. *Critical Reviews in Food Science and Nutrition, 61*(2), 177–192.

Gupta, S. C., Patchva, S., & Aggarwal, B. B. (2013). Therapeutic roles of curcumin: Lessons learned from clinical trials. *AAPS Journal, 15*(1), 195–218.

Health Canada. (2018). Natural health products regulations. *Health Canada Guidance Document.*

Husen, A., & Iqbal, M. (2021). Plant-based potential nutraceuticals for improving the human immune system. In *Traditional herbal therapy for the human immune system* (Ed. Husen, A., pp. 1–11). Boca Raton, FL: CRC Press. https://doi.org/10.1201/9781003137955-1

Invest India. (2023). Indian nutraceuticals industry: Growth and opportunities. *Government of India, Invest India Reports.*

Jaglan, P., Buttar, H. S., Kaur, G., Garg, V. K., & Télessy, I. G. (2025). Bioeconomic model for the extraction of bioactive compounds for nutraceuticals and phytomedicines from fruits and vegetables waste: A systematic review. *Molecular Medicine and Biomedical Research in the Era of Precision Medicine*, 975–988.

Jarouliya, U., Keservani, R. K., Kesharwani, R. K., Patel, V. K., & Bharti, A. D. (Eds.). (2024). *Immune-boosting nutraceuticals for better human health: Novel applications* (1st ed.). Apple Academic Press. https://doi.org/10.1201/9781003371069

Jayan, L. S., & Wani, K. M. (2025). Utilization of industrial fruits and vegetables waste streams for the production of 3D printed and synbiotic foods. *Journal of Food Measurement and Characterization*, 1–21.

Kalra, E. K. (2014). Nutraceutical – Definition and introduction. *AAPS PharmSciTech, 5*(3), 27–28.

Kaur, K. (2016). Functional nutraceuticals: Past, present, and future. *Neutraceuticals, 4*, 41–78.

Kavitha Shree, G. G., Arokiamary, S., Kamaraj, M., & Aravind, J. (2025). Biorefinery approaches for converting fruit and vegetable waste into sustainable products. *International Journal of Environmental Science and Technology, 22*(8), 7211–7230.

Kussmann, M., Abe Cunha, D. H., & Berciano, S. (2023). Bioactive compounds for human and planetary health. *Frontiers in Nutrition, 10*, 1193848.

Kussmann, M., & Fay, L. B. (2008). Nutrigenomics and personalized nutrition: Science and concept. *Personalized Medicine, 5*(5), 447–455. https://doi.org/10.2217/17410541.5.5.447

Martirosyan, D. M., & Singh, J. (2015). A new definition of functional food by FFC: What makes a new definition unique? *Functional Foods in Health and Disease, 5*(6), 209–223.

Meléndez-Martínez, A. J., Böhm, V., Borge, G. I. A., Cano, M. P., Fikselová, M., Gruskiene, R. & O'Brien, N. M. (2021). Carotenoids: Considerations for their use in functional foods, nutraceuticals, nutricosmetics, supplements, botanicals, and novel foods in the context of sustainability, circular economy, and climate change. *Annual Review of Food Science and Technology, 12*(1), 433–460.

Mokoena, M. P., Mutanda, T., & Olaniran, A. O. (2016). Perspectives on the probiotic potential of lactic acid bacteria from African traditional fermented foods and beverages. *Food & Nutrition Research, 60*(1), 29630.

Mordor Intelligence. (2023). *Global nutraceuticals market report, 2023–2028.* https://www.mordorintelligence.com/industry-reports/global-nutraceuticals-market-industry

Nagpal, D., Nema, S., Nagpal, S., Pandey, M. M., Kaushik, D., & Kathuria, H. (2025). Management and prevention of neurodegenerative disorders: Can antioxidant-rich dietary interventions help?. *Antioxidants, 14*(9), 1078.

Nasri, H., Baradaran, A., Shirzad, H., & Rafieian-Kopaei, M. (2014). New concepts in nutraceuticals as alternative for pharmaceuticals. *International Journal of Preventive Medicine, 5*(12), 1487–1499.

Pai, S., Hebbar, A., & Selvaraj, S. (2022). A critical look at challenges and future scopes of bioactive compounds and their incorporation in the food, energy, and pharmaceutical sectors. *Environmental Science and Pollution Research, 29*(24), 35518–35541.

Pandey, M., Verma, R. K., & Saraf, S. A. (2010). Nutraceuticals: New era of medicine and health. *Asian Journal of Pharmaceutical and Clinical Research, 3*(1), 11–15.

Prabhakar, P., Verma, D. K., & Banerjee, M. (2025). Safety and toxicity of nutraceuticals: Concerns, challenges and global regulatory aspects. In *Food bioactives and nutraceuticals: Dietary medicine for human health and well-being* (pp. 437–462). Singapore: Springer Nature Singapore.

Puri, V., Nagpal, M., Singh, I., Singh, M., Dhingra, G. A., Huanbutta, K., . . . & Sangnim, T. (2022). A comprehensive review on nutraceuticals: Therapy support and formulation challenges. *Nutrients, 14*(21), 4637.

Qureshi, I., Habib, M., Bashir, K., Jan, K., & Jan, S. (2024). Introduction to functional foods and nutraceuticals. In *Functional foods and nutraceuticals: Chemistry, health benefits and the way forward* (pp. 1–15). Cham: Springer International Publishing.

Radha, A., Ahluwalia, V., Rai, A. K., Varjani, S., Awasthi, M. K., Sindhu, R., . . . & Kumar, V. (2024). The way forward to produce nutraceuticals from agri-food processing residues: Obstacles, solutions, and possibilities. *Journal of Food Science and Technology*, *61*(3), 429–443. https://doi.org/10.1007/s13197-023-05729-9

Rane, B. R., Amkar, A. J., Patil, V. S., Vidhate, P. K., & Patil, A. R. (2024). Opportunities and challenges in the development of functional foods and nutraceuticals. *Formulations, Regulations, and Challenges of Nutraceuticals*, 227–254.

Rao, T. J. M., Kesharwani, R. K., Keservani, R. K., & Sharma, A. K. (Eds.). (2024). *Formulations, regulations, and challenges of nutraceuticals*. CRC Press.

Riar, C. S., & Panesar, P. S. (2024). Bioactive compounds and nutraceuticals: Classification, potential sources, and application status. In *Bioactive compounds and nutraceuticals from dairy, marine, and nonconventional sources* (pp. 3–60). Apple Academic Press.

Rishi, P., Thakur, R., & Vij, S. (2020). Nutraceuticals: Current regulatory scenario and future prospects. *International Journal of Pharmaceutical Sciences and Research*, *11*(3), 1000–1010.

Ristori, S., Bertoni, G., Bientinesi, E., & Monti, D. (2025). The role of nutraceuticals and functional foods in mitigating cellular senescence and its related aspects: A key strategy for delaying or preventing aging and neurodegenerative disorders. *Nutrients*, *17*(11), 1837.

Saiwal, N., Dahiya, M., & Dureja, H. (2019). Nutraceutical insight into vegetables and their potential for nutrition-mediated healthcare. *Current Nutrition & Food Science*, *15*(5), 441–453.

Santini, A., Cammarata, S. M., Capone, G., Ianaro, A., Tenore, G. C., Pani, L., & Novellino, E. (2018). Nutraceuticals: opening the debate for a regulatory framework. *British Journal of Clinical Pharmacology*, *84*, 659–672. https://doi.org/10.1111/bcp.13496.

Santini, A., & Novellino, E. (2018). Nutraceuticals: Beyond the diet before the drugs. *Current Opinion in Food Science*, *22*, 145–152.

Shan, F., Liu, L., Li, L., Wang, W., Bi, Y., & Li, M. (2025). Management, safety, and efficacy evaluation of nutraceutical and functional food: A global perspective. *Comprehensive Reviews in Food Science and Food Safety*, *24*(4), e70222.

Shao, J. W., Jiang, J. L., Zou, J. J., Yang, M. Y., Chen, F. M., Zhang, Y. J., & Jia, L. (2020). Therapeutic potential of ginsenosides on diabetes: From hypoglycemic mechanism to clinical trials. *Journal of Functional Foods*, *64*, 103630. https://doi.org/10.1016/j.jff.2019.103630

Shimizu, T. (2012). Health claims on functional foods: The Japanese perspective. *Nutrition Research Reviews*, *25*(2), 242–247.

Shinde, N., Bangar, B., Deshmukh, S., & Kumbhar, P. (2014). Nutraceuticals: A review on current status. *Research Journal of Pharmacy and Technology*, *7*(1), 110–113.

Shirodkar, S., Khanchey, F., & Babre, N. (2025). Role of functional foods and nutraceuticals in nutrition and health. *Academia Nutrition and Dietetics*, *2*(2).

Singh, A., Singh, J., Parween, G., Khator, R., & Monga, V. (2024). A comprehensive review of apigenin, a dietary flavonoid: Biological sources, nutraceutical prospects, chemistry, and pharmacological insights and health benefits. *Critical Reviews in Food Science and Nutrition*, 1–37.

Vu, N. T. (2025). Current approaches to food waste valorization for nutraceuticals. In *Sustainable pharmaceutical product development and optimization processes: From eco-design to supply chain integrity* (pp. 307–338). Singapore: Springer Nature Singapore.

Xiang, J., Mlambo, R., Shaw, I., Seid, Y., Shah, H., He, Y., . . . & He, B. (2023). Cryopreservation of bioflavonoid-rich plant sources and bioflavonoid-microcapsules: Emerging technologies for preserving bioactivity and enhancing nutraceutical applications. *Frontiers in Nutrition*, *10*, 1232129.

Yadav, S., Malik, K., Moore, J. M., Kamboj, B. R., Malik, S., Malik, V. K., . . . & Bishnoi, D. K. (2024). Valorisation of agri-food waste for bioactive compounds: Recent trends and future sustainable challenges. *Molecules*, *29*(9), 2055.

Yang, S., Fan, Y., Cao, Y., Wang, Y., Mou, H., & Sun, H. (2024). Technological readiness of commercial microalgae species for foods. *Critical Reviews in Food Science and Nutrition*, *64*(22), 7993–8017.

11 Importance of Vegetables in Regular Diet

Simran Panda, Anushka Pandey, Adnan Alam, Tanushree Tandon, Kavita Srivastava, and Tanvi Jain

11.1 INTRODUCTION

Vegetables are recurrent agricultural crops, with certain divisions such as roots, stalks, flowers, fruits, and leaves, which can be consumed entirely or partially, cooked or raw (Welbaum, 2015). Vegetables are universally regarded as a critical constituent of a balanced and healthy diet. For centuries, human societies across different religions have relied on vegetables as a primary food source for their growth and development, nutriment, and medicinal purposes. Contemporary nutritional science has already confirmed what traditional wisdom long suggested: vegetables are not only nutrient-dense but also carry bioactive compounds which protect against a wide range of chronic diseases. Their consumption has been consistently associated with improved overall health outcomes, prevention of malnutrition, and reduced mortality from NCDs (Boeing et al., 2012). Vegetables are particularly important because they provide essential macronutrients and micronutrients in appropriate quantities without contributing excessive energy or fat. Unlike animal-based foods, vegetables are naturally low in saturated fat and cholesterol, while being high in dietary fibre, vitamins (such as vitamin A, vitamin C, vitamin K, and folate), and minerals (such as potassium, magnesium, and iron). Moreover, vegetables are rich in phytochemicals, including carotenoids, flavonoids, and glucosinolates, which act as antioxidants, modulate immune responses, and protect against oxidative stress and inflammation (Liu, 2013; Dixit, 2023; Generalić Mekinić et al., 2025). These bioactive compounds contribute to a long-term disease prevention, making vegetables a cornerstone of functional foods. Globally, inadequate consumption of vegetables is a major public health concern. The WHO (WHO, 2020b) recommended a minimum daily intake of 400 grams of fruits and vegetables combined; however, dietary surveys suggested that most populations fall short of these requirements. Low intake is particularly observable in urbanized societies, where processed and energy-dense foods often replace fresh production. Insufficient vegetable consumption has been linked to an increased risk of CVDs, obesity, type 2 diabetes, gastrointestinal disorders, and certain cancers (Aune et al., 2017). In fact, WHO estimates that low fruit and vegetable intake accounts for nearly 2.8 million deaths annually, underlining the urgent need to promote vegetable-rich diets.

From an epidemiological standpoint, populations with high vegetable intake tend to exhibit lower rates of chronic disease. For example cruciferous vegetables such as broccoli, cabbage, and cauliflower contain glucosinolates, which yield compounds with anticancer properties. Leafy green vegetables like spinach and kale are abundant in lutein and zeaxanthin, which are associated with eye health and prevention of AMD (Mila et al., 2024). Tomatoes, rich in lycopene, are known to support cardiovascular health and reduce the risk of prostate cancer (Boeing et al., 2012). Such findings highlight the diversity of health-promoting compounds provided by different categories of vegetables.

Vegetables are also central to sustainable dietary patterns such as the Mediterranean diet and the DASH (Dietary Approaches to Stop Hypertension) diet, both of which are widely endorsed for

DOI: 10.1201/9781003606710-11

their cardioprotective and anti-obesity benefits (Aune et al., 2017). In addition to personal health, these plant-forward dietary patterns contribute to planetary well-being by reducing environmental footprints associated with excessive meat consumption (Willett et al., 2019; Mateus et al., 2024). Thus, vegetables are critical not only for human nutrition but also for sustainable food systems and climate resilience.

The cultural and medicinal value of vegetables further enhances their importance. Traditional medicine systems such as Ayurveda, Traditional Chinese Medicine (TCM), and Unani have long recommended vegetables as natural remedies for ailments ranging from digestive issues to metabolic disorders. Garlic, for instance, has been valued for its antimicrobial and cardioprotective properties, while bitter gourd is used in the management of diabetes. Modern scientific research increasingly validates these traditional claims, revealing the therapeutic potential of vegetables as both preventive and adjunct treatments (Slavin & Lloyd, 2012; Moskwa et al., 2023).

Despite such well-established benefits, several barriers limit vegetable consumption, including socioeconomic constraints, seasonal availability, and lack of awareness. Urbanization and rapid lifestyle changes have also shifted dietary preferences toward processed foods, leading to a nutritional gap that contributes to both undernutrition and overnutrition worldwide. Bridging this gap requires coordinated action – ranging from agricultural innovation and improved supply chains to public health policies and consumer education (WHO, 2020b).

11.2 CATEGORIZATION OF VEGETABLES

There are more than 10,000 plant species approximately which are used as vegetables in the world. These plants can be categorized by considering a democratic set of characteristics. It is important for food scientists, dieticians, and researchers to sub-categorize the vegetables by considering health and nutrition. These categories will be more useful if they are based on similarities in the composition of nutritional content in food (Pennington & Fisher, 2010).

Based on the specific nutritional values and the portion of the plant used for nourishment, all vegetables can be categorized.

11.2.1 Green Vegetables

11.2.1.1. Green Leafy Vegetables

This section includes spinach, lettuce, curly lettuce, chard, purslane, etc. They consist of important minerals like iron and calcium, vitamins (A, C, and riboflavin), and other fibre sources. Freshly cultivated leaves contain more vitamin C than mature plants of green vegetables. The green exterior portion of cabbage and lettuce is richer in iron, vitamins, and calcium than their inner leaves.

11.2.1.2 Stalk Vegetables

The most commonly available stalk vegetables are asparagus and celery. Stalk vegetables comprise vitamins and minerals in accordance with the green colour. Asparagus is specifically a good source of folic acid and other folate-based compounds which are responsible for producing various important receptors present inside the body.

11.2.1.3 Fruit and Flower Vegetables

These include vegetables which resemble flowers, like cauliflower and broccoli, which are frequently eaten. Iron, phosphorus, vitamins A and C, and riboflavin are all essential components of broccoli. Additionally, cauliflower is a significant source of vitamins A and C. They can be used as cooked vegetables or as raw ingredients in salads. The most popular fruits and vegetables are tomatoes and all peppers. Vitamin C is abundant in both.

11.2.2 Root Vegetables

11.2.2.1 Root, Tuber, and Bulb Vegetables

This group of vegetables includes potatoes, carrots, beets, turnips, onions, and radishes. The precursor to vitamin A, β-carotene, is abundant in orange and yellow mixtures. One of the best examples of a root vegetable is the onion, which also includes vitamin C.

11.2.2.2 Legumes

This group comprises soy beans, peas, and plants derived from legumes. Saponin and soluble fibres like pectin and β-glucans are abundant in this category. Various categories in nutritional standards are based on the nutritional content available in various nations, and subgroups within these classifications may vary from nation to nation. For example the basic food guidelines used in the United States (Basic 7 and Basic 4 Food Groups and Food Guide Pyramid) are more closely linked to citrus fruits for vitamin C and dark green leafy and dark orange/yellow group vegetables for β-carotene. The 2010 USDA (U.S. Department of Agriculture) My-pyramid food guide included green leafy vegetables and broccoli, other leafy vegetables, legumes, unique vegetables (dark orange, tomato, Allium vegetables, etc.), and additional vegetables. The Turkey Nutrition Guide 2015, published by the Turkish Ministry of Health, categorizes vegetables into five main groups based on colour and nutritional characteristics. These include dark green leafy vegetables such as spinach, chard, lettuce, and parsley; other green vegetables like broccoli, okra, peas, and zucchini; red–orange–blue–purple vegetables including tomatoes, carrots, beets, and eggplants; white vegetables such as onions, cabbage, cauliflower, and mushrooms; and starchy vegetables like potatoes and corn. This classification emphasizes dietary diversity and encourages the inclusion of various coloured vegetables for balanced nutrition (Schmidt & Oliveira, 2023; Salesse et al., 2024).

11.3 IMPORTANCE OF VEGETABLES

Vegetables hold a fundamental position in human nutrition and overall well-being. Beyond being a daily dietary component, they are also important in addressing public health concerns such as malnutrition, micronutrient deficiencies, and NCDs. From providing essential vitamins and minerals to reducing the risk of lifestyle-related disorders, vegetables play a key role in keeping people healthy. In today's world, where fast food and processed diets are common, including vegetables in daily meals has become even more important (Boeing et al., 2012; Slavin & Lloyd, 2012).

11.3.1 Role in Fighting Malnutrition and Deficiencies

In many parts of the world, people suffer from malnutrition and lack of important nutrients. Vegetables are a natural and affordable solution to this problem. Leafy vegetables like spinach, amaranth, and kale are packed with iron, folate, vitamin A, and vitamin K, which are important for blood health, bones, and immunity (Singh et al., 2020). Likewise, carrots and pumpkins provide vitamin A, which prevents night blindness and supports immunity. Broccoli and cabbage contain folate that is essential for pregnant women, as it reduces the chances of birth defects in babies (Steinmetz & Potter, 1996). Primal micronutrients provided by vegetables includes vitamin A which is mostly present in carrots and leafy greens; vitamin C is primarily available in cabbage, bitter gourd, and squash; and B vitamins present in legumes, calcium, iron, and iodine, each of which is required for the growth, immune system functioning, and overall holistics development of the body and thus protecting the same from of deficiency-related disorders. A Holocene study analysed alteration in the abundance, diversity, and composition of crop commodities in the food supplies of over 142 countries, comparing data from 1961 to 2009 (Khoury et al., 2014). Thus, eating a variety of vegetables helps prevent nutrient deficiencies and supports overall health.

11.3.2 Contribution of Vegetables in Chronic Disease Prevention

One of the most extensively studied benefits of vegetables is their protective role against mostly all the chronic diseases. Many studies and meta-analyses revealed that diets rich in vegetables are connected with a significantly lower risk of CVDs, diabetes, obesity, and several types of cancer (Aune et al., 2017). Non-starchy vegetables like broccoli, bitter gourd, and okra help regulating the blood sugar levels by improving insulin sensitivity and reducing glycaemic load. Cruciferous vegetables contain glucosinolates which break down into biologically active compounds with anti-carcinogenic properties. Similarly, tomatoes are rich in lycopene, which has been linked to reduced prostate and breast cancer risk (Boeing et al., 2012). Vegetables like beetroot and spinach contain natural nitrates that lower blood pressure and improve blood circulation. Potassium-rich vegetables also reduce the risk of stroke (Lai et al., 2023; Chen et al., 2024; Wang et al., 2025). Figure 11.1 explains the important role of vegetables in promoting human health.

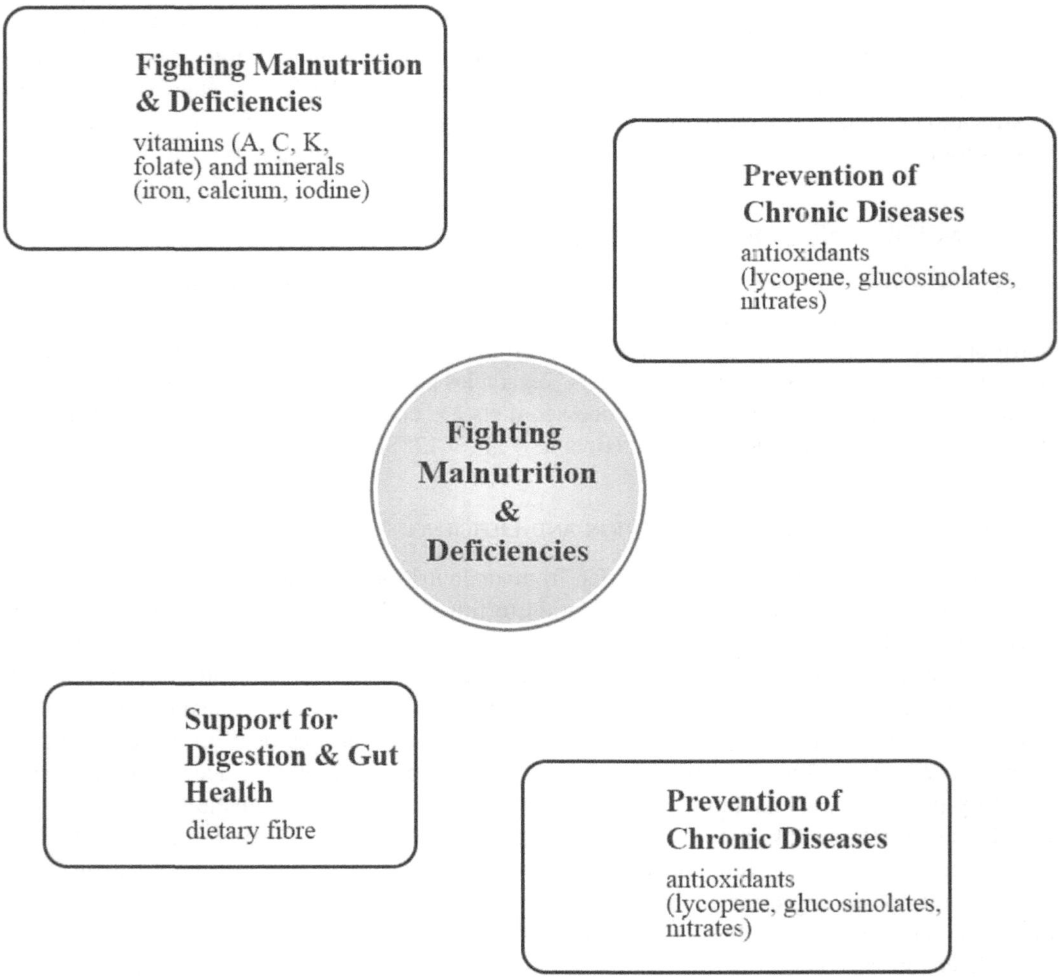

FIGURE 11.1 Importance of vegetables.

11.3.3 Support for Digestion and Gut Health

Vegetables are an excellent source of dietary fibre. Such fibres helps in better digestion by preventing constipation, control blood sugar by slowing down glucose absorption, and enhance gut health by feeding good bacteria in the intestine (Makki et al., 2018). For example carrots, beans, peas, and leafy greens help maintain a healthy gut microbiome. A balanced gut supports not just digestion but also immunity and mental health.

11.3.4 Economic and Cultural Importance

Vegetables not only nourish populations but also contribute significantly to the global economy. Smallholder farmers in developing countries rely heavily on vegetable cultivation as a source of income and livelihood. Urban agriculture initiatives which focus on vegetable production have been promoted as sustainable strategies to improve local food security (FAO, 2019). Culturally, vegetables are a part of every traditional diet. For example the Mediterranean diet uses plenty of tomatoes, cucumbers, and leafy greens whereas Asian diets include gourds, roots, and cruciferous vegetables. This cultural importance makes vegetables acceptable and easy to include in diets across the world. Figure 11.2 describes the economic and cultural importance of vegetables.

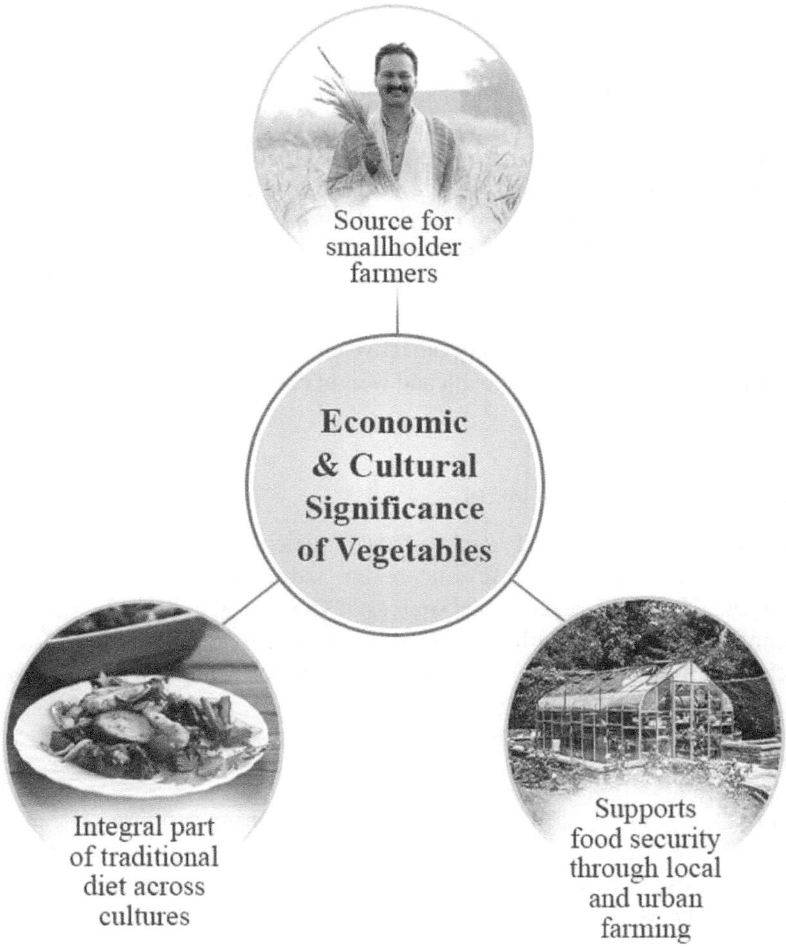

FIGURE 11.2 Economic and cultural significance of vegetables.

11.3.5 Global Dietary Recommendations

Global dietary recommendations consistently stated that adults should consume a minimum of 400 grams of fruits and vegetables per day (five portions) of fruits and vegetables per day, not including starchy vegetables (WHO, 2020b). This range is being hugely endorsed by stellar organizations like the WHO and the FAO which reflected the consensus on the importance of vegetables for preventing chronic disease and supporting optimal nutrition for all human beings. Studies have shown that individuals who select nutrient-rich food despite cost and convenience barrier demonstrate a form of nutritional resilience which helps in maintaining a healthier diet under limiting conditions (Aggarwal et al., 2016).

11.4 NUTRITIONAL COMPOSITION OF VEGETABLES

Vegetables are rich in a wide range of nutrients which are essential for human health. Unlike processed foods, they naturally provide vitamins, minerals, dietary fibre, antioxidants, and bioactive compounds that support growth, immunity, and disease prevention. The nutritional composition of vegetables varies depending on type, colour, and part of the plant (leaves, roots, fruits, or stems). Because of this diversity, nutritionists recommend eating a wide variety of vegetables every day to cover all nutrient needs (Slavin & Lloyd, 2012; Singh et al., 2020).

11.4.1 Macronutrients in Vegetables

Vegetables contribute carbohydrates, proteins, negligible fats, and dietary fibre. Carbohydrates are the main energy component, especially in starchy vegetables like potatoes and peas, while leafy greens provide fewer carbohydrates but more fibre (FAO, 2020). Protein levels in vegetables are modest but significant in leafy greens such as spinach and amaranth (Odhav et al., 2007). Fats are generally low, except in avocado and olives, which provide heart-healthy fatty acids. Dietary fibre, both soluble and insoluble, regulates digestion, blood glucose, and cholesterol (Anderson et al., 2009). Additionally, vegetables contain 70–95% water, contributing to hydration and satiety (Gupta et al., 2015). Recent research highlights that vegetables produce more than just macronutrients, which are also important sources of bioactive compounds that help to maintain a metabolic balance, and contribute to better heart health and weight control (Suresh Kumar, 2024; Nurzyńska-Wierdak et al., 2025).

11.4.2 Micronutrients in Vegetables

Many vegetables have been extensively analysed and have demonstrated, through research, the asset of incorporating micronutrients like iron, zinc, boron, and manganese in relation to growth, yield, and quality. Vegetables serve as exceptional sources of vitamins and minerals, which are crucial for metabolism, immunity, and overall health (Sandeep et al., 2019). Table 11.1 gives an overview of major phytochemicals found in commonly consumed vegetables, along with their compound types and biological functions.

Vitamins

- **Vitamin A:** Found in carrots, spinach, and pumpkin; supports vision, skin, and immunity.
- **Vitamin C:** Present in tomatoes, bell peppers, and broccoli; acts as an antioxidant and boosts immunity.

TABLE 11.1
Major Phytochemicals in Vegetables and Their Biological Functions

Vegetable	Major Phytochemical	Compound Type	Biological Function	References
Carrot	β-Carotene	Carotenoid	Antioxidant, vision protection	Ikram et al. (2024)
Tomato	Lycopene	Carotenoid	Anticancer	Balali et al. (2025),
Spinach	Lutein, zeaxanthin	Xanthophylls	Eye protection, anti-inflammatory	Mrowicka et al. (2022)
Broccoli	Sulforaphane	Isothiocyanate	Detoxification, anticancer	Moskwa et al. (2023),
Onion	Quercetin	Flavonoid	Anti-inflammatory, immune support	Aghababaei and Hadidi (2023)
Garlic	Allicin	Organosulphur	Antimicrobial, cardioprotective	Bhatwalkar et al. (2021)

- **Vitamin K:** Abundant in leafy greens like kale and spinach; important for blood clotting and bone health.
- **Folate (Vitamin B9):** Found in cabbage, peas, and spinach; vital for DNA synthesis and pregnancy.

Minerals

- **Iron:** Present in spinach, amaranth, and beans; important for red blood cell production.
- **Calcium:** Found in broccoli, kale, and okra; strengthens bones and teeth.
- **Potassium:** Abundant in potatoes, beetroot, and tomatoes; regulates blood pressure.
- **Magnesium:** Found in leafy greens; supports muscle and nerve function.

11.4.3 Antioxidants and Phytochemicals

Antioxidants refer to a group of compounds which help prevent or slow down the oxidation of cellular components. The body's antioxidant defence system comprises both naturally synthesized antioxidants and those obtained through the diet. Consuming a variety of dietary antioxidants is considered essential for protecting cells from oxidative damage, regulating inflammation, and reducing the risk of various diseases. Besides key nutrients such as β-carotene, vitamins C and E, and selenium, plants and plant-derived foods also provide numerous phytochemicals which exhibit strong antioxidant properties. Apart from vitamins and minerals, vegetables contain phytochemicals (plant-based bioactive compounds) which protect against diseases. Other important constituents like carotenoids, commonly found in vegetables such as carrots, pumpkin, and sweet potato, are well known for their ability to protect vision and reduce the risk of certain cancers through their antioxidant activity. Flavonoids which are mostly found in onions, kale, and broccoli reduce inflammation and prevent heart diseases. Glucosinolates found naturally in cruciferous vegetables like cabbage and cauliflower play a crucial role in detoxification and have anti-carcinogenic potential. Lutein and zeaxanthin, found in kale and spinach, protect against age-related eye diseases. Lycopene, predominantly present in tomato and watermelon, lowers the risk of prostate and breast cancers. These constituents act as antioxidants, neutralizing free radicals and preventing cell damage (Boeing et al., 2012; Nurzyńska-Wierdak et al., 2025).

11.4.4 Water and Fibre Content

Many vegetables are high in water content, and they can help you stay hydrated. Mostly all the vegetables contain 70–95% water, making them excellent for hydration. They are also high in dietary fibre, which helps in improving digestion and prevents constipation. It also helps control cholesterol and blood sugar levels. They are big supporter of the survival of healthy gut bacteria. This makes vegetables an essential part of a healthy digestive system (Makki et al., 2018).

11.5 USES OF VEGETABLES IN HUMAN HEALTH

Vegetables play a central role in maintaining human health due to their rich content of vitamins, minerals, dietary fibre, and bioactive compounds. They are considered one of the most cost-effective sources of nutrition which not only prevent nutrient deficiencies but also reduce the risk of several chronic diseases. In modern nutritional science, vegetables are recognized as functional foods, as they provide health benefits beyond basic nutrition (Slavin & Lloyd, 2012). Following subsections highlight the different ways vegetables contribute to human health.

11.5.1 Vegetables for Cardiovascular Issues

Vegetables are very essential for both the heart and blood vessels. Green leafy vegetables like spinach and lettuce have natural nitrates which help in relaxing the blood vessels and support in keeping the blood pressure under control. Tomatoes are rich in lycopene which reduces bad cholesterol and protects the heart from damage. Likewise, carrots having beta-carotene help in keeping arteries clean and thus prevent blockage issues. Broccoli, cabbage, and Brussels sprouts contain compounds which reduce inflammation and lower the risk of stroke and heart attack. Many studies show that people who eat four to five servings of vegetables daily have better heart health compared to people who eat very few vegetables. So, including vegetables in daily meals is one of the easiest and most natural ways to protect the heart (Sharma & Kumar, 2016).

11.5.2 Diabetes and Obesity Control

Vegetables are naturally low in calories and high in fibre, helping in weight management by increasing satiety. Vegetables play a crucial role in preventing and management of diabetes and obesity through their various roles which affect the glucose metabolism, insulin sensitivity, inflammation reduction, and body weight regulation. Non-starchy vegetables such as cucumbers, lettuce, and spinach have a low glycaemic index, making them ideal for blood sugar control (Jaceldo-Siegl et al., 2019). Vegetables are rich in dietary fibre, antioxidants, and phytochemicals which improve glycaemic control and insulin sensitivity.

11.5.3 Immune System and Anti-Inflammatory Benefits

Inflammation is related to an increased probability of a variety of chronic diseases. A diet high in vegetables and fruits helps to reduce the chronic inflammation, as both are flush sources of antioxidants and other biologically active compounds, which may improve the overall immune function (Hosseini et al., 2018). Most vegetables are rich in vitamins C and E, which support to enhance the immune responses. Flavonoids and polyphenols act as natural anti-inflammatory compounds. Onions, garlic, and ginger have antimicrobial properties and are widely studied for their ability to reduce infections.

11.5.4 Bone and Eye Health

Regular intake of vegetables plays an important role in keeping bones strong and maintaining good eyesight. Leafy green vegetables like kale and spinach provide essential nutrients such as calcium, magnesium, and vitamin K needed for healthy bone formation and to lower the risk of osteoporosis. Vitamin K helps in the process of binding calcium to bone proteins through the activation of osteocalcin, which supports bone density and overall skeletal strength (Weaver et al., 2016). Carotenoids such as lutein and zeaxanthin found mostly in spinach, broccoli, and corn gather in retina thus protecting the eye from harmful blue light. Recent research has further demonstrated that higher dietary intake of these carotenoids is associated with improved visual performance, better contrast sensitivity, and reduced progression of retinal degeneration (Becerra et al., 2023; Johnson, 2021). They also act as antioxidants, reducing damage to eye cells and age-related problems like cataract and macular degeneration (Ma & Lin, 2019; Landrum & Nolan, 2013). Beta-carotene from carrots and sweet potatoes is converted into vitamin A which is essential for maintaining the health of the cornea and night vision (Tanumihardjo, 2011).

11.5.5 Vegetables and Mental Heath

Eating vegetables is not only important for physical health but also required for a good mental health. Nutrients such as folate, magnesium, and potassium support nerve function and help in the production of neurotransmitters which control the mood and emotions. Studies showed that people who consume more fruits and vegetables often feel happier and have a lower risk of depression and anxiety. Various studies reported that people who increased their vegetable intake had better mental well-being and lower stress levels. Another study found that eating at least five portions of fruits and vegetables was linked to a more positive mood and better life satisfaction (Shams-Rad et al., 2022). Recent research has further strengthened these findings, suggesting that plant-based diets high in green leafy and cruciferous vegetables can improve cognitive performance, reduce the risk of depression, and protect against neurodegenerative disorders such as Alzheimer's disease (Brookie et al., 2024; Nguyen et al., 2023).

11.6 LIMITATIONS IN VEGETABLE CONSUMPTION

Vegetables are widely recognized as essential components of a healthy diet due to their rich supply of vitamins, minerals, fibre, and phytochemicals. Despite these benefits, their consumption across the globe remains below the recommended levels suggested by the WHO and FAO, which is at least 400 g/day per adult. Several limitations and barriers ranging from production challenges to cultural, economic, and personal factors contribute to this dietary gap.

11.6.1 Effects of Climate Change on Vegetable Cultivation

- High temperature: Elevated temperatures can precipitate substantial reductions in tomato productivity, primarily due to diminished fruit set and the resultant production of smaller, inferior-quality fruits.
- Low temperature: Low temperatures, including chilling and freezing injuries, can adversely affect all plant species. Damage resulting from exposure to temperatures above 0°C (ranging from 0°C to 10°C) is classified as chilling injury, while brief exposure to temperatures below 0°C results in freeze injury.
- Flooding: Many vegetables exhibit a heightened sensitivity to flooding, particularly those with shallow root systems. Under waterlogged conditions, roots struggle to acquire oxygen as soil air is supplanted by inundating water. This suffocation severely impedes root respiration, thereby disrupting their essential functions of nutrient and water absorption. Consequently, this leads to root necrosis, rendering the plants susceptible to infections by soil-borne pathogens.
- Salinity: The threat posed by escalating soil salinity is significant for vegetable production, particularly in irrigated croplands, which fulfil 40% of the global food demand (FAO,

2001). In arid and hot environments, elevated rates of evapotranspiration result in considerable water loss from the soil, leaving behind salts that hinder the plant's capacity to absorb water.

- Soil erosion engenders detrimental effects that are often challenging to quantify. The erosion of soil from the root zone leads to the dislodging of plants, which is particularly harmful to many vegetables (such as tomatoes, eggplants, and chilies). The extent of damage to vegetable crops due to storms and their subsequent recovery is contingent upon various factors, including the specific type of vegetable, its growth stage, the prevailing weather conditions immediately following the storm, and the presence of pathogenic organisms. Prolonged hot and humid conditions after a storm exacerbate the situation, significantly increasing the incidence of diseases, particularly bacterial infections.

11.6.2 Socio-Economic Barriers

Socio-economic factors play a major role in shaping vegetable consumption patterns across the world. In many low- and middle-income countries, fresh vegetables can be expensive compared to cheaper calorie-dense foods such as rice, bread, and processed snacks. This price disparity limits access to nutrient-rich diets, particularly among economically disadvantaged populations. Income inequality further exacerbates the issues as families with limited resources prioritize staple foods over vegetables, leading to nutrient deficiencies. Urbanization has also significantly influenced eating habits, especially among the poor. Urban poor populations may lack access to fresh vegetables and instead rely on packaged, preserved, or fried foods. Recent studies highlight that improving affordability, establishing urban gardens, and subsidizing vegetable production could help bridge this socio-economic nutrition gap (Chen & Miller, 2013; Rahman & Willott, 2025).

11.6.3 Agricultural and Post-Harvest Limitations

Agricultural and post-harvest limitations significantly affect the consistent availability, quality, and safety of vegetables worldwide. Many vegetables are seasonal, making them unavailable year-round. The lack of efficient storage systems and cold chain infrastructure exacerbates post-harvest loss thus causing spoilage, reducing both availability and affordability. Another major concern is the excessive use of chemical pesticides and fertilizers, which also discourages consumers due to safety concerns. Moreover, climate change poses an escalating threat to vegetable cultivation, as rising temperatures, erratic rainfall, and water scarcity adversely impact yields and diminish the nutritional content of crops (Zhao et al., 2023; Verma et al., 2025).

11.6.4 Cultural and Behavioural Barriers

Cultural and behavioural barriers play a major role in influencing vegetable consumption patterns across different populations. Many communities prefer cereals and meat over vegetables, which usually leads to an imbalanced diet. Children and many adolescents often dislike the text and texture of certain vegetables and have a perception that vegetables are less filling which discourages their inclusion in daily meals. Lack of nutrition education results in an underestimation of the role of vegetables in disease prevention. Cultural cooking practices also influence the nutritional value of vegetables. Overcooking vegetables destroys heat-sensitive vitamins like vitamin C and folate, reducing nutritional benefits. To address these barriers, health campaigns and educational initiatives focusing on the cultural importance and health benefits of vegetables are necessary (Miller et al., 2021; Yue et al., 2024).

11.6.5 Health-Related Barriers

Health-related barriers also play a significant role in limiting vegetable consumption among different populations. Some individuals experience bloating or discomfort from high-fibre vegetables, leading to avoidance. Additionally, food safety concerns contribute to hesitation towards fresh product consumption. Unhygienic washing or irrigation with contaminated water can lead to foodborne illnesses (e.g. E. coli, Salmonella). Another, though less common, barrier is food allergies or sensitivities as some people develop allergic reactions to certain vegetables like celery, spinach, or carrots, further reducing their intake options. Public health initiatives and technological interventions in post-harvest management can also reduce health-related risks, ensuring that vegetables remain both safe and accessible (de Oliveira et al., 2023).

11.6.6 Policy and Market Barriers

Policy and market barriers significantly influence vegetable consumption and availability across regions. In many countries, governments often subsidize cereal crops rather than vegetables, making vegetables relatively costlier. Moreover, marketing of junk food increases and processed food industries advertise aggressively, while vegetable promotion remains minimal. Another limitation is the lack of strong institutional support for vegetable-based nutrition programmes. Only a few countries implement mandatory vegetable-based meals in school feeding schemes. Strengthening agricultural policies, regulating food marketing, and integrating vegetable-based nutrition into public programmes are crucial steps to bridge this policy gap and promote healthier diets (Gómez & Ricketts, 2013; FAO, 2023).

11.6.7 Psychological and Lifestyle Barriers

Psychological and lifestyle barriers play a major role in reducing vegetable consumption, particularly in urban and working populations. Busy lifestyles encourage fast-food consumption instead of cooking fresh vegetables. The preparation of vegetables includes washing, peeling, chopping, and cooking and is often perceived as time-consuming and inconvenient, especially among working adults and students with limited time for meal preparation. Additionally, social and peer influences strongly affect food choices where children and young adults often imitate peers who prefer fast foods over vegetables. Encouraging meal-prep culture, improving access to pre-cut or ready-to-cook vegetables, and promoting positive social campaigns around healthy eating could help overcome these barriers (Pollard et al., 2022; Hartmann & Siegrist, 2020).

11.7 CONCLUSION

Vegetables are indispensable for maintaining health and preventing disease, as they supply essential nutrients and protective compounds which strengthen immunity, aid digestion, and enhance overall well-being. Regular consumption lowers the risk of CVD, diabetes, obesity, and various cancers, making them a cornerstone of preventive healthcare. Beyond individual benefits, vegetables play an important role in ensuring food and nutrition security, supporting rural economies, and promoting sustainable food systems with reduced environmental footprints. Despite these advantages, global consumption continues to lag behind recommended levels due to affordability, limited access, and lifestyle changes. Advances in biotechnology, hydroponics, and vertical farming can enhance availability, nutrient density, and year-round supply, while policies, education, and awareness campaigns are needed to overcome barriers. Recognizing vegetables as vital resources rather than optional foods is essential for improving public health, safeguarding ecosystems, and securing a sustainable global future.

REFERENCES

Aggarwal, A., Rehm, C. D., Monsivais, P., & Drewnowski, A. (2016). Importance of taste, nutrition, cost and convenience in relation to diet quality: Evidence of nutrition resilience among US adults using National Health and Nutrition Examination data. *American Journal of Health Promotion*, *33*(6), 869–875.

Aghababaei, F., & Hadidi, M. (2023). Recent advances in potential health benefits of quercetin. *Pharmaceuticals (Basel, Switzerland)*, *16*(7), 1020.

Anderson, J. W., Baird, P., Davis, R. H., Ferreri, S., Knudtson, M., Koraym, A., Waters, V., & Williams, C. L. (2009). Health benefits of dietary fiber. *Nutrition Reviews*, *67*(4), 188–205.

Aune, D., Giovannucci, E., Boffetta, P., Fadnes, L. T., Keum, N., Norat, T., et al. (2017). Fruit and vegetable intake and the risk of cardiovascular disease, total cancer and all-cause mortality: A systematic review and dose–response meta-analysis of prospective studies. *International Journal of Epidemiology*, *46*(3), 1029–1056.

Balali, A., Fathzadeh, K., Askari, G., & Sadeghi, O. (2025). Dietary intake of tomato and lycopene, blood levels of lycopene, and risk of total and specific cancers in adults: A systematic review and dose-response meta-analysis of prospective cohort studies. *Frontiers in Nutrition*, *12*, 1516048.

Becerra, M., Garcia, R., & Lopez, P. (2023). Global vegetable consumption trends and nutritional challenges: A policy perspective. *Public Health Nutrition*, *26*(5), 812–820.

Bhatwalkar, S. B., Mondal, R., Krishna, S. B. N., Adam, J. K., Govender, P., & Anupam, R. (2021). Antibacterial properties of organosulfur compounds of garlic (*Allium sativum*). *Frontiers in Microbiology*, *12*, 613077. https://doi.org/10.3389/fmicb.2021.613077. PMID: 34394014; PMCID: PMC8362743.

Boeing, H., Bechthold, A., Bub, A., Ellinger, S., Haller, D., Kroke, A., et al. (2012). Critical review: Vegetables and fruit in the prevention of chronic diseases. *European Journal of Nutrition*, *51*(6), 637–663.

Brookie, K. L., Best, G. I., & Conner, T. S. (2024). Increasing fruit and vegetable consumption improves psychological well-being: A randomized controlled trial. *British Journal of Nutrition*, *132*(2), 234–243.

Chen, E., & Miller, G. E. (2013). Socioeconomic status and health: Mediating and moderating factors. *Annual Review of Clinical Psychology*, *9*(1), 723–749. https://doi.org/10.1146/annurev-clinpsy-050212-185634

Chen, Y., Wu, Q., & Zhao, X. (2024). Macronutrient balance in vegetables and their role in human metabolic health. *Journal of Food Biochemistry*, *48*(1), e14562.

de Oliveira, L. M., Santos, R. F., & Souza, M. C. (2023). Policy gaps and food system inequities in low-vegetable consumption regions. *Global Food Security*, *39*, 100688.

Dixit, V. (2023). Functional foods: Exploring the health benefits of bioactive compounds. *Frontiers in Nutrition*, *10*, 5546753.

Food and Agriculture Organization (FAO). (2001). *The state of the world's land and water resources for food and agriculture (SOLAW)*. FAO of the United Nations. https://www.fao.org

Food and Agriculture Organization of the United Nations (FAO). (2019). *The state of food and agriculture: Moving forward on food loss and waste reduction*. FAO.

Food and Agriculture Organization of the United Nations (FAO). (2020). *Food balance sheets*. FAO.

Food and Agriculture Organization of the United Nations (FAO). (2023). *FAO food composition database*. https://www.fao.org/infoods/infoods/tables-and-databases/en/

Generalić Mekinić, I., Skroza, D., Ljubenkov, I., Katalinić, V., & Šimat, V. (2025). Bioactive compounds in foods: New and novel sources from vegetables and fruits. *Foods*, *14*(9), 1617.

Gómez, M. I., & Ricketts, K. D. (2013). Food value chain transformations in developing countries: Selected hypotheses on nutritional implications. *Food Policy*, *42*, 139–150. ISSN 0306-9192. https://doi.org/10.1016/j.foodpol.2013.06.010

Gupta, S., Lakshmi, A. J., Manjunath, M. N., & Prakash, J. (2015). Analysis of nutrient and antinutrient content of underutilized green leafy vegetables. *LWT – Food Science and Technology*, *38*(4), 339–345.

Hartmann, C., & Siegrist, M. (2020). Development and validation of the Eating Motivation Survey: Identifying factors influencing eating habits. *Appetite*, *144*, 104464.

Hosseini, B., Berthon, B. S., Saedisomeolia, A., Starkey, M. R., Collison, A., Wark, P. A. B., & Wood, L. G. (2018). Effects of fruit and vegetable consumption on inflammatory biomarkers and immune cell populations: A systematic literature review and meta-analysis. *The American Journal of Clinical Nutrition*, *108*(1), 136–155.

Ikram, A., Rasheed, A., Khan, A., Khan, R., Ahmad, M., Bashir, R., & Mohamed, M. (2024). Exploring the health benefits and utility of carrots and carrot pomace: A systematic review. *International Journal of Food Properties*, *27*, 180–193.

Jaceldo-Siegl, K., Estevez, D., Fraser, G. E., Hayes-Bautista, D., Flores, H., Jordan, M., et al. (2019). Plant-based diets in Hispanic/Latino adult Adventists in the United States and their association with body mass index. *American Journal of Health Promotion*, *33*(6), 869–875.

Johnson, P. (2021). Barriers to vegetable consumption: A global overview. *Nutrition Policy Review*, *12*(3), 210–225.

Khoury, C. K., Bjorkman, A. D., Dempewolf, H., Ramirez-Villegas, J., Guarino, L., Jarvis, A., Rieseberg, L. H., & Struik, P. C. (2014). Increasing homogeneity in global food supplies and the implications for food security. *Proceedings of the National Academy of Sciences*, *111*(11), 4001–4006.

Lai, H. S., Wang, J., & Lin, Q. (2023). Macronutrient quality in vegetables and its effect on human metabolism. *Journal of Food Science and Human Health, 12*(2), 145–153.

Landrum, J. T., & Nolan, J. M. (Eds.). (2013). *Carotenoids and Retinal Disease*. CRC.

Liu, R. H. (2013). Health-promoting components of fruits and vegetables in the diet. *Advances in Nutrition, 4*(3), 384S–392S.

Ma, L., & Lin, X. M. (2019). Effects of lutein and zeaxanthin on visual function and pathology in age-related macular degeneration: A review. *Nutrients, 12*(1), 1–13.

Makki, K., Deehan, E. C., Walter, J., & Bäckhed, F. (2018). The impact of dietary fiber on gut microbiota in host health and disease. *Cell Host & Microbe, 23*(6), 705–715.

Mateus, A. R. S., et al. (2024). Unveiling the potential of bioactive compounds in vegetable and fruit by-products and their applications toward the food chain. *Food Chemistry: X, 22*, 101743.

Mila, K. J., et al. (2024). Nutritional composition, bioactive compounds and antioxidant activity of indigenous vegetables. *Scientific Reports, 14*, 78625.

Miller, M. A., Bates, S., Ji, C., & Cappuccio, F. P. (2021). Systematic review and meta-analyses of the relationship between short sleep and incidence of obesity and effectiveness of sleep interventions on weight gain in preschool children. *Obesity Reviews, 22*(2), e13113. https://doi.org/10.1111/obr.13113

Ministry of Health, Turkey. (2015). *Turkey nutrition guide*. Republic of Turkey Ministry of Health.

Moskwa, J., et al. (2023). Vegetables as a source of bioactive compounds with health-promoting properties: Focus on skin protection. *Nutrients, 15*(16), 3594.

Mrowicka, M., Mrowicki, J., Kucharska, E., & Majsterek, I. (2022). Lutein and zeaxanthin and their roles in age-related macular degeneration-neurodegenerative disease. *Nutrients, 14*(4), 827.

Nguyen, Y. H. T., Tran, T. Q., Hoang, D. T., Tran, T. M. T., & Thang Nguyen, T. T. (2023). Does land fragmentation affect household livelihood choice in Vietnam? *Economic Alternatives, 2*, 256–272. https://doi.org/10.37075/EA.2023.2.02

Nurzyńska-Wierdak, R., Suresh, P., & Kaul, V. (2025). Nutritional and functional roles of macronutrients and bioactive compounds in vegetables: An updated perspective. *Frontiers in Plant Science, 16*, 1423378.

Odhav, B., Beekrum, S., Akula, U., & Baijnath, H. (2007). Preliminary assessment of nutritional value of traditional leafy vegetables in KwaZulu-Natal, South Africa. *Journal of Food Composition and Analysis, 20*(5), 430–435.

Pennington, J. A. T., & Fisher, R. A. (2010). Food component profiles for fruit and vegetable subgroups. *Journal of Food Composition and Analysis, 23*(5), 411–418.

Pollard, C. M., Pulker, C. E., Meng, X., Kerr, D. A., & Scott, J. A. (2022). Consumer attitudes and barriers to healthy eating: A review of recent evidence. *Public Health Nutrition, 25*(3), 593–605.

Rahman, I., & Willott, C. (2025). Social, economic and ecological drivers of tuberculosis disparities in Bangladesh: Implications for health equity and sustainable development policy. *Challenges, 16*(3), 37. https://doi.org/10.3390/challe16030037

Salesse, F., Eldridge, A. L., Mak, T. N., & Gibney, E. R. (2024). A global analysis of portion size recommendations in food-based dietary guidelines. *Frontiers in Nutrition, 11*, 1476771. https://doi.org/10.3389/fnut.2024.1476771

Sandeep, G., Swaminathan, V., Paramaguru, P., & Janaki, D. (2019). Effect of foliar application of micronutrients on growth, yield, and quality of annual moringa (Moringa oleifera Lam.) var. PKM-1. *Journal of Pharmacognosy and Phytochemistry, 8*(3), 363–367.

Schmidt, H. D. O., & Oliveira, V. R. D. (2023). Overview of the incorporation of legumes into new food options: An approach on versatility, nutritional, technological, and sensory quality. *Foods, 12*(13), 2586. https://doi.org/10.3390/foods12132586

Shams-Rad, S., Talepasand, S., & Hasani, J. (2022). Fruit and vegetable consumption and mental well-being: A meta-analysis of observational studies. *Nutrients, 14*(15), 3057.

Sharma, U., & Kumar, P. (2016). Extent of deficiency, crop responses and future challenges. *International Journal of Advanced Research, 4*(4), 1402–1406.

Singh, R., Verma, R., & Prakash, S. (2020). Nutritional importance of green leafy vegetables: A review. *Journal of Current Research in Food Science, 3*(1), 23–30.

Steinmetz, K. A., & Potter, J. D. (1996). Vegetables, fruit, and cancer prevention: A review. *Journal of the American Dietetic Association, 96*(10), 1027–1039.

Suresh Kumar, D. (2024). Ecosystem services based approaches in agricultural policy making. *Indian Journal of Agricultural Economics, 79*(1), 141–143.

Tanumihardjo, S. A. (2011). Vitamin A: Biomarkers of nutrition for development. *The American Journal of Clinical Nutrition, 94*(2), 658S–65S. https://doi.org/10.3945/ajcn.110.005777. Epub 2011 Jun 29. PMID: 21715511; PMCID: PMC3142734.

U.S. Department of Agriculture (USDA). (2010). *Dietary Guidelines for Americans, 2010*. U.S. Government Printing Office. https://www.dietaryguidelines.gov/

Verma, S., Kumar, S., Kumar, V., Yadav, S., Singh, P., Saurabh, A., Kumar, A., & Jain, M. (2025). Recent advancements and innovations in post-harvest handling, storage, and technology for vegetables: A review. *Archives of Current Research International*, *25*, 161–180.

Wang, H., Egun, I. L., Gan, T., Huang, Z., Liu, H., Gustave, W., & Zhang, X. (2025). Enhancing vegetable yield and quality with biochar: Prospects and challenges. *Frontiers in Sustainable Food Systems*, *9*, 1621366. https://doi.org/10.3389/fsufs.2025.1621366

Weaver, C. M., Gordon, C. M., Janz, K. F., Kalkwarf, H. J., Lappe, J. M., Lewis, R., O'Karma, M., Wallace, T. C., & Zemel, B. S. (2016). The National Osteoporosis Foundation's position statement on peak bone mass development and lifestyle factors: A systematic review and implementation recommendations. *Osteoporosis International*, *27*(4), 1281–1386.

Welbaum, G. E. (2015). Vegetable production and practices. In *IARC Handbooks of Cancer Prevention: Fruit and Vegetables* (Vol. 8, pp. 1–15). International Agency for Research on Cancer.

Willett, W., Rockström, J., Loken, B., Springmann, M., Lang, T., Vermeulen, S., et al. (2019). Food in the anthropocene: The EAT–lancet commission on healthy diets from sustainable food systems. *The Lancet*, *393*(10170), 447–492.

World Health Organization (WHO). (2020a). *Healthy Diet: Fact Sheet*. https://www.who.int/news-room/fact-sheets/detail/healthy-diet

World Health Organization (WHO). (2020b). *Increasing Fruit and Vegetable Consumption to Reduce the Risk of Noncommunicable Diseases*. WHO. https://www.who.int/news-room/fact-sheets/detail/healthy-diet

Yue, Z., Yuan, K., Seki, M., Agake, S.-I., Matsumura, K., Okita, N., Naoi, W., Naoi, K., Toyota, K., Tanaka, H., Sugihara, S., Yasuda, M., & Ohkama-Ohtsu, N. (2024). Comparative analysis of Japanese soils: Exploring power generation capability in relation to bacterial communities. *Sustainability*, *16*(11), 4625. https://doi.org/10.3390/su16114625

Zhao, H., Yang, Y., Yang, C., Song, R., & Guo, W. (2023). Evaluation of spatial resolution on crop disease detection based on multiscale images and category variance ratio. *Computers and Electronics in Agriculture*, *207*, 107743. https://doi.org/10.1016/j.compag.2023.107743

12 Nutrigenomics
A Pathway to a Healthy Life

Saloni Malik, Raviraj B. Barot, Pradeep Pushparaj Selvadoss, Sachidanand Singh, and Nisha Singh

12.1 INTRODUCTION TO NUTRIGENOMICS

Nutrigenomics, the study of interaction of genes and nutrients, aims to know how dietary elements impact on the expression of genes and also how the genetic variance modulates the reaction of an individual to food and well-being in general (Hosseiniara & Hosseini Zijoud, 2024). To create individualized dietary recommendations and improve health outcomes, this emerging discipline integrates molecular biology, genomics, bioinformatics, artificial intelligence, machine learning, microbiome research, and nutritional science. A crucial part of nutritional genomics, which also includes nutrigenetics, the study of genetic variations in nutrient metabolism, is nutrigenomics. The overall aim of nutrigenomics is to determine how one can alter their diet depending on her genetic predisposition to prevent and control such chronic and communicable diseases such as cancer, diabetes, obesity, and CVD. With continued developments in omics technologies and personalized nutrition, nutrigenomics is expected to change dietary advice and public health interventions in the coming decades. 'Let food be thy medicine and medicine be thy food', wrote Hippocrates, the founder of Western medicine, a millennium ago. The promotion of health and the study of gene-food connections were made possible in large part by 'the Human Genome Project (National Human Genome Research Institute, 2013)'. This chapter's goal is to conduct a comprehensive review to assess the modulation of nutrient-containing substances and their interactions with genomic stability and gene expression regulation. Additionally, each person's distinct genetic composition affects their dietary preferences and nutritional needs, which differ depending on environmental and regional circumstances. This emphasizes the concepts of nutrigenomics and customized nutrition.

As Figure 12.1 describes, the goal of the transdisciplinary 'Nutrigenetic and Nutrigenomic Research and Training Unit (N2RTU)' implementation is in Malaysia. It is implemented to advance precision nutrition via integrative research, translational application, and capacity building (Dhanapal et al., 2022). The framework is divided into five fundamental domains:

1. The Research Core: It focuses on using omics technologies (such as metabolomics and transcriptomics) to uncover diet-responsive biomarkers, genomic and epigenomic profiling, and cellular models and clinical trials to study gene–diet interactions.
2. The Data Integration and Bioinformatics Core: To comprehend intricate diet–gene relationships, it integrates multi-omics datasets. Uses machine learning and artificial intelligence to create predictive models for customized nutrition while preserving population genetic data specific to a particular area.
3. Training and Capacity Building: To create a skilled workforce and foster interdisciplinary partnerships, it offers structured educational programmes in omics and bioinformatics, including diplomas, workshops, and practical training.
4. Translational and Clinical Integration: Incorporates nutrigenomic data into clinical practice and dietary guidelines, including genetic screening, nutritional counselling,

DOI: 10.1201/9781003606710-12

and assessment of tailored nutrition treatments to lower the risk of noncommunicable diseases.

5. The Policy and Public Health Interface: It guarantees ethical and legal conformity in genomic research and application, incorporates genetic data into nutritional surveillance, and offers scientific support for national nutrition policy.

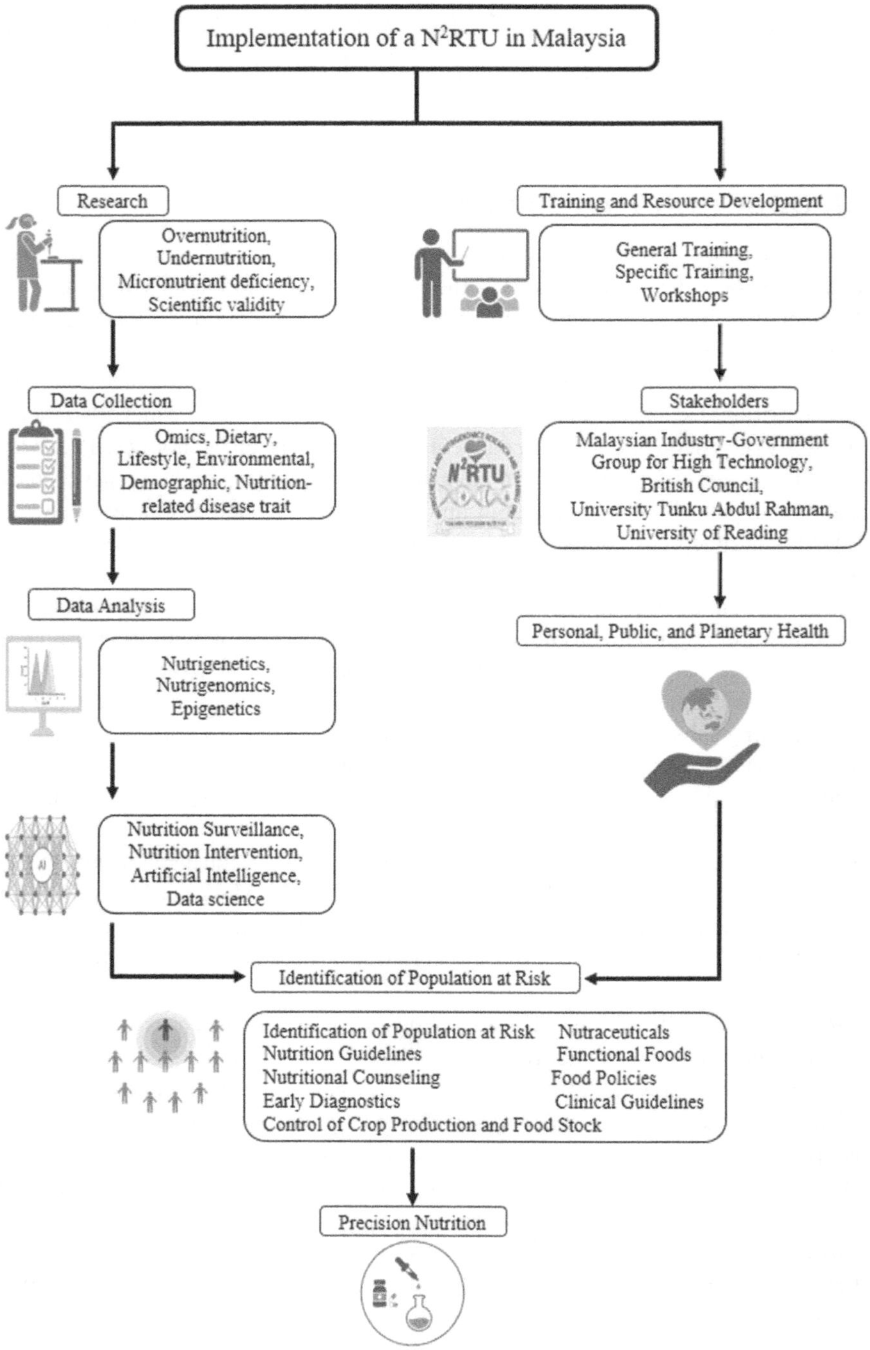

FIGURE 12.1 Unit (N^2RTU) in Malaysia, outlining the integration of research, data analysis, stakeholder training, and policy development to support precision nutrition and improve public health outcomes.

12.2 UNDERSTANDING OF NUTRIGENOMICS

Some investigators argue that the impact of nutrition on gene expression is only the beginning because dietary components interact at every level between gene and protein or metabolic product(s), including signal transduction and protein post-translational modifications (Haro et al., 2019). The linkage among diet and cancer or other disorders is well established (Saha et al., 2024). Modern progress in genomics, particularly nutrigenomics, transcriptomics (the examination of the level of expression of mRNAs in a given cell population), metabolomics, the study of the unique chemical signatures left by particular cellular processes, and epigenomics, the omics study of epigenetic elements, has substantially advanced the investigation of nutrient modulation in the carcinogenesis pathway. Nutrient-modulated carcinogenesis research is directed towards the mechanisms by which nutrients modify DNA damage and repair mechanisms, DNA methylation, which influences cellular phenotypes and gene expression, antioxidant rearrangement and oxidative stress, target receptors, signal transduction pathways, cell cycle controls and checkpoints, apoptosis, and antiangiogenic processes. Nutritional genomics, proteomics, and metabolomics enable researchers, for the first time, to simultaneously investigate the biological effects of dietary components at the levels of gene expression and cellular function (Satrio et al., 2024).

Overall, transcription factors are the key mechanisms through which nutrients regulate gene expression (Saha et al., 2024). The nuclear receptor super family of transcription factors is one of the most significant groups of nutrient sensors, consisting of 48 numbers of the human genome. Many receptors that bind nutrients and their metabolites belong to this superfamily. The nuclear receptors and peroxisomes proliferator-activated receptor (PPAR) binds fatty acids and retinoid X receptor which further binds cholesterol metabolites, heterodimerize and also bind to certain nucleotide sequences named response elements within the promoter regions of the genes (Pérez et al., 2025). The binding of the ligands triggers a conformational change in the nuclear receptors, leading to the dissociation of the corepressors and association of the coactivator proteins that stimulate transcription. Such transcription factors are used as nutritional sensors in metabolically active organs like the liver, stomach, and adipose tissue which in turn regulate the transcripts of selected genes to changes in the nutritional status. Nuclear receptors regulate many diverse biological functions such as nutrient metabolism, embryonic development, cell proliferation, and cellular differentiation.

12.2.1 Gene–Nutrient Interaction

Nutrigenetics, which examines how genetic variation affects nutrient metabolism, utilization, food sensitivities, and individual dietary requirements, and nutrigenomics, which examines how nutrients impact genome function, including gene expression, epigenetic modifications, mutation rates, developmental programming, and even long-term genome evolution, are the two subfields that make up nutritional genomics (Mansour et al., 2024). Nutrient-driven genomic effects contribute to genetic diversity among populations, while genetic variation modulates nutritional responsiveness. These interactions are reciprocal. This interactive interface highlights the intricacies of human nutrition and its effects on health, disease susceptibility, and the formulation of customized dietary recommendations.

12.2.1.1 Nutrient–Gene Interactions

Specifically, three types of nutritional gene interactions occur:

1. Direct interactions: When nutrients engage with a receptor, they can act as transcription factors, binding to DNA and inducing gene expression.
2. Epigenetic interactions: Nutrients can change the structure of DNA (or histone proteins in chromatin), altering gene expression over time.
3. Genetic variations: Common genetic differences, such as single-nucleotide polymorphisms (SNPs), can affect gene expression and activity.

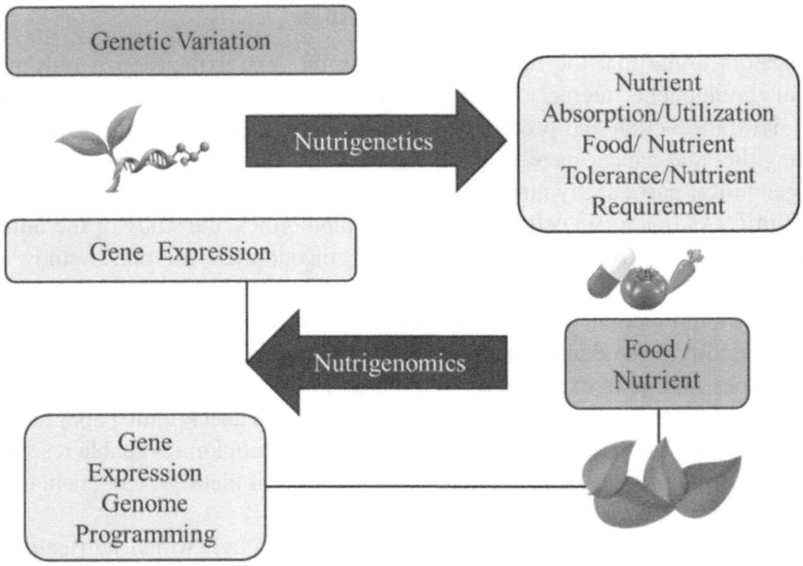

FIGURE 12.2 Overview of nutrition-genome interactions.

All of these systems can alter metabolism and dietary requirements. Therefore Figure 12.2 describes the influence of genetic variation in nutrient responses i.e. Nutrigenetics while Nutrigenomics explain the way nutrients modulate gene expression.

12.2.2 Mechanism of Gene Expression Regulation

The development of mammals depended heavily on their capacity to quickly adapt to changes in their nutritional environment while maintaining a metabolism that met all their needs and continually produced ATP. Quick yet durable reactions that simultaneously modify a wide range of interconnected metabolic processes are required for adaptation to fluctuations in food availability (energy and specific nutrients). This is mainly caused by hormones that can be classified by their chemical composition (e.g. steroids or peptide hormones) and/or the way and length of their activity.

12.2.3 Role of Epigenetics in Nutrigenomics

Epigenomics is the study of epigenetic changes in the cell/tissue at any specific time (Richter et al., 2024). The epigenome is composed of chemical substances that change or mark the genome, indicating what a cell is capable of doing and where and when it can do it. These markers are known as epigenetic marks. As the cells divide, they transfer these epigenetic marks that will be inherited by future generations. The immediate environment (nutrition, medications, and other factors) alters these signals, and they define the phenotype to the environment in which they occur (Richter et al., 2024; Carlberg, 2019; Virolainen et al., 2023). When analysing the influence of diet on the expression of genes, the substances that are immediately associated with hormonal secretion and action, including proteins, carbohydrates, lipids, and calcium, should be dissociated with the dietary components that do not seem to be directly integrated into these hormonal pathways and mechanisms, including specific vitamins or minerals (Bravo-Ruiz et al., 2021; Tang et al., 2025). However, when it comes to hormone-mediated modifications in cellular metabolism, different patterns (concentrations) of specific metabolites or nutrients within a cell can also affect gene expression by interacting with specific targets such as nuclear receptors and response elements. Apart from creating a long-lasting state of deficiency, the lack of a necessary nutrient can also indirectly change hormone activity and secretion, with intricate direct and indirect consequences on gene expression (Bravo-Ruiz et al., 2021; Tang et al., 2025; Lal et al., 2022).

Epigenomics is the study of all epigenetic changes occurring in a cell or tissue at a specific moment (Richter et al., 2024). Chemicals that alter or mark the genome in a way that can show what a cell is capable of doing, as well as where and when it can do it, make up the epigenome. We refer to these indicators as epigenetic marks. These epigenetic markers are transferred from one generation to the next during cell division. The phenotype is determined by the signals that are been altered by genotype in the surrounding environment (diet, drugs, and other factors, for instance) (Richter et al., 2024; Carlberg, 2019; Bordoni et al., 2021).

Chromosome structure can be altered by epigenetic processes, leading to changes in gene expression (Farsetti et al., 2023). Histone acetylation and DNA methylation are two examples. The nutritional enzyme, DNMT that is known to transfer a methyl group of S-adenosylmethionine to certain positions on the DNA, is catalysed, and it is responsible for chromatin remodelling, which is intimately linked to DNA methylation (Richter et al., 2024). 'Folic acid, vitamins B6, B12, B2, choline, and methionine' are all metabolized by S-adenosylmethionine (Tiwari et al., 2022). These deficits increase the risk of developing NTCD by altering DNA methylation and disrupting carbon metabolism. The transcription-related gene is suppressed by DNA hypermethylation, while hepatocellular and prostate malignancies are associated with hypomethylation (Lal et al., 2022).

12.3 KEY NUTRIENTS AND THEIR IMPACT ON GENE EXPRESSION

12.3.1 Omega-3 Fatty Acids and Inflammation

The anti-inflammatory properties of the so-called omega-3 polyunsaturated fatty acids (PUFAs) are potent (particularly, the action of docosahexaenoic acid (DHA) and eicosapentaenoic acid (EPA)) with regard to anti-inflammatory effects. By modulating transcription factors such as nuclear factor-kappa B (NF-κB) and peroxisome PPARs, these fatty acids are incorporated into cell membranes and affect gene expression.

A. Restriction of NF-κB by omega-3 fatty acids leads to transcription of pro-inflammatory cytokines genes including TNF-α, IL-6, and COX-2 that have less transcription, which reduces systemic inflammation.
B. When 'EPA/DHA activates PPAR-γ', genes related to adipogenesis and cytokine production are downregulated, which promotes an anti-inflammatory state.
C. Omega-3s also affect epigenetic processes, such as histone acetylation and DNA methylation, which further alter immunological and inflammatory gene networks. Because of these benefits, omega-3 fatty acids are essential for treating long-term inflammatory conditions like cancer, heart disease, and autoimmune diseases.

12.3.2 Folate and DNA Methylation

One-carbon metabolism, which is essential for DNA synthesis, repair, and methylation, depends heavily on folate (vitamin B9). It contributes methyl groups to the methylation of cytosine residues in CpG islands, a significant epigenetic alteration that controls gene expression.

A. The conversion of homocysteine to methionine, which is subsequently transformed into S-adenosylmethionine (SAM), the universal methyl donor for DNA, RNA, and histone methylation, requires folate.

B. Global DNA hypomethylation, a condition linked to genomic instability; oncogene activation; and an elevated risk of colorectal, cervical, and breast malignancies, may result from a folate deficiency.
C. On the other hand, in certain situations of folate excess, hypermethylation of tumour suppressor gene promoters may occur, highlighting the significance of a balanced folate intake.

Furthermore, DNA methylation profiles and disease susceptibility can be notably modified by polymorphisms of the MTHFR gene that encode the enzyme methylenetetrahydrofolate reductase that participates in folate metabolism (Mansour et al., 2024; Raghubeer & Matsha, 2021).

12.3.3 Vitamin D3 and Immune Regulation

Through its interaction with the 'vitamin D receptor (VDR), a nuclear receptor that binds to vitamin D response elements (VDREs) in the promoter regions of target genes, vitamin D3 (cholecalciferol)' controls gene expression.

A. A VDR activation regulates about 1,000 genes, many of which are involved in immune function, cell cycle control, apoptosis, and differentiation.
B. In immune cells, vitamin D3 regulates both innate and adaptive immunity.
C. It increases the production of antimicrobial peptides such as cathelicidin.
D. It maintains immunological tolerance by promoting regulatory T cell activity and suppressing 'pro-inflammatory cytokines (such as IL-6, IL-17, and TNF-α)'.
E. Through gene regulation, vitamin D3 has been shown in cancer biology to prevent proliferation and trigger apoptosis in a variety of cancer cell types.
   Vitamin D deficiency is associated with autoimmune diseases, cancer, and increased susceptibility to infections, highlighting its gene-regulatory role in maintaining immune homeostasis.

12.3.4 Other Bioactive Compounds

Several other bioactive dietary compounds exert epigenetic and gene regulatory effects, impacting health and disease risk:

1. Curcumin is a polyphenol present in turmeric, and it affects chromatin remodelling by inhibiting histone acetyltransferases (HATs) and activating histone deacetylases (HDACs). Additionally, it inhibits transcription factors such as STAT3 and NF-κB (Kassem et al., 2023).
2. Resveratrol: Sirtuins (SIRT1), NAD-dependent deacetylases, which regulate lifespan-related, metabolic, and cancer-suppressive gene expression, are activated by resveratrol that is found in grapes and red wine.
3. SFN: This is a compound in cruciferous vegetables, which encourages the expression of antioxidant and detoxifying genes via stimulating Nrf 2, a transcription factor, which provides protection against oxidative stress and carcinogens.
4. Epigallocatechin gallate (EGCG): It is a green tea catechin that is known to act on HDACs and DNMTs, which subsequently influence the reactivation of tumour-suppressor genes and silencing of genes.

   These compounds demonstrate how dietary interventions can modulate gene expression and epigenetic regulation, offering potential strategies for disease prevention and therapy.

12.4 HEALTH BENEFITS OF NUTRIGENOMICS-BASED DIETARY INTERVENTIONS

12.4.1 Obesity and Metabolic Syndrome

Genes that affect weight management can be categorized in various ways. The first group relies on genetically controlled systems and processes. This includes the equilibrium of body weight, physical activity, hunger, insulin signalling, adipocyte formation, mitochondrial activities, lipid turnover, thermogenesis, and energy efficiency.

Genes are then grouped according to how they regulate metabolic activities. The following genes are among the polymorphisms in this group that have been connected to genetically mediated differences in dietary weight loss techniques (Peña-Romero et al., 2017):

I. 'Controlling energy intake (e.g. MC3R, MC4R, POMC, LEP, LEPR, and FTO)
II. Lipid metabolism and adipogenesis (e.g. PLIN1, APOA5, LIPC, FABP2)
III. Thermogenesis (e.g. ADBRs, UCPs)
IV. Adipocytokine production (e.g. ADIPOQ, IL6)
V. Transcription factors (e.g. PPARG, TCF7L2, CLOCK)
VI. Type II diabetes and nutrigenomics'

12.4.2 Cardiovascular Disease

Cardiovascular illnesses (CVDs) are defined by the WHO as diseases due to abnormalities of the heart and blood vessels, such as cerebrovascular disease, coronary heart disease (heart attacks), stroke, peripheral arterial disease, rheumatic heart disease, congenital heart disease, increased blood pressure (hypertension), and heart failure. They will remain, according to projections, the leading cause of morbidity and mortality in the twenty-first century in the next decades (Kassem et al., 2023). A healthy lifestyle may help avoid this disease, which could eventually reduce its prevalence in the general population. Given that the vast majority of chronic diseases can be avoided with dietary and hygiene practices, nutrition is undoubtedly a crucial element.

Over the years, numerous interventional and epidemiological studies have been conducted. Despite progress, the mechanisms underlying individual differences in the pathophysiology and presentation of CVD remain unknown. Genetics is a significant factor in CVDs, as several genes and their variants can influence CVD risk in different ways.

Dietary consumption, genetic variations, and environmental factors can influence the total risk of CVDs. If enough information is available about the connection between specific genetic variants, food, and CVD risk, it would be possible to provide people with dietary advice based on their genotype. In addition to preserving health and fitness, this would increase life expectancy (Kassem et al., 2023; Gianfredi et al., 2025).

12.4.3 Type 2 Diabetes

The complicated metabolic disease known as T2DM is caused by a combination of low-grade inflammation, decreased pancreatic β-cell activity, and insulin resistance. The critical role of both innate and adaptive immune processes in the onset and progression of 'type 2 diabetes', especially in obese individuals, has been highlighted by recent discoveries in immune metabolism. Changes in immune cell populations contribute to a pro-inflammatory milieu in visceral adipose tissue, which is an essential immunological hub. Anti-inflammatory M2 macrophages preserve tissue equilibrium in a lean physiological state. Still, obesity causes a shift toward M1 macrophages that are pro-inflammatory and releases cytokines such as TNF-α, IL-6, and IL-1 β molecules that are known to interfere with insulin receptor signalling pathways (Bravo-Ruiz et al., 2021; Felisbino et al., 2021). Inflammation and metabolic dysregulation are further exacerbated by alterations in the adaptive immune compartment, characterized by decreased regulatory T cells and increased infiltration of 'CD8 T cells and B cells'. These immune-mediated disruptions affect hepatic, muscular, and pancreatic organs, as well as adipose depots, exacerbating insulin resistance and disrupting glucose homeostasis (Bravo-Ruiz et al., 2021). In particular, β-cell dysfunction and death are associated with activation of NLRP3 inflammasome and IL-1β secretion in pancreatic islets. Two of the interventions that have proven effective in reducing systemic inflammation and enhancing metabolic outcomes in individuals with type 2 diabetes include IL-1 receptor blockage and planned lifestyle changes (Felisbino et al., 2021).

12.4.4 Cancer Prevention

Cancer is a complicated illness that is impacted by environmental, genetic, and epigenetic variables, including diet. Poor eating habits are associated with about 35% of cancer cases, and appropriate nutrition may be able to prevent up to 70% of these cases. Nutrient deficiencies can impair DNA repair and antioxidant defences, thereby increasing genomic instability. Individual vulnerability to dietary carcinogens (such as heterocyclic aromatic amines in cooked red meat) and hormone-dependent malignancies is influenced by genetic polymorphisms, such as those in N-acetyltransferase (NAT2) and hormone-related genes. Plant-based foods contain antioxidants and phytoestrogens (such as lignin and isoflavones) that modify gene expression and lessen oxidative DNA damage (Ionescu et al., 2021). By modulating inflammation, the immune response, and carcinogen elimination, omega-3 fatty acids and other phytochemicals present in fruits and vegetables have a protective effect. Through gene-nutrient interactions, signalling pathway modifications, and reduced oxidative stress, nutrigenomics research underscores the importance of bioactive food substances in cancer prevention.

Oesophageal cancer in the world is represented mostly by oesophageal squamous cell carcinoma (ESCC), approximately by 60–70%. Along with genetic susceptibility and infectious pathogens, a massive epidemiological investigation carried out in China established that food intake was a critical modifiable risk factor in the growth and advancement of ESCC (Ionescu et al., 2021; Jiang et al., 2025). The study found that a much higher risk of ESCC was associated with regular intake of pickled foods and extremely hot beverages, both common in traditional Asian diets. Furthermore, compared with people who consume little meat, greater consumption of red meat and processed meat was associated with 57% and 55% higher risks, respectively. On the other hand, a high intake of fruits and vegetables was linked to a 30% lower incidence of ESCC, underscoring the preventive role of plant-based foods rich in fibre, antioxidants, and phytochemicals. Increased vulnerability to ESCC has been linked to nutritional deficits, especially in β-carotene, vitamin E, and selenium. These micronutrient supplements have been shown to lower ESCC-related mortality by 17%. Additionally, it has been demonstrated that moderate riboflavin (vitamin B2) supplementation, a micronutrient found in eggs, green leafy vegetables, and organ meats such as liver and kidney, reduces the incidence of ESCC, prevents recurrence, and improves patient prognosis. These results highlight the potential of micronutrient optimization in oesophageal cancer therapy and prevention (Ionescu et al., 2021).

Tobacco use and Helicobacter pylori (H. pylori) infection have been closely linked to stomach cancer, and there is increasing evidence linking certain dietary factors to the disease's genesis. High consumption of pickled and salty foods has been linked to an elevated risk of stomach cancer, according to a new systematic review and meta-analysis. Chronic alcohol use was also found to be a contributory factor, with a 19% increased risk seen even in those who had stopped drinking. This is probably because chronic alcohol exposure causes irreversible damage to the liver and mucosa. Conversely, no significant correlation was found between coffee or green tea consumption and stomach cancer. Significantly, eating a lot of fruits and vegetables was linked to a 48% and 62% lower risk of stomach cancer, respectively, indicating a preventive benefit of plant-based diets strong in fibre and antioxidants (Wang et al., 2023).

As the third-most common cause of cancer-related deaths, CRC is a significant global health burden that is strongly associated with modifiable lifestyle variables. Tobacco use, infections, and especially diet and obesity are substantial causes of CRC.

The development of CRC is strongly linked to high consumption of red and processed meats. This is ascribed to the following:

I. Red meat contains heme iron that is build up in the colon, damages the mucosa, and modifies the gut microbiota.
II. Oncogenes such as b-catenin, K-Ras, and APC can have their heterocyclic amines (HCAs) mutated by high-temperature cooking.
III. N-nitroso compounds (NOCs) are found in processed meat and damage DNA and promote the formation of cancer.
IV. IGF-1 elevation and cell division by PI3K/Akt pathway need hyperinsulinemia and insulin resistance.
V. Elevated leptin and a decreased adiponectin disrupt cell signalling.
VI. The chronic inflammation which develops a tumour-promoting environment is being mediated by the cytokines such as TNF-α, IL-6, and CRP.

Role of key micronutrients in cancer prevention and immune modulation is as follows.

12.4.4.1 Vitamin C

Vitamin C, also known as ascorbic acid, is frequently obtained through diet and supplementation. It is a water-soluble antioxidant that is essential for protecting cells from oxidative damage, a significant factor in cancer development. It stops lipid peroxidation, directly scavenges ROS, and maintains the redox integrity of other antioxidants such as glutathione and vitamin E. By preventing oxidative DNA lesions, such as 8-oxoguanine and strand breaks, vitamin C preserves genomic integrity. Furthermore, vitamin C regulates immunological function by increasing natural killer (NK) cell activity and T-cell proliferation, both of which are critical for tumour immunosurveillance. It has shown dose-dependent anticancer activity: low doses may promote cell growth, whereas high doses induce apoptosis, particularly in melanoma cells. The pro-oxidant impact of excessive buildup in cells may cause adverse effects. Furthermore, vitamin C increases the effectiveness of chemotherapy and reduces the required dosages. Its capacity to generate hydrogen peroxide, which builds up and encourages the death of cancer cells, is one suggested method. It has also been shown to reduce the death rate from breast cancer.

12.4.4.2 Vitamin A

Retinol, retinal, and provitamin A carotenoids are components of vitamin A. It controls epigenetic methylation, which in turn controls cell proliferation and differentiation, and functions as a free radical scavenger, minimizing oxidative stress-induced DNA damage. It protects premalignant oral lesions; reduces the risk of lung, head, and neck malignancies; and reduces chemotherapy-induced mucositis. Retinoids provide enhanced protection against oral carcinogenesis when combined with agents such as bexarotene. Clinical applications of retinoids include differentiation treatment and chemoprevention, particularly for acute promyelocytic leukaemia. However, high-dose β-carotene supplementation has been linked to a higher risk of lung cancer among smokers and populations exposed to asbestos, suggesting that demographic context, source, and dosage are important considerations. Liver, dairy products, and eggs are natural sources of vitamin A, and carrots, sweet potatoes, and dark leafy greens are rich sources of provitamin A carotenoids.

12.4.4.3 Vitamin D

Oral, breast, prostate, and pancreatic cancers are among the many cancers that vitamin D protects against. It influences immunological responses and gene expression by interacting with vitamin D receptors (VDRs) (Starska-Kowarska, 2023). It prevents angiogenesis and cell division in pancreatic cancer. It causes programmed cell death in oral tumours and enhances the results of radiation therapy for breast cancer. It is expressed in breast tissue and, following ligand binding, modifies gene transcription in breast cancer. Specific VDR gene polymorphisms, such as FokI (rs2228570), may change receptor function and affect BC susceptibility. Also, VDR and retinoid X receptor (RXR) heterodimerize to bind vitamin D response elements (VDREs) in the promoters of genes, which regulate the expression of the oestrogen receptor (ERα) in certain tumour subtypes (Torres et al., 2024). Vitamin D affects the expression of critical BC-related genes, including CYP24A1, which degrades vitamin D, and CYP27B1, which activates vitamin D. Inhibiting CYP24A1 may restore vitamin D's tumour-suppressive action; however, overexpression of this gene has been associated with a poor prognosis (Torres et al., 2024).

12.4.4.4 Folic Acid

A synthesized, stable version of folate (vitamin B9), folic acid (pteroylmonoglutamic acid), is necessary for one-carbon metabolism. Folic acid is crucial for the methylation of proteins, RNA, and DNA, which is essential for genome stability. These processes may be hampered, and tumour development may be promoted by mutations in the MTHFR enzyme, which converts folate to its active form (Mansour et al., 2024). Concerns about its dual pro- and anti-carcinogenic potential persist, despite some evidence supporting its preventive function against CRC and its improvement of lung cancer treatment. Adequate folate levels help prevent chromosomal instability, abnormal methylation, and DNA strand breaks, all of which are indicators of carcinogenesis. Preserving promoter methylation in tumour suppressor genes promotes regular gene expression. Large-scale randomized trials have not demonstrated an increased risk of colorectal, prostate, or general cancer, despite concerns that high folic acid intake may promote the growth of pre-existing neoplasms.

12.4.4.5 Vitamins and Viral Defence

Vitamins significantly enhance host immunity, particularly during viral infections like COVID-19.

I. Vitamin D improves immunity by strengthening physical barriers, reducing cytokine release, and modulating both cellular and humoral responses.
II. Vitamin A boosts immune function through NK cell regulation, phagocytosis, and cytokine modulation (TNF-α, IL-2).
III. Vitamins C, E, B1, and B6 have all been shown to aid in recovery and treatment during viral infections, supporting their inclusion in therapeutic protocols.

12.5 FUTURE PROSPECTIVE AND CHALLENGES

By modulating gene expression through dietary interventions, the discipline of nutrigenomics offers new pathways for illness prevention, management, and personalized healthcare, marking a paradigm shift in biomedical science. The application of nutrigenomic information to clinical and public health practice is still complex, despite tremendous advancements. This chapter discusses the prospects and related difficulties, with a focus on new tools and technical developments that will make this transition easier.

12.5.1 Advancement in Personalized Medicine

In the context of nutrigenomics, personalized medicine refers to tailoring nutritional therapies according to a person's metabolic, transcriptomic, genetic, and epigenomic profiles. By focusing on particular pathways linked to illness susceptibility and progression, it seeks to maximize health outcomes. Genetic variations in nutrient absorption, metabolism, and response may include SNPs in enzyme-coding (e.g. MTHFR, FTO, APOE) or nutrient-transporter genes (Chiarelli et al., 2024). The development of customized dietary strategies for disease prevention, such as lowering cardiovascular risk, controlling metabolic syndrome, and modifying cancer risk, is made possible by combining this information with phenotypic data (Felisbino et al., 2021; Chiarelli et al., 2024).

However, there are obstacles to adoption in clinical settings, including the need for robust population-based studies, biomarker validation, ethical considerations in handling genomic data, and patient adherence. Furthermore, the interpretation of nutrigenomic data is complicated by multifactorial effects that necessitate systems-level research, such as the gut microbiome, environment, and lifestyle (Fischer, 2024). As shown in Figure 12.3, a continuous loop where individual genetic variation drives personalized medicine, enabling targeted therapies for each patient.

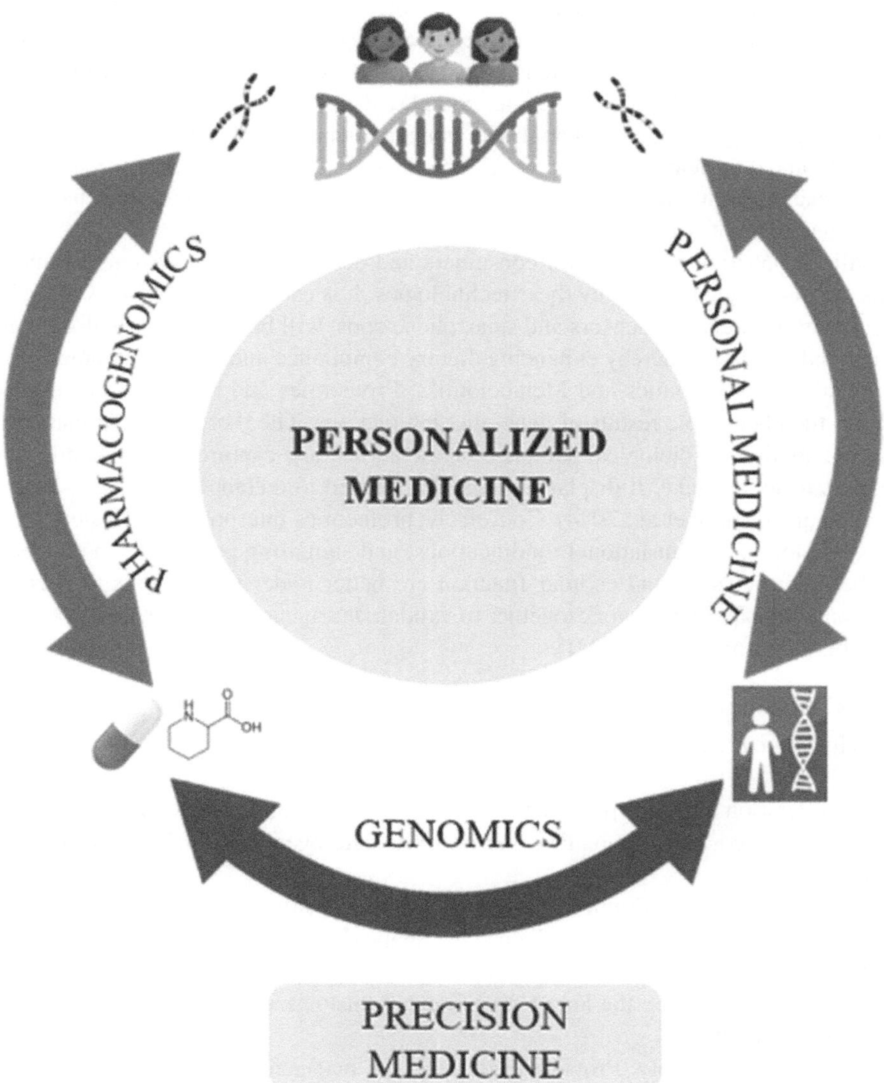

FIGURE 12.3 Schematic representation of personalized medicine.

12.5.2 Technology Advances in Nutrigenomics

Recent technological developments have dramatically increased the ability to study nutrient-gene interactions at previously unheard-of scale and resolution. These developments have enabled the evaluation of molecular responses to dietary treatments in a high-throughput, integrative, and dynamic manner.

Omics and high-throughput genomics technologies: With the advent of DNA microarrays, next-generation sequencing (NGS), and epigenome-wide association studies (EWAS), the analysis of genomic landscapes is currently changing (Shaman, 2024). Techniques like DNA and RNA microarrays have been primarily used to study stress-responsive mechanisms in plants (Nandhan et al., 2025). Recent advances showed that the same is used in personalized nutrition research (Singar et al., 2024). These resources make it easier to identify genetic variants and epigenetic changes linked to nutritionally significant health outcomes. Researchers can thoroughly profile gene expression variations in response to specific foods or dietary patterns when combined with transcriptomic tools such as RNA sequencing (Shaman, 2024).

Moreover, a systems biology perspective is enabled by integrative omics techniques such as proteomics, metabolomics, and lipidomics. The metabolic pathways, regulatory networks, and feedback mechanisms involved in diet-related disease processes can be clarified using these techniques.

AI in the study of nutrigenomics: Large, complex datasets produced by multi-omics research are increasingly being managed and analysed using artificial intelligence (AI), including machine learning (ML) and deep learning methods. AI techniques make it easier to identify new biomarkers, predict how each person will respond to dietary changes, and group populations based on nutrigenomic profiles (Phugat & Goel, 2025).

Personalized nutrition platforms for consumers and decision-support systems for medical professionals are also made possible by these technologies. It is expected that AI-driven digital health solutions, including wearable sensors and smartphone apps, will be crucial for real-time nutritional monitoring and feedback, thereby enhancing dietary compliance and health outcomes.

The Function of Proteomics and Metabolomics: Proteomics and metabolomics offer essential insights into the phenotypic results of gene–diet interactions. The thorough examination of small-molecule metabolites in biological samples, or metabolomics, captures the dynamic physiological responses to food intake. It displays biomarkers related to metabolic health, disease risk, and nutritional status (Catussi et al., 2024). Conversely, proteomics interprets diet-related alterations in protein expression, post-translational modifications, and signalling pathways. The molecular connections between nutrition and cellular function are better understood thanks to these analyses. Proteomics and metabolomics work together to validate nutrigenomic results and advance customized nutrition (Al-Amrani et al., 2021).

12.5.2.1 Challenges

Despite technological advances, several challenges remain:

1. Data Integration and Interpretation: Standardized procedures and sophisticated bioinformatics tools are needed to manage and integrate heterogeneous data from different omics platforms.
2. Ethical and Regulatory Frameworks: Privacy, data security, and informed consent are ethical issues that are brought up by the use of genetic data in nutrition.
3. Clinical Translation and Public Health Implementation: To determine causality, create guidelines, and evaluate the long-term effects of customized nutrition, randomized clinical trials are required.
4. Accessibility and Equity: Providing fair access to nutrigenomic services is still a significant public health issue, especially in low- and middle-income nations.

12.6 CONCLUSIONS

Nutrigenomics is currently being accepted as an essential component in healthcare that provides specific dietary remedies to individuals and communities across their genetic profiles. It finds a proactive application in managing and minimizing the vulnerability of chronic diseases like obesity, diabetes, and heart disease and even supporting the prevention of cancer through informing decisions about functional foods and nutraceuticals. Genetic testing is assisting in clinical practice to make more accurate choices about supplementation, including folate or personalized levels of fat, and population health agencies are also embracing the application of genomics to create effective interventions. Surpassing the distance that existed between basic science and patient care, nutrigenomics is rapidly integrating into hospitals, clinics, and community health, which is why the concept of precision nutrition is becoming a real possibility to lots of people today. Therefore, gene-nutrient interactions can be studied more thoroughly with the help of technological progress in bioinformatics and omics technologies that provide opportunities to treat specific nutritional disorders. Though the challenges are still evident, more efficient genotype-guided diets through continuous research ensure the improvement of disease prevention and health promotion on a more individual level.

REFERENCES

Al-Amrani, S., Al-Jabri, Z., Al-Zaabi, A., Alshekaili, J., & Al-Khabori, M. (2021). Proteomics: Concepts and applications in human medicine. *World Journal of Biological Chemistry*, *12*(5), 57.

Bordoni, L., Petracci, I., Zhao, F., Min, W., Pierella, E., Assmann, T. S., . . . & Gabbianelli, R. (2021). Nutrigenomics of dietary lipids. *Antioxidants*, *10*(7), 994.

Bravo-Ruiz, I., Medina, M. Á., & Martínez-Poveda, B. (2021). From food to genes: Transcriptional regulation of metabolism by lipids and carbohydrates. *Nutrients*, *13*(5), 1513.

Carlberg, C. (2019). Nutrigenomics of vitamin D. *Nutrients*, *11*(3), 676. https://doi.org/10.3390/nu11030676

Catussi, B. L. C., Turco, E. G. L., Pereira, D. M., Teixeira, R. M. N., Castro, B. P., & Massaia, I. F. D. (2024). Metabolomics: Unveiling biological matrices in precision nutrition and health. *Clinical Nutrition ESPEN*, *64*, 314–323.

Chiarelli, R., Caradonna, F., & Naselli, F. (2024). Autophagy and nutrigenomics: A winning team against chronic disease and tumors. *Frontiers in Nutrition*, *11*, 1409142.

Dhanapal, A. C. T. A., Wuni, R., Ventura, E. F., Chiet, T. K., Cheah, E. S. G., Loganathan, A., Quen, P. L., Appukutty, M., Noh, M. F. M., Givens, I., & Vimaleswaran, K. S. (2022). Implementation of nutrigenetics and nutrigenomics research and training activities for developing precision nutrition strategies in Malaysia. *Nutrients*, *14*(23), 5108. https://doi.org/10.3390/nu14235108

Farsetti, A., Illi, B., & Gaetano, C. (2023). How epigenetics impacts on human diseases. *European Journal of Internal Medicine*, *114*, 15–22.

Felisbino, K., Granzotti, J. G., Bello-Santos, L., & Guiloski, I. C. (2021). Nutrigenomics in regulating the expression of genes related to type 2 diabetes mellitus. *Frontiers in Physiology*, *12*, 699220.

Fischer, F. (2024). The interplay of nutrition, genetics, and lifestyle in the prevention and management of chronic diseases: Insights from modern research. *Public Health Spectrum*, *1*(1).

Gianfredi, V., Nucci, D., Pennisi, F., Maggi, S., Veronese, N., & Soysal, P. (2025). Aging, longevity, and healthy aging: The public health approach. *Aging Clinical and Experimental Research*, *37*(1), 1–12.

Haro, D., Marrero, P. F., & Relat, J. (2019). Nutritional regulation of gene expression: Carbohydrate-, fat- and amino acid-dependent modulation of transcriptional activity. *International Journal of Molecular Sciences*, *20*(6), 1386.

Hosseiniara, S. M., & Hosseini Zijoud, S. S. (2024). Nutrigenomics: A promising frontier in chronic disease prevention. *Journal of Preventive and Complementary Medicine*, *3*(4), 195–200. https://doi.org/10.22034/jpcm.2024.496108.1196

Ionescu, V. S., Popa, A., Alexandru, A., Manole, E., Neagu, M., & Pop, S. (2021). Dietary phytoestrogens and their metabolites as epigenetic modulators with impact on human health. *Antioxidants*, *10*(12), 1893.

Jiang, W., Zhang, B., Xu, J., Xue, L., & Wang, L. (2025). Current status and perspectives of esophageal cancer: A comprehensive review. *Cancer Communications*, *45*(3), 281–331.

Kassem, N. M., Abdelmegid, Y. A., El-Sayed, M. K., Sayed, R. S., Abdel-Aalla, M. H., & Kassem, H. A. (2023). Nutrigenomics and microbiome shaping the future of personalized medicine: A review article. *Journal of Genetic Engineering and Biotechnology*, *21*(1), 134. https://doi.org/10.1186/s43141-023-00599-2

Lal, M. K., Sharma, E., Tiwari, R. K., Devi, R., Mishra, U. N., Thakur, R., . . . & Sahu, S. K. (2022). Nutrient-mediated perception and signalling in human metabolism: A perspective of nutrigenomics. *International Journal of Molecular Sciences*, *23*(19), 11305.

Mansour, S., Alkhaaldi, S. M., Sammanasunathan, A. F., Ibrahim, S., Farhat, J., & Al-Omari, B. (2024). Precision nutrition unveiled: gene–nutrient interactions, microbiota dynamics, and lifestyle factors in obesity management. *Nutrients*, *16*(5), 581.

Nandhan, N. S., Saikrishna, M., Yadav, M., Barot, M., Parveen, S., Modi, A., & Husen, A. (2025). Abiotic stress induced regulation of antioxidant enzyme genes using microarray techniques. In *Role of Antioxidants in Mitigating Plant Stress* (pp. 391–403). Academic Press.

National Human Genome Research Institute (2013). *Talking Glossary of Genetic Terms*. http://www.genome.gov/Research Funding/.

Peña-Romero, A. C., Navas-Carrillo, D., Marín, F., & Orenes-Piñero, E. (2017). The future of nutrition: Nutrigenomics and nutrigenetics in obesity and cardiovascular diseases. *Critical Reviews in Food Science and Nutrition*, *58*(17), 3030–3041. https://doi.org/10.1080/10408398.2017.1349731

Pérez, A. R., Bottasso, O. A., & Santucci, N. E. (2025). Immune-endocrine crossroads: The impact of nuclear receptors in Tuberculosis and Chagas disease. *Frontiers in Endocrinology*, *16*, 1538376.

Phugat, S., & Goel, P. (2025). Review on Advancement of AI in Nutrigenomics. In *Artificial Intelligence (AI) in Cell and Genetic Engineering* (Methods in Molecular Biology, Vol. 2952, pp. 429–444). Springer Nature, Cham, Switzerland. https://doi.org/10.1007/978-1-0716-4690-8_23

Raghubeer, S., & Matsha, T. E. (2021). Methylenetetrahydrofolate (MTHFR), the one-carbon cycle, and cardiovascular risks. *Nutrients*, *13*(12), 4562.

Richter, E., Patel, P., Babu, J. R., Wang, X., & Geetha, T. (2024). The importance of sleep in overcoming childhood obesity and reshaping epigenetics. *Biomedicines*, *12*(6), 1334.

Saha, B., AT, R., Adhikary, S., Banerjee, A., Radhakrishnan, A. K., Duttaroy, A. K., & Pathak, S. (2024). Exploring the relationship between diet, lifestyle and gut microbiome in colorectal cancer development: A recent update. *Nutrition and Cancer*, *76*(9), 789–814.

Satrio, R. D., Fendiyanto, M. H., & Miftahudin, M. (2024). Tools and techniques used at global scale through genomics, transcriptomics, proteomics, and metabolomics to investigate plant stress responses at the molecular level. In *Molecular Dynamics of Plant Stress and Its Management* (pp. 555–607). Singapore: Springer Nature Singapore.

Shaman, J. A. (2024). The future of pharmacogenomics: Integrating epigenetics, nutrigenomics, and beyond. *Journal of Personalized Medicine*, *14*(12), 1121. https://doi.org/10.3390/jpm14121121

Singar, S., Nagpal, R., Arjmandi, B. H., & Akhavan, N. S. (2024). Personalized nutrition: Tailoring dietary recommendations through genetic insights. *Nutrients, 16*(16), 2673. https://doi.org/10.3390/nu16162673

Starska-Kowarska, K. (2023). Role of vitamin D in head and neck cancer – immune function, anti-tumour effect, and its impact on patient prognosis. *Nutrients*, *15*(11), 2592.

Tang, J., Krushelnycky, L., Shaqo, A., & Cho, C. E. (2025). A comprehensive review of nutritional influences on the serotonergic system. *Advances in Nutrition*, 100524.

Tiwari, D., Rani, A., & Jha, H. C. (2022). Homocysteine and folic acid metabolism. In *Homocysteine Metabolism in Health and Disease* (pp. 3–36). Singapore: Springer Nature Singapore.

Torres, A., Cameselle, C., Otero, P., & Simal-Gandara, J. (2024). The impact of vitamin D and its dietary supplementation in breast cancer prevention: An integrative review. *Nutrients*, *16*(5), 573. https://doi.org/10.3390/nu16050573

Virolainen, S. J., VonHandorf, A., Viel, K. C., Weirauch, M. T., & Kottyan, L. C. (2023). Gene–environment interactions and their impact on human health. *Genes & Immunity*, *24*(1), 1–11.

Wang, Y., Liu, B., Han, H., Hu, Y., Zhu, L., Rimm, E. B., . . . & Sun, Q. (2023). Associations between plant-based dietary patterns and risks of type 2 diabetes, cardiovascular disease, cancer, and mortality–a systematic review and meta-analysis. *Nutrition Journal*, 22(1), 46.

13 Industrially Important Vegetables and Their Phytochemicals

Hiral Desai, Anil Patani, Sachidanand Singh, and Megha Barot

13.1 INTRODUCTION

Vegetables provide macronutrients, micronutrients, and a variety of bioactive substances known as phytochemicals (Sharma & Kaushik, 2021). They are essential to a healthy diet. Even though they are not essential nutrients, these organic plant substances are essential for maintaining health and averting persistent illnesses. Vegetables are key sources of vitamins like vitamins K, C, and A, and minerals like potassium, magnesium, iron, dietary fibre, and various macronutrients needed for metabolic functions (Ebabhi & Adebayo, 2022). Eating vegetables regularly is linked to a lower chance of getting heart disease, diabetes, obesity, and several types of cancer. Their wide range of nutrients makes vegetables essential for overall health and important for long-term well-being (Boeing et al., 2012). Vegetables are gaining more importance for their industrial uses as well as their nutritional value. The culinary, pharmaceutical, cosmetic, and nutraceutical industries all use phytochemicals found in vegetables (Kawatra et al., 2023). For example garlic contains allicin, which has antibacterial and medicinal properties (Choo et al., 2020). Natural antioxidants such as carotenoids and polyphenols serve as food preservatives and health-boosting additives. Phytochemicals are natural compounds which give plants colour, flavour, and disease resistance. Phytochemicals can be broadly categorized into a number of groups, including glucosinolates, alkaloids, carotenoids, flavonoids, and polyphenols, based on their various chemical structures and biological functions (Nigar et al., 2025). Leafy vegetables are rich in such phytochemicals and can become a reliable alternative to other plant- or plant part-based sources such as medicinal plants (Kumar et al., 2023), fruits (Wan et al., 2025), edible flowers (Książkiewicz et al., 2025), and seaweeds (Barot et al., 2016, 2019).

13.2 CLASSIFICATION OF PHYTOCHEMICALS

13.2.1 Saponins

Based on the nature of the aglycone, saponins – glycosides found in plants that contain sugar moieties and sapogenin – are divided into two categories: triterpenoids and steroidal saponins (Hamdi et al., 2020). These are prevalent in legumes such as common beans (*Phaseolus vulgaris* L.), pigeon peas (*Cajanus cajan* L.), garden peas (*Pisum sativum* L.), and black grams (*Vigna mungo* L.).

13.2.2 Polyphenols

Polyphenols are a group of naturally occurring compounds which have phenolic structures. This family has four primary subgroups: phenolic acids, stilbenes, lignans, and flavonoids. Artichokes (*Cynara cardunculus var. scolymus* L.), spinach (*Spinacia oleracea* L.), broccoli (*Brassica oleracea*

DOI: 10.1201/9781003606710-13

var. italica L.), chicory (*Cichorium intybus* L.), flax (*Linum usitatissimum* L.), onions (*Allium cepa* L.), apples (*Malus domestica* L.), plums (*Prunus subg. Prunus* L.), pears (*Pyrus* L.), grapes (*Vitis vinifera* L.), and cherries (*Prunus avium* L.) are rich sources of polyphenols.

13.2.3 Carotenoids

Brilliant orange, red, and yellow pigments known as carotenoids are found in photosynthetic bacteria, algae, and plants. Fruits are particularly rich in carotenoids, but they are also abundant in vegetables like sweet potatoes, carrots, pumpkins, and spinach. The round purple turnip (*Brassica rapa* var. rapa L.), fenugreek (*Trigonella foenum-graecum* L.), spinach (*Spinacia oleracea* L.), cabbage (*Brassica oleracea* L.), daikon radish (*Raphanus sativus* L.), oranges (*Citrus sinensis* L.), parsley (*Petroselinum crispum* L.), tomatoes (*Solanum lycopersicum* L.), Carrots (*Daucus carota* L.), and green leafy vegetables are rich sources of them. Carotenoids which are frequently found include lutein, zeaxanthin, lycopene, fucoxanthin, β-cryptoxanthin, β-carotene, and α-carotene (Quitério et al., 2022). In most fruits and vegetables, β-carotene is the most abundant carotenoid, which is followed by α-carotene; tangerines, persimmons, and oranges contain β-cryptoxanthin, a major carotenoid; green leafy vegetables primarily contain lutein and zeaxanthin; tomatoes contain lycopene; and brown algae contain fucoxanthin.

13.2.4 Alkaloids

Alkaloids are compounds which contain nitrogen and are mostly derived from amino acids. They are extensively utilized in both medicine and pest control and are renowned for their potent pharmacological properties. Despite being less prevalent in vegetables than in medicinal plants, some species, like eggplant (Solanum melongena), contain significant alkaloids with antioxidant and antimicrobial qualities, such as solanine and solasodine. When used in small amounts, these substances may have therapeutic benefits in addition to aiding in plant defence (Ghosh et al., 2022).

13.2.5 Glucosinolates

Glucosinolates are sulphur-containing substances with anticancer and detoxifying qualities which are mostly present in cruciferous vegetables like kale, broccoli, and cabbage. They yield indoles and physiologically active isothiocyanates when hydrolysed by enzymes (Baldelli et al., 2025).

13.3 INDUSTRIALLY IMPORTANT VEGETABLES AND THEIR PHYTOCHEMICALS

More than being just a source of vitamins and macronutrients, vegetables are nutritious matrices that contain a variety of secondary metabolites, or phytochemicals, which have significant health benefits and commercial value. Phenolics, carotenoids, glucosinolates, alkaloids, saponins, betalains, and terpenoids are among the compounds which contribute to colour, flavour, and plant defence. When ingested by humans, these compounds have antimicrobial, chemopreventive, antioxidant, and anti-inflammatory qualities (Hossain et al., 2025). Here Table 13.1 presents various phytochemicals and their properties.

13.4 EXTRACTION TECHNIQUES OF VEGETABLE PHYTOCHEMICALS

Extraction is the process of removing the desired components from the original substance. The solvents used to extract phytochemicals can be separated into two categories: organic solvents like cyclohexane, hexane, benzene, acetone, methyl acetate, ethyl acetate, chloroform, methanol, butanol, etc., and green solvents like deep eutectic solvents, supercritical CO_2, isopropanol, acetic acid,

TABLE 13.1
Major Phytochemicals of Vegetables and Their Properties

Phytochemicals	Source	Properties	References
Lycopene	Tomatoes	Antioxidant	Fernandez-Pan et al. (2024)
Alpha and beta-carotene	Carrots and green leafy vegetables	Antioxidant, immune modulator	Chandra et al. (2012)
Glucosinolates and indoles	Broccoli, kale	Immune modulator	Chandra et al. (2012)
Allyl sulphides	Onion and garlic	Anti-neoplastic, blood purifier, anti-bacterial, antioxidant	Chandra et al. (2012)
Indoles and isothiocyanates	Broccoli, cabbage, cauliflower, brussels, sprouts	Immune modulator, antioxidant	Chandra et al. (2012)
Isoflavones	Soybeans, tofu	Anti-carcinogenic, decrease cholesterol levels (LDL)	Chandra et al. (2012)
Lignans	Soybeans, flax seed	Soybeans, flax seed	Chandra et al. (2012)
Lutein	Kale, spinach	Antioxidant	Eom et al. (2023)
Anthocyanins	red cabbage	Anti-inflammatory, antioxidant	Zhao et al. (2024)
Quercetin	onion	Anti-inflammatory, antioxidant	Xu et al. (2024)

ionic liquids, fatty oils, glycerol, ethanol, and water. Because green solvents are non-toxic, renewable, recyclable, and biodegradable (Kumar et al., 2023). To separate phytochemicals from vegetables, traditional extraction techniques like maceration, Soxhlet extraction, and solvent extraction have been widely employed (Figure 13.1). However, these methods may break down thermolabile compounds and often require high volumes of solvent and lengthy processing periods. Modern green extraction technologies have become effective, sustainable, and selective replacements for these drawbacks (Chemat et al., 2022).

13.4.1 Microwave-Assisted Extraction

High-frequency electromagnetic waves known as microwaves are found in the electromagnetic spectrum between radiofrequency and far-infrared (IR). Their wavelengths range from 1 mm to 1 cm, and their frequencies range from 0.3 to 300 GHz (Irfan et al., 2021). This extraction method extracts plant metabolites using strong microwaves and solvents. MAE has also been used to analyse food ingredients, such as fatty acids from vegetable and animal fats, pectin from leftover mango peel, flavour composition, and amino acids from vegetables (Kumar et al., 2023). Microwave devices run on two frequencies: 2,450 MHz and 915 MHz. The dipoles align and randomly rotate 4.9×10^9 times/sec. at 2,450 MHz (Akhtar et al., 2019). The solution is heated as a result of the forced molecular movement and molecular friction. Because MAE heats the sample solvent mixture quickly, it can be used to quickly extract a variety of acetone–hexane, analytes, including those that are thermally unstable. For solvents like dichloromethane–acetone and acetone, the temperature reached during extraction in closed-vessel MAE is two to three times the solvent's boiling point, which improves

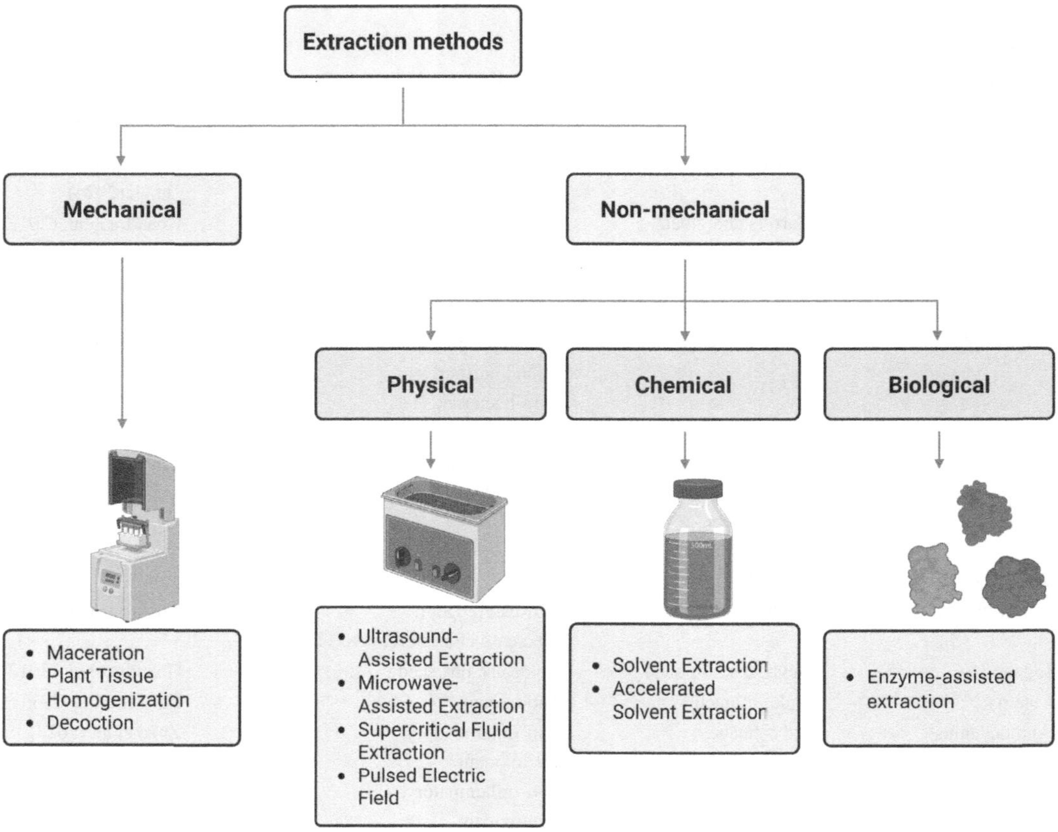

FIGURE 13.1 Different methods of phytochemical extraction.

sample matrix's extraction efficiencies. The effectiveness of MAE depends on a number of factors, including the material of the sample, the components that are obtained, the solvent characteristics, and their dielectric constants. MAE's drawback is that it may harm heat-sensitive polyphenols like anthocyanins as well as polyphenols with a lot of hydroxyl-type substituents. Thus, it is used sparingly when utilizing polymeric polyphenols like tannins and anthocyanins (Shital et al., 2018).

13.4.2 Ultrasound-Assisted Extraction

Acoustic or ultrasonic cavitation serves as the foundation for ultrasonic extraction, also known as UAE. Acoustic interaction happens when high-power ultrasonic waves are coupled into the slurry using a probe-type ultrasonic processor. High and low pressure cycles are created as the waves pass through the liquid. Cavitation occurs in the UAE due to the mechanical energy produced by ultrasonic waves (Shital et al., 2018). This completes extraction and produces high-quality extracts quickly and affordably (Yousuf et al., 2018). Ultrasound applications in the food industry can be classified as either high-intensity low-frequency ultrasound (20 kHz $\leq$ f $\leq$ 100 kHz) or low-intensity high-frequency ultrasound (f > 100 kHz). Low-intensity ultrasound does not alter the material's chemical or physical properties, but high-intensity electromagnetic waves produce significant temperature and pressure gradients caused by bubble cavitation, which disturbs the matrix. High shear forces, pressure differentials, and extremely high temperatures and heating/cooling rates are all caused by acoustic/ultrasonic cavitation. Sonoporation, cell disruption, particle breakdown, erosion, and surface peeling are all consequences of cavitation bubbles imploding on a solid's surface due to micro-jets and inter-particular collision. Both micro-mixing and macro-turbulence are produced when cavitation bubbles in liquid media implode. According to sample type, UAE can extract 85–97% of its material in 10–60 minutes at temperatures between 20 and 80°C (Medina-Torres et al., 2017). Important variables in the UAE include the solvent's nature and properties, extraction time, temperature, power, and gas presence. UAE yield is initially increased by temperature, but it decreases as the temperature rises above 45 °C. Various phytochemicals can be extracted by the UAE process using a variety of solvents, including acetone, alcohols, acidified water, and water. For the extraction of pectin, acidified water is ideal (Kumar et al., 2021). The use of UAE for collecting phytochemicals associated with complex matrixes like algae is limited by inadequate power management, which leads to poor energy transmission inside the extracted vessel (Carreira-Casais et al., 2021).

13.4.3 Enzyme-Assisted Extraction

Enzymes are powerful stimulants which can be utilized to produce, modify, or remove complex bioactive substances from the environment. Enzyme-based extraction relies on the inherent ability of enzymes to catalyse reactions with exact specificity and regioselectivity. It might be a good substitute for conventional extraction techniques based on solvents and provides a green approach which lessens environmental issues. The three most commonly used enzymes for removing bioactive substances are cellulases, hemicelluloses, and pectinases. The cell wall is hydrolysed by enzymes, which also alter the structural integrity of material from plants. Most enzyme-assisted extraction is conducted between 33 and 67°C, and extraction periods vary from 20 to 220 minutes. The extract's extraction efficiency can range from 38 to 52% (Pontillo et al., 2021).

Because of their constituent type, concentrations, enzymatic activity, and composition, enzymes have a major influence on hydrolytic properties. As the temperature rises, the extraction media's viscosity falls, increasing the solubility of bioactive compounds because activity of enzyme is correlated with temperature. In the food processing industry, pectic enzymes are widely used to extract pectin, flavours, pigments, and essential oils from plant parts. For oil extraction, the most often used enzymes are cellulase, α-amylase, and pectinase. The quality of oils treated with enzymes is generally better than that of oils extracted with hexane. One technological barrier to extracting a large number of samples is the high cost of enzymes. The extraction yield of phytochemicals is limited because the enzyme combinations currently on the market cannot fully degrade plant cell walls. Environmental factors like temperature, nutrient availability, and the percentage of dissolved oxygen also affect the rates at which enzymes are extracted (Puri et al., 2012).

13.4.4 Pressurized Liquid Extraction

Pressurized liquid extraction (PLE) is a rapid and effective method of using liquid solvents to extract analytes from a solid matrix at high temperatures and pressures. It has excellent mass transfer and solubility qualities (Alvarez-Rivera et al., 2020). PLE is also known as high-pressure solvent extraction (HSPE), accelerated fluid extraction (ASE), and enhanced solvent extraction (ESE). In order to facilitate extraction, PLE uses high pressure to keep solvents in a liquid state past their boiling point. This method's high temperature and pressure drastically lower the amount of solvent needed and the extraction period. The extraction duration in PLE is between 5 and 20 minutes, and the extraction accuracy is between 95 and 100%. Temperatures between 50 and 200°C and pressures between 50 and 300 psi are used for extraction (Kumar et al., 2023). PLE requires significantly less solvent and extraction time than Soxhlet. As CO_2 is not an effective solvent for removing polar compounds, PLE is a viable substitute for SFE in the extraction of bioactive compounds from the ocean sponge Cacospongia mycofijiensis. Because organic solvents are used in place of CO_2, this extraction technique is regarded as an organic variant of supercritical fluid extraction. Because this method requires very little solvent, it is also referred to as a 'green' extraction method. The extraction temperature, solvent surface tension, and pressure all affect extraction efficiency. Better penetration and extraction rates are made possible by the use of high pressure, high temperatures, and low solvent surface tension. Ionic surfactant solutions in PLE have recently been replaced by non-ionic surfactant solutions (Alvarez-Rivera et al., 2020). Allergens such as paraben, bronidox, and triclosan – which are also used as cosmetic preservatives – have been effectively extracted using PLE. The main drawbacks of PLE are the presence of interferents, low analyte selectivity, high extract dilutions (when employing multi-cycle extraction), and the need for advanced machinery (Alara et al., 2021).

13.5 INDUSTRIAL APPLICATIONS OF PHYTOCHEMICALS

13.5.1 Food Industry

Due to their safety and health-promoting qualities, natural pigments like betalains, carotenoids, and chlorophylls are being used more and more as alternatives to synthetic colourants like tartrazine and allura red. Betanin, which comes from beetroot, is a natural red dye used in dairy products, candies, and drinks. Products like margarine, baked goods, and beverages use carotenoids like lutein and β-carotene from spinach and carrots as colourants and fortifiers (Martinez et al., 2024).

13.5.2 Pharmaceutical Applications

Phytochemicals derived from vegetables have shown a variety of pharmacological effects, for example anti-inflammatory, anticancer, and antioxidant properties. Broccoli contains glucoraphanin hydrolysed to produce SFN, which is sold as a nutraceutical supplement for cancer prevention and detoxification (Andrés et al., 2025). Formulations aimed at metabolic and CVDs use polyphenolic extracts from spinach, onions, and garlic. Furthermore, several countries have approved tomato extracts high in lycopene as natural supplements for skin and prostate health (Pawase et al., 2024).

13.5.3 Cosmetic Industry

Phytochemicals including polyphenols, carotenoids, and flavonoids are used in skincare products because of their anti-ageing and antioxidant qualities (Singh et al., 2025). For example lutein and chlorophyll-rich spinach and broccoli extracts prevent photooxidative damage and improve skin radiance (Bernstein et al., 2016). Additionally, betalain extracts are used as safe, biodegradable colourants in natural hair and lip care products.

13.5.4 Health and Economic Importance

Vegetable phytochemicals serve the dual purposes of advancing public health and the national and global economies. Vegetables are vital for health because they contain dietary fibre, vitamins, minerals, and a range of bioactive secondary metabolites which support the operation of vital physiological systems. Consuming phytochemical-rich vegetables regularly has been associated with decreased consequences of chronic illnesses like inflammatory diseases, obesity, cancer, heart disease, and type 2 diabetes. Phytochemicals with potent anti-inflammatory and antioxidant properties, such as phenolic acids, flavonoids, carotenoids, glucosinolates, and betalains, help the body scavenge free radicals, alter cell signalling pathways, and reduce oxidative stress. Increased immunity, better metabolic health, and a delayed onset of age-related disorders are all associated with diets rich in vegetables, including broccoli, spinach, tomatoes, carrots, and beetroot, according to recent clinical and epidemiological research (Pawase et al., 2024; Martinez et al., 2024). Through specific biochemical processes, phytochemicals also aid in the prevention of diseases. For example the function of SFN, which is derived from the glucosinolates in broccoli, in controlling detoxification enzymes and stopping cancer cells from proliferating has been extensively researched. Because of its potent antioxidant properties, lycopene, a carotenoid mostly found in tomatoes, is linked to a lower risk of cardiovascular and prostate disorders. Similar to this, betalains from beetroot have hepatoprotective and anti-inflammatory properties and are becoming more widely acknowledged for their potential in clinical nutrition and functional food formulations (Andrés et al., 2025).

Vegetables rank among the most valuable components of the global agricultural supply chain in terms of economic value. FAO (2023) states that the growing demand for functional foods, growing consumer awareness of wellness and health, and expansion in the plant-based and nutraceutical product industries are all contributing factors to the global vegetable market's continued growth. The growth of the phytochemical extraction industries has produced high-value markets for antioxidants, dietary supplements, natural pigments, medicinal formulations, and ingredients for cosmetics. As an illustration, lycopene which was extracted from tomato processing waste is currently marketed for use in skincare and dietary supplements. Because consumers are choosing clean-label products, beetroot's betalain pigments have emerged as significant natural alternatives to synthetic food dyes. High amounts of by-products, including peels, pomace, stems, and leaves, are produced by vegetable processing industries. Significant amounts of phytochemicals that can be extracted and sold are present in these residues. This strategy lowers environmental waste while simultaneously boosting industrial profitability and promoting circular economy models. Carrot pomace is utilized to create β-carotene powders, and onion peels are utilized to extract quercetin, which is frequently utilized in pharmaceutical and nutraceutical formulations (Thomsen & Asalf, 2023).

13.6 CHALLENGES AND FUTURE PROSPECTS

Vegetable-derived phytochemicals are gaining attention in science and industry, but a number of obstacles still prevent widespread use. Variations in phytochemical content brought on by variations in cultivar genetics, soil composition, climate, and post-harvest handling are among the main causes for concern. It is challenging to maintain uniform quality and standardization in commercial products because of this variability. Furthermore, a lot of phytochemicals are susceptible to oxidative degradation and are thermolabile, which lowers their bioactivity during processing, storage, and transit. Although stabilization methods like nanoencapsulation and microencapsulation present promising alternatives, they raise manufacturing costs, which puts small-scale producers at a financial disadvantage (Pawase et al., 2024). The yield and stability of phytochemicals could be greatly increased in the future by developments in precision agriculture, metabolic engineering, and plant biotechnology. The biosynthesis of specific compounds can be improved by breeding initiatives and CRISPR-based genome editing, while environmental variability can be reduced by vertical farming and CEA. Additionally, it will be more crucial than ever to value the by-products of vegetable processing in order to advance circular bioeconomy models which support both economic profitability and environmental sustainability (Suresh et al., 2025).

13.7 CONCLUSION

Industrially significant vegetables are essential bioresources which are abundant in phytochemicals with potent antimicrobial, antioxidant, and therapeutic properties. They are much more than just dietary staples. They are essential to the food, cosmetic, pharmaceutical, and nutraceutical industries due to their wide range of compounds, which are organosulphur metabolites, glucosinolates, flavonoids, and carotenoids. Furthermore, turning vegetable waste into valuable functional ingredients provides a sustainable way to cut down on agro-waste. The full industrial and health-promoting potential of these phytochemical-rich vegetables will only be realized through advanced research and creative processing techniques, forming a more robust, sustainable, and circular bioeconomy.

REFERENCES

Akhtar, I., Javad, S., Yousaf, Z., Iqbal, S., & Jabeen, K. (2019). Review: Microwave assisted extraction of phytochemicals an efficient and modern approach for botanicals and pharmaceuticals. *Pakistan Journal of Pharmaceutical Sciences*, *32*, 223–230.

Alara, O. R., Abdurahman, N. H., & Ukaegbu, C. I. (2021). Extraction of phenolic compounds: A review. *Current Research in Food Science*, *4*, 200–214.

Alvarez-Rivera, G., Bueno, M., Ballesteros-Vivas, D., Mendiola, J. A., & Ibañez, E. (2020). Pressurized liquid extraction. In *Liquid-Phase Extraction*. Amsterdam, The Netherlands: Elsevier, pp. 375–398.

Andrés, C. M. C., Pérez de la Lastra, J. M., Munguira, E. B., Juan, C. A., & Pérez-Lebeña, E. (2025). The multifaceted health benefits of broccoli – A review of glucosinolates, phenolics and antimicrobial peptides. *Molecules*, *30*(11), 2262.

Baldelli, S., Lombardo, M., D'Amato, A., Karav, S., Tripodi, G., & Aiello, G. (2025). Glucosinolates in human health: Metabolic pathways, bioavailability, and potential in chronic disease prevention. *Foods*, *14*(6), 912.

Barot, M., Kumar, N. J. I., & Kumar, R. N. (2016). Bioactive compounds and antifungal activity of three different seaweed species Ulva lactuca, Sargassum tenerrimum and Laurencia obtusa collected from Okha coast, Western India. *Current Research in Food Science*, *4*(4), 284–289.

Barot, M., Nirmal Kumar, J. I., & Kumar, R. N. (2019). An evaluation of the nutritional composition of seaweeds as potential source of food and feed. *National Academy Science Letters*, *42*(6), 459–464.

Bernstein, P. S., Li, B., Vachali, P. P., Gorusupudi, A., Shyam, R., Henriksen, B. S., & Nolan, J. M. (2016). Lutein, zeaxanthin, and meso-zeaxanthin: The basic and clinical science underlying carotenoid-based nutritional interventions against ocular disease. *Progress in Retinal and Eye Research*, *50*, 34–66.

Boeing, H., Bechthold, A., Bub, A., Ellinger, S., Haller, D., Kroke, A., . . . & Watzl, B. (2012). Critical review: Vegetables and fruit in the prevention of chronic diseases. *European Journal of Nutrition*, *51*(6), 637–663.

Carreira-Casais, A., Otero, P., Garcia-Perez, P., Garcia-Oliveira, P., Pereira, A. G., Carpena, M., . . . & Prieto, M. A. (2021). Benefits and drawbacks of ultrasound-assisted extraction for the recovery of bioactive compounds from marine algae. *International Journal of Environmental Research and Public Health*, *18*(17), 9153.

Chandra, S., Sah, K., Bagewadi, A., Keluskar, V., Shetty, A., Ammanagi, R., & Naik, Z. (2012). Additive and synergistic effect of phytochemicals in prevention of oral cancer. *European Journal of General Dentistry*, *1*(3), 142–147.

Chemat, F., Abert-Vian, M., & Fabiano-Tixier, A. S. (2022). Green extraction of natural products: Concept and recent advances. *Trends in Analytical Chemistry*, *154*, 116664.

Choo, S., Chin, V. K., Wong, E. H., Madhavan, P., Tay, S. T., Yong, P. V. C., & Chong, P. P. (2020). Antimicrobial properties of allicin used alone or in combination with other medications. *Folia Microbiologica*, *65*(3), 451–465.

Ebabhi, A., & Adebayo, R. (2022). Nutritional values of vegetables. *Vegetable Crops-Health Benefits and Cultivation*, 1839699485.

Eom, J. W., Lim, J. W., & Kim, H. (2023). Lutein induces reactive oxygen species-mediated apoptosis in gastric cancer AGS cells via NADPH oxidase activation. *Molecules*, *28*(3), 1178.

FAO. (2023). *The State of Food and Agriculture 2023 – Revealing the True Cost of Food to Transform Agrifood Systems*. https://doi.org/10.4060/cc7724en

Fernandez-Pan, I., Horvitz, S., Ibañez, F. C., Arroqui, C., Beriain, M. J., & Virseda, P. (2024). Extra-virgin olive oil enriched with lycopene: From industrial tomato by-products to consumer. *Food Science & Nutrition*, *12*(8), 5815–5823.

Ghosh, R., Badavath, V. N., Chowdhuri, S., & Sen, A. (2022). Identification of alkaloids from Terminalia chebula as potent SARS-CoV-2 main protease inhibitors: An in silico perspective. *ChemistrySelect*, *7*(14), e202200055.

Hamdi, M., Feki, A., Bardaa, S., Li, S., Nagarajan, S., Mellouli, M., . . . & Nasri, R. (2020). A novel blue crab chitosan/protein composite hydrogel enriched with carotenoids endowed with distinguished wound healing capability: In vitro characterization and in vivo assessment. *Materials Science and Engineering: C*, *113*, 110978.

Hossain, M. S., Wazed, M. A., Asha, S., Amin, M. R., & Shimul, I. M. (2025). Dietary phytochemicals in health and disease: Mechanisms, clinical evidence, and applications – a comprehensive review. *Food Science & Nutrition*, *13*(3), e70101.

Irfan, M., Suprajaa, P. S., Praveen, R., & Reddy, B. M. (2021). Microwave-assisted one-step synthesis of nanohydroxyapetite from fish bones and mussel shells. *Materials Letters*, *282*, 128585.

Kawatra, A., Gupta, S., Dhankhar, R., Singh, P., & Gulati, P. (2023). Application of phytochemicals in therapeutic, food, flavor, and cosmetic industries. In *Phytochemical Genomics: Plant Metabolomics and Medicinal Plant Genomics*. Singapore: Springer Nature Singapore, pp. 85–108.

Książkiewicz, M., Karczewska, M., Nawrot, F., Korybalska, K., & Studzińska-Sroka, E. (2025). Traditionally used edible flowers as a source of neuroactive, antioxidant, and anti-inflammatory extracts and bioactive compounds: A narrative review. *Molecules*, *30*(3), 677.

Kumar, A., P, N., Kumar, M., Jose, A., Tomer, V., Oz, E., . . . & Oz, F. (2023). Major phytochemicals: Recent advances in health benefits and extraction method. *Molecules*, *28*(2), 887.

Kumar, K., Srivastav, S., & Sharanagat, V. S. (2021). Ultrasound assisted extraction (UAE) of bioactive compounds from fruit and vegetable processing by-products: A review. *Ultrasonics Sonochemistry*, *70*,

Martinez, R. M., Melo, C. P., Pinto, I. C., Mendes-Pierotti, S., Vignoli, J. A., Verri, W. A., & Casagrande, R. (2024). Betalains: A narrative review on pharmacological mechanisms supporting the nutraceutical potential towards health benefits. *Foods*, *13*(23), 3909.

Medina-Torres, N., Ayora-Talavera, T., Espinosa-Andrews, H., Sánchez-Contreras, A., & Pacheco, N. (2017). Ultrasound assisted extraction for the recovery of phenolic compounds from vegetable sources. *Agronomy*, *7*(3), 47.

Nigar, S., Shimul, I. M., Hossain, M. S., Sultana, R., Asha, S., & Huq, A. O. (2025). Comparative analysis on phytonutrient properties of different fig varieties (Ficus spp.). *Food Chemistry Advances*, *6*, 100878.

Pawase, P. A., Goswami, C., Shams, R., Pandey, V. K., Tripathi, A., & Rustagi, S. (2024). A conceptual review on classification, extraction, bioactive potential and role of phytochemicals in human health. *Future Foods*, *9*, 100313.

Pontillo, A. R. N., Papakosta-Tsigkri, L., Lymperopoulou, T., Mamma, D., Kekos, D., & Detsi, A. (2021). Conventional and enzyme-assisted extraction of rosemary leaves (Rosmarinus officinalis L.): Toward a greener approach to high added-value extracts. *Applied Sciences*, *11*(8), 3724.

Puri, M., Sharma, D., & Barrow, C. J. (2012). Enzyme-assisted extraction of bioactives from plants. *Trends in Biotechnology*, *30*(1), 37–44.

Quitério, E., Grosso, C., Ferraz, R., Delerue-Matos, C., & Soares, C. (2022). A critical comparison of the advanced extraction techniques applied to obtain health-promoting compounds from seaweeds. *Marine Drugs*, *20*(11), 6

Sharma, M., & Kaushik, P. (2021). Vegetable phytochemicals: An update on extraction and analysis techniques. *Biocatalysis and Agricultural Biotechnology*, *36*, 102149.

Shital Patil, S. P., Ranveer, R. C., Debaje, P. P., Kadam, J. H., & Sahoo, A. K. (2018). *Ultrasound Assisted Extraction of Curcumin*, *37*, 250–252. https://www.arccjournals.com/journal/asian-journal-of-dairy-and-food-research/DR-1356

Singh, H., Kamal, Y. T., Pandohee, J., Mishra, A. K., Biswas, A., Mohanto, S., . . . & Chopra, H. (2025). Dietary phytochemicals alleviate the premature skin aging: A comprehensive review. *Experimental Gerontology*, *199*, 112660.

Suresh, A., Koodal, S. J., Pant, U., & Palariya, P. (2025). Future prospects and challenges in vegetable biotechnology. *Shabbir Ahmed Jyotika Purohit Kiran Mary Kandir Anirudha Chattopadhyay Anubhav Singh*, *16*.

Thomsen, M. G., & Asalf, B. (2023, October). Precipitation, drought, and soil factors affecting carrot quality. In *III International Symposium on Carrot and Other Apiaceae 1393* (pp. 1–6).

Wan, C., Langyan, S., Echeverría, J., Devkota, H. P., Tewari, D., Moosavi, M. A., Ezzat, S. M., Perez-Vazquez, A., Fraga-Corral, M., Cravotto, G., & Prieto, M. A. (2025). Edible fruits and berries as a source of functional polyphenols: Current scene and future perspectives. *Phytochemistry Reviews*, *24*(2), 1197–1225.

Xu, K., Qu, L., Li, H., Ren, X., Yan, N., & Fu, X. (2024). Functional properties of dietary quercetin in cardiovascular health and disease. *Food Frontiers*, *5*(5), 1951–1967.

Yousuf, O., Gaibimei, P., & Singh, A. (2018). Ultrasound assisted extraction of oil from soybean. *International Journal of Current Microbiology and Applied Sciences*, *7*, 843–852.

Zhao, Y., Wang, L., Huang, Y., Evans, P. C., Little, P. J., Tian, X., . . . & Xu, S. (2024). Anthocyanins in vascular health and disease: Mechanisms of action and therapeutic potential. *Journal of Cardiovascular Pharmacology*, *84*(3), 289–302.

14 Vegetable-Based Pharmaceuticals and Nutraceutical Products Available in Market

Pragati Khare, Noopur Khare, Tanvi Jain, and Sachidanand Singh

14.1 INTRODUCTION

'Any substance that is a food or a part of a food and provides medical or health benefits, including the prevention and treatment of diseases' is one definition of the phrase 'nutritraceutical', which is derived from the words 'nutrition' and 'pharmaceutical'. According to this definition, dietary supplements including functional foods, therapeutic foods, and phytochemicals may fall under the category of nutraceuticals. A wide range of molecules originating from plants which have anti-inflammatory, anti-carcinogenic, anti-mutagenic, and antioxidant qualities are referred to as phytochemicals (Husen, 2024a,2024b; Husen & Pant, 2024a,2024b; Pant et al., 2025; Pant & Husen, 2025). Because of their medicinal qualities, nutraceuticals have a long history and important biomedical uses (Husen, 2021, 2022, 2023a, 2023b). Numerous studies conducted to date have demonstrated the many benefits nutraceuticals offer in both cancer prevention and treatment, as well as in enhancing general health. Dr Michaelsporn coined the term 'chemoprevention', which describes using a natural treatment to lower the incidence and death rate of cancer. This is predicated on the idea that prevention, not cure, is the key to lowering the occurrence of a disease. A new field of oncology research called 'chemoprevention' aims to stop cancer by using natural substances. Chemoprevention has been demonstrated to be essential in preventing tumour invasion and metastasis. In order to produce synergistic anticancer benefits, a number of nutraceuticals are being employed either alone or in conjunction with traditional therapy as adjuvant therapy. The natural origin, safety, and health benefits of vegetable-based medications and nutraceutical products have made them extremely popular in recent years. These items, which come from plant sources, are used for general health, illness prevention, and therapeutic purposes.

14.2 VEGETABLE-BASED PHARMACEUTICALS

Bioactive substances generated from vegetables and other plant sources are referred to as plant-derived medications or vegetable-based pharmaceuticals. These substances are used to treat a variety of illnesses in both traditional and modern medicines because of their therapeutic qualities. Numerous pharmaceutical firms conduct thorough research and clinical testing before developing medications derived from plants.

Some well-known plant-based pharmaceuticals have made significant contributions to modern medicine. Artemisinin, derived from *Artemisia annua*, is widely used for treating malaria due to its potent antimalarial properties. Paclitaxel, obtained from the Pacific yew tree (*Taxus brevifolia*), is

 DOI: 10.1201/9781003606710-14

TABLE 14.1
Vegetable-Based Pharmaceuticals

Bioactive Compound	Vegetable Source	Therapeutic Use	Reference
Curcumin	Turmeric (*Curcuma longa*)	Anti-inflammatory, antioxidant	AlAli et al. (2021)
Lycopene	Tomatoes	Cardiovascular protection, cancer prevention	AlAli et al. (2021)
Sulforaphane	Broccoli	Detoxification, anticancer	AlAli et al. (2021)
Beta-carotene	Carrots, sweet potatoes	Eye health, immune function	AlAli et al. (2021)
Allicin	Garlic (*Allium sativum*)	Antimicrobial, cardiovascular support	Choi and Ko (2017)
Gingerol	Ginger (*Zingiber officinale*)	Antiemetic, anti-inflammatory	Choi and Ko (2017)
Polyphenols	Green tea (*Camellia sinensis*)	Antioxidant, neuroprotective	Bicknell et al. (2023)
Glycyrrhizin	Liquorice (*Glycyrrhiza glabra*)	Anti-ulcer, anti-inflammatory	Pastorino et al. (2018)
Kavalactones	Kava root (*Piper methysticum*)	Anxiolytic, sedative	Pastorino et al. (2018)
Apigenin	Chamomile (*Matricaria chamomilla*)	Antispasmodic, anti-inflammatory	Srivastava et al. (2010)

a powerful anticancer drug commonly used in chemotherapy. Morphine, extracted from the opium poppy (*Papaver somniferum*), serves as a potent painkiller for managing severe pain. Additionally, resveratrol, found in grapes and peanuts, is recognized for its cardiovascular benefits, including antioxidant and anti-inflammatory effects which promote heart health (Chaachouay & Zidane, 2024) as seen in Table 14.1.

14.2.1 Nutraceutical Products

Nutraceuticals are food-derived products which provide medicinal or health benefits beyond basic nutrition. These products can be functional foods, dietary supplements, or herbal extracts which support overall health and prevent diseases. Some common vegetable-based nutraceuticals play a crucial role in promoting health and preventing diseases. Curcumin, found in turmeric, is widely recognized for its strong anti-inflammatory and antioxidant properties, making it valuable in managing chronic conditions. Lycopene, a red pigment present in tomatoes, supports heart health and is linked to reduced cancer risk. SFN, abundant in broccoli, is known for its detoxifying effects and potential in cancer prevention (AlAli et al., 2021).

14.2.2 Market Availability and Demand

The demand for vegetable-based pharmaceuticals and nutraceuticals is growing due to increasing consumer awareness of natural and plant-based remedies. These products are available in various forms such as capsules, powders, teas, extracts, and fortified foods. Many pharmaceutical and nutraceuticals companies have introduced innovative products which cater to specific health concerns like immunity boosting, weight management, and chronic disease prevention (Puri et al., 2022).

14.3 HISTORICAL BACKGROUND

Food has been used for medical purposes for thousands of years, and many cultures still employ particular foods as part of their traditional medical procedures. For instance, herbs and other natural materials have long been employed in traditional Chinese medicine for their therapeutic properties. However, the present notion of nutraceuticals dates back to the 1980s. Since then, academics looking into the health effects of food ingredients as well as the food and supplement businesses have adopted this term. Conversely, functional foods have a longer history; for ages, people from different cultures have used particular foods for their health benefits. Herbs and spices have been utilized for their anti-inflammatory and antioxidant qualities, while fermented foods like kefir and yoghurt have been used for their probiotic qualities. In the 1980s, the Japanese government recognized some meals as having health benefits beyond basic nourishment and started to encourage their use, giving rise to the idea of functional foods (Wang et al., 2025). As people look for natural solutions to enhance their health and wellness, nutraceuticals and functional foods are both seeing tremendous increases in popularity today. According to Grand View Research, the global nutraceutical market is expected to grow from its 2019 valuation of USD 382.51 billion to USD 722.49 billion by 2027.

14.3.1 Possible Health Advantages

Beyond just basic nutrition, nutraceuticals and functional foods may offer a host of additional health advantages. Functional foods and nutraceuticals play a significant role in promoting overall health and preventing disease, particularly in areas such as cardiovascular, digestive, immune, cognitive, and bone health. Numerous studies have demonstrated that omega-3 fatty acids, which are naturally found in fatty fish and also available as supplements, contribute to improved cardiovascular health by reducing the risk of heart disease. Similarly, garlic and plant sterols – found in products like margarine and dairy – have been shown to effectively reduce cholesterol levels, further supporting heart health (Vignesh et al., 2024). Digestive health benefits are largely attributed to probiotics, which are beneficial bacteria found in fermented foods and supplements; these microorganisms enhance gut function and reduce the risk of certain gastrointestinal disorders. Prebiotics, such as those found in onions and bananas, act as dietary fibre that supports the growth of healthy gut bacteria, thereby improving gut microbiota balance. Functional foods and nutraceuticals also support the immune system; for example elderberries and vitamin C from citrus fruits have been shown to boost immune responses and help prevent infections. In the domain of cognitive function, nutraceuticals like ginkgo biloba and omega-3 fatty acids have demonstrated benefits in enhancing memory and mental alertness while reducing the risk of cognitive decline associated with ageing. Bone health also benefits from the calcium and vitamin D found in dairy products and supplements, which are crucial in maintaining bone density and preventing osteoporosis. Furthermore, many functional foods and nutraceuticals possess anti-inflammatory properties which may reduce the risk of chronic diseases such as arthritis and certain cancers. However, it is important to note that the health benefits of all functional foods and nutraceuticals are not fully substantiated; additional clinical research is needed to validate these claims. Moreover, individual responses may vary depending on factors like genetics, lifestyle, and existing health conditions, highlighting the need for personalized dietary guidance (Obayomi et al., 2024).

14.4 SAFETY AND EFFICACY CONSIDERATIONS

Although functional foods and nutraceuticals may offer health advantages, it's crucial to evaluate their efficacy and safety. Functional foods and nutraceuticals offer promising health benefits, but several important concerns must be considered to ensure their safe and effective use. One major issue is security, as these products are not subject to the same rigorous regulations as pharmaceutical drugs. This means their safety may not be thoroughly evaluated before they become available to consumers. Consequently, individuals should consult healthcare professionals before taking supplements, especially because some may cause adverse effects or interact negatively with prescription medications. Regarding efficacy, while certain nutraceuticals have proven health benefits, the quality and dosage of these products can vary significantly. Some may not contain the advertised levels of active ingredients, and many manufacturer claims may lack solid scientific evidence. Individual variation also plays a crucial role; genetic factors, health status, and lifestyle can influence how a person responds to these products. Therefore, what benefits one person may not necessarily help another. Additionally, interactions with medications or other supplements can pose risks. Healthcare professionals should always be consulted to avoid unwanted interactions. Another concern is contamination and adulteration; some products have been found to contain harmful substances like heavy metals or undisclosed additives. To minimize risks, it is essential to choose supplements and functional foods from reputable sources. Overall, while functional foods and nutraceuticals can contribute to improved health, they should be used cautiously, with professional guidance and careful consideration of quality and personal health factors (Dixit et al., 2023; Fernandes et al., 2024).

14.5 NUTRACEUTICALS IN VARIOUS DISEASES

As shown in Figure 14.1, nutraceuticals can prevent and treat a variety of diseases and health problems by improving health, well-being, and immunity. Nutraceuticals can be used to treat a number of illnesses, some of which are covered here.

14.5.1 Nutraceuticals in CVDs

Nutraceutical intervention can be used to treat cardiovascular problems because they have the highest risk factors of any disease. It has been demonstrated that nutraceuticals can effectively cure cardiovascular conditions. The heart and blood vessels are the main organs affected by cardiovascular disorders (CVDs). CVDs are one of the main causes of death, accounting for about 30% of fatalities globally each year. Arrhythmias, congestive heart failure, angina, hypertension, and

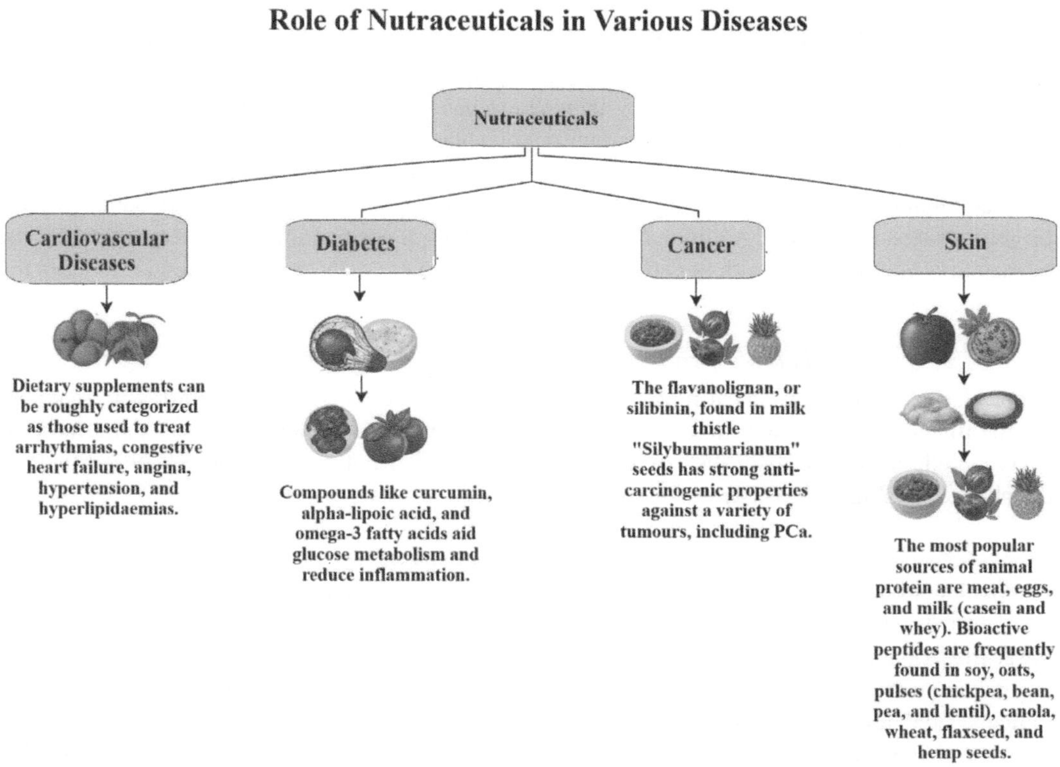

FIGURE 14.1 Nutraceuticals in various diseases.

hyperlipidaemia can all be treated with dietary supplements. Additionally, they have been demonstrated to be beneficial for risk management and CVD prevention (Alissa & Ferns, 2012).

14.5.1.1 Allicin and Alliin

Atherosclerosis and ischaemic heart disease are associated with elevated blood cholesterol and plasma triglyceride levels. The antihyperlipidemic herb Allium sativum reduces endogenous cholesterol synthesis and eliminates significant amounts of cholesterol and its metabolites from the faeces. As a result, the HDL to LDL ratio is more favourable. Allicin and alliin can affect cholesterol levels if stomach acids are able to do so. Garlic supplementation's impact on serum cholesterol was assessed in 13 placebo-controlled trials with 781 participants. Garlic has some inherent antihypertensive qualities in addition to its antihyperlipidemic ones (Sun et al., 2018).

14.5.1.2 Omega-3 Fatty Acids

PUFAs, another name for omega-3 fatty acids, are derived from marine sources. DHA and marine omega-3 eicosapentaenoic acid (EPA) can be used to prevent and treat CVDs. One study found that taking fish oil supplements reduced the death rate by 29% over a two-year period in the diet and reinfarction experiment (DART), a randomized trial that involved 2,033 men following myocardial infarction.

Consuming fish oil led to a significant 20% reduction in overall mortality, a 45% reduction in unplanned deaths, and a 30% reduction in CVD fatalities. According to recent clinical trial research, omega-3 fatty acids improve the health of people with atherosclerosis-related plaque development and help reduce the incidence of cardiac arrhythmias. Omega-3 fatty acids help cure arrhythmias by extending the relative refractory period and enhancing the electrical stability of heart cells (Banaszak et al., 2024).

14.5.1.2.1 Soy Isoflavones

The antihyperlipidemic, antihypertensive, anti-hyperglycaemic, antioxidant, anticancer, anti-inflammatory, anti-obesity, and neuroprotective properties of soy proteins and soy isoflavones make them valuable nutrients with potentially therapeutic benefits that lend a biological plausibility to observational associations. Clinical trial data clearly show that soy protein ingestion lowers human serum cholesterol levels. Furthermore, the U.S. Food and Drug Administration (USFDA) has demonstrated that postmenopausal women's blood pressure decreased when they consumed 25 g of soy proteins or isoflavone daily. Additionally, soy proteins have a positive impact on serum lipid levels, particularly in individuals with hypercholesterolaemia. A diet having less saturated fat reduces the incidence of coronary heart disease among study participants (Jheng et al., 2020).

14.5.1.2.2 Proteins, Peptides, and Amino Acids

A correlation exists between cardiovascular illnesses and hypertension. The mainstay of treatment for the illness has been ACE (angiotensin-converting enzyme) inhibitors; however, these medications include adverse effects such as hypotension, increased potassium, reduced renal function, coughing, and skin rashes (Herman et al., 2023).

14.5.1.2.3 Antioxidant Vitamins

Antioxidants have been considered as possible supplements for long-term conditions like cancer and heart disease. Through their ability to counteract the harmful effects of free radicals, they lower the oxidation of LDL cholesterol. Large amounts of antioxidant vitamins are found in fruits, vegetables, fish, and fixed oils. These vitamins either trap or stop the production of oxygen-free radicals. Epidemiologic studies of CHD patients who consume high amounts of antioxidants have shown that they reduce the incidence of morbidity and death. Antioxidant vitamins C and E found in supplements aid in the prevention of CHD. β-Carotene supplementation is not advised, nevertheless, as it may have negative effects (Rychter et al., 2022).

14.5.1.2.4 Phytosterols

Phytosterols and cholesterol are structurally similar. They often compete for absorption in the small intestine. Wood pulp, seeds, nuts, cereals, and vegetable oils are examples of natural sources of these. Phytosterol consumption leads to lower blood LDL levels, reduced cholesterol absorption, and increased LDL uptake by the liver. Consuming plant sterols can reduce LDL levels by up to 15%, according to research. Plant sterols come from natural grains including sunflower, soy, and maize. Consuming 2–3 grams of plant sterols or stanols daily has been shown in numerous trials to reduce LDL cholesterol levels by up to 20%, while individual variances are important (Poli et al., 2021).

14.5.2 Nutraceuticals in Cancer

Hair loss, nausea, vomiting, diarrhoea, and severe side effects are just a few of the adverse consequences of traditional cancer therapies like radiation therapy and chemotherapy. Because some cancers are exceedingly resistant to systemic cytotoxic chemotherapy and radiation, these treatments are not very helpful in increasing patient survival. In this instance, a variety of combination medications offer a successful cancer treatment choice. In a similar vein, several plants and natural supplements have been demonstrated to mitigate the adverse effects of radiation and chemotherapy. To reduce side effects and boost treatment effectiveness, these should be used in addition to chemotherapy or radiation therapy.

Cell proliferation causes cancer, which requires therapy, and the nutraceutical industry is evolving to satisfy consumer demands. It is possible to trace the evolution of the nutraceutical industry as it started with health promotion and has since advanced to illness prevention. Numerous herbs and phytochemicals which are now in use have been demonstrated to be successful nutraceuticals in slowing the growth of tumours, lessening the negative effects of radiation and chemotherapy, and increasing the sensitivity of these treatments. Pharmacologically, they are also safe. Caponio and his associates evaluated the effects of the phenolic compounds found in Aglianico Grape pomace (GP) on CRC cell lines at various stages of development after subjecting the cells to an in-vitro digesting model (Anand et al., 2023). Aglinico GP extract was found to have a substantial effect on cell division, apoptosis, and other cellular processes. Bax, the Bax/Bcl-2 ratio, and caspase-3 were all significantly elevated in both HT29 and SW480 cells. UHPLC-DAD analysis revealed that anthocyanins, phenolic acids, and flavonoids were primarily responsible for the elevated TPC and antioxidant activity in the Aglianico GP-digested extract. In both CRC cell lines, apigenin and chrysin dramatically decreased clone number, migration, and invasion at 25 μM while increasing apoptosis. Additionally, apigenin and chrysin significantly decreased p-AKT and p-P38. The tumour-inhibiting actions of apigenin and chrysin were effectively diminished by a P38 agonist known as anisomycin. Together, apigenin (25 μM) and chrysin (25 μM) prevented the proliferation and spread of CRC cells by blocking the P38-MAPK/AKT pathway (Caponio et al., 2022).

14.5.2.1 Curcumin (Diferuloyl-Methane) from Turmeric (Curcuma Longa)

Curcumin is one of the best nutraceuticals for cancer treatment. Preclinical studies on curcumin show that it inhibits carcinogenesis in a number of cancer types, including pancreatic, colorectal, prostate, gastric, and hepatic cancer, at every stage, including angiogenesis, metastasis, and proliferation. It is significantly more successful in treating cancer when combined with chemotherapy and radiation (Zoi et al., 2021).

14.5.2.2 Ginger

It has been demonstrated that the adverse effects of radiation and chemotherapy can be mitigated by ginger, a nutraceutical with antioxidant, anti-inflammatory, and antimutagenic properties. These characteristics give ginger its advantageous radio-protector properties. Ginseng and its polysaccharides can help lessen the negative effects of cancer treatment regimens, which have been found to reduce the likelihood of cancer recurrence by 50%. Ginsenoside Rf has also been demonstrated to help cancer patients lower their morphine dosages (Ayustaningwarno et al., 2024).

14.5.2.3 Genistein

Genistein is a potent isoflavone that may help prevent cancer. Several compounds only exhibit anticancer effects at higher concentrations than may be obtained from regular dietary intake, according to an in-vitro study. We feel that the mechanism of distribution is an important factor that needs to be considered in in-vivo research and clinical trials because it is difficult to achieve the desired effect at the tumour location. When developing a treatment, the natural substances' non-toxicity is an important factor to take into account. Nevertheless, it has been found that some drugs, such as genistein, work better when administered early in life (Ayustaningwarno et al., 2024).

14.5.3 Nutraceuticals in the Treatment of Prostate Cancer (PCa)

The most prevalent kind of cancer and the second-leading cause of cancer-related deaths among male Americans is prostate cancer. Men of African and American heritage have been found to have the greatest prostate cancer mortality rates when compared to men of European descent. Despite the fact that current treatment approaches are quite strong and successful, resistance to the treatment, metastasis, and many other issues could arise in the future. Therefore, to get over these significant obstacles and offer appropriate care and treatment of this illness, more potent and non-toxic therapeutic techniques are needed. As a result, there are a number of safe and effective nutraceuticals that may be used to treat PCa. It is clear that the use of nutraceuticals may aid in the development of more precise medication designs, lessen the toxicities linked to chemotherapy, lower disease resistance, and potentially treat both localized and advanced stages of the disease. The following section discusses a few nutraceuticals which may be utilized as treatment (McHugh et al., 2022).

14.5.3.1 Silibinin

The flavanolignan, or silibinin, found in milk thistle 'Silybummarianum' seeds has strong anticarcinogenic properties against a variety of tumours, including PCa. Significant anticancer potential of silibinin in the treatment of PCa has been demonstrated in a pre-clinical animal model, and bioavailability studies have also been assessed in phase II clinical trials (Wallace et al., 2005).

14.5.3.2 Soy Isoflavones

Soy isoflavones, which belong to the polyphenolic flavonoid family, are utilized extensively in Asian and African cuisines and are mostly found in soybeans, red clover, kudzu root, etc. Because soy isoflavones have a significant impact on insulin and inflammatory signalling, clinical investigations have shown some advantages against PCa. Isoflavones have also been reported to exhibit immunomodulatory qualities in the plasma of patients with asymptomatic biochemically recurrent PCa. It was discovered that Chinese patients who had high plasma genistein levels had a 69% lower chance of developing PCa in the future (Kim, 2021). Conversely, compared to individuals receiving milk protein therapy, soy protein (alcohol-washed) consumption reduced tissue levels of Bax and PCNA. According to a recent study by Zhu and his colleagues, glycation with soy soluble polysaccharide at greater degrees of glycation (DG) boosted the soy protein isolate's potential cholesterol-lowering properties.

14.5.4 Nutraceuticals for Skin Treatment

The skin, the largest organ in the body, protects the body from various infections, UV radiation, and substances which cause sensitivity. Because skin is essential for protecting the body, it can undergo changes that are detrimental to human health, such as inflammation, photoaging, and immunological dysfunction. Nutraceuticals may be used to treat skin-related diseases and prevent or slow down premature skin ageing. Carotenoids, plant extracts, bioactive peptides, bioactive polysaccharides, etc., are examples of these nutraceuticals. Supplementing with these products lowers signs of ageing and shields against UV radiation, according to several human trials (Yousef et al., 2017).

14.5.4.1 Bioactive Peptides

Combinations of two or more amino acids generate peptide moieties, which are small molecules with a molecular weight of less than 3 kDa. Some of these molecules may have significant biological functions; these are known as bioactive peptides. Numerous dietary proteins, both plant and animal, have been found to contain bioactive peptides. The most popular sources of animal protein are meat, eggs, and milk (casein and whey). Bioactive peptides are frequently found in soy, oats, pulses (chickpea, bean, pea, and lentil), canola, wheat, flaxseed, and hemp seeds (Singh et al., 2015). Because of their higher solubility and bioavailability, peptides utilized for cosmetic purposes are typically produced from collagen and are used in nutraceutical formulations. VERISOL®, which includes bioactive collagen peptide (BCP), has usually been the subject of controlled studies where it has been detected. In this trial, participants received VERISOL® and a placebo for eight weeks, and measurements of skin wrinkles were taken both before and after the therapy. After eight weeks of treatment, it was found that BCP significantly reduced the volume of eye wrinkles as compared to a placebo. Along with this, BCP use was associated with higher levels of fibrillin, elastin, and procollagen type 1. As a result, this treatment decreased wrinkles and had positive effects on the synthesis of skin matrix (Wang et al., 2022).

Pork origin and Peptan F: Other collagen peptide-containing nutraceuticals derived from fish, such as Peptan P, are used to slow down the ageing process by efficiently preserving the moisture content of the skin's layers. A recent study found that using the nutraceutical product Celergen® improved skin qualities without increasing the risk of oxidative damage, demonstrating that it is a safe and effective supplement (Rajabimashhadi et al., 2023).

14.5.4.2 Bioactive Polysaccharides

These are sugar-based polymers with structural and energy-storage properties. They are found in a variety of monosaccharide combinations, physicochemical characteristics, and structural types, including plants, fungi, mammals, and prokaryote species. The marine-derived glycosaminoglycans are the most beneficial of them for nutraceutical compositions. These formulations, which are dietary supplements for skin care and contain certain protein fractions in addition to glycosaminoglycans, were tested on humans using Imedeen® DermOne®. Zinc gluconate and vitamin C, which are important for skin maintenance, were included in the supplement along with the protein. Ten women received treatment with 500 mg of Imedeen® for 90 days during the trial. Dryness, brittleness of the hair and nails, wrinkles, and mottling were among the parameters assessed. All of these indicators showed improvement after 90 days, and observations confirmed the flexibility and thickness of the skin (Chen et al., 2024).

14.5.4.3 Bioactive Botanical Extracts

These extracts are complex blends of several substances with different origins and structures. They have been used and reviewed for a long time. One of the most important natural substances with cosmetic uses is polyphenol, which comes in a wide variety of families and structures. These micronutrients are derived from plants and can be obtained through diet. These are helpful as support therapies for both illness prevention and disease improvement. The bioavailability of different polyphenols varies greatly, and the polyphenols which are most prevalent in our diet have the highest concentration of active metabolites in their target tissues. Depending on the extraction process and families, their percentage and composition change. These are used to create Pycnogenol®, a formulation that is high in procyanidins (B1, B2, B3, B7 C1, and C2), flavonoids, and catechins. It also contains phenolic acids including caffeic and ferulic acids. Because of its antidiabetic, anti-inflammatory, and antioxidant qualities, it has also been shown to have a number of benefits for the cardiovascular system and to decrease cholesterol (Sridhar et al., 2021).

14.5.4.4 Carotenoids

Algae, photosynthesizing microorganisms, and a variety of plants all naturally contain these pigments. Their tetraterpenoid structure is linear. Fruits, vegetables, and other natural foods include them. Probiotics and carotenoids have been shown to reduce UV-induced skin damage and to modulate early UV-effect skin indicators. The photoprotective effects of a carotenoid combination supplement containing lutein, β-carotene, and β-carotene have been demonstrated. In a similar vein, beta-carotene, lutein, and lycopene carotenoids are said to provide erythema prevention. It functions as a cofactor for the production of collagen fibres and prevents fibroblasts from producing elastin, which is a common occurrence in photodamaged skin. It works in concert with its transformation mechanism when combined with vitamin E. The primary lipophilic antioxidant is vitamin E, which is present as tocopherols. It inhibits the lipid peroxidation of PUFAs by binding with peroxyl radicals. Furthermore, there is ample evidence to support its use in reducing sunburn, atopic dermatitis, photodamage, etc. (Mostafa & Hashem, 2024).

14.5.5 Nutraceuticals as Specialized Medical Products

Dietary meals and supplements used for specific medicinal objectives are considered specialized medical products under the legal foundation as seen in Table 14.2. In addition to the many national protocols issued, usually by the 'Ministry of Agriculture' and/or 'Ministry of Health' of various nations worldwide, these dietary supplements should be regulated in accordance with regulatory bodies like the 'European Food Safety Authority' and the 'U.S. Food and Drug Administration'. Non-specific biological therapies called nutraceuticals are used to control symptoms, prevent cancer, and improve wellness (Ruthsatz et al., 2022).

TABLE 14.2
List of Nutraceuticals with Health Benefits

Nutraceuticals	Nutrients	Health Benefits	References
Water-soluble vitamins	Vitamin B1	Neurological function	Sridhar et al. (2021)
	Vitamin B2	Energy metabolism	
	Vitamin B3	Brain function	
	Vitamin B6	Convert proteins to energy	
	Vitamin B12	Metabolism of fats, proteins, carbohydrates	
	Vitamin C	Wound healing, antioxidant	
	Folic acid	Formation of RBCs	
Fat-soluble vitamins	Vitamin A	Cancer, skin disorder	Ruthsatz et al. (2022)
	Vitamin D	Formation of bones and teeth	
	Vitamin E	Antioxidant	
	Vitamin K	Blood clotting	
Minerals	Calcium	Maintaining bone strength	Chen et al. (2024)
	Iron	Oxygen transport	
	Copper	Iron absorption	
	Iodine	Functioning of thyroid gland	
	Magnesium	Healthy nerve and muscle function	
Herbals	Garlic	Antibacterial	Mostafa and Hashem (2024)
	Ginger	Carminative	
	Green tea	Antioxidant	
	Ginseng	Adaptogen	
	Aloe vera	Wound healing	

14.6 CHALLENGES IN THE FORMULATION OF NUTRACEUTICALS AND DIETARY SUPPLEMENT

The high melting point, poor water solubility, and chemical instability of active ingredients provide challenges when creating nutraceuticals. For instance curcumin, oil-soluble vitamins, carotenoids, and omega-3 fatty acids are all highly nutritious but poorly soluble. Formulating these as innovative delivery systems is therefore one feasible strategy. They are expensive because of these new distribution methods. Therefore, work is required to make these formulations economical. The high melting point of nutraceuticals presents another difficulty in their formulation (Santini et al., 2018). For instance the high melting points of carotenoids, fatty alcohols, and phytosterols might make formulation unstable. Thus, making a solid dispersion or dissolving it in a high-grade solvent and adding it to meals as suspended nanocrystals are one conceivable strategy. But once more, the problem is that it results in poor stability and shelf life, an unpleasant look, and an unpleasant mouthfeel and odour, all of which have an impact on market value and consumer demand. Thus, the development of cost-effective technology is necessary (Henao-Ardila et al., 2024).

Furthermore, certain, selective bacterial strains are required for the creation of probiotics. Choosing the right strain and then incorporating it into meals are current difficulties. In addition to their careless handling, the application of any mismatched bacterial or poisonous cultures might have disastrous results. Solid dose formulation and process design for drug goods and nutrition items which are similar but have different purposes and regulatory requirements are additional factors to take into account in addition to these difficulties (Walker et al., 2015). Last but not least, developing dose forms for dietary supplements and nutraceuticals which are appropriate for various ageing population groups – particularly children and older adults – presents a difficulty. This is because members of this group have difficulties swallowing solid dosages (such as tablets or capsules) due to dysphagia. Therefore, while administering nutraceuticals and dietary supplements, advanced dosage forms including orodispersible tablets, fast-dissolving films, and easy-swallowing gels – which are typically employed in pharmaceutical applications – must be taken into account (Prajapati et al., 2020).

14.7 APPROACHES TO DEAL WITH FORMULATION CHALLENGES

Maceration, percolation, Soxhlet extraction, MAE, and counter current extraction are some of the several extraction techniques (Bommakanti et al., 2023). Plant extracts, herbal concentrates, fruit, vegetable, and speciality concentrates, as well as fungi and microorganisms which are added to feedstock to create concentrates (a process known as bio-fermentation), are examples of natural bioactive compounds. Concentrates may contain production-related excipients (e.g. spray-dried carriers) (Morgan et al., 2006).

There have been reports of several nanotechnology-prepared supplement formulations or 'nano formulations', which have improved bioavailability, decreased adverse effects, and shielded the active ingredients from deterioration. Preclinical and clinical research have shown that the polyphenols derived from dietary sources as nutraceuticals have a significant positive impact on the symptoms of disease. Their decreased bioavailability, which results from lower gastrointestinal absorption in the upper tract due to hydrophobicity, presence in foods in the form of polymers or glycosylation, and tight binding with food matrices are one of the most frequent drawbacks. The body's bio-accessibility of polyphenols is reduced as a result of all these circumstances. Thus, research and development methods based on food processing, in addition to nanoformulations, enzymatic therapy, and probiotic combination therapies, have shown to be beneficial in overcoming this obstacle (Stéphane et al., 2022), as seen in Figure 14.2.

14.7.1 Liposomes and Nanoemulsions

Liposomes and nanoemulsions have special qualities, and nanoliposomes can be utilized both to promote health and to avoid diseases. Lipid-based nanocarriers, or nanophytosomes, are one of

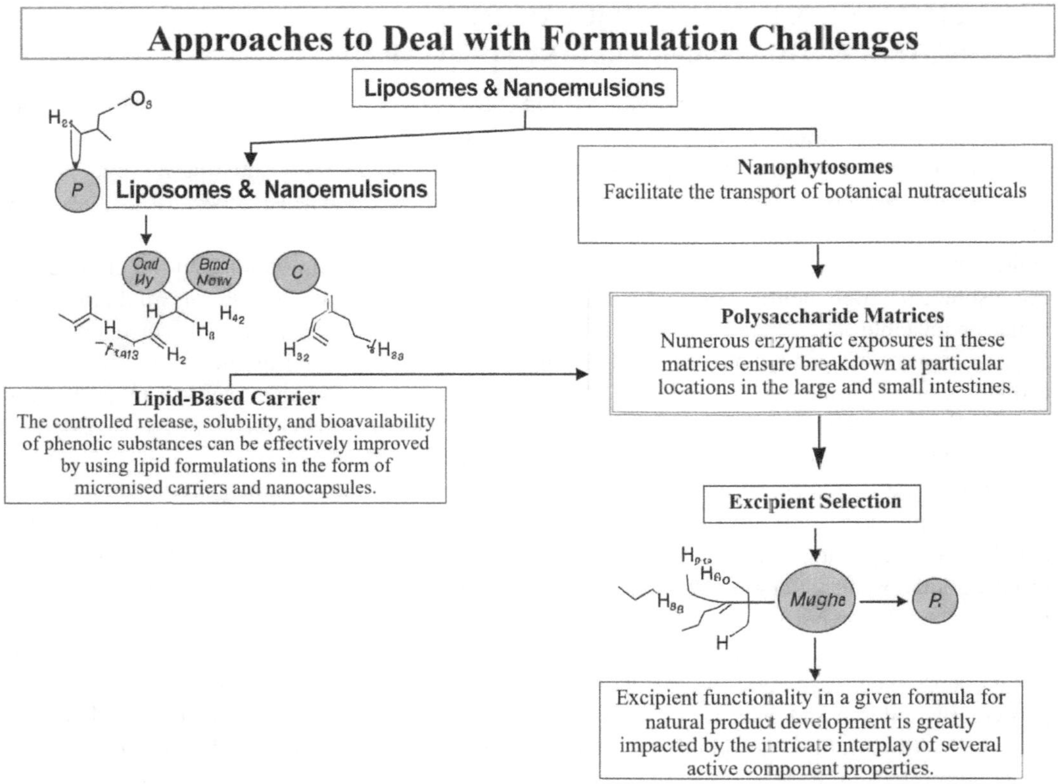

FIGURE 14.2 Approaches to deal with formulation challenges.

the most recent examples which facilitate the transport of botanical nutraceuticals. They could be applied to the creation of innovative functional beverages and food items in a variety of culinary products (Rudzińska et al., 2024).

A study found that rutin's chemical and physical stability increased when rutin complexes assumed the structure of phytosomes. These phytosome complexes of rutin were produced with the help of phosphatidylcholine (PC), which was able to mask the undesirable properties of rutin. They had a molar ratio (rutin:PC) of 1:3, a particle size of less than 100 nm, and a 99% encapsulation efficiency. These were known as phytosomes or phosphatidylcholine (PC)–rutin complexes (Hooresfand et al., 2015).

14.7.2 Lipid-Based Carriers

The controlled release, solubility, and bioavailability of phenolic substances can be effectively improved by using lipid formulations in the form of micronized carriers and nanocapsules. For instance because of their physical stability, β-carotene nanocapsules (>300 nm) exhibited very little change throughout storage, indicating that they can be employed extensively as both nutraceutical goods and functional meals and beverages (Subramanian, 2021).

14.7.3 Polysaccharide Matrices

Numerous enzymatic exposures in these matrices ensure breakdown at particular locations in the large and small intestines. When applied as a coating for nanoparticles, they can successfully postpone the nonspecific release of bioactive substances contained inside until the coating is exposed to the environment in which it was designed to be released. To increase the oral bioavailability, these coated nanoparticles may be able to target different GI tract organs (Visan & Cristescu, 2023).

14.7.4 Excipient Selection

The next strategy for addressing nutraceutical formulation issues is to adjust formulation parameters through sensible excipient selection. In terms of nutrition, the finished formula needs to be strong enough to handle the different physical properties of natural ingredients in a complicated recipe. The manufacturer and materials must adhere to internal safety and quality standards as well as performance criteria. The usefulness of one formulation may be the dysfunctionality of another. Only in relation to a specific formulation and manufacturing method can excipient functioning be accurately evaluated. Excipient functionality in a given formula for natural product development is greatly impacted by the intricate interplay of several active component properties. Sometimes, excipients which appear to be equivalent in function are not (Helal et al., 2019).

14.8 SAFETY AND QUALITY CONTROL OF NUTRACEUTICALS

Users consume nutraceuticals, which are sold as over-the-counter products, as supplements. As a result, their safety is crucial because failure to do so could have fatal consequences. Contamination, intentional or unintentional adulteration, or deceptive labelling are the most frequently reported problems. Three distinct detection techniques can be used to show adulteration: (1) the presence of an undeclared drug, (2) a component's deviation from its typical level (content), and (3) the likelihood of a profile. Intentional or unintentional adulteration are also possible. Various circumstances can lead to unintentional adulteration. For instance contamination with fertilizers, heavy metals, fertilizers, or microbiological agents may occur throughout the various stages of plant growth, during the formulation and manufacture of nutraceuticals or during storage. Other potential sources of adulteration include artificial substances, substitute species, dust, pollens, insects, rodents, parasites, mould, fungi, bacteria, poisons, and heavy metals. Contamination of this kind can result in infections or even more serious problems including gastritis and its consequences, liver damage, and even life-threatening diseases. Thus, quality control of raw materials and final products is necessary and can be established by the specifications given in specific monographs, in addition to microbiological control and the stability of the active component or compounds (Ronis et al., 2018). Intentional adulteration in supplements or herbal treatments might have serious negative effects. Usually unreported, it happens with synthetic substances. It typically takes place with the goal of changing the pharmacological reaction and yields financial gains. Nutraceuticals from plants are frequently sourced in extremely limited quantities, and extract processing is expensive and time-consuming. Regulatory bodies are therefore prohibited from allowing certain adulterations.

14.9 CONCLUSIONS

Nutraceuticals are substances which offer physiological benefits or protection against long-term illnesses. Nutraceuticals can be used to maintain the body's structure or function, prevent chronic diseases, enhance health, and slow down the ageing process. The potential health, safety, and nutritional benefits of nutraceuticals have garnered a lot of attention recently. In recent studies, these medications have shown promising advantages in a range of issues. The development of oxidative stress-related disease-modifying indications for diseases such as obesity, Alzheimer's, cardiovascular, cancer, diabetes, ophthalmic, immunological, inflammatory, and Parkinson's diseases has taken a great deal of effort.

Nutraceuticals are products that are used not just for feeding but also as medicines. A nutraceutical product could be any chemical which provides protection against chronic disease or has physiological effects. Nutraceuticals can be used to maintain the body's structure or function, prevent chronic diseases, enhance health, and slow down the ageing process. The potential health, safety, and nutritional benefits of nutraceuticals have garnered a lot of attention recently. In recent studies, these medications have shown promising advantages in a range of issues. In the current review, a lot of effort has gone into providing innovative concepts about nutraceuticals based on their indications for modifying diseases. Emphasis has been placed on the presenting of herbal nutraceuticals which are helpful in treating oxidative stress-related hard-to-cure ailments like obesity, cancer, diabetes, eye, immunological, inflammatory, Parkinson's, Alzheimer's, and CVDs. For recent articles addressing different aspects of nutraceuticals as a pharmaceutical alternative, we searched scientific databases such as Medline, PubMed, and Google Scholar. The terms used included nutraceutical, allergy, cardiovascular, cancer, diabetic, ophthalmic, immunological, inflammatory, or Parkinson.

REFERENCES

AlAli, M., Alqubaisy, M., Aljaafari, M. N., AlAli, A. O., Baqais, L., Molouki, A., & Lim, S. H. E. (2021). Nutraceuticals: Transformation of conventional foods into health promoters/disease preventers and safety considerations. *Molecules*, *26*(9), 2540.

Alissa, E. M., & Ferns, G. A. (2012). Functional foods and nutraceuticals in the primary prevention of cardiovascular diseases. *Journal of Nutrition and Metabolism*, *2012*(1), 569486.

Anand, U., Dey, A., Chandel, A. K. S., Sanyal, R., Mishra, A., Pandey, D. K., . . . & de la Lastra, J. M. P. (2023). Cancer chemotherapy and beyond: Current status, drug candidates, associated risks and progress in targeted therapeutics. *Genes & Diseases*, *10*(4), 1367–1401.

Ayustaningwarno, F., Anjani, G., Ayu, A. M., & Fogliano, V. (2024). A critical review of Ginger's (Zingiber officinale) antioxidant, anti-inflammatory, and immunomodulatory activities. *Frontiers in Nutrition*, *11*, 1364836.

Banaszak, M., Dobrzyńska, M., Kawka, A., Górna, I., Woźniak, D., Przysławski, J., & Drzymała-Czyż, S. (2024). Role of Omega-3 fatty acids eicosapentaenoic (EPA) and docosahexaenoic (DHA) as modulatory and anti-inflammatory agents in noncommunicable diet-related diseases–Reports from the last 10 years. *Clinical Nutrition ESPEN*, *63*, 240–258.

Bicknell, B., Liebert, A., Borody, T., Herkes, G., McLachlan, C., & Kiat, H. (2023). Neurodegenerative and neurodevelopmental diseases and the gut-brain axis: The potential of therapeutic targeting of the microbiome. *International Journal of Molecular Sciences*, *24*(11), 9577.

Bommakanti, V., Puthenparambil Ajikumar, A., Sivi, C. M., Prakash, G., Mundanat, A. S., Ahmad, F., & Rana, S. S. (2023). An overview of herbal nutraceuticals, their extraction, formulation, therapeutic effects and potential toxicity. *Separations*, *10*(3), 177.

Caponio, G. R., Cofano, M., Lippolis, T., Gigante, I., De Nunzio, V., Difonzo, G., . . . & Notarnicola, M. (2022). Anti-proliferative and pro-apoptotic effects of digested aglianico grape pomace extract in human colorectal cancer cells. *Molecules*, *27*(20), 6791.

Chaachouay, N., & Zidane, L. (2024). Plant-derived natural products: A source for drug discovery and development. *Drugs and Drug Candidates*, *3*(1), 184–207.

Chen, Y. J., Sui, X., Wang, Y., Zhao, Z. H., Han, T. H., Liu, Y. J., & Ye, Z. H. (2024). Preparation, structural characterization, biological activity, and nutritional applications of oligosaccharides. *Food Chemistry: X*, *22*, 101289.

Choi, J. H., & Ko, C. M. (2017). Food and drug interactions. *Journal of Lifestyle Medicine*, *7*(1), 1.

Dixit, V., Joseph Kamal, S. W., Bajrang Chole, P., Dayal, D., Chaubey, K. K., Pal, A. K., . . . & Bachheti, R. K. (2023). Functional foods: Exploring the health benefits of bioactive compounds from plant and animal sources. *Journal of Food Quality*, *2023*(1), 5546753.

Fernandes, F. A., Carocho, M., Prieto, M. A., Barros, L., Ferreira, I. C., & Heleno, S. A. (2024). Nutraceuticals and dietary supplements: Balancing out the pros and cons. *Food & Function*, *15*(12), 6289–6303.

Helal, N. A., Eassa, H. A., Amer, A. M., Eltokhy, M. A., Edafiogho, I., & Nounou, M. I. (2019). Nutraceuticals' novel formulations: The good, the bad, the unknown and patents involved. *Recent Patents on Drug Delivery & Formulation*, *13*(2), 105–156.

Henao-Ardila, A., Quintanilla-Carvajal, M. X., & Moreno, F. L. (2024). Emulsification and stabilisation technologies used for the inclusion of lipophilic functional ingredients in food systems. *Heliyon*, *10*(11).

Herman, L. L., Padala, S. A., Ahmed, I., & Bashir, K. (2023). Angiotensin-converting enzyme inhibitors (ACEI). In *StatPearls*. StatPearls Publishing.

Hooresfand, Z., Ghanbarzadeh, S., & Hamishehkar, H. (2015). Preparation and characterization of rutin-loaded nanophytosomes. *Pharmaceutical Sciences*, *21*(3), 145–151.

Husen, A. (2021). *Traditional Herbal Therapy for the Human Immune System.* CRC Press. https://doi.org/10.1201/9781003137955

Husen, A. (2022). *Herbs, Shrubs and Trees of Potential Medicinal Benefits.* CRC Press. https://doi.org/10.1201/9781003205067

Husen, A. (2023a). *Exploring Poisonous Plants: Medicinal Values, Toxicity Response and Other Potential Uses.* CRC Press. https://doi.org/10.1201/b23017

Husen, A. (2023b). *Antidiabetic Medicinal Plants and Herbal Treatments.* CRC Press. https://doi.org/10.1201/b23347

Husen. A. (2024a). *Antimalarial Medicinal Plants.* Taylor & Francis Group. https://doi.org/10.1201/9781003378396

Husen, A. (2024b). *Medicinal Spices and Condiment Crops.* Taylor & Francis Group. https://doi.org/10.1201/9781003387046

Husen, H., & Pant, M. (2024a). *Exploring Medicinal Orchids.* Taylor & Francis Group. https://doi.org/10.1201/b23375

Husen, H., & Pant, M. (2024b). *Tissue Culture Techniques and Medicinal Plants: Enhancing Propagation and Production.* Taylor & Francis Group. https://doi.org/10.1201/b23374

Jheng, H. F., Hayashi, K., Matsumura, Y., Kawada, T., Seno, S., Matsuda, H., . . . & Goto, T. (2020). Anti-inflammatory and antioxidative properties of isoflavones provide renal protective effects distinct from those of dietary soy proteins against diabetic nephropathy. *Molecular Nutrition & Food Research*, *64*(10), 2000015.

Kim, I. S. (2021). Current perspectives on the beneficial effects of soybean isoflavones and their metabolites for humans. *Antioxidants*, *10*(7), 1064.

McHugh, J., Saunders, E. J., Dadaev, T., McGrowder, E., Bancroft, E., Kote-Jarai, Z., & Eeles, R. (2022). Prostate cancer risk in men of differing genetic ancestry and approaches to disease screening and management in these groups. *British Journal of Cancer*, *126*(10), 1366–1373.

Morgan, S. L., & Baggott, J. E. (2006). Medical foods: Products for the management of chronic diseases. *Nutrition Reviews*, *64*(11), 495–501.

Mostafa, H. S., & Hashem, M. M. (2024). Microalgae as a source of carotenoids in foods, obstacles and solutions. *Phytochemistry Reviews*, 1–43.

Obayomi, O. V., Olaniran, A. F., & Owa, S. O. (2024). Unveiling the role of functional foods with emphasis on prebiotics and probiotics in human health: A review. *Journal of Functional Foods*, *119*, 106337.

Pant, M., & Husen, A. (2025). *Exploring Medicinal Bamboos.* Taylor & Francis Group. https://doi.org/10.1201/9781003485933

Pant, M., Sohrab, S. S., & Husen, A. (2025). *Genetic Improvement and Conservation Practice of Medicinal Plants.* Taylor & Francis Group. https://doi.org/10.1201/9781003482659

Pastorino, G., Cornara, L., Soares, S., Rodrigues, F., & Oliveira, M. B. P. (2018). Liquorice (Glycyrrhiza glabra): A phytochemical and pharmacological review. *Phytotherapy Research*, *32*(12), 2323–2339.

Poli, A., Marangoni, F., Corsini, A., Manzato, E., Marrocco, W., Martini, D., & Visioli, F. (2021). Phytosterols, cholesterol control, and cardiovascular disease. *Nutrients*, *13*(8), 2810.

Prajapati, R. N., & Kumar, S. (2020). The role, scope, health benefits and market growth of nutraceuticals: An overview. *Current Research in Pharmaceutical Sciences*, *10*(3), 30–42.

Puri, V., Nagpal, M., Singh, I., Singh, M., Dhingra, G. A., Huanbutta, K., & Sangnim, T. (2022). A comprehensive review on nutraceuticals: Therapy support and formulation challenges. *Nutrients*, *14*(21), 4637.

Rajabimashhadi, Z., Gallo, N., Salvatore, L., & Lionetto, F. (2023). Collagen derived from fish industry waste: Progresses and challenges. *Polymers*, *15*(3), 544.

Ronis, M. J., Pedersen, K. B., & Watt, J. (2018). Adverse effects of nutraceuticals and dietary supplements. *Annual Review of Pharmacology and Toxicology*, *58*(1), 583–601.

Rudzińska, M., Grygier, A., Knight, G., & Kmiecik, D. (2024). Liposomes as carriers of bioactive compounds in human nutrition. *Foods*, *13*(12), 1814.

Ruthsatz, M., Chen, J., Wu, C., & Morck, T. (2022). Foods for special medical purposes/medical foods: A global regulatory synopsis. *Regulatory Affairs Professionals Society*, 1–36.

Rychter, A. M., Hryhorowicz, S., Słomski, R., Dobrowolska, A., & Krela-Kaźmierczak, I. (2022). Antioxidant effects of vitamin E and risk of cardiovascular disease in women with obesity–a narrative review. *Clinical Nutrition*, *41*(7), 1557–1565.

Santini, A., Cammarata, S. M., Capone, G., Ianaro, A., Tenore, G. C., Pani, L., & Novellino, E. (2018). Nutraceuticals: Opening the debate for a regulatory framework. *British Journal of Clinical Pharmacology*, *84*(4), 659–672.

Singh, S., & Devi, M. B. (2015). Vegetables as a potential source of nutraceuticals and phytochemicals: A review. *International Journal of Pharmaceutical Sciences*, *5*(2), 1–14.

Sridhar, A., Ponnuchamy, M., Kumar, P. S., Kapoor, A., Vo, D. V. N., & Prabhakar, S. (2021). Techniques and modeling of polyphenol extraction from food: A review. *Environmental Chemistry Letters*, *19*(4), 3409–3443.

Srivastava, J. K., Shankar, E., & Gupta, S. (2010). Chamomile: A herbal medicine of the past with a bright future. *Molecular Medicine Reports*, *3*(6), 895–901.

Stéphane, F. F, Jules, B. K., Batiha, G. E., Ali, I., Bruno, L. N. (2022). Compounds from medicinal plants. *Natural Medicinal Plants*, *11*, 147.

Subramanian, P. (2021). Lipid-based nanocarrier system for the effective delivery of nutraceuticals. *Molecules*, *26*(18), 5510.

Sun, Y. E., Wang, W., & Qin, J. (2018). Anti-hyperlipidemia of garlic by reducing the level of total cholesterol and low-density lipoprotein: A meta-analysis. *Medicine*, *97*(18), e0255.

Vignesh, A., Amal, T. C., Sarvalingam, A., & Vasanth, K. (2024). A review on the influence of nutraceuticals and functional foods on health. *Food Chemistry Advances*, *5*, 100749.

Visan, A. I., & Cristescu, R. (2023). Polysaccharide-based coatings as drug delivery systems. *Pharmaceutics*, *15*(9), 2227.

Walker, R. M., Decker, E. A., & McClements, D. J. (2015). Physical and oxidative stability of fish oil nanoemulsions produced by spontaneous emulsification: Effect of surfactant concentration and particle size. *Journal of Food Engineering*, *164*, 10–20.

Wallace, S. N., Carrier, D. J., & Clausen, E. C. (2005). Batch solvent extraction of flavanolignans from milk thistle (Silybum marianum L. Gaertner). *Phytochemical Analysis: An International Journal of Plant Chemical and Biochemical Techniques*, *16*(1), 7–16.

Wang, L., Wang, N., Zhang, W., Cheng, X., Yan, Z., Shao, G., & Fu, C. (2022). Therapeutic peptides: Current applications and future directions. *Signal Transduction and Targeted Therapy*, *7*(1), 48.

Wang, L., Zhu, X., Liu, H., & Sun, B. (2025). Medicine and food homology substances: A review of bioactive ingredients, pharmacological effects and applications. *Food Chemistry*, *463*, 141111.

Yousef, H., Alhajj, M., & Sharma, S. (2017). Anatomy, skin (integument), epidermis.

Zoi, V., Galani, V., Lianos, G. D., Voulgaris, S., Kyritsis, A. P., & Alexiou, G. A. (2021). The role of curcumin in cancer treatment. *Biomedicines*, *9*(9), 1086.

15 Cultivated Vegetables versus Wild Vegetables
A Nutrient Perspective

Mahmoud A.A. Ali, Assiya Ansabayeva, Mikhail Puchkov, Mostafa Abdelkader, and Karim M. Hassan

15.1 INTRODUCTION

Around the world, rural people frequently use wild plants (WPs) and mushrooms, which are also sold in urban marketplaces, especially in Africa (Ambrose-Oji, 2009; Uusiku et al., 2010; Cernansky, 2015). WPs offer various food sources, leading to improvements in food security and poverty reduction due to their low costs and high positive consumer effect (Weinberger & Pichop, 2009; Yang & Keding, 2009). In rural areas, WPs are crucial as they are affordable and accessible to families. This allows for a more varied diet and serves as a source of suitable macro- and micronutrients, leading to lower susceptibility to diseases (Shackleton & Shackleton, 2004; Shackleton et al., 2007; Garekae & Shackleton, 2020). WPs can tolerate environmental changes relatively well, providing local people with an alternative food source during a crisis (Ohiokpehai, 2003). Depending on the type, CVs and WPs will synergize biodiversity and enhance sustainability. This will lead to adapting to local conditions and increasing diversification, resulting in a low carbon footprint and reduced chemical inputs. These are some of the traits that make indigenous food systems mainly appealing.

Wild vegetables (WVs) generally have smaller seeds with limited variation (Table 15.1), whereas CVs exhibit larger seeds (Kluyver et al., 2017; Garibaldi et al., 2021; Hernández-Terán et al., 2017). WPs are smaller and slower-growing, while domesticated ones have greater size, faster growth rates, and big aboveground biomass (De Casas et al., 2024; Lei et al., 2024). WVs always have smaller leaves, whereas domesticated varieties have developed larger leaves which contribute significantly to overall plant size (Milla & Matesanz, 2017; Lei et al., 2024). Productivity of WVs is lower, while domestication has led to maximizing productivity (Garibaldi et al., 2021; Lei et al., 2024; De Casas et al., 2024). Wild maintain high genetic diversity, in contrast to domesticated ones which have undergone genetic changes due to the selective process (Smýkal et al., 2018; Meyer et al., 2012; Coyne et al., 2020).

WVs show good tolerance to environmental stresses, whereas domesticated plants perform optimally under favourable conditions and exhibit reduced stress tolerance (Matesanz & Milla, 2018; Gómez-Fernández et al., 2024). Resistance to herbivores and pathogens is more pronounced in wild species, while domesticated vegetables tend to be more susceptible to pests and diseases (Martínez-Romero et al., 2020). Wild species often maintain more beneficial microbial associations and fewer pathogens, while domesticated crops experience altered microbiomes with reduced mutualistic relationships (Shi et al., 2019; Martín-Robles et al., 2020). WVs contain higher bioactive and nutrient compounds, whereas selection for high yield led to a reduction in nutritional quality (Fernandez et al., 2021; Smýkal et al., 2018).

Vegetable domestication is one of the most transformative developments in human history, as it links plant selection to nutrition and human health (Harris, 2007). The shift from collecting WPs to domesticated crops began about 10,000 years ago. This process is known as the Neolithic transition

DOI: 10.1201/9781003606710-15

TABLE 15.1
Comparison Between Wild and Domesticated Vegetable Crops

Trait/Aspect	Wild Vegetables	Domesticated Vegetables	Citations
Seed Size	Smaller, less variation	Larger (20% to 2.5× in vegetables; up to 15× in grains)	Kluyver et al. (2017), Garibaldi et al. (2021), Hernández-Terán et al. (2017)
Plant Size/ Biomass	Smaller, slower-growing	Larger, faster-growing, more aboveground biomass	Milla and Matesanz (2017), De Casas et al. (2024), Lei et al. (2024)
Leaf Size	Smaller	Larger leaves, contributing to plant size	Milla and Matesanz (2017), Lei et al. (2024)
Yield (Seed/ Fruit)	Lower total yield	Higher yield, greater total seed/fruit mass	Garibaldi et al. (2021), Lei et al. (2024), De Casas et al. (2024)
Genetic Diversity	High	Reduced due to selection and bottlenecks	Smýkal et al. (2018), Meyer et al. (2012), Coyne et al. (2020)
Phenotypic Plasticity	More tolerant to environmental stresses, more stable under variable conditions	Higher performance in optimal conditions, but less tolerant to stressors	Matesanz and Milla (2018), Gómez-Fernández et al. (2024)
Defensive Traits	Higher (more resistance to herbivores/pathogens)	Lower, more susceptible to pests	Fernandez et al. (2021), Coyne et al. (2020)
Microbiome/ Soil Interactions	More beneficial symbionts, fewer pathogens	Altered microbiome, more pathogens, fewer mutualists	Martínez-Romero et al. (2020), Shi et al. (2019), Martín-Robles et al. (2020)
Nutritional Quality	Higher in wild species	Lower due to selection for yield	Coyne et al. (2020), Fernandez et al. (2021), Smýkal et al. (2018)

from hunting to agriculture, which considered the beginning of organized farming systems and constituted transformation in human socioeconomic organization (Bhargava & Srivastava, 2019). Humans selected plants for desired parameters such as taste, reduced bitterness, and yield. Across the years, these processes developed important genetics and biochemical traits which distinguish domesticated crops from their WPs (Alseekh et al., 2021).

Domestication has significantly enhanced crop productivity and food security, leading to substantial changes in the nutritional profile of plants. We can find several WV crops which have lower yield, but they also have higher levels of minerals, vitamins, and bioactive compounds (Bvenura & Sivakumar, 2017). On the other hand, domesticated vegetables are often selected and bred for specific traits such as productivity and nutritional quality. Numerous studies have confirmed that vegetables, including wild relatives of legumes, leafy greens, and root vegetables, commonly contain a broader spectrum of phytochemical compounds and demonstrate greater antioxidant capacity compared to cultivated varieties (Mohammed Abdus Satter et al., 2016; Sánchez-Mata et al., 2012). These works indicate that the stresses driven by domestication have inadvertently altered plants' domestication and have influenced and altered human diets over the years. Early hunters consumed a diverse array of WPs rich in active compounds, which support a diverse stomach microbiota and adaptive metabolic capacity (Rampelli et al., 2025). Currently, a limited number of cultivated vegetable crops dominate the human diet, which has contributed to a reduction in dietary diversity and an increase in nutrient deficiencies, often referred to as 'hidden hunger' (Sánchez-Mata et al., 2012; Yilmaz & Yilmaz, 2025). Investigating the nutritional progression of vegetable crops shapes the fundamental relationship among farming, nutritional composition, and human health.

An in-depth understanding of WVs and CVs is vital for comprehending the effects of domestication and environmental alteration on the nutritional status and quality of our food. WV crops, often classified as non-cultivated, abandoned, or indigenous species, exhibit pronounced genetic and phytochemical diversity which does not exist in current farming systems. Wild vegetable crops (WVCs) are adapted to native habitat conditions; typically require minimal inputs; and contain a rich source of nutrients, antioxidants, and bioactive compounds which contribute to achieving food security and sustainability (Knez et al., 2023).

Although WVs often contain high amounts of specific phytochemicals and antioxidant compounds compared to CVs, cultivation also enhances other valuable compounds, such as sugars, organic acids, and beta-carotene (Botella et al., 2024). The difference in nutritional contents between WVs and CVs plants highlights the importance of investigating these distinct genetic resources for the development of functional foods and nutraceutical compounds (Ceccanti et al., 2020). Furthermore, abandoned indigenous vegetables, especially in regions such as Sub-Saharan Africa, have significant potential for improving dietary diversity, enhancing the nutritional quality of food, and boosting sustainability in the current climate change-affected agricultural landscape (Ndlovu et al., 2024).

CVs benefit from adjusting growth conditions and selective breeding programmes which enhance productivity and yield uniformity. This often reduces the biosynthesis of specific secondary metabolites. Secondary metabolites are essential for plant defence systems and also have health-promoting properties for the human diet (Abdelkader et al., 2024a, 2024b). Breeding programmes often aim to select for traits such as yield, taste, shape, hardness, and appearance. Maximizing these traits resulted in reduced levels of secondary metabolites, which are crucial for plant tolerance to abiotic and biotic stressors (Suliman et al., 2024). This has an adverse effect on plant health and the nutritional content of consumed vegetables (Ku et al., 2020).

The reintegration of WVs into our food system offers various benefits. It may enhance biodiversity conservation, nutrient intake, and sustainable agriculture. Wild species naturally contain higher concentrations of secondary metabolites, which confer high resistance to environmental stressors and a higher nutritional profile (Botella et al., 2024). Many studies have shown that domestication reduces the content of secondary metabolites, making them less tolerant to environmental stressors. Additionally, the investigations revealed that domestication decreases the concentration of bioactive compounds beneficial to human health. Reintegration of diverse plant species helps improve these issues, contributing to resilient agroecosystems and better nutritional outcomes in the face of environmental challenges.

Additionally, the nutritional attributes of WVs have a positive effect on human health. WVs may be utilized in crop breeding programmes to restore genes associated with nutrient concentration and stress tolerance (Ebert, 2020). This method aligns with global initiatives such as the Convention on Biological Diversity and FAO's cross-cutting initiative on biodiversity for food and nutrition. These initiatives highlight the status of WVs species in boosting food security and improving nutrition quality (Rokka et al., 2025). Therefore, studying the nutrient variations between WVs and CVs enhances understanding of the relationships between plants and human health. Also, this will inform various approaches for developing resilient, nutritionally sustainable, and adaptive food systems.

15.2 NUTRITIONAL COMPARISON OF CULTIVATED (CVS) AND WILD VEGETABLES (WVS)

Both types of vegetables play a crucial role in human health, serving as a source of essential nutrients and bioactive compounds. These substances are vital for maintaining physiological activity and reducing disease risk. However, scaling evidence suggests meaningful compositional content differences between WVs and CVs species, primarily due to genetic selection, environmental conditions, and domestication (Disciglio et al., 2017; Ku et al., 2020). These differences remain poorly understood (Kissanga et al., 2021).

Macronutrients, such as those commonly found in CVs, contribute reasonable quantities of energy; nonetheless, their lipid, protein, and carbohydrate profiles vary among species. WVs often exhibit higher protein levels and dietary fibre content than CVs, indicating that their adaptive mechanisms to harsh environmental conditions may also enhance traits that are vital for survival and nutritional value (Hundare, 2025; Li et al., 2020). For example findings reveal that indigenous African leafy vegetables commonly have higher crude protein content than cultivated species. For instance African leafy vegetables like *Amaranthus hybridus*, *Urena lobata*, and *Cleome gynandra* have been described to show crude protein levels fluctuating roughly from 19.28% to 30% on a dry weight, with some wild leafy vegetable varieties reaching 20–28 g/100 g dry weight (Kissanga et al., 2021; Lewu & Kambizi, 2014).

This is very often moderately higher than the protein concentrations typically observed in widely cultivated leafy vegetables. However, specific studies on cultivated leafy vegetables may also show considerable protein level on a dry weight matter, with green leafy vegetables like spinach, parsley, and coriander evaluated at 25.13% to 33.25% dry weight matter (Opazo-Navarrete et al., 2021). Correspondingly, advanced fibre content in WVs promotes slower digestion and enhanced regulation of glycaemia, improving their value in operating diet-related chronic diseases (Figure 15.1).

Micronutrients, such as those found in WVs, often exhibit greater performance than those in domestic vegetables in terms of mineral and vitamin concentrations. This is mainly evident for magnesium (Mg), calcium (Ca), iron (Fe), zinc (Zn), and vitamin C, all of which are typically found in greater amounts in wild species (García-Herrera & De Cortes Sánchez-Mata, 2016). For example wild amaranths and wild spider plants have been tested to include up to two to three times more Ca and Fe than cultivated spinach (Uusiku et al., 2010). β-Carotene, known as a precursor to vitamin A, is also richer in wild leafy vegetables due to significant exposure to sunlight stress, which stimulates carotenoid biosynthesis (Agea et al., 2014).

An alternative critical trend indicates a balance between high yield quantity and decreased nutrient levels. Domesticating vegetables has commonly involved selecting genotypes with larger edible organs and faster growth rates. Selecting crops according to these parameters frequently dilutes the concentration of nutrients through a 'yield dilution effect' (Singh & Van Der Knaap, 2022; Thapa et al., 2024). Current high-yield vegetable crops accumulate excess water and carbohydrates. On

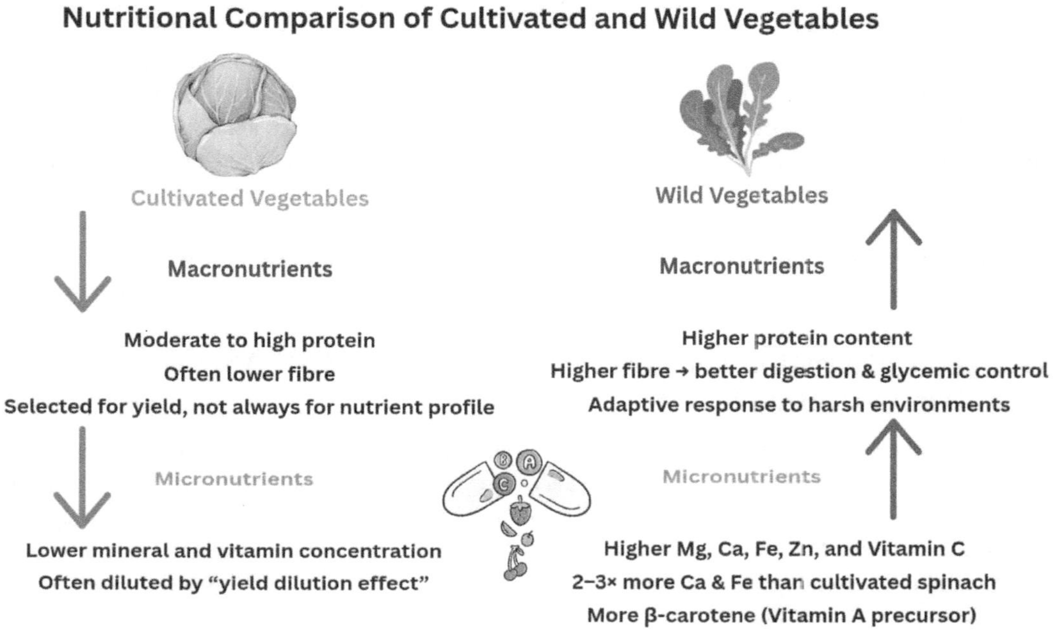

FIGURE 15.1 Comparative nutrient profile of cultivated and wild vegetables.

the other hand, minor minerals and vitamins were accumulated in comparison to their wild types. In this manner, although cultivated vegetables provide a greater quantity of food, WVs may offer a higher nutritional quality.

Many studies from various regions demonstrated the nutritional advantages of WVs over domesticated ones. For example, in South Africa, numerous wild leafy vegetables, such as Chenopodium album, Sonchus asper, Solanum nigrum, and *Urtica urens*, have been reported to possess superior nutritional properties compared with cultivated vegetables (Afolayan & Jimoh, 2009). The researchers analysed the leaves of these plants for their nutritional value, antinutritive elements, and polyphenolic composition. The protein content of their leaves ranged from 13.25% to 26.44%, while the fibre and mineral (ash) levels were approximately 16–23% and 13–27%, respectively. *U. urens* had the most significant levels of Ca, potassium (K), phosphorus (P), and Zn. In addition, a high concentration of Fe was detected in *S. asper.* These estimates were found to be comparable with or superior to those of domesticated vegetables such as cabbage, lettuce, and spinach. Additionally, the total phenolic concentration increased from 4.58 to 10.53 mg/g. Moreover, anti-nutritional compounds such as phytate, saponins, and alkaloids were found in vegetables at relatively low levels. The available minerals in WVs make them effective and important providers in human diets, particularly for low-income populations (Afolayan & Jimoh, 2009).

Brassicaceae vegetables are well known for their rich phytochemical compounds. A previous study explored the phenolic content, antioxidant activities, and anticancer potential of domestic Croatian and wild cabbage. The researchers identified sinapic acid as the predominant phenolic acid, followed by ferulic acid and salicylic acid. A comparative analysis revealed that wild cabbage had a notably higher content of sinapic acid. Kale had a higher content of kaempferol and quercetin. Both extracts have demonstrated significant antioxidant activity and exhibited antiproliferative properties against HeLa cells without impacting normal human skin fibroblasts. Wild cabbage extract exhibited an antiproliferative effect at a lower concentration of 10 μg/mL. This is possibly due to its higher content of sinapic acid (Lučić et al., 2023). These findings highlight the potential of *Brassica* species as a source of bioactive composites with antioxidant activity plus anticancer properties, demanding advanced evaluation of their therapeutic applications.

These studies emphasize that WVs are not only high in nutritional value but also socially and environmentally valuable. Their incorporation into the food systems may improve nutritional diversity and reinforce sustainable agricultural practices.

15.3 BIOACTIVE COMPOUNDS IN WILD VERSUS DOMESTICATED VEGETABLES

Bioactive compounds are recognized as phytochemicals which naturally act as secondary metabolites in plant tissues. Phytochemicals play a significant function in human health, primarily through their antioxidant, anti-inflammatory, and disease-defensin properties (Lučić et al., 2023; Sánchez-Mata et al., 2012). Each group of phytochemicals, such as polyphenols, flavonoids, and alkaloids, has unique functional and nutritional values (El-Beltagi et al., 2023).

Polyphenols (PPs), including phenolic acids, tannins, and flavonoids, are commonly found in vegetable parts (Abdelkader et al., 2024c). PP acts as a powerful antioxidant by scavenging ROS and chelating heavy metal ions, thereby reducing oxidative stress (Tawfik et al., 2022; Rao & Zheng, 2025). For example wild chicory (*Cichorium intybus*) exhibits a higher phenolic content and antioxidant activity compared to domesticated lettuce (Stanciu et al., 2019). Additionally, dandelion (Taraxacum officinale) exhibited a higher content of PP and antioxidant capacity (Miłek et al., 2019). This higher content is frequently related to exposing WPs to natural environmental stresses such as ultraviolet radiation, temperature fluctuations, and drought. Environmental stressors accelerate the biosynthesis of phenolic compounds.

Flavonoid substances are known for their varied biological properties. Flavonoids have anti-inflammatory, antimicrobial, and cardioprotective effects. Wild *Amaranthus spinosus* had notably higher total flavonoid contents than cultivated *Amaranthus* species, indicating its strong antioxidant contents (Sarker & Oba, 2019). *Cleome gynandra* (wild species) had higher total flavonoid contents and antioxidant capacity than CVs (Rjeibi et al., 2017). Wild *C. gynandra* exhibits significantly higher total flavonoid and antioxidant activity than domestic spinach and cabbage, with stated estimates surpassing 250 mg quercetin equivalents per 100 g DM (dry matter), twice that of domestic *B. oleracea* (Moyo et al., 2018).

Carotenoids are precursors to vitamin A and play an important role in the pigmentation of the upper surface. β-Carotene, lutein, and zeaxanthin are essential for maintaining eyesight and supporting the immune system. WVs have higher carotenoid levels compared to their domesticated species (Botella et al., 2024). Wild *Solanum nigrum* contains 50–115 mg β-carotene/100 g DM, which is significantly superior to kale or lettuce. Elevated lutein levels contribute to its vitamin A capacity (Raju et al., 2007). β-Carotene was measured in A. *hybridus* at 27–32 mg β-carotene/100 g DM (Mziray et al., 2000). These wild, leafy vegetables provide a high amount of vitamin A, thereby helping to alleviate vitamin A deficiency.

Alkaloids are correlated with plant defence systems and bitterness. Alkaloids also enhance human health in small quantities, exhibiting antimicrobial and neuroprotective effects. Wild *Solanum nigrum*, *Momordica charantia*, and *Capsicum frutescens* contain alkaloids which have therapeutic roles in traditional medicine (Narayana, 2021). Domestication has reduced alkaloid concentration to increase sweetness and enhance safety. As a result, domestication has altered the biochemical composition of CVs, leading to a decrease in bioactive compounds. Selective breeding, in general, prioritized agronomic traits (yield, texture, colour, and taste) over phytochemical value, thereby consciously declining the health-promoting potential of many domesticated varieties (Smýkal et al., 2018).

The nutrient dilution effect occurs when excessive yields result in reduced nutrients (Kaspari & Welti, 2024; Ku et al., 2020). This was found in various vegetable crops, including tomato (Abdelkader et al., 2019; Dixon et al., 2024), and carrot (Ellison, 2019). Recent varieties contain less amounts of vitamin C, polyphenols, and β-carotene than their wild relatives (Medina-Lozano et al., 2021; Smýkal et al., 2018). The CEA diminishes environmental stressors (temperature fluctuations, water shortage, and nutrient deficiency), requiring additional reduction in plants' needs to generate defence-associated stress compounds (Gruda et al., 2025). On the other hand, WPs in nature create higher concentrations of phytochemicals as responses to stress (Chávez-Arias et al., 2022). Current breeding programmes reestablish bioactive traits from the WPs to restore lycopene, flavour, and flavonoid profiles in tomato (Adithya et al., 2025) and develop glucosinolate concentrations in domestic *Brassica* species (Garcia et al., 2025).

Therefore, domestication has increased crop productivity and food availability; it has also reduced the phytochemical content of domestic vegetables. Comprehension of these evolutionary trade-offs is crucial for informing future breeding programmes and conservation strategies which aim to develop both dietary and functional food qualities.

15.4 HEALTH BENEFITS OF WILD VEGETABLES COMPARED TO COMMONLY CULTIVATED VARIETIES

WVs offer a rare source of nutritional and bioactive substances which contribute to enhanced health benefits compared to most domesticated ones. General nutritional and clinical studies have discovered that the higher levels of vitamins, minerals (Ca, K, P, Fe, and Zn), and phytochemicals (polyphenols, flavonoids, carotenoids, and glucosinolates) in wild types help to enhance antioxidant and anti-inflammatory potential (Sánchez-Mata et al., 2012; Uusiku et al., 2010). Phytochemicals stabilize free radicals, decrease lipid peroxidation, and modulate detoxification enzymes, leading to protecting cellular structure from oxidative damage (Narayana, 2021).

Epidemiological evidence has also found a relationship between the consumption of WVs and a decrease in chronic diseases. Cardiovascular disorders, diabetes, obesity, and certain cancers

(Bayang et al., 2025; Delzenne et al., 2025) are among these diseases. Increasing fibre content and reducing the glycaemic index of many WVs improve glucose metabolism and lipid content. Rich in antioxidant activity, they reinforce the immune system and promote cardiovascular health. Bitter melon has hypoglycaemic and hypolipidaemic effects attributed to its bioactive alkaloids and saponins (Malik et al., 2025). Wild *Brassica* species are rich in polyphenols, exhibiting chemoprotective activities associated with glucosinolates and isothiocyanates. This alters carcinogen metabolism and triggers apoptotic mechanisms in cancer cells (Alqaraawi & Almujaydil, 2025). Incorporating WVs into modern nutrition improves dietary quality and reduces the risk of chronic disease. Rich in bioactive compounds, WVs are practical components of sustainable agriculture for promoting human health and biodiversity conservation (Figure 15.2).

A comparison of vegetable and fruit consumption in ten European nations against the WHO's recommendations for minimum daily intake per person indicated that all countries, except Greece, exhibited vegetable consumption below the recommended minimum (Trichopoulou et al., 2001). Additionally, the number of people who did not consume vegetables was higher than the number of people who did not consume fruits. This suggests that more efforts are needed to encourage people to eat more WVs (Table 15.2). In fact, they have been a significant part of African diets for hundreds of years, providing millions of people with the vitamins, minerals, and nutrients they need to maintain their health and ward off dangerous diseases (Grivetti & Ogle, 2000). A recent assessment found that at least 1,000 of the more than 45,000 plant species in Sub-Saharan Africa might be eaten as green leafy vegetables (Flyman & Afolayan, 2006). The Plant Resources of Tropical Africa indicated that 6,376 plants were helpful to people. About 397 of them are considered WVs (Adebooye & Opabode, 2004). This list is likely too short, as local people use many more plants as food than are documented in the literature (Padulosi et al., 2002).

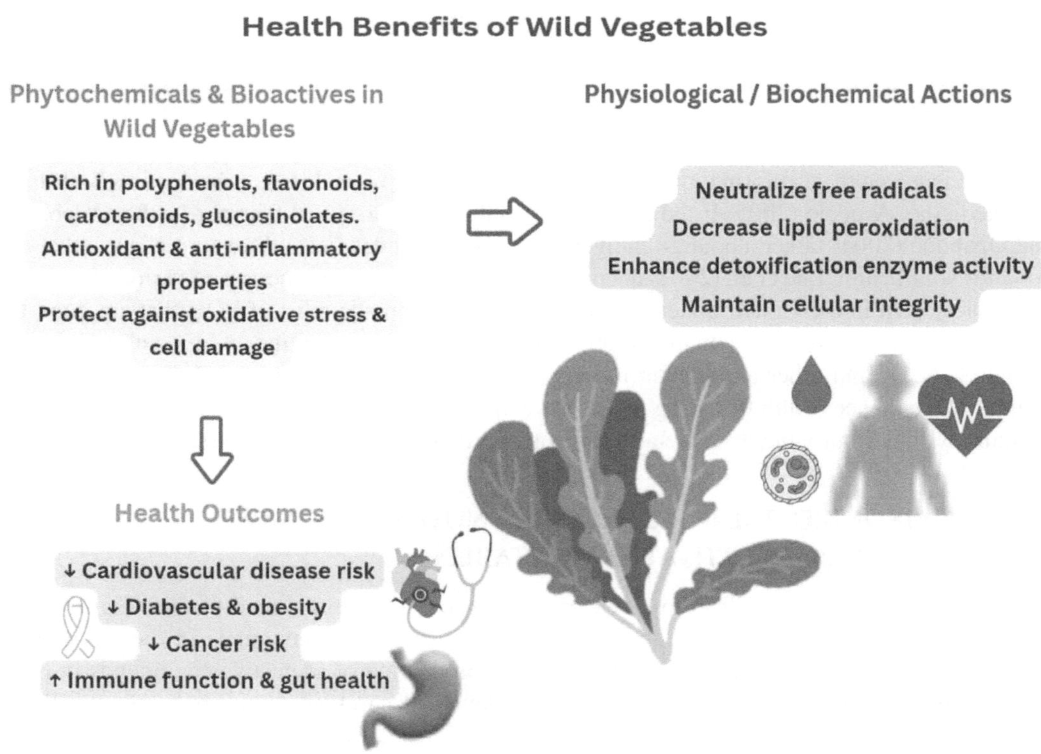

FIGURE 15.2 Health benefits of wild vegetables.

TABLE 15.2
Examples of Some Wild Vegetable Species Around the World

Scientific Name	Common Name	Edible Use	Nutritional Benefits	Citation
Diplazium esculentum	Pakis, Pahu	Young shoots	Enriched in vitamin B9	Indonesian Food Composition Data (2020)
Ipomoea aquatica	Kangkung air	Leaves and stems as a vegetable	High in Fe and provitamin A	Indonesian Food Composition Data (2020)
Eremurus spectabilis	Foxtail lily	Shoots, buds, and young leaves	High in mineral content and antioxidants.	Tosun et al. (2012)
Scolymus hispanicus	Golden thistle	Roots and young leaves	Enriched in dietary fibre, Mg, and Ca.	García-Herrera et al. (2014)
Tacca leontopetaloides	Kai Niue	Root starch	Main source of carbohydrates, and it also contains vitamin C.	Wardah et al. (2017)
Malva sylvestris	Mallow, Tibi	Leafy vegetable	Strong antioxidant capabilities; rich in phenols, flavonoids, and carotenoids	Barros et al. (2010)
Nasturtium offcinale	Watercress	Leafy vegetable	High in vitamin K, vitamin A, vitamin C, vitamin B6, Mn, and Ca	BFN Project (2020)
Chorchorus olitorius	Jute mallow	Leaves used as vegetables	Rich in beta-carotene, vitamin C, folic acid, Ca, and Fe	FAO/Government of Kenya (2018)
Amaranthus tortuosus	Amaranth, Ekichabo	Leaves used as vegetables	High in proteins, fibres, calcium, and vitamin C	FAO/Government of Kenya (2018)
Cleome gynandra	Spider plant	Leaves used as vegetables	Rich in beta-carotene, vitamin C, folic acid, Ca, and Fe.	FAO/Government of Kenya (2018)
Caulerpa racemosa	Nama	Sea vegetable	Provides proteins, fibre, minerals, vitamins, and polyunsaturated fatty acids.	de Gaillande et al. (2017)

There is a wealth of information on how wild, traditional vegetables can be used as both food and medicine. Ayodele (2005) listed 28 leafy vegetables that are good for your health, Adebooye and Opabode (2004) talked about 24 indigenous leafy vegetables in southwestern Nigeria that are good for your health, Lee et al. (2003) documented the antibacterial properties of various fruits and vegetables eaten in the United States, and Ogle et al. (2003) made a list of WVs that are good for the health and for eating in Vietnam. Additionally, WVs contain a significant amount of Vitamins A and C. These and other antioxidant micronutrients found in fresh fruits and vegetables (Szeto et al., 2002) enhance health by aiding in the prevention of cancer and hypertension, stimulating the immune system, improving drug metabolism, and facilitating tissue regeneration (Walingo, 2005).

15.5 ENVIRONMENTAL INFLUENCES ON NUTRIENT CONTENT IN WILD AND CULTIVATED VEGETABLES

The nutrient status of vegetable crops is clearly influenced by environmental factors (Figure 15.3). These factors impact the minerals, vitamins, and phytochemical compounds in the crops (Šola et al., 2025). Sunlight and temperature stimulate the production of protective compounds in plant tissues. These external factors directly impact the levels of carotenoids and polyphenols. These compounds enhance the antioxidant activities of the wild species (Jangpangi et al., 2025; Hassan et al., 2024; Abou El-Nasr et al., 2025). In contrast, managing irrigation and fertilization of the cultivated vegetables produces limited stress-induced phytochemicals. Moreover, stressors such as

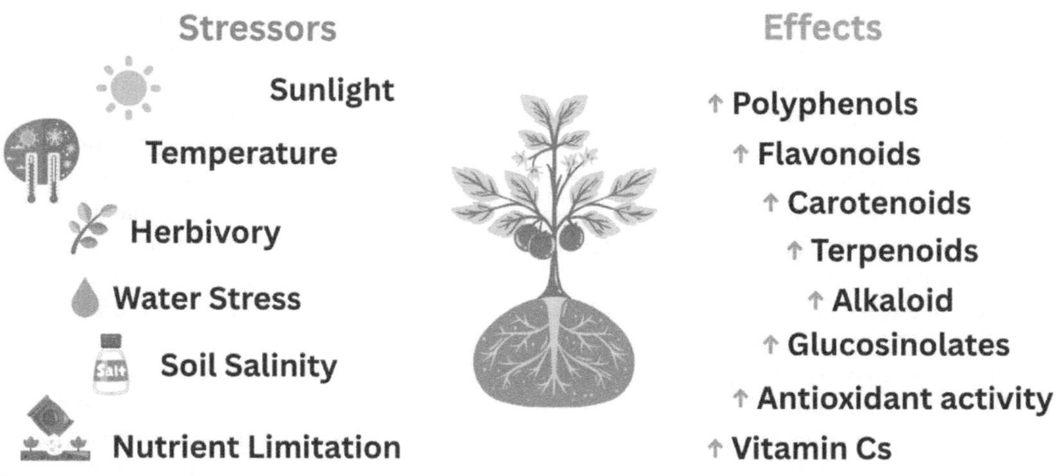

FIGURE 15.3 Environmental stressors impacting the nutrient composition of wild and cultivated vegetables.

drought, salinity, herbivory, and competition in natural habitats stimulate defence system-related metabolic mechanisms, particularly the phenylpropanoid and isoprenoid pathways, which contribute to high levels of flavonoids and terpenoids (Alkhateeb et al., 2024; El-Sayed et al., 2022).

Domesticated vegetables, implementing sustainable agriculture practices (organic practice, intercropping, combining, slight tillage, and decreased agrochemical inputs), may enhance soil fertility and phytochemical content (De Pasquale et al., 2023). Organic agriculture practices, for instance, expose vegetable crops to mild stress and restricted nutrient availability, which often stimulates the synthesis of secondary metabolites beneficial to human well-being (Mohd Din et al., 2017). For example previous study observed how diverse soil amendments such as chicken manure, vermicompost Vermi, horse manure, sewage sludge, elemental inorganic fertilizer Inorg, organic fertilizer Org, and biochar, affect the level of glucosinolates, vitamin C, phenols compounds, plus reducing sugars in three varieties of turnips (Purple Top White Globe PTWG, Scarlet Queen Red SQR, and Tokyo Cross TC), arugula, and mustard greens cultivated under natural field environments. The findings of this study revealed that mustard greens exhibited higher levels of glucosinolates (974 µg/g fresh shoots) than arugula (651 µg/g fresh shoots). Additionally, the TC variety of turnip showed the highest levels of glucosinolates, vitamin C, and sugars.

Furthermore, improving the soil with sewage sludge, chicken manure, and horse manure substantially boosted the vitamin C level in mustard shoots by 82%, 90%, and 31%, respectively, and the total phenols by 77%, 70%, and 36%, respectively. These findings underscore the importance of soil amendments in enhancing the phytochemical composition of *Brassica* crops (De Pasquale et al., 2023). Adjusting the environmental factors which influence nutrients in WVs and CVs is a crucial step in building sustainable food systems and conserving environmental integrity.

15.6 FUTURE PERSPECTIVES: INTEGRATING WILD VEGETABLES INTO MODERN DIETS

The integration of WVs into our food system holds significant potential for enhancing nutrient intake. This may improve nutritional diversity and biodiversity. WV relatives of cultivated vegetable crops can be utilized as a gene bank. WVs genes can enhance nutrient contents, disease resistance, and tolerance to environmental stressors (Bal et al., 2025). Employing these genetic resources in breeding programmes may help reestablish missing nutritional traits and bioactive compound characteristics. Successful integration of rich lycopene from *Solanum pimpinellifolium* into current tomatoes and rich glucosinolate from wild *Brassica* species into modern varieties is evidence (Adithya et al., 2025; Garcia et al., 2025). Preserving and reintroducing WVs in recent farming practices contribute to agrobiodiversity preservation. This approach can help achieve resilient food systems in response to climate change and environmental degradation.

Raising the intake of WVs requires new strategies which integrate awareness with action. Availability of WVs and policy support can play a crucial role in this matter. Public health and educational initiatives can also promote the awareness of the nutritional value and medicinal benefits of WVs. Food innovation and food processing for value enhancement may make them more appealing to modern customers (Borelli et al., 2020). Combining these WVs with school food programmes and local markets can improve food security, particularly in regions where nutrient deficiencies are prevalent.

Policy frameworks which promote sustainable vegetable harvesting, organic practices, organic certification, and sustainable trade practices may further enhance the economic sustainability of WVs. These approaches promote ecological conservation. Cooperation between scientists, farmers, consumers, and policymakers is vital in integrating WVs into our nutrition programmes. This direction aligns with the Sustainable Development Goals (SDGs) in relation to health and biodiversity. WVs' revalorization presents numerous benefits for biodiversity, making WVs a vital component of the emerging sustainable agriculture and resilient food approaches, alongside domesticated species.

15.7 CONCLUSION

This chapter highlights the biochemical and ecological differences between wild and domestic vegetables. Additionally, it underscores how the domestication and management of environmental factors impact nutrient composition. Various studies indicate that WVs contain higher levels of micronutrients, phytochemical compounds, and antioxidant activity compared to their domesticated counterparts. These variations are attributed to the nutrient dilution effect. Selective plant breeding programmes aimed at increasing yield and taste can lead to a decrease in active compounds. Both dietary and clinical data prove the health benefits of WVs. Their higher phytochemical content diminishes the risks of chronic diseases (CVD, diabetes, and certain cancers). The bioactive compounds in WVs play vital roles in managing oxidative stress and promoting metabolic health. WVs are a valuable genetic source for enhancing the diversity and nutrient content in breeding programmes. Adding WV species to modern meal plans could make food security better. Sustainable harvesting and planting in the area may help improve the quality of food and the stability of the environment. Recognizing the nutritional value of WPs can boost food security and increase diversity. Combining old and new farming methods can help keep wild species alive and include them in plans.

REFERENCES

Abdelkader, M. M., Elkhawaga, F. A., Suliman, A. A., Puchkov, M., Kuranova, K. N., Mahmoud, M. H., & Abdelkader, M. F. (2024b). Understanding the regular biological mechanism of susceptibility of tomato plants to low incidences of blossom-end rot. *Horticulturae*, *10*(6), 648.

Abdelkader, M. M., Puchkov, M. Y., Lysakov, M. A., Loktionova, E. G., & Suliman, A. A. (2019). Effect of crezacin and humic acid on growth and physiological aspects of tomato plants (Lycopersicon esculentum). *Journal of Applied Horticulture*, *21*(1), 61–66.

Abdelkader, M. M., Suliman, A. A., Salem, S. S., Assiya, A., Voronina, L., Puchkov, M., . . . & Abdelkader, M. F. (2024a). Studying the combined impact of salinity and drought stress-simulated conditions on physio-biochemical characteristics of lettuce plant. *Horticulturae*, *10*(11), 1186.

Abdelkader, M. M., Zargar, M., Bayat, M., Pakina, E., Shehata, A. S., & Suliman, A. A. (2024c). Biogenic nano-fertilizers as a sustainable approach to alleviate nitrate accumulation and enrich quality traits of vegetable crops. *Horticulturae*, *10*(8), 789.

Abou El-Nasr, M. K., Hassan, K. M., Abd-Elhalim, B. T., Kucher, D. E., Rebouh, N. Y., Ansabayeva, A., Abdelkader, M., Ali, M. A., & Nasser, M. A. (2025). The emerging roles of nanoparticles in managing the environmental stressors in horticulture crops – a review. *Plants*, *14*(14), 2192. https://doi.org/10.3390/plants14142192

Adebooye, O. C., & Opabode, J. T. (2004). Status of conservation of the indigenous leaf vegetables and fruits of Africa. *African Journal of Biotechnology*, *3*(12), 700–705.

Adithya, A., Indu Rani, C., Savitha, B. K., Murugan, M., Sudha, M., & Prabhu, M. (2025). Harnessing the crop wild relatives in genetic improvement of eggplant, potato and tomato. *Genetic Resources and Crop Evolution*, *72*(3), 2613–2629. https://doi.org/10.1007/S10722-024-02165-2/METRICS

Afolayan, A. J., & Jimoh, F. O. (2009). Nutritional quality of some wild leafy vegetables in South Africa. *International Journal of Food Sciences and Nutrition*, *60*(5), 424–431. https://doi.org/10.1080/09637480701777928

Agea, J. G., Kimondo, J. M., Woiso, D. A., Okia, C. A., Obaa, B. B., Isubikalu, P., Bosco, J., Okullo, L., Obua, J., & Teklehaimanot, Z. (2014). Proximate composition, vitamin C and beta-carotene contents of fifteen selected leafy wild and semi-wild food plants (WSWFPs) from Bunyoro-Kitara Kingdom, Uganda. *Journal of Natural Products and Plant Resources*, *4*(3).

Alkhateeb, O. A., Ali, M. A. A., Abdou, A. H., Abdelaal, K., & El-Azm, N. A. I. A. (2024). Integrating soil mulching and subsurface irrigation for optimizing deficit irrigation effectiveness as a water-rationing strategy in tomato production. *Notulae Botanicae Horti Agrobotanici Cluj-Napoca*, *52*(1), 13514–13514. https://doi.org/10.15835/NBHA52113514

Alqaraawi, S. S., & Almujaydil, M. S. (2025). Nutritional properties and potential therapeutic uses of Brassica nigra. *Discover Food*, *5*(1), 1–17. https://doi.org/10.1007/S44187-025-00635-9

Alseekh, S., Scossa, F., Wen, W., Luo, J., Yan, J., Beleggia, R., Klee, H. J., Huang, S., Papa, R., & Fernie, A. R. (2021). Domestication of crop metabolomes: Desired and unintended consequences. *Trends in Plant Science*, *26*(6). https://doi.org/10.1016/j.tplants.2021.02.005

Ambrose-Oji, B. (2009). Urban food systems and African indigenous vegetables: Defining the spaces and places for African indigenous vegetables in urban and peri-urban agriculture. In *African Indigenous Vegetables in Urban Agriculture*, eds. C. Shackleton, A. Drescher, and M. Pasquini (London: Earthscan), 1–33.

Ayodele, A. E. (2005). The Medicinally important leafy vegetables of Southwestern Nigeria. *Ethnobotanical Leaflets*, *2005*(1), 16.

Bal, S., Chattopadhyay, A., & Hazra, P. (2025). Utilization of crop wild relatives in vegetable breeding programs could enhance crop adaptation to challenging environments. In (pp. 73–99). https://doi.org/10.1007/978-981-96-2413-3_4

Barros, L., Carvalho, A. M., & Ferreira, I. C. F. R. (2010). Leaves, flowers, immature fruits and leafy flowered stems of Malva sylvestris: A comparative study of the nutraceutical potential and composition. *Food and Chemical Toxicology*, *48*, 1466–1472.

Bayang, J. P., Touwang, C., Mamoudou, H., Woudam, E. S., & Koubala, B. B. (2025). Variation of nutrients and bioactive compounds of five wild edible leafy vegetables from far north region of Cameroon. *Food Chemistry Advances*, *6*, 100849. https://doi.org/10.1016/J.FOCHA.2024.100849

BFN Project. *BFN Species Database*. Available online: http://www.b4fn.org/resources/species-database/ (accessed on 29 July 2020).

Bhargava, A., & Srivastava, S. (2019). Human civilization and agriculture. In *Participatory Plant Breeding: Concept and Applications*, 1–27. https://doi.org/10.1007/978-981-13-7119-6_1

Borelli, T., Hunter, D., Powell, B., Ulian, T., Mattana, E., Termote, C., Pawera, L., Beltrame, D., Penafiel, D., Tan, A., Taylor, M., & Engels, J. (2020). Born to eat wild: An integrated conservation approach to secure wild food plants for food security and nutrition. *Plants*, *9*(10), 1299. https://doi.org/10.3390/PLANTS9101299

Botella, M. Á., Hellín, P., Hernández, V., Dabauza, M., Robledo, A., Sánchez, A., Fenoll, J., & Flores, P. (2024). Chemical composition of wild collected and cultivated edible plants (Sonchus oleraceus L. and Sonchus tenerrimus L.). *Plants*, *13*(2), 269. https://doi.org/10.3390/PLANTS13020269/S1

Bvenura, C., & Sivakumar, D. (2017). The role of wild fruits and vegetables in delivering a balanced and healthy diet. *Food Research International*, *99*, 15–30. https://doi.org/10.1016/J.FOODRES.2017.06.046

Ceccanti, C., Landi, M., Incrocci, L., Pardossi, A., Venturi, F., Taglieri, I., Ferroni, G., & Guidi, L. (2020). Comparison of three domestications and wild-harvested plants for nutraceutical properties and sensory profiles in five wild edible herbs: Is domestication possible? *Foods*, *9*(8), 1065. https://doi.org/10.3390/FOODS9081065

Cernansky, R. (2015). The rise of Africa's super vegetables. *Nature*, *522*, 146–148. https://doi.org/10.1038/522146a

Chávez-Arias, C. C., Ramírez-Godoy, A., & Restrepo-Díaz, H. (2022). Influence of drought, high temperatures, and/or defense against arthropod herbivory on the production of secondary metabolites in maize plants. A review. *Current Plant Biology*, *32*, 100268. https://doi.org/10.1016/J.CPB.2022.100268

Coyne, C., Kumar, S., Wettberg, E., Marques, E., Berger, J., Redden, R., Ellis, T., Brus, J., Zablatzká, L., & Smýkal, P. (2020). Potential and limits of exploitation of crop wild relatives for pea, lentil, and chickpea improvement. *Legume Science*, *2*. https://doi.org/10.1002/leg3.36

De Casas, R., Martín-Brull, I., Milla, R., & Ocaña-Calahorro, F. (2024). Selection for domestication favored taxa characterized by fast growth and tolerance of high intraspecific density. *Plants, People, Planet*. https://doi.org/10.1002/ppp3.10596

de Gaillande, C., Payri, C., Remoissenet, G., & Zubia, M. (2017). Caulerpa consumption, nutritional value and farming in the Indo-Pacific Region. *Journal of Applied Phycology*, *29*, 2249–2266.

Delzenne, N. M., Bindels, L. B., Neyrinck, A. M., & Walter, J. (2025). The gut microbiome and dietary fibres: Implications in obesity, cardiometabolic diseases and cancer. *Nature Reviews Microbiology*, *23*(4), 225–238. https://doi.org/10.1038/S41579-024-01108-Z;SUBJMETA

De Pasquale, C., Moore, N. J., López Núñez, R., & Antonious, G. F. (2023). The impact of organic, inorganic fertilizers, and biochar on phytochemicals content of three brassicaceae vegetables. *Applied Sciences*, *13*(15), 8801. https://doi.org/10.3390/APP13158801

Disciglio, G., Tarantino, A., Frabboni, L., Gagliardi, A., Giuliani, M. M., Tarantino, E., & Gatta, G. (2017). Qualitative characterization of cultivated and wild edible plants: Mineral elements, phenols content and antioxidant capacity. *Italian Journal of Agronomy*, *12*(4), 1036. https://doi.org/10.4081/IJA.2017.1036

Dixon, M. M., Afkairin, A., Davis, J. G., Chitwood-Brown, J., Buchanan, C. M., Ippolito, J. A., Manter, D. K., & Vivanco, J. M. (2024). Tomato domestication rather than subsequent breeding events reduces microbial associations related to phosphorus recovery. *Scientific Reports*, *14*(1), 1–15. https://doi.org/10.1038/S41598-024-60775-3;SUBJMETA

Ebert, A. W. (2020). The role of vegetable genetic resources in nutrition security and vegetable breeding. *Plants*, *9*(6). https://doi.org/10.3390/plants9060736

El-Beltagi, H. S., El-Sayed, S. M., Abdelhamid, A. N., Hassan, K. M., Elshalakany, W. A., Nossier, M. I., Alabdallah, N. M., Al-Harbi, N. A., Al-Qahtani, S. M., Darwish, D. B. E., & Abbas, Z. K. (2023). Potentiating biosynthesis of alkaloids and polyphenolic substances in Catharanthus roseus plant using κ-carrageenan. *Molecules*, *28*(8), 3642. https://doi.org/10.3390/molecules28083642

Ellison, S. (2019). *Carrot Domestication*, 77–91. https://doi.org/10.1007/978-3-030-03389-7_5

El-Sayed, S. M., Hassan, K. M., Abdelhamid, A. N., Yousef, E. E., Abdellatif, Y. M., Abu-Hussien, S. H., Nasser, M. A., Elshalakany, W. A., Darwish, D. B. E., Abdulmajeed, A. M., & Alabdallah, N. M. (2022). Exogenous paclobutrazol reinforces the antioxidant and antimicrobial properties of lavender (Lavandula officinalis L.) oil through modulating its composition of oxygenated terpenes. *Plants*, *11*(12), 1607. https://doi.org/10.3390/plants11121607

FAO/Government of Kenya. (2018). *Kenya Food Composition Tables*. Food and Agricultural Organization of the United Nations: Nairobi, Kenya.

Fernandez, A., Sáez, A., Quintero, C., Gleiser, G., & Aizen, M. (2021). Intentional and unintentional selection during plant domestication: Herbivore damage, plant defensive traits and nutritional quality of fruit and seed crops. *The New Phytologist*. https://doi.org/10.1111/nph.17452

Flyman, M. V., & Afolayan, A. J. (2006). The suitability of wild vegetables for alleviating human dietary deficiencies. *South African Journal of Botany*, *72*(4), 492–497.

Garcia, G., Treccarichi, S., Arena, D., Ben Ammar, H., Maggioni, L., & Branca, F. (2025). Capturing the Brassica oleracea L. wild relatives diversity for improving nutraceutical traits of cole crops. *Genetic Resources and Crop Evolution*, *72*(7), 8855–8871. https://doi.org/10.1007/S10722-025-02479-9/METRICS

García-Herrera, P., & De Cortes Sánchez-Mata, M. (2016). The contribution of wild plants to dietary intakes of micronutrients (II): Mineral elements. In *Mediterranean Wild Edible Plants: Ethnobotany and Food Composition Tables*, 141–171. https://doi.org/10.1007/978-1-4939-3329-7_7/FIGURES/24

García-Herrera, P., Sánchez-Mata, M. C., Cámara, M., Fernández-Ruiz, V., Díez-Marqués, C., Molina, M., & Tardío, J. (2014). Nutrient composition of six wild edible Mediterranean Asteraceae plants of dietary interest. *Journal of Food Composition and Analysis*, *34*, 163–170.

Garekae, H., & Shackleton, C. M. (2020). Foraging wild food in urban spaces: The contribution of wild foods to urban dietary diversity in South Africa. *Sustainability*, *12*, 678. https://doi.org/10.3390/su12020678

Garibaldi, L., Aizen, M., Sáez, A., Gleiser, G., Strelin, M., & Harder, L. (2021). The influences of progenitor filtering, domestication selection and the boundaries of nature on the domestication of grain crops. *Functional Ecology*. https://doi.org/10.1111/1365-2435.13819

Gómez-Fernández, A., Aranda, I., & Milla, R. (2024). Early human selection of crops' wild progenitors explains the acquisitive physiology of modern cultivars. *Nature Plants*. https://doi.org/10.1038/s41477-023-01588-6

Grivetti, L. E., & Ogle, B. M. (2000). value of traditional foods in meeting macro-and micronutrient needs: The wild plant connection. *Nutrition Research Reviews*, *13*(1), 31–46.

Gruda, N. S., Samuolienė, G., Dong, J., & Li, X. (2025). Environmental conditions and nutritional quality of vegetables in protected cultivation. *Comprehensive Reviews in Food Science and Food Safety*, *24*(2), e70139. https://doi.org/10.1111/1541-4337.70139

Harris, D. (2007). Agriculture, cultivation and domestication: Exploring the conceptual framework of early food production. In *Rethinking Agriculture: Archaeological and Ethnoarchaeological Perspectives*. Routledge. https://doi.org/10.4324/9781315421018

Hassan, K. M., Ajaj, R., Abdelhamid, A. N., Ebrahim, M., Hassan, I. F., Hassan, F. A., Alam-Eldein, S. M., & Ali, M. A. (2024). Silicon: A powerful aid for medicinal and aromatic plants against abiotic and biotic stresses for sustainable agriculture. *Horticulturae*, *10*(8), 806.

Hernández-Terán, A., Wegier, A., Benítez, M., Lira, R., & Escalante, A. (2017). Domesticated, genetically engineered, and wild plant relatives exhibit unintended phenotypic differences: A comparative meta-analysis profiling rice, canola, maize, sunflower, and pumpkin. *Frontiers in Plant Science*, *8*. https://doi.org/10.3389/fpls.2017.02030

Hundare, A. (2025). Wild versus cultivated: Assessing the nutritional superiority of a diploid potato species solanum okadae for future food security. *Molecular Nutrition & Food Research*, e70236. https://doi.org/10.1002/MNFR.70236

Indonesian Food Composition Data. Available online: https://www.panganku.org/en-EN/tentang_kami (accessed on 14 July 2020).

Jangpangi, D., Patni, B., Chandola, V., & Chandra, S. (2025). Medicinal plants in a changing climate: understanding the links between environmental stress and secondary metabolite synthesis. *Frontiers in Plant Science*, *16*, 1587337. https://doi.org/10.3389/FPLS.2025.1587337/FULL

Kaspari, M., & Welti, E. A. R. (2024). Nutrient dilution and the future of herbivore populations. *Trends in Ecology and Evolution*, *39*(9), 809–820. https://doi.org/10.1016/j.tree.2024.05.001

Kissanga, R., Sales, J., Moldão, M., Alves, V., Mendes, H., Romeiras, M. M., Lages, F., & Catarino, L. (2021). Nutritional and functional properties of wild leafy vegetables for improving food security in Southern Angola. *Frontiers in Sustainable Food Systems*, *5*, 791705. https://doi.org/10.3389/FSUFS.2021.791705/BIBTEX

Kluyver, T., Jones, G., Pujol, B., Bennett, C., Mockford, E., Charles, M., Rees, M., & Osborne, C. (2017). Unconscious selection drove seed enlargement in vegetable crops. *Evolution Letters*, *1*, 64–72. https://doi.org/10.1002/evl3.6

Knez, M., Ranić, M., & Gurinović, M. (2023). Underutilized plants increase biodiversity, improve food and nutrition security, reduce malnutrition, and enhance human health and well-being. Let's put them back on the plate! *Nutrition Reviews*, *82*(8), 1111. https://doi.org/10.1093/NUTRIT/NUAD103

Ku, Y. S., Contador, C. A., Ng, M. S., Yu, J., Chung, G., & Lam, H. M. (2020). The effects of domestication on secondary metabolite composition in legumes. *Frontiers in Genetics*, *11*, 581357. https://doi.org/10.3389/FGENE.2020.581357

Lee, Y. L., Cesario, T., Wang, Y., Shanbrom, E., & Thrupp, L. (2003). Antibacterial activity of vegetables and juices. *Nutrition*, *19*(11–12), 994–996.

Lei, Z., Li, Z., Zhang, W., He, D., & Zhang, Y. (2024). From wild to cultivated crops: general shift in morphological and physiological traits for yield enhancement following domestication. *Crop and Environment*. https://doi.org/10.1016/j.crope.2024.03.001

Lewu, M. N., & Kambizi, L. (2014). Comparative nutritional assessment of selected leafy vegetables. *Journal of Food, Agriculture and Environment*, *12*(2).

Li, X., Yadav, R., & Siddique, K. H. M. (2020). Neglected and underutilized crop species: The key to improving dietary diversity and fighting hunger and malnutrition in Asia and the Pacific. *Frontiers in Nutrition*, *7*. https://doi.org/10.3389/fnut.2020.593711

Lučić, D., Pavlović, I., Brkljačić, L., Bogdanović, S., Farkaš, V., Cedilak, A., Nanić, L., Rubelj, I., & Salopek-Sondi, B. (2023). Antioxidant and antiproliferative activities of kale (Brassica oleracea L. Var. acephala DC.) and wild cabbage (Brassica incana Ten.) polyphenolic extracts. *Molecules*, *28*(4), 1840. https://doi.org/10.3390/MOLECULES28041840/S1

Malik, M. M., Malik, S., & Malik, I. (2025). Hypoglycemic and lipid-lowering effects of bitter melon (Momordica charantia) in type 2 diabetic patients. *Global Health Dynamics*, *1*(1), 20–24. https://doi.org/10.64229/P5WQ9942

Martínez-Romero, E., Aguirre-Noyola, J., Taco-Taype, N., Martínez-Romero, J., & Zúñiga-Dávila, D. (2020). Plant microbiota modified by plant domestication. *Systematic and Applied Microbiology*, *43*(5), 126106. https://doi.org/10.1016/j.syapm.2020.126106

Martín-Robles, N., García-Palacios, P., Rodríguez, M., Rico, D., Vigo, R., Sánchez-Moreno, S., De Deyn, G., & Milla, R. (2020). Crops and their wild progenitors recruit beneficial and detrimental soil biota in opposing ways. *Plant and Soil*, *456*, 159–173. https://doi.org/10.1007/s11104-020-04703-0

Matesanz, S., & Milla, R. (2018). Differential plasticity to water and nutrients between crops and their wild progenitors. *Environmental and Experimental Botany*, *145*, 54–63. https://doi.org/10.1016/j.envexpbot.2017.10.014

Medina-Lozano, I., Bertolín, J. R., & Díaz, A. (2021). Nutritional value of commercial and traditional lettuce (Lactuca sativa L.) and wild relatives: Vitamin C and anthocyanin content. *Food Chemistry*, *359*. https://doi.org/10.1016/j.foodchem.2021.129864

Meyer, R., Duval, A., & Jensen, H. (2012). Patterns and processes in crop domestication: An historical review and quantitative analysis of 203 global food crops. *The New Phytologist*, *196*(1), 29–48. https://doi.org/10.1111/j.1469-8137.2012.04253.x

Miłek, M., Marcinčáková, D., & Legáth, J. (2019). Polyphenols content, antioxidant activity, and cytotoxicity assessment of taraxacum officinale extracts prepared through the micelle-mediated extraction method. *Molecules*, *24*(6), 1025. https://doi.org/10.3390/MOLECULES24061025

Milla, R., & Matesanz, S. (2017). Growing larger with domestication: A matter of physiology, morphology or allocation? *Plant Biology*, *19*(3), 475–483. https://doi.org/10.1111/plb.12545

Mohammed Abdus Satter, M., Khan, M. M. R. L., Jabin, S. A., Abedin, N., Islam, M. F., & Shaha, B. (2016). Nutritional quality and safety aspects of wild vegetables consume in Bangladesh. *Asian Pacific Journal of Tropical Biomedicine*, *6*(2), 125–131. https://doi.org/10.1016/J.APJTB.2015.11.004

Mohd Din, A. R. J., Cheng, K. K., & Sarmidi, M. R. (2017). Assessment of compost extract on yield and phytochemical contents of Pak Choi (*Brassica rapa* cv. *chinensis*) grown under different fertilizer strategies. *Communications in Soil Science and Plant Analysis*, *48*(3), 274–284. https://doi.org/10.1080/00103624.2016.1269793

Moyo, M., Amoo, S. O., Aremu, A. O., Gruz, J., Šubrtová, M., Jarošová, M., Tarkowski, P., & Doležal, K. (2018). Determination of mineral constituents, phytochemicals and antioxidant qualities ofcleome gynandra, compared tobrassica oleracea and beta vulgaris. *Frontiers in Chemistry*, *5*. https://doi.org/10.3389/fchem.2017.00128

Mziray, R. S., Imungi, J. K., & Jaruri, E. G. (2000). Changes in ascorbic acid, beta-carotene and sensory properties in sundried and stored Amaranthus hybridus vegetables. *Ecology of Food Nutrition*, *39*(6). https://doi.org/10.1080/03670244.2000.9991630

Narayana, C. K. (2021). Phytochemicals in vegetables and their therapeutic properties. *Phytochemicals in Vegetables and Their Therapeutic Properties*. https://doi.org/10.1201/9781003245308

Ndlovu, M., Scheelbeek, P., Ngidi, M., & Mabhaudhi, T. (2024). Underutilized crops for diverse, resilient and healthy agri-food systems: A systematic review of sub-Saharan Africa. *Frontiers in Sustainable Food Systems*, *8*, 1498402. https://doi.org/10.3389/FSUFS.2024.1498402/BIBTEX

Ogle, B. M., Tuyet, H. T., Duyet, H. N., & Dung, N. N. X. (2003). Food, feed or medicine: The multiple functions of edible wild plants in Vietnam. *Economic Botany*, *57*(1), 103–117.

Ohiokpehai, O. (2003). Promoting the nutritional goodness of traditional food products. *Pakistan Journal of Nutrition*, *2*, 267–270. https://doi.org/10.3923/pjn.2003.267.270

Opazo-Navarrete, M., Burgos-Díaz, C., Soto-Cerda, B., Barahona, T., Anguita-Barrales, F., & Mosi-Roa, Y. (2021). Assessment of the nutritional value of traditional vegetables from southern chile as potential sources of natural ingredients. *Plant Foods for Human Nutrition (Dordrecht, Netherlands)*, *76*(4), 523. https://doi.org/10.1007/S11130-021-00935-2

Padulosi, S., Hodgkin, T., Williams, J. T., & Haq, N. (2002). Underutilized crops: Trends, challenges and opportunities in the 21st century. In *Managing Plant Genetic Diversity. Proceedings of an International Conference, Kuala Lumpur, Malaysia, 12–16 June 2000* (pp. 323–338). Wallingford: CABI Publishing.

Raju, M., Varakumar, S., Lakshminarayana, R., Krishnakantha, T. P., & Baskaran, V. (2007). Carotenoid composition and vitamin A activity of medicinally important green leafy vegetables. *Food Chemistry*, *101*(4). https://doi.org/10.1016/j.foodchem.2006.04.015

Rampelli, S., Pomstra, D., Barone, M., Fabbrini, M., Turroni, S., Candela, M., & Henry, A. G. (2025). Consumption of only wild foods induces large scale, partially persistent alterations to the gut microbiome. *Scientific Reports*, *15*(1), 1–13. https://doi.org/10.1038/S41598-025-00319-5;SUBJMETA

Rao, M. J., & Zheng, B. (2025). The role of polyphenols in abiotic stress tolerance and their antioxidant properties to scavenge reactive oxygen species and free radicals. *Antioxidants*, *14*(1), 74. https://doi.org/10.3390/ANTIOX14010074

Rjeibi, I., Ben Saad, A., Ncib, S., Souid, S., & Alimi, H. (2017). Characterization of Amaranthus spinosus collected from different regions: Phytochemical and biological properties. *Journal of Food Biochemistry*, *41*(5). https://doi.org/10.1111/jfbc.12397

Rokka, S., El Bilali, H., Borelli, T., Calabrese, G. J., Lecci, S., Otieno, G., Tietiambou, F. R. S., Bteich, M. R., Grazioli, F., Kabre, B., Seindira, M., Karambiri, S., Nanema, J., Bakasso, Y., Guimbo, I. D., Dambo, L., Nouhou, B., Hima, H. M., Ouedraogo, N., . . . & Acasto, F. (2025). Current policies for promoting neglected and underutilized crop species in Burkina Faso and Niger. *International Journal of Agricultural Sustainability*, *23*(1). https://doi.org/10.1080/14735903.2025.2459541

Sánchez-Mata, M. C., Loera, R. D. C., Morales, P., Fernández-Ruiz, V., Cámara, M., Marqués, C. D., Pardo-de-Santayana, M., & Tardío, J. (2012). Wild vegetables of the Mediterranean area as valuable sources of bioactive compounds. *Genetic Resources and Crop Evolution*, *59*(3), 431–443. https://doi.org/10.1007/S10722-011-9693-6/TABLES/4

Sarker, U., & Oba, S. (2019). Nutraceuticals, antioxidant pigments, and phytochemicals in the leaves of Amaranthus spinosus and Amaranthus viridis weedy species. *Scientific Reports*, *9*(1). https://doi.org/10.1038/s41598-019-50977-5

Shackleton, C. M., & Shackleton, S. E. (2004). The importance of non-timber forest products in rural livelihood security and as safety nets: A review of evidence from South Africa. *South African Journal of Science*, *100*, 658–664. Available online: https://core.ac.uk/download/pdf/145031088.pdf

Shackleton, C. M., Shackleton, S. E., Buiten, E., & Bird, N. (2007). The importance of dry woodlands and forests in rural livelihoods and poverty alleviation in South Africa. *Forest Policy and Economics*, *9*, 558–577. https://doi.org/10.1016/j.forpol.2006.03.004

Shi, S., Chang, J., Tian, L., Nasir, F., Ji, L., Li, X., & Tian, C. (2019). Comparative analysis of the rhizomicrobiome of the wild versus cultivated crop: insights from rice and soybean. *Archives of Microbiology*, *201*, 879–888. https://doi.org/10.1007/s00203-019-01638-8

Singh, J., & Van Der Knaap, E. (2022). Unintended consequences of plant domestication. *Plant and Cell Physiology*, *63*(11), 1573–1583. https://doi.org/10.1093/PCP/PCAC083

Smýkal, P., Nelson, M. N., Berger, J. D., & Von Wettberg, E. J. B. (2018). The impact of genetic changes during crop domestication. *Agronomy*, *8*(7), 119. https://doi.org/10.3390/AGRONOMY8070119

Šola, I., Poljuha, D., Pavičić, I., Jurinjak Tušek, A., & Šamec, D. (2025). Climate change and plant foods: The influence of environmental stressors on plant metabolites and future food sources. *Foods*, *14*(3), 416. https://doi.org/10.3390/FOODS14030416

Stanciu, G., Rotariu, R., Popescu, A., & Tomescu, A. (2019). Phenolic and mineral composition of wild chicory grown in Romania. *Revista de Chimie*, *70*(4). https://doi.org/10.37358/rc.19.4.7087

Suliman, A. A., Elkhawaga, F. A., Zargar, M., Bayat, M., Pakina, E., & Abdelkader, M. (2024). Boosting resilience and efficiency of tomato fields to heat stress tolerance using cytokinin (6-benzylaminopurine). *Horticulturae*, *10*(2), 170.

Szeto, Y. T., Tomlinson, B., & Benzie, I. F. (2002). Total antioxidant and ascorbic acid content of fresh fruits and vegetables: Implications for dietary planning and food preservation. *British Journal of Nutrition*, *87*(1), 55–59.

Tawfik, E., Ahmed, M. F., Albalawi, D. A., Aljuaid, B. S., Darwish, D. B. E., Mahmoud, S. F., Hassan, K. M., Ibrahim, M. F., & Abdel Razik, A. B. (2022). Molecular identification of zantedeschia culture with determination of its morphometric and metabolic activities for Mediterranean acclimatization. *Plants*, *11*(17), 2311. https://doi.org/10.3390/plants11172311

Thapa, M., Liu, L., Barkla, B. J., Kretzschmar, T., Rogiers, S. Y., & Rose, T. J. (2024). Accumulation patterns of anthocyanin and γ-oryzanol during black rice grain development. *PLoS ONE*, *19*(5), e0302745. https://doi.org/10.1371/JOURNAL.PONE.0302745

Tosun, M., Ercisli, S., Ozer, H., Turan, M., Polat, T., Ozturk, E., Padem, H., & Kilicgun, H. (2012). Chemical composition and antioxidant activity of foxtail lily (*Eremurus spectabilis*). *Acta Scientiarum Polonorum. Hortorum Cultus*, *11*, 145–153.

Trichopoulou, A., Naska, A., & Vasilopoulou, E. (2001). Guidelines for the intake of vegetables and fruit: The Mediterranean approach. *International Journal for Vitamin and Nutrition Research*, *71*(3), 149–153.

Uusiku, N. P., Oelofse, A., Duodu, K. G., Bester, M. J., & Faber, M. (2010). Nutritional value of leafy vegetables of sub-Saharan Africa and their potential contribution to human health: A review. *Journal of Food Composition and Analysis*, *23*(6), 499–509. https://doi.org/10.1016/J.JFCA.2010.05.002

Walingo, K. M. (2005). Role of vitamin C (ascorbic acid) on human health-a review. *African Journal of Food, Agriculture, Nutrition and Development*, *5*(1).

Wardah, Sambas, E. N., Ridwan, & Ariani, D. (2017). Starch product of wild plants species Jalawure (*Tacca leontopetaloides* L.) Kuntze as the source of food security in the south coastal West Java. *IOP Conference Series: Materials Science and Engineering*, *193*, 012035.

Weinberger, K., & Pichop, G. N. (2009). Marketing of African indigenous vegetables along urban and peri-urban supply chains in sub-Saharan Africa. In *African Indigenous Vegetables in Urban Agriculture*, eds. C. Shackleton, A. Drescher, and M. Pasquini (London: Earthscan), 225–244.

Yang, R. Y., & Keding, G. B. (2009). Nutritional contributions of important African indigenous vegetables. In *African Indigenous Vegetables in Urban Agriculture*, eds. C. Shackleton, A. Drescher, and M. Pasquini (London: Earthscan), 105–143.

Yilmaz, H., & Yilmaz, A. (2025). Hidden hunger in the age of abundance: The nutritional pitfalls of modern staple crops. *Food Science & Nutrition*, *13*(2), e4610. https://doi.org/10.1002/FSN3.4610

16 From Conventional Science to Industrial Processing

Cooking Methods of Vegetables for Various Health Benefits

Loredana Dumitrașcu and Iuliana Aprodu

16.1 INTRODUCTION

Although all surveys regarding food consumption indicate that 98–99% of the population consume meat, in the last years, an increasing interest in diets low in animal-based products and ultra-processed foods was noticed (Kent et al., 2022). As a result, the dietary patterns that primarily include different types of vegetables, fruits, legumes, whole grains nuts, and seeds, specific to vegans, lacto-ovo vegetarians or pescatarians, became more common (Esquivel, 2022). The increasing popularity of the plant-rich diets was assigned to the potential health-related benefits and the low impact on the environment (Alcorta et al., 2021).

16.1.1 Classification of Vegetables

Vegetables are the edible parts of the plants, which, unlike fruits, are normally eaten as salads, appetizers, or part of the main course of a meal. Considering the plant tissues and organs consumed, Oke and Paliyath (2006) classified the vegetables into three groups: earth vegetables including roots (sweet potatoes, carrots, radish, turnip) tubers (potatoes), corns (taro), and bulbs (onions, garlic, scallion); herbage vegetables including leaves (cabbage, spinach, beet greens, lettuce, mustard greens, parsley, romaine), stalks (celery, rhubarb), flower buds (cauliflower, broccoli, artichokes, mushrooms), sprouts and shoots (asparagus, bamboo shoots); and fruit vegetables including legumes (peas and beans), cereals (sweet corn), different types of vines (cucumber, pumpkin, squash, zucchini), berry (tomato, eggplant), and tree (avocado, breadfruit) fruits (Figure 16.1). Because of their insignificant contribution to the overall nutritional components, herbs and spices are not included in this classification.

The vegetables are also classified for dietary guidance purpose, helping the people to easily select the items that meet their needs. The groups and subgroups of vegetables defined in the food guides of many countries vary also depending on the availability (Pennington & Fisher, 2009).

Considering the nutrient composition, including the macronutrients, 9 vitamins (vitamin A, vitamin E, vitamin C, thiamine, riboflavin, niacin, vitamin B6, folate, and vitamin B12) and 8 minerals (calcium, phosphorus, magnesium, iron, zinc, copper, sodium, and potassium), the most consumed 36 vegetables were classified, in the MyPyramid, into 5 subgroups (Marcoe et al., 2006; Pennington & Fisher, 2009):

- dark-green vegetables including broccoli, romaine, spinach, collard greens, mustard greens, kale, and turnip greens; are good sources of calcium, potassium, magnesium, fibre, vitamin A, vitamin C, and vitamin E; out of the vegetables of this group, broccoli is the most consumed (37.7%);

DOI: 10.1201/9781003606710-16

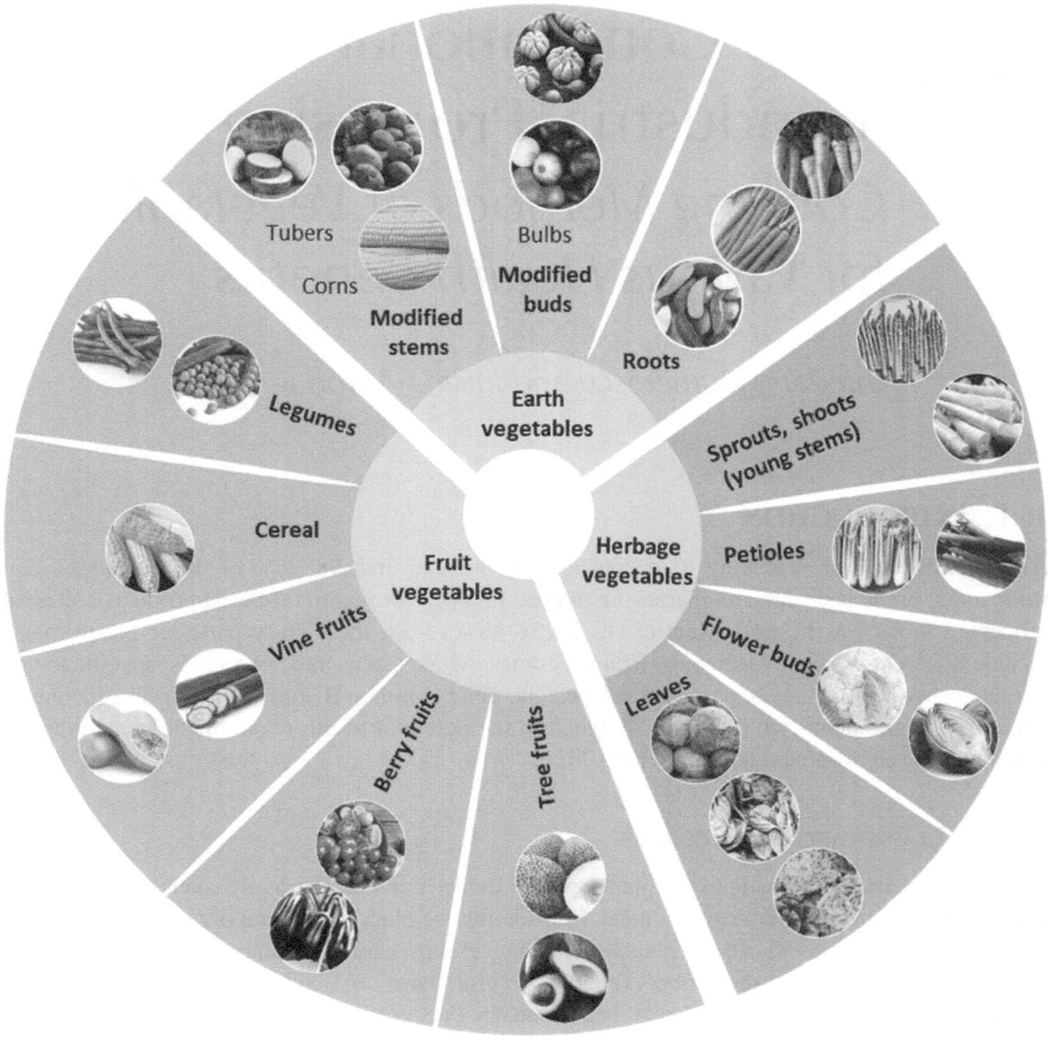

FIGURE 16.1 Classification of vegetables based on the edible parts.

- orange vegetables including carrots, sweet potato, winter squash, and pumpkin; are good sources of fibre, vitamin A, and vitamin E;
- dry beans and peas including pinto beans, white beans, soybeans, kidney beans, black beans, lentils, chickpeas, lima beans, cowpeas, and split peas; are good sources of calcium, potassium, magnesium, fibre, and vitamin E; pinto beans are the most consumed (44.3%);
- starchy vegetables including potato, corn, and green peas; are good sources of potassium; potatoes are the most consumed (83%);
- other vegetables including tomato, lettuce, cucumber, pepper, onion, celery, cabbage, green beans, mushroom, bean sprouts, summer squash, and cauliflower; onions are the most consumed (19.2%).

Pennington and Fisher (2009) indicated that vegetables belong to the following botanic families: Amaryllidaceae (garlic, leek, onion, scallion); Asteraceae/Compositae (artichoke, endive, Jerusalem artichoke, lettuce, romaine); Chenopodiaceae (beet, spinach); Convolvulaceae (sweet potato), Cruciferae/Brassica (broccoli, Brussels sprouts, cabbage, cauliflower, Chinese broccoli, Chinese cabbage, collards, kale, mustard greens, radish, rutabaga, turnip); Cucurbitaceae (squash, cucumber, pumpkin, zucchini); fungi (mushrooms); Gramineae/Poaceae (sweet corn); Lauraceae (avocado); Leguminosae (blackeye peas, green peas, jicama, kidney beans, lentils, lima beans, mung beans, navy beans, pigeon peas, pinto beans, green snap beans, snowpeas, soybeans); Liliaceae (asparagus); Malvaceae (okra), Portulacacea (rhubarb); Solanaceae (eggplant, sweet green pepper, sweet red pepper, potato, tomato); and Umbelliferae (carrot, celery, parsley, parsnip). This classification does not factor the consumed parts of the plants.

16.1.2 Chemical Composition of Vegetables

Vegetables have a central role in the plant-rich diets, providing important amounts of nutrients and phytochemicals with key roles in supporting the overall body functions and human health.

Various vegetables provide important amounts of macronutrients, dietary fibres, vitamins, and minerals. The composition of the legumes varies significantly with the cultivar and growth-related conditions, such as soil type, climate, and environmental factors (Fabbri & Crosby, 2016).

The major component of most vegetables is water, which accounts for over 70% of the fresh weight. The wet vegetables are therefore highly perishable compared to the dry crops like beans.

The protein content of the vegetables is usually lower than 3.5%, except for dried beans and peas (Oke & Paliyath, 2006; Fabbri & Crosby, 2016). In many cases, the plant proteins are deficient in one or several essential amino acids. Methionine is the limiting amino acid in soybeans, legumes have low amounts of methionine and cysteine, whereas grains are low in lysine (Oke & Paliyath, 2006; Mariotti & Gardner, 2019). In order to balance the amino acid profile, it is therefore highly recommended to combine different plant proteins in the diet. For instance a varied diet combining the leafy vegetables, legumes, and grains ensures the intake of all 20 amino acids, including the 9 essential amino acids.

The content of carbohydrates, including sugars and polysaccharides like starch, cellulose, hemicellulose, pectin, and lignin, varies significantly with the vegetable type, ranging from 2.8% in lettuce to 27% in sweet potatoes (Oke & Paliyath, 2006). Starch is the major carbohydrate especially in the roots and tubers, which are storage organs for plants, ensuring most of the energy in the diet. Corn, white potatoes, dried beans, and peas are grouped as starchy vegetables as they have high contents of starch (Slavin, 2013; Fabbri & Crosby, 2016). The dark green vegetables, on the other hand, have small or no starch, being rich in dietary fibres, whereas small sweet potatoes are rich in sucrose (Slavin, 2013). The leaves, stem, and stalk vegetables are usually good sources of soluble and insoluble dietary fibres (1–5.4%) (Septembre-Malaterre et al., 2018; Slavin, 2013), with important roles in preventing chronic diseases, including cardiovascular and inflammatory diseases, diabetes, obesity, and different types of cancer. Most beneficial health responses are mediated by the gut microbiota, which can ferment the vegetable fibres (Cui et al., 2019).

Lipids are minor components in vegetables, being mainly located in the cuticles and cell membranes or are part of the intracellular oleosomes. The lipid fractions include mono-, di-, and triacylglycerol; free fatty acids; free and esterified sterols; and phosphor- and glycolipids (Fabbri & Crosby, 2016). The essential fatty acids are among the most important components of the vegetable lipids. Vidrih et al. (2009) indicated contents of fatty acids in vegetables ranging from 0.5% in red cabbage to 4% in tarragon, the poly- and mono-unsaturated fatty acid accounting for 45–81% and 1–25% of the total content, respectively. The fatty acids profile of many vegetables meets the current nutritional recommendations regarding the n-6n/n-3 ratio, which should be lower than 10/1, in order to be effective in coronary heart disease prevention. Green vegetables like tarragon, courgette, lettuce, rucola, broccoli, mangold, kale, cucumber, radish sprouts, and French beans have a suitable n-6n/n-3 ratio. Tarragon, green lettuce, mangold, zucchini, and broccoli are particularly rich in α-linolenic acid (C18:3), which is a precursor of the long chain poly-unsaturated fatty acid (Vidrih et al., 2009). Shiitake, champignons, oyster mushroom, carrot, and fennel have high levels of C18:2; radish sprouts and chickpea have high levels of C18:1; while French beans, kale, and artichoke are rich in C16:1 (Vidrih et al., 2009). Phytosterols are structurally similar to cholesterol and are found in low amounts (51–370 ppm) in fresh vegetables like Brussels sprouts, cauliflower, broccoli, and dill (Piironen et al., 2003). Avocado, on the other hand, has been reported to have the highest amount of phytosterols of 752 ppm. Sitosterol is the predominant sterol in all vegetables (43–86%), except for cucumber and spinach, where Δ7-sterols were found to be more abundant (Piironen et al., 2003).

The level of vitamins in vegetables depends on the variety, anatomical parts or organs, maturity at harvesting, and growing practices (Septembre-Malaterre et al., 2018). The dark green leaves, which host the most intense metabolic activity of the plant, accumulate important amounts of carotenoids (beta-carotene, alpha-carotene, lutein, zeaxanthin), vitamin C, vitamin K, folate, and flavonoids, whereas the orange legumes' fruits, roots, and tubers accumulate alpha-carotene and beta-carotene (Pennington & Fisher, 2009; Pennington & Fisher, 2010). Several examples of vegetables which provide over 50% of the dietary reference intake (DRI) for vitamin C are the dark green leafy vegetables like spinach, tomatoes, and cabbages (broccoli, cabbage, cauliflower, Brussels sprouts, Chinese broccoli, and Chinese cabbage), while the vegetables providing 50% of the DRI for vitamin K are lettuce, spinach, cabbages, and bulbs from Allium family, namely garlic, leek, onion, scallion (Pennington & Fisher, 2010; Septembre-Malaterre et al., 2018). Peas and beans are rich in water-soluble vitamins like thiamine (B1), riboflavin (B2), and niacin (Fabbri & Crosby, 2016). In particular, the legumes and dark green leafy vegetables provide over 25% of the DRI for folate, whereas the vitamin B6 is mainly found in dark green leafy vegetables, Allium family bulbs, cabbages, and legumes, which provide over 10% of the DRI (Pennington & Fisher, 2009; Pennington & Fisher, 2010; Septembre-Malaterre et al., 2018).

Vegetables are also good sources of macrominerals and trace elements. The dark green leafy vegetables are rich in iron (10% DRI), copper (10% DRI), manganese (10% DRI), potassium, calcium, magnesium, and zinc; cabbages and lettuce are good sources of manganese; legumes provide copper (25% DRI), iron (10% DRI), zinc (10% DRI), manganese (10% DRI), magnesium, calcium, phosphorus, potassium, sodium, cobalt, sulphur, and fluorine (Pennington & Fisher, 2010; Fabbri & Crosby, 2016). Anyway, because of the high contents of fibres and phytates and oxalates in the vegetables, the intestinal absorption of certain minerals might be limited (Oke & Paliyath, 2006).

In addition to the macro- and micronutrients, vegetables also provide important amounts of phytochemicals with protective roles against various chronic diseases, such as cardiovascular or cancer, owing to their antioxidant properties. Among the compounds with antioxidant activity reported in various vegetables, of particular importance are various vitamins, carotenoids, phenolic acids, flavonoids (flavanones, flavones, flavonols, and anthocyanins), and other polyphenols (Pennington & Fisher, 2010; Fabbri & Crosby, 2016; Septembre-Malaterre et al., 2018). The biological activity and bioavailability of these compounds depend on the chemical structure. Legumes and bulbs from Allium family accumulate the highest concentrations of flavonols among vegetables, whereas the dark green leafy vegetables are rich in bioactive compounds with antioxidant activity like beta-carotene, lutein, zeaxanthin, flavones, and flavonols. The most important anthocyanidins providers among vegetables are tomatoes and other red vegetables, followed by vegetables from cabbage family (Pennington & Fisher, 2010). Moreover, glucosinolates and isothiocyanates were reported to exert anticancer activity (Verkerk et al., 2009).

Some vegetables also contain anti-nutrients at a level which can hinder the optimum utilization of various nutrients from vegetables or even cause poisoning. Therefore, it is crucially important to apply proper preparation methods, such as thermal treatment and removal of certain parts of the plants, which effectively reduce the anti-nutrient levels. Several examples of anti-nutrients found in vegetables are oxalates found in high amounts in greed leafy vegetables; alkaloids (solanine in potatoes), nitrites; phytates, tannins, and saponins identified in Leguiminosae family; and protease inhibitors in peas (Fabbri & Crosby, 2016; Lo et al., 2018).

16.2 HEALTH-RELATED BENEFITS OF THE PLANT-BASED DIETS

There are important differences between the plant-based and conventional omnivore diets in terms of energy values, macro- and micronutrients intake, and presence of diverse phytochemicals with important physiological functions. Vegetables provide important amounts of essential nutrients and bioactive compounds, contributing to the health status of the individuals. The main health benefits associated with vegetables intake are mainly attributed to their high contents of dietary fibres, vitamins, minerals, and phytochemicals (Behera et al., 2024). The health benefits should not be linked to a specific compound. Therefore, balancing the diet through integration of a large variety of vegetables is of high importance in order to ensure the existence of a wide range of nutrients and phytochemicals.

The most frequently indicated health benefits associated with the plant-based diets consist of the improvement of metabolic parameters which indicate a reduced risk of T2DM, obesity-related dysfunctions, vascular disease different types of cancers, and chronic low-grade inflammation (Medawar et al., 2019; Kent et al., 2022; Esquivel, 2022). The main pathways through which the plant-based diets improve the inflammatory status in humans, help better control of the blood sugar level, and affect neurotransmitter metabolism are by direct dietary intake or by positively influencing the gut activity (Medawar et al., 2019).

Because of the high dietary fibre intake, the plant-based diets might help preventing the coronary diseases and many types of cancer, support the digestive health by balancing the gut microbiome, lower the risk of diabetes through regulating the glycaemic response, and help with weight management (Fabbri & Crosby, 2016; Behera et al., 2024). The balanced gut microbiome influences the brain functions, being further linked to healthy ageing (Medawar et al., 2019). Storey and Anderson (2014) found positive effects on heart health when consuming 14 g dietary fibre per 1,000 kcal in the diet. A review study of Medawar et al. (2019) on the beneficial impact of plant-based diet revealed weight loss and important metabolic improvements both on healthy individuals and on people dealing with T2DM and obesity. The plant-based diets are usually low in calories, explaining the weight loss effect. Regarding the improvement of the metabolic status of the individuals participating in various studies, the plant-based diet ensured important reduction of the glycated haemoglobin, low-density (LDL), high-density lipoproteins (HDL), and total cholesterol (TC) levels (Medawar et al., 2019).

Only several examples of vegetables ensuring unique health benefits are presented here. The dark green leafy vegetables include high amounts of vitamins A, C, K, and folate, which are vital for energy metabolism, immune system, vision functions, and bone health (Behera et al., 2024). As previously detailed, vegetables are rich in bioactive compounds with antioxidant, anti-inflammatory, and anti-carcinogenic properties. Because of the high contents of phytochemicals, the consumption of green leafy vegetables was also linked to reduced inflammatory activity and cardiovascular health (Behera et al., 2024).

Tomatoes and other red vegetables like red peppers, beetroots, and red potatoes contain high amounts of lycopene, ellagic acid, quercetin, hesperidin, etc., and were reported to have good scavenging activity of the harmful free radicals, efficiently lower the blood pressure and LDL cholesterol levels, and to reduce the risk of cancer (Behera et al., 2024).

Unlike other vegetables, cruciferous vegetables, such as broccoli, Brussels sprouts, cauliflower, cabbage, and kale, contain glucosinolates and isothiocyanates and have preventive properties against lung, gastrointestinal, prostate, breast, and chemically induced cancers (Behera et al., 2024).

A major drawback of the plant-based diets consists in the risk of having specific nutrient deficiencies. The restrictive plant-based diets might result in deficiency of vitamin B12 and vitamin D, which are naturally found in the animal-derived products (Medawar et al., 2019; Kent et al., 2022). The low intake of vitamin B12 was linked to neurological and cognitive health depreciation (Medawar et al., 2019).

Although many clinical trials have been completed, Medawar et al. (2019) reported that there is inconclusive evidence regarding the link between the plant-based diet and the cognitive and mental effects.

16.3 DOMESTIC COOKING AND PROCESSING OF VEGETABLES

Vegetables can be eaten in many ways and, depending on consumer preferences, can be consumed raw or dried, cooked in domestic space, or processed at industrial level. A preliminary step in cooking or processing vegetables is represented by the preparation method. Frequently, this step includes trimming, washing, soaking, peeling, cutting, chopping, or dicing (Fabbri & Crosby, 2016). Preparation methods applied to vegetables are depicted in Figure 16.2. Some preparation methods such as washing do not alter the bioavailability of the bioactive compounds, whereas operations like cutting, peeling, or chopping can decrease the phytochemical content (Fabbri & Crosby, 2016). Most of the preparation methods, except for washing, change the texture of the vegetables, which results in the activation of enzymatic and non-enzymatic reactions. Methods like cutting damage the onion tissue which promotes the release of an enzyme called allinase that in contact with sulphur compounds increases the flavour and pungency (Fabbri & Crosby, 2016). Soaking and peeling of vegetables containing chemical substances used for controlling vegetable diseases represent an efficient way of minimizing the risk exposure and protecting human health (Chang et al., 2023). Soaking of peas and beans prior to cooking allows reducing the cooking time and the anti-nutrients' content.

Cooking vegetables at household level is known as domestic cooking, while cooking at industrial level is known as processing. The most common domestic cooking methods applied for vegetables are presented in Table 16.1. All of them are based on heat treatment in the forms of steaming, boiling, frying, or baking with the aim of improving the sensory attributes of vegetables and assure food safety (Zhao et al., 2019). At industrial level, processing of vegetables can be performed by using conventional or/and novel technologies (Table 16.2). Most of the technologies have the main purpose of assuring the food safety of the vegetable products while minimizing the impact on the nutrients and texture. The extent to which these technologies exert a negative impact on health-promoting components is dependent on the applied technology and operational conditions.

Blanching uses steaming or boiling vegetables for a short period of time followed by rapid cooling. The method reduces the activity of some quality deterioration enzymes such as peroxidase and PPO and the microbial load, while preserving or enhancing the nutritional properties of vegetables (Xiao et al., 2017).

Conventional technologies such as pasteurization contribute to increasing the preservation of the vegetable products by eliminating the vegetative spoilage microorganisms and non-spore-forming bacteria (Srivastava & Sit, 2025). Sterilization is applied to vegetable foods with a prolonged shelf life. However, it is well known that these processing technologies reduce the content of health-promoting compounds like polyphenols, carotenoids, vitamins, or minerals.

TABLE 16.1
Domestic Cooking Methods Most Used for Vegetables

Method	Principle
Boiling	Cooking in boiling liquid
Frying	Cooking in hot oil
Roasting	Cooking in a pan in a hot oven
Simmering	Cooking in liquid below the boiling point
Sauteing	Cooking in a thin layer of hot oil
Steaming	Cooking vegetables in a suspended basket over simmering liquid in a covered pot
Stewing	Sauteing followed by simmering
Microwaving	Cooking using microwaves
Pressure cooking	Cooking with a pressure cooker
Sous-vide	Cooking under vacuum using specific plastic pouches, at a set temperature

FIGURE 16.2 Examples of preparation methods applied to vegetables.

Unlike pasteurization and sterilization, freezing uses low temperatures which reduce the metabolic processes, microbiological load, and qualitative degradation reactions (Neri et al., 2020). Although freezing pretreatments cause the damage of cell structure with negative consequences on the quality and nutrient profile of vegetables, some studies indicate that freezing could exert a positive impact on the quality and functionality, as the frozen state promotes the release of anthocyanins and phenolic acids. The application of blanching prior to freezing and the use of cryoprotectants are common strategies applied to preserve the quality attributes of vegetables during freezing (Neri et al., 2020).

Drying extends the shelf life of the vegetable material by reducing the water activity, which, on the other hand, causes physical, biological, and chemical changes with impact on the quality of the dried product (Oliveira et al., 2016).

Frying is used in both domestic and industrial conditions to prepare or preserve vegetable meals with superior sensory attributes (Srivastava & Sit, 2025). On the other hand, the high temperature applied during frying has a negative effect on many thermolabile compounds. This limitation, however, could be mitigated by applying pre-frying procedures like drying, vacuum frying, ultrasound, microwave, or high-pressure processing (Srivastava & Sit, 2025).

Numerous innovative technologies have been promoted over the last years due to their ability of obtaining safe and high-quality products with prolonged shelf life in a sustainable manner. The most innovative technologies that can be applied for processing vegetables are high pressure processing, ohmic heating, pulsed electric fields, microwaves, ultrasound, infrared heating, pulsed light, radio frequency, ionizing radiation, and ozone treatment (Table 16.2). Some of these technologies although are producing heat, the level is below the threshold of conventional pasteurization, making them attractive for obtaining safe and 'fresh like' vegetables products with health-promoting properties (Srivastava & Sit, 2025).

TABLE 16.2
Conventional and Novel Processing Methods for Vegetables

Method	Principle	Reference
	Conventional	
Blanching	Adding moisture at 70–100°C, for a short time	Srivastava and Sit (2025)
Pasteurization	Heat treatment at 65–90°C, few seconds to 30 minutes	Srivastava and Sit (2025)
Sterilization/Canning	Flash heating over 100°C, in a container	Srivastava and Sit (2025)
Drying	Reducing the water content using heat treatment	Srivastava and Sit (2025)
Freezing	Phase transition through heat removal	Srivastava and Sit (2025)
Frying	Rapid dehydration using hot oil	Srivastava and Sit (2025)
	Novel	
Pulsed electric fields	Electric fields of high intensity (5–50 kV/cm) create transmembrane potential, leading to electroporation ensuring irreversible inhibition of vegetative of microbial cells	Srivastava and Sit (2025)
High-pressure processing	Hydrostatic pressure (100 to 1,000 MPa) for short time at 20–60°C improves vegetables preservation	Denoya et al. (2021)
Ohmic heating	Alternative electric current through two electrodes to semi-conductive food materials	Srivastava and Sit (2025)
Ultrasound	Ultrasound generates acoustic cavitation and streaming that promote mass transfer and microbial inactivation	Lepaus et al. (2023)
Microwave	Conversion of electromagnetic energy into heat energy in a very short time	Chen et al. (2024)
Radio frequency	Radio frequency (10–50 MHz) generates electromagnetic heating	Altemimi et al. (2019)
Pulsed light	The use of light energy in concentrated form (0.01–50 J/cm^2) for a short time	Birwal et al. (2019)
Ultraviolet light	The use of UV light (100–400 nm) has germicidal effects	Koutchma (2009)
Infrared frequency	The use of electromagnetic spectrum between 0.75 and 1,000 μm	Aboud et al. (2019)
Ionization radiation	The use of ions of beta or gamma rays for microbial inactivation	Munir and Federighi (2020)
Ozone treatment	Ozone generates hydroxyl radicals that improve vegetables' preservation	Glowacz and Rees (2016)

Electric fields processing, like ohmic heating and pulsed electric fields, are green technologies that apply alternative electric current through two electrodes to semi-conductive food materials. The electric fields processing accelerates the inactivation of bacteria and enzymes, therefore enabling the use of lower processing temperatures than conventional heating (Ramaswamy et al., 2014). In foods with high conductivity, ohmic heating could induce overheating. The best results in terms of quality and efficiency are obtained by combining ohmic heating with blanching, fermentation, dehydration, and pasteurization/sterilization (Ramaswamy et al., 2014). The application of pulsed electric fields enhances the extraction of bioactive compounds, juice yield, and the shelf life of the vegetable products, being considered an excellent technology for both preserving the health-promoting components and assuring the safety of the products.

High pressure processing is a non-thermal technology known as cold pasteurization which subjects liquid and solid vegetables to pressures between 400 and 600 MPa at temperatures below 45°C, being capable of preserving the sensorial and nutritional value, while inactivating most of the pathogens and vegetative spoilage organisms (EFSA BIOHAZ Panel, 2022).

Ultrasound processing uses acoustic waves at frequencies higher than 20 kHz, providing enhanced quality and retention of vegetable characteristics in terms of texture, nutritional value, and organoleptic properties (Bhargava et al., 2021).

Pulsed light processing uses wavelengths from ultraviolet (100 nm) to near infrared (1100 nm), being one of the modern methods applied for disinfecting the surface of vegetables (Srivastava & Sit, 2025).

Plasma technology can be applied for preserving the quality and assuring the safety of vegetables. This effect is correlated with the limited penetration depth of the active substance plasma (Zeng et al., 2025).

Microwaves technology uses electromagnetic waves between 915 MHz and 2,450 MHz; capable of inner penetration and faster heating; and being used in drying, blanching, thawing, or baking of vegetables. Compared to conventional technologies, it has the potential to better preserve the health-promoting components (Xiao et al., 2017).

Many studies have shown that combining conventional with novel technologies represents a promising alternative for minimizing the nutrients loss, reduction of antinutritional factors, or enhancing the nutritional profile of the vegetable products (Oliveira et al., 2016).

16.4 EFFECTS OF DOMESTIC COOKING AND INDUSTRIAL PROCESSING ON NUTRITIONAL PROPERTIES OF VEGETABLES

16.4.1 Minimization of Nutrients Loss

Cooking methods and conditions play an essential role in the preservation of vegetable nutrients. Domestic cooking methods which use severe temperatures and prolonged cooking time cause the loss of most valuable nutrients. However, in many cases, applying mild cooking procedures could minimize the nutrients' loss (Table 16.3). Steam blanching of carrots, sweet potatoes, and collard greens followed by drying minimized the β-carotene loss between 54.8 and 98.1% (Mdziniso et al., 2006). Oloyede et al. (2021) tested the effect of three domestic cooking methods (steaming, microwave, and stir frying) on the glucosinolates content of cabbage. Steaming for 2 min retained about 97% of glucosinolates, while in stir frying, the amount of glucosinolates decreased by about 70%. Usually, it is assumed that fresh vegetables have a higher antioxidant activity compared to frozen vegetables (Danesi & Bordoni, 2008). Danesi and Bordoni (2008) tested the antioxidant activity of fresh and frozen vegetables (carrots, zucchini, tomatoes, green beans, peas, and yellow pepper) cooked at domestic level by steaming, boiling, and microwave. Cooking decreased more the antioxidant activity of fresh green vegetables than of those in frozen form, whereas for yellow/red vegetables, an opposite effect during cooking on antioxidant activity was measured. Among the tested cooking methods, boiling caused the highest loss of antioxidant activity.

At the industrial level, pressure-assisted thermal sterilization (preheating at 98 °C for 5 min and pressurizing at 600 MPa, 90 °C for 5 min) applied to pumpkin, butternut squash, beetroot, and purple potatoes reduced the vitamin C loss between 3 and 14%, while chlorophyll and β-carotene did not show important changes (Al-Ghamdi et al., 2020). Microwave-assisted thermal sterilization was reported to provide a higher level of nutrients' retention and stability compared with conventional thermal processing. For example Albahr et al. (2024) showed that microwave-assisted thermal sterilization was able to retain by about 74% of anthocyanins in cabbage puree, 80% of betalains red beetroot, and up to 100% of β-carotene in carrot puree. Used in combination with packages with high gas barriers, microwave-assisted thermal sterilization can be applied to preserve vegetable nutrients for prolonged storage.

Ultrasound osmotic dehydration used as pretreatment prior to freezing was able to provide a higher retention of ascorbic acid in radishes compared with osmotic dehydration (Xu et al., 2014). The effect was correlated with the ultrasound ability of reducing osmotic dehydration time and drip loss. Patras et al. (2011) tested the impact of blast freezing and blanching on the antioxidant activity,

TABLE 16.3
Examples of Cooking/Processing Methods That Minimize the Nutrient Loss of Vegetables

Cooking/ Processing Method	Conditions	Vegetable	Effect	Reference
Blanching and drying	Steaming for 1.5–3 min	• Carrots • Sweet potatoes • Collard greens	Minimization of β-carotene loss between 54.8 and 98.1%	Mdziniso et al. (2006)
Domestic cooking: Stir frying Microwave Steaming	• Stir frying: 90 s in olive oil at 100°C • Microwave treatment: 900 W, 60% power, 3 min • Steaming: Hot steam for 2 min	• Cabbage	Mild steaming enabled the highest retention of glucosinolates.	Oloyede et al. (2021)
Domestic cooking: Steaming Boiling Microwave	• Steaming: 30 min • Boiling for 20 min in salty water • Microwave treatment: 700 W, 5 min	• Carrots, • Zucchini, • Tomatoes, • Green beans • Peas • Yellow pepper	The antioxidant activity of zucchini, peas, and green beans was higher when cooked in frozen form and lower for tomatoes, zucchini and yellow pepper.	Danesi and Bordoni (2008)
Blanching and freezing	• Immersion in water at 95°C for 3 min. • Blast freezing at −30°C for 3 min.	• Broccoli • Carrots	Blanching was able to totally maintain vitamin C than unblanched variants.	Patras et al. (2011)
Pressure-assisted thermal sterilization (PATS)	• Preheating the purees at 98°C for 5 min and pressurizing at 600 MPa, 90 °C for 5 min	Pumpkin, butternut squash, pea, beetroot, and purple potato purees	Vitamin C showed 3–14% reduction. Total chlorophyll and β-carotene did not vary significantly.	Al-Ghamdi et al. (2020)
Microwave-assisted thermal sterilization	• 915 MHz, 124°C for • 3.12–3.8 s	• Carrot puree • Beetroot puree • Cabbage puree	Reduction of anthocyanins in cabbage puree by about 26%. β-Carotene of carrot puree was totally preserved. The content of betalains was reduced by 20%.	Albahr et al. (2024)
Ultrasound-assisted osmotic dehydration (UOD) and freezing	• 250 W, 30°C • Immersion freezing at −20°C	• Radish	Better ascorbic retention than osmotic dehydration	Xu et al. (2014)

phenols and ascorbic acid content of broccoli, carrots, and green beans. The application of blanching prior to freezing allowed preserving the antioxidant activity of the vegetables, due to enzymes' inactivation. Storage of blanched green beans for 14 weeks in a home-type freezer at −27°C did not result in the phenolic content and antioxidant activity decreasing; however, the vitamin C started to decrease after 7 weeks of storage (Neri et al., 2020).

16.4.2 Nutritional Enhancement

Domestic cooking and industrial processing can improve the nutritional attributes of vegetable products as summarized in Table 16.4. Steam cooking of broccoli is a friendly process for preserving the

TABLE 16.4
Examples of Nutritional Enhancement of Vegetables by Different Cooking/Processing Methods

Cooking/Processing Method	Conditions	Vegetable	Effect	Reference
Domestic Blanching	• Steam blanching, 10 min	Broccoli	• 90% increase of β-carotene • 52% increase of polyphenols • 23% increase of α-tocopherol; • 44% increase of γ-tocopherol • 25% increase of vitamin E	Gliszczynska-Swiglo et al. (2006)
Domestic blanching	• Hot steam, 2 min • Internal cabbage temperature 75–80°C	Cabbage	• Steaming increased the formation of isothiocyanates by 23-fold	Oloyede et al. (2021)
Boiling Steaming Microwaving	• Boiling: 100°C, 10 min • Steaming: hot steam, 10 min • Microwave: power 1,000 W, 10 min	Broccoli	*Boiling* • 63.8% increase of β-carotene • 88% increase of lutein • 30.4% increase of γ-tocopherol • 47% increase of α-tocopherol *Steaming* • 93.9% increase of β-carotene • 88.7% increase of lutein • 37.4% increase of γ-tocopherol • 91.3% increase of α-tocopherol *Microwaving* • 90.6% increase of β-carotene • 84.1 % increase of lutein • 45.3% increase of γ-tocopherol • 72.2% increase of α-tocopherol	Hwang and Kim (2013)
Blanching and sonication	• Hot water and acidified water at 100°C, 4 min • 70% amplitude, 2 min	Carrot juice	• 16% increase of chlorogenic acid • 47.36% increase of lycopene content • 19% increase of lutein content	Jabbar et al. (2014)
Ohmic heating	• 4–11 V/cm, 30 min, 0–100°C	Tomato by product	• 77% increase in the recovery of rutin	Coelho et al. (2019)
Ultrasound	• 60% amplitude for 10 min • 20% amplitude for 20 min	Tomatoes after post harvesting	• Increase of lycopene, total soluble polyphenols, and rutin after ultrasound treatment • Ultrasound promoted the biosynthesis of bioactive nutrients	Tovar-Pérez et al. (2020)
Fermentation	• 60 h at 37°C with *Lactobacillus plantarum* and *Lactobacillus pentosus*	Broccoli puree	• The phenolic content was 1.5-fold higher than in unfermented variant	Hou et al. (2023)
Fermentation	• 120 h at 37°C with *Lactobacillus plantarum*	Carrot juice	• Fermentation increased the antioxidant activity of the juice and flavour	Zhang et al. (2019)
Fermentation	• 72 h at 45°Cwith *Lactobacillus curvatus*	Red beetroot	• Fermentation improved the antioxidant activity and antidiabetic potential	Daliri et al. (2023)

health-promoting constituents of some vegetables. For example, Gliszczynska-Swiglo et al. (2006) studied the impact of water and steam blanching of domestically cooked broccoli on vitamin C, polyphenols, carotenoids, tocopherols, and glucosinolates content. Steam cooking maintained the initial vitamin C level and increased the levels of polyphenols, glucosinolates, β-carotene, lutein, and α- and γ-tocopherols compared to fresh broccoli. The increase was associated with enhanced availability of phytochemicals for extraction. Steaming of cabbage for 2 min resulted in the highest production of isothiocyanate, a compound responsible for inhibiting the development of cancer cells (Oloyede et al., 2021). When compared to boiling, steaming and microwaving increased to a greater extent the extraction of carotenoids and tocopherols of broccoli (Hwang & Kim, 2013). Combining blanching with ultrasound can improve the health-related nutrients of carrot juice as suggested by Jabbar et al. (2014). The authors reported that the application of blanching and ultrasound treatment showed a significant increase of total chlorogenic acid, carotenoids, lycopene, and lutein compared to control. The increase of chlorogenic acid was correlated with the release of bound form of chlorogenic acid due to cavitation induced during sonication which caused the disruption of cell wall. Ohmic heating of tomato by products caused the permeabilization of the cellular tissue promoting the recovery of bioactive compounds (Coelho et al., 2019).

The content of secondary metabolites of vegetables after post-harvest can be improved without impacting their quality. Tovar-Pérez et al. (2020) applied ultrasound treatment on tomatoes submerged in distilled water after postharvest and evaluated the quality and nutritional profile for a period of six days. Ultrasound treatment induced a significant increase of polyphenols, lycopene, and rutin in both the pulp and the skin at the beginning and after six days of post-harvest evaluation period.

Fortification is another strategy for achieving the nutritional enhancement of vegetables. Calcium fortification of eggplants, oyster mushroom, and carrot samples using vacuum impregnation was proposed as an alternative to developing functional vegetables (Gras et al., 2003). The fortification of carrot juice with sodium selenate was recently reported by Mezeyova et al. (2024). The average selenium content after pasteurization in fortified carrot juice was 11-fold higher than in unfortified sample. This form of selenium is thermostable, it bounds to proteins and can be used totally by the human body; thus, lower volumes of carrot juice are necessary to drink for achieving the daily recommended dose of selenium.

The nutritional profile of vegetables can be enhanced through biotechnological processes in which microorganisms and enzymes are involved in producing new metabolites. For example lactic acid bacteria fermentation represents one of the most applied technologies for maintaining and increasing the safety, sensorial, and nutritional properties of vegetables, while being cheap and environmentally friendly (Dissanayake et al., 2025). The fermentation of broccoli puree with two strains of *Lactobacillus* for 60 h doubled the phenolic content compared to raw broccoli (Hou et al., 2023), whereas the fermentation of red beetroot pulp with *Lactobacillus curvatus* promoted the formation of dihydromyricetin, which cannot be found in raw red beetroot (Daliri et al., 2023). The fermentation of carrot juice for 120 h at 37°C improved the antioxidant activity and sensorial properties (Zhang et al., 2019), while in pumpkin juice, the fermentation with *Lactobacillus paracasei* doubled the content of riboflavin compared with unfermented variant (Sun et al., 2022). Germination activates endogenous enzymes with excellent potential to boost the bioactivity of vegetables by producing new metabolites that provide protection against many diseases (Gunathunga et al., 2024).

16.4.3 Reduction of the Antinutritional Factors

Vegetables represent an important source of essential micro and macronutrients. On the other hand, there have been some health concerns as the nutritional profile of vegetables can be limited by the presence of antinutritional factors such as lectins, oxalates, phytates, phytoestrogens, and tannins (Petroski & Minich, 2020; Gunathunga et al., 2024; Ohanenye et al., 2020) (Figure 16.3). Lectins are

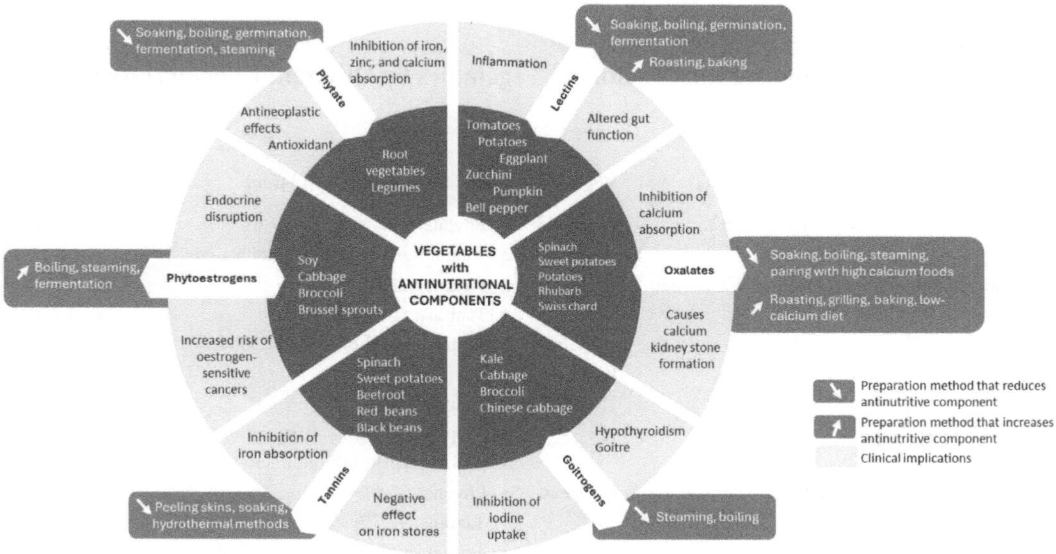

FIGURE 16.3 Antinutritive components of vegetables, health implications, and preparation methods which reduce or increase the antinutritive potential.

found in small amounts in vegetables, high concentrations being found in raw legumes. Bean lectins can be eliminated by soaking, autoclaving, and boiling for at least 35 min at 97°C (Patterson et al., 2016) (Table 16.5) and in lentils and peas by fermentation for 72 h and 24 h (Petroski & Minich, 2020; Patterson et al., 2016)). Oxalates can be found in spinach, sweet potatoes, Swiss chard, and rhubarb, with high concentrations being reported in leaves, while smaller content were reported in root vegetables (Petroski & Minich, 2020). As oxalate is water soluble, cooking methods like boiling and steaming could reducing oxalate content. Boiling of Swiss chard leaves, spinach, rhubarb, carrots, and broccoli was reported to be more efficient than steaming, whereas boiling of potatoes had a higher effect on reducing the oxalate content than baking and steaming (Chai & Liebman, 2005). Soaking followed by boiling can reduce the oxalate content of vegetables up to 77% (Akhtar et al., 2011). Glucosinolates are dietary goitrogens found mainly in vegetables of *Brassica* family which have the potential to prevent cancer; however, goitrin (produced from progoitrin) and thiocyanate (resulted from degradation of indol glucosinolate) could induce adverse effects of the thyroid. Thus, for population with iodine deficiency, methods like steaming, boiling, and microwaving can reduce the total glucosinolates content in vegetables like broccoli (Hwang & Kim, 2013). Phytoestrogens like isoflavones, lignans, stilbenes, and coumestrol are well known for their health effects. In vegetables, phytoestrogens are found in reduced amounts and in high amounts in soy; however, in some cases, isoflavones and other phytoestrogens could cause the growth of oestrogen-sensitive cancers (Petroski & Minich, 2020). Methods like boiling, steaming, pressure steaming, and fermentation with *Lactobacillus* and *Bifidobacterium* (Figure 16.3) can increase the content of beta-glucosides and aglycones, making them bioavailable (Petroski & Minich, 2020). Phytic acid or phytate is an anti-nutritional component which is found in high amounts in legumes like peas, lentils chickpeas, beans, and in negligible amounts in some root vegetables (Popova & Mihaylova, 2019). Phytic acid presence can be associated with the reduced bioavailability of iron, zinc, and calcium (López-Moreno et al., 2022). Methods such as soaking, boiling, germination, fermentation, and microwave are frequently employed to reduce the concentration of phytates (Table 16.5). Tannins are well recognized for the health-promoting benefits; yet, some studies indicated tannins' chelator potential of inhibiting the absorption of iron, copper, and zinc (Popova & Mihaylova, 2019).

TABLE 16.5
The Reduction of Antinutritional Factors of Vegetables by Different Cooking/Processing Methods

Cooking/ Processing Method	Conditions	Vegetable	Reduction Effect	Reference
Boiling	• 97°C, 35 min	Dry beans	• Phytate: 65% • Tannins and trypsin inhibitor activity: 70%	Patterson et al. (2016)
Pressure cooking	• 121°C, 15 psi, 30 min	Dry beans	• Phytate: 68% • Tannins: 50% • Trypsin-inhibitor activity: 100%	Patterson et al. (2016)
Soaking	• 12 h, 30°C	Kidney beans	• Phytate: 6% • Tannis: 24.2% • Trypsin-inhibitor activity: 5.48% • Chymotrypsin-inhibitor activity: 15.1%	Patterson et al. (2016)
Fermentation	• 24 h, room temperature	Dry peas	• Phytate: 67.8% • Tannins: 92.7% • Trypsin-inhibitor activity: 41.6%	Patterson et al. (2016)
Microwaves	• 1200 W, 15 min	Peas	• Phytate: 63.8% • Tannins: 83.6% • Trypsin-inhibitor activity: 100%	Patterson et al. (2016)
Soaking and ohmic heating	• Soaking: 12 h • 60 V, 110°C, 15 min	Black beans	• Trypsin-inhibitor activity: 57.44%	Lima-Becerra et al. (2021)
Domestic boiling, steaming, and microwaving	• Boiling: 100°C, 10 min • Steaming: hot steam, 10 min • Microwave: power 1,000 W, 10 min	Broccoli	• Total glucosinolates after boiling: 71.3% • Total glucosinolates after steaming: 67.7% • Total glucosinolates after microwaving: 64.64%	Hwang and Kim (2013)

Preparation methods like peeling and soaking combined with hydrothermal methods can reduce tannins content of vegetables, improving the bioavailability of proteins and minerals (Ojo, 2022).

16.5 CONCLUSIONS

Vegetables are an important reservoir of micro and micronutrients, attractive for their taste, colour, texture, and being an essential component for a healthy diet. Vegetable-based diets contribute to preventing several chronic diseases and to reduce micronutrient deficiencies. Although many vegetables are consumed fresh, frequently, they are cooked or processed to ensure the safety and increase their sensorial attractiveness and stability. Through cooking and processing, positive and negative changes in the nutritional profile occur. Therefore, it is important to know for consumers the most critical steps in cooking and processing that will optimize the nutrition quality of vegetables while ensuring food safety.

REFERENCES

Aboud, S. A., Altemimi, A. B., Al-HiIphy, R. S., Yi-Chen, L., & Cacciola, F. (2019). A comprehensive review on infrared heating applications in food processing. *Molecules*, *24*, 4125. https://doi.org/10.3390/molecules24224125

Akhtar, M. S., Israr, B., Bhatty, N., & Ali, A. (2011). Effect of cooking on soluble and insoluble oxalate contents in selected Pakistani vegetables and beans. *International Journal of Food Properties*, *14*(1), 241–249. https://doi.org/10.1080/10942910903326056

Albahr, Z., Promsorn, J., Tang, Z., Ganjyal, G. M., Tang, J., et al. (2024). Storage and thermal stability of selected vegetable purees processed with microwave-assisted thermal sterilization. *Food Innovation and Advances*, *3*(2), 99–110. https://doi.org/10.48130/fia-0024-0010.

Alcorta, A., Porta, A., Tárrega, A., Alvarez, M. D., & Vaquero, M. P. (2021). Foods for plant-based diets: Challenges and innovations. *Foods*, *10*(2), 293.

Al-Ghamdi, S., Sonar, C. R., Patel, J., Albahr, Z., & Sablani, S. S. (2020). High pressure-assisted thermal sterilization of low-acid fruit and vegetable purees: Microbial safety, nutrient, quality, and packaging evaluation. *Food Control*, *114*, 107233. https://doi.org/10.1016/j.foodcont.2020.107233

Altemimi, A., Aziz, S. N., Al-Hilphy, A. R. S., Lakhssassi, N., Watson, D. G., & Ibrahim, S. A. (2019). Critical review of radio-frequency (RF) heating applications in food processing. *Food Quality and Safety*, *3*(2), 81–91. https://doi.org/10.1093/fqsafe/fyz002

Behera, S. R., Pandey, R., Dhathri, M., & Bisht, A. (2024). Nutritional value and health benefits of vegetables. In: Jawadagi, R., Hadimani, H., Evoor, S., & Kishorkumar, G. K. (Eds.). *Vegetable Science Frontiers: Insights of Olericulture*. Stella International Publication, Haryana, India.

Bhargava, N., Mor, R. S., Kumar, K., & Sharanagat, V. S. (2021). Advances in application of ultrasound in food processing: A review. *Ultrasonics Sonochemistry*, *70*, 105293. https://doi.org/10.1016/j.ultsonch.2020.105293

Birwal, P., Deshmukh, G. P., & Ravindra, M. R. (2019). Nonthermal processing of dairy beverages. In: *Milk-based Beverages*. Elsevier Inc. https://doi.org/10.1016/B978-0-12-815504-2.00012-8

Chai, W., & Liebman, M. (2005). Effect of different cooking methods on vegetable oxalate content. *Journal of Agricultural and Food Chemistry*, *53*(8), 3027–3030. https://doi.org/10.1021/jf048128d

Chang, J., Dou, L., Ye, Y., & Zhang, K. (2023). Reduction in the residues of penthiopyrad in processed edible vegetables by various soaking treatments and health hazard evaluation in China. *Foods*, *12*, 892. https://doi.org/10.3390/foods1204089

Chen, F., Xu, B., Cui, W., Wang, Y., Wan, F., & Cheng, A. (2024). Effects of microwave treatment on vegetable oil quality & biological activities. *Trends in Food Science and Technology*, *153*(1), 104748. https://doi.org/10.1016/j.tifs.2024.104748

Coelho, M., Pereira, R., Rodrigues, A. S., Teixeira, J. A., & Pintado, M. E. (2019). Extraction of tomato by-products' bioactive compounds using ohmic technology. *Food and Bioproducts Processing*, *117*, 329–339. https://doi.org/10.1016/j.fbp.2019.08.005

Cui, J., Lian, Y., Zhao, C., Du, H., Han, Y., Gao, W., Xiao, H., & Zheng, J. (2019). Dietary fibers from fruits and vegetables and their health benefits via modulation of gut microbiota. *Comprehensive Reviews in Food Science and Food Safety*, *18*(5), 1514–1532.

Daliri, E. B.-M., Balnionytė, T., Stankevičiūtė, J., Lastauskienė, E., Meškys, R., & Burokas, A. (2023). High temperature lacto-fermentation improves antioxidant and antidiabetic potentials of Lithuanian red beet-root. *LWT*, *185*, Article 115122. https://doi.org/10.1016/j.lwt.2023.115122

Danesi, F., & Bordoni, A. (2008). Effect of home freezing and Italian style of cooking on antioxidant activity of edible vegetables. *Journal of Food Science*, *73*(6), 109–112. https://doi.org/10.1111/j.1750-3841.2008.00826.x

Denoya, G. I., Colletti, A. C., Vaudagna, S. R., & Polenta, G. A. (2021). Application of non-thermal technologies as a stress factor to increase the content of health-promoting compounds of minimally processed fruits and vegetables. *Current Opinion in Food Science*, *42*, 224–236. https://doi.org/10.1016/j.cofs.2021.06.008

Dissanayake, I. H., Tabassum, W., Alsherbiny, M., Chang, D., Li, C. G., & Bhuyan, D. J. (2025). Lactic acid bacterial fermentation as a biotransformation strategy to enhance the bioavailability of phenolic antioxidants in fruits and vegetables: A comprehensive review. *Food Research International*, *209*, 116283. https://doi.org/10.1016/j.foodres.2025.116283

EFSA Biohaz Panel (EFSA Panel on Biological Hazards), Koutsoumanis, K., Alvarez-Ordóñez, A., Bolton, D., Bover-Cid, S., Chemaly, M., Davies, R., De Cesare, A., Herman, L., Hilbert, F., Lindqvist, R., Nauta, M., Peixe, L., Ru, G., Simmons, M., Skandamis, P., Suffredini, E., Castle, L., Crotta, M., Grob, K., Milana, M. R., Petersen, A., Roig Sagués, A. X., Vinagre Silva, F., Barthélémy, E., Christodoulidou, A., Messens, W., & Allende, A. (2022). Scientific opinion on the efficacy and safety of high-pressure processing of food. *EFSA Journal*, *20*(3), 7128. https://doi.org/10.2903/j.efsa.2022.7128

Esquivel, M. K. (2022). Nutrition benefits and considerations for whole foods plant-based eating patterns. *American Journal of Lifestyle Medicine*, *16*(3), 284–290.

Fabbri, A. D. T., & Crosby, G. A. (2016). A review of the impact of preparation and cooking on the nutritional quality of vegetables and legumes. *International Journal of Gastronomy and Food Science*, *3*, 2–11. https://doi.org/10.1016/j.ijgfs.2015.11.001

Gliszczynska-Swiglo, A., Ciska, E., Pawlak-Lemanska, K., Chmielewski, J., Borkowski, T., Tyrakowska, B. (2006). Changes in the content of health promoting compounds and antioxidant activity of broccoli after domestic processing. *Food Additives & Contaminants*, *23*, 1088–1098.

Glowacz, M., & Rees, D. (2016). Exposure to ozone reduces postharvest quality loss in red and green chilli peppers. *Food Chemistry*, *210*, 305–310. https://doi.org/10.1016/j.foodchem.2016.04.119

Gras, M. L., Vidal, D., Betoret, N., Chiralt, A., & Fito, P. (2003). Calcium fortification of vegetables by vacuum impregnation: Interactions with cellular matrix. *Journal of Food Engineering*, *56*(2–3), 279–284. https://doi.org/10.1016/S0260-8774(02)00269-8

Gunathunga, C., Senanayake, S., Jayasinghe, M. A., Brennan, C. S., Truong, T., Marapana, U., & Chandrapala, J. (2024). Germination effects on nutritional quality: A comprehensive review of selected cereals and pulses changes. *Journal of Food Composition and Analysis*, *128*, 106024. https://doi.org/10.1016/j.jfca.2024.106024

Hou, F., Cai, Y., & Wang, J. (2023). Antioxidant capacity changes and untargeted metabolite profile of broccoli during lactic acid bacteria fermentation. *Fermentation*, *9*(5), 474. https://doi.org/10.3390/fermentation9050474

Hwang, E. S., & Kim, G. H. (2013). Effects of various heating methods on glucosinolate, carotenoid and tocopherol concentrations in broccoli. *International Journal of Food Sciences and Nutrition*, *64*(1), 103–111. https://doi.org/10.3109/09637486.2012.704904

Jabbar, S., Abid, M., Hu, B., Wu, T., Hashim, M. M., Lei, S., Zhu, X., & Zeng, X. (2014). Quality of carrot juice as influenced by blanching and sonication treatments. *LWT*, *55*(1), 16–21. https://doi.org/10.1016/j.lwt.2013.09.007

Kent, G., Kehoe, L., Flynn, A., & Walton, J. (2022). Plant-based diets: A review of the definitions and nutritional role in the adult diet. *Proceedings of the Nutrition Society*, *81*(1), 62–74.

Koutchma, T. (2009). *Advances in Ultraviolet Light Technology for Non-Thermal Processing of Liquid Foods*, 138–155. https://doi.org/10.1007/s11947-008-0178-3

Lepaus, B. M., Valiati, B. S., Machado, B. G., Domingos, M. M., Silva, M. N., Faria-Silva, L., Bernardes, P. C., Oliveira, D. da S., & de São José, J. F. B. (2023). Impact of ultrasound processing on the nutritional components of fruit and vegetable juices. *Trends in Food Science and Technology*, *138*(June), 752–765. https://doi.org/10.1016/j.tifs.2023.07.002

Lima-Becerra, I., María, B.-A., Dorantes-Campuzano, F., Mojica, L., Loarca-Piña, G., Morales-Sánchez, E., Ramírez-Jiménez, A. K., & Gaytán-Martínez, M. (2021). Ohmic heating as an emerging technology for the improvement of the techno-functional properties of common bean flour. *Biology and Life Sciences Forum*, *6*, 95. https://doi.org/10.3390/Foods2021-11000

Lo, D., Hsin-I, W., WanJen, W., & RayYu, Y. (2018). Anti-nutrient components and their concentrations in edible parts in vegetable families. *CABI Reviews*, *2018*, 1–30.

López-Moreno, M., Garcés-Rimón, M., & Miguel, M. (2022). Antinutrients: Lectins, goitrogens, phytates and oxalates, friends or foe? *Journal of Functional Foods*, *89*. https://doi.org/10.1016/j.jff.2022.104938

Marcoe, K., Juan, W., Yamini, S., Carlson, A., & Britten, P. (2006). Development of food group composites and nutrient profiles for the MyPyramid Food Guidance System. *Journal of Nutrition Education and Behavior*, *38*(6), S93–S107.

Mariotti, F., & Gardner, C. D. (2019). Dietary protein and amino acids in vegetarian diets – a review. *Nutrients*, *11*(11), 2661.

Mdziniso, P., Hinds, M. J., Bellmer, D. D., Brown, B., & Payton, M. E. (2006) Physical quality and carotene content of solar-dried green leafy and yellow succulent vegetables. *Plant Foods for Human Nutrition*, *61*(1), 13–21. https://doi.org/10.1007/s11130-006-0003-y

Medawar, E., Huhn, S., Villringer, A., & Veronica Witte, A. (2019). The effects of plant-based diets on the body and the brain: A systematic review. *Translational Psychiatry*, *9*(1), 226.

Mezeyova, I., Mezey, J., Slosar, M., Hegedusova, A., & Rosa, R. (2024). Selenization and its impact on quantitative and qualitative parameters of carrot juice. *Food Bioscience*, *59*. https://doi.org/10.1016/j.fbio.2024.103933

Munir, M. T., & Federighi, M. (2020). Control of foodborne biological hazards by ionizing radiations. *Foods*, *9*, 878. https://doi.org/10.3390/foods9070878

Neri, L., Faieta, M., Di Mattia, C., Sacchetti, G., Mastrocola, D., & Pittia, P. (2020). Antioxidant activity in frozen plant foods: Effect of cryoprotectants, freezing process and frozen storage. *Foods*, *9*(12), 1–35. https://doi.org/10.3390/foods9121886

Ohanenye, I. C., Tsopmo, A., Ejike, C. E. C. C., & Udenigwe, C. C. (2020). Germination as a bioprocess for enhancing the quality and nutritional prospects of legume proteins. *Trends in Food Science and Technology*, *101*, 213–222. https://doi.org/10.1016/j.tifs.2020.05.003

Ojo, M. A. (2022). Tannins in foods: Nutritional implications and processing effects of hydrothermal techniques on underutilized hard-to-cook legume seeds-a review. *Preventive Nutrition and Food Science*, *27*(1), 14–19. https://doi.org/10.3746/pnf.2022.27.1.14

Oke, M., & Paliyath, G. (2006). Biochemistry of vegetable processing. In: Y. H. Hui (Ed.). *Food Biochemistry and Food Processing*. Blackwell Publishing.

Oliveira, S. M., Brandão, T. R. S., & Silva, C. L. M. (2016). Influence of drying processes and pretreatments on nutritional and bioactive characteristics of dried vegetables: A review. *Food Engineering Reviews*, *8*(2), 134–163. https://doi.org/10.1007/s12393-015-9124-0

Oloyede, O. O., Wagstaff, C., & Methven, L. (2021). The impact of domestic cooking methods on myrosinase stability, glucosinolates and their hydrolysis products in different cabbage (*Brassica oleracea*) accessions. *Foods*, *10*, 2908. https://doi.org/10.3390/foods10122908

Patras, A., Tiwari, B. K., & Brunton, N. P. (2011). Influence of blanching and low temperature preservation strategies on antioxidant activity and phytochemical content of carrots, green beans and broccoli. *LWT*, *44*(1), 299–306. https://doi.org/10.1016/j.lwt.2010.06.019

Patterson C. A., Curran, J., & Der, T. (2016). Effect of processing on antinutrient compounds in pulses. *Cereal Chemistry*, *94*(1), 2–10. https://doi.org/10.1094/CCHEM-05-16-0144-FI

Pennington, J. A., & Fisher, R. A. (2009). Classification of fruits and vegetables. *Journal of Food Composition and Analysis*, *22*, S23–S31.

Pennington, J. A., & Fisher, R. A. (2010). Food component profiles for fruit and vegetable subgroups. *Journal of Food Composition and Analysis*, *23*(5), 411–418.

Petroski, W., & Minich, D. M. (2020). Is there such a thing as "anti-nutrients"? A narrative review of perceived problematic plant compounds. *Nutrients*, *12*, 2929. https://doi.org/10.3390/nu12102929

Piironen, V., Toivo, J., Puupponen-Pimiä, R., & Lampi, A. M. (2003). Plant sterols in vegetables, fruits and berries. *Journal of the Science of Food and Agriculture*, *83*(4), 330–337.

Popova, A., Mihaylova, D. (2019). Antinutrients in plant base foods: A review. *The Open Biotechnology Journal*, *13*, 68–76. https://doi.org/10.2174/1874070701913010068

Ramaswamy, H. S., Marcotte, M., Sastry, S., & Abdelrahim, K. (2014). Ohmic heating in food processing. In: *Ohmic Heat: Food Process*. http://doi.org/10.1201/b16605

Septembre-Malaterre, A., Remize, F., & Poucheret, P. (2018). Fruits and vegetables, as a source of nutritional compounds and phytochemicals: Changes in bioactive compounds during lactic fermentation. *Food Research International*, *104*, 86–99.

Slavin, J. L. (2013). Carbohydrates, dietary fiber, and resistant starch in white vegetables: Links to health outcomes. *Advances in Nutrition*, *4*(3), 351S–355S.

Srivastava, P. K., & Sit, N. (2025). A review on fruit and vegetable processing using traditional and novel methods. *Future Postharvest and Food*, *2*(1), 4–26. https://doi.org/10.1002/fpf2.12046

Storey, M., & Anderson, P. (2014). Income and race/ethnicity influence dietary fiber intake and vegetable consumption. *Nutrition Research*, *34*(10), 844–850.

Sun, X., Zhang, Y., Li, F., Jiao, X., Ma, D., Zhang, L., Yang, B., Zhao, J., Han, J., & Li, Q. (2022). Effects of lactic acid bacteria fermentation on chemical compounds, antioxidant capacities and hypoglycemic properties of pumpkin juice. *Food Bioscience*, *50*, Article 102126. https://doi.org/10.1016/j.fbio.2022.102126

Tovar-Pérez, E. G., Aguilera-Aguirre, S., López-García, U., Valdez-Morales, M., Ibarra-Zurita, A. K., Ortiz-Basurto, R. I., & Chacón-López, A. (2020). Effect of ultrasound treatment on the quality and contents of polyphenols, lycopene and rutin in tomato fruits. *Czech Journal of Food Sciences*, *38*(1), 20–27. https://doi.org/10.17221/189/2019-CJFS

Verkerk, R., Schreiner, M., Krumbein, A., Ciska, E., Holst, B., Rowland, I., De Schrijver, R., Hansen, M., Gerhäuser, C., Mithen, R. & Dekker, M. (2009). Glucosinolates in Brassica vegetables: The influence of the food supply chain on intake, bioavailability and human health. *Molecular Nutrition & Food Research*, *53*(S2), S219–S219.

Vidrih, R., Filip, S., & Hribar, J. (2009). Content of higher fatty acids in green vegetables. *Czech Journal of Food Sciences*, *27*(1), S125.

Xiao, H. W., Pan, Z., Deng, L. Z., El-Mashad, H. M., Yang, X. H., Mujumdar, A. S., Gao, Z. J., & Zhang, Q. (2017). Recent developments and trends in thermal blanching – A comprehensive review. *Information Processing in Agriculture*, *4*(2), 101–127. https://doi.org/10.1016/j.inpa.2017.02.001

Xu, B., Zhang, M., Bhandari, B., & Cheng, X. (2014). Influence of ultrasound-assisted osmotic dehydration and freezing on the water state, cell structure, and quality of radish (*Raphanus sativus* L.) cylinders. *Drying Technology: An International Journal*, *32*(15), 1803–1811. https://doi.org/10.1080/07373937.2014.947427

Zeng, X., Wang, H., Zhongqi, F., Li, Q., Li, H., Xi, Y., Jiang, W., Cao, J., Liu, B., & Li, J. (2025). Recent advances in the applications of different technologies in the processing of ready-to-eat fruit and vegetable products to improve their freshness, quality and safety: A review. *Food Control*, *178*(January). https://doi.org/10.1016/j.foodcont.2025.111503

Zhang, X., Duan, W., Zou, J., Zhou, H., Liu, C., & Yang, H. (2019). Flavor and antioxidant activity improvement of carrot juice by fermentation with Lactobacillus plantarum WZ-01. *Journal of Food Measurement and Characterization*, *13*(4), 3366–3375. https://doi.org/10.1007/s11694-019-00260-y

Zhao, C., Liu, Y., Lai, S., Cao, H., Guan, Y., San Cheang, W., Liu, B., Zhao, K., Miao, S., Riviere, C., Capanoglu, E., & Xiao, J. (2019). Effects of domestic cooking process on the chemical and biological properties of dietary phytochemicals. *Trends in Food Science and Technology*, *85*(January), 55–66. https://doi.org/10.1016/j.tifs.2019.01.004

17 Best Farming and Cultivation Method for Vegetables to Enhance Their Nutritional Value

Noopur Khare, Pragati Khare, Tanvi Jain, Sachidanand Singh, Megha Barot, and Azamal Husen

17.1 INTRODUCTION

With increasing awareness of the link between diet and health, the nutritional quality of vegetables has gained significant attention. However, the nutrient composition of vegetables is not solely determined by their genetic makeup; rather, it is greatly influenced by farming and cultivation practices. Different agricultural methods can either enhance or degrade the nutritional value of vegetables, making it imperative to adopt best farming practices that optimize their health benefits (Javed et al., 2019). This chapter explores the importance of nutritional quality in vegetables, the key factors that affect their nutrient composition, and an overview of various farming and cultivation techniques that can help maximize their nutritional potential.

17.1.1 Importance of Nutritional Quality in Vegetables

Vegetables are a fundamental component of a balanced diet, supplying essential micronutrients such as vitamin A, vitamin C, folate, potassium, calcium, and iron. They also provide antioxidants, phytochemicals, and dietary fibre that contribute to disease prevention and overall well-being (Kumar et al., 2020). The nutritional quality of vegetables depends on several intrinsic and extrinsic factors, including their variety, growing conditions, and post-harvest handling. For instance leafy greens such as spinach and kale are rich in iron and vitamin K, while carrots and bell peppers are excellent sources of beta-carotene and vitamin C. Root vegetables like sweet potatoes and beets are high in complex carbohydrates and fibre. The presence of phytochemicals like flavonoids, carotenoids, and polyphenols further enhances the health benefits of vegetables by providing antioxidant and anti-inflammatory properties (Sharma et al., 2021).

Given the increasing consumer preference for nutrient-rich foods, there is a growing need for agricultural practices which not only maximize yield but also improve the nutritional density of vegetables. Traditional farming systems often prioritize quantity over quality, leading to the depletion of essential nutrients in the soil and a decline in vegetable nutritional value. By adopting best farming and cultivation methods, farmers can ensure that vegetables retain their optimal nutrient profile while also promoting environmental sustainability and soil health (Chaudhari et al., 2024).

17.1.2 Factors Impacting Nutritional Value of Vegetables

Several factors influence the nutrient composition of vegetables, including genetic, environmental, and agronomic factors as seen in Figure 17.1.

DOI: 10.1201/9781003606710-17

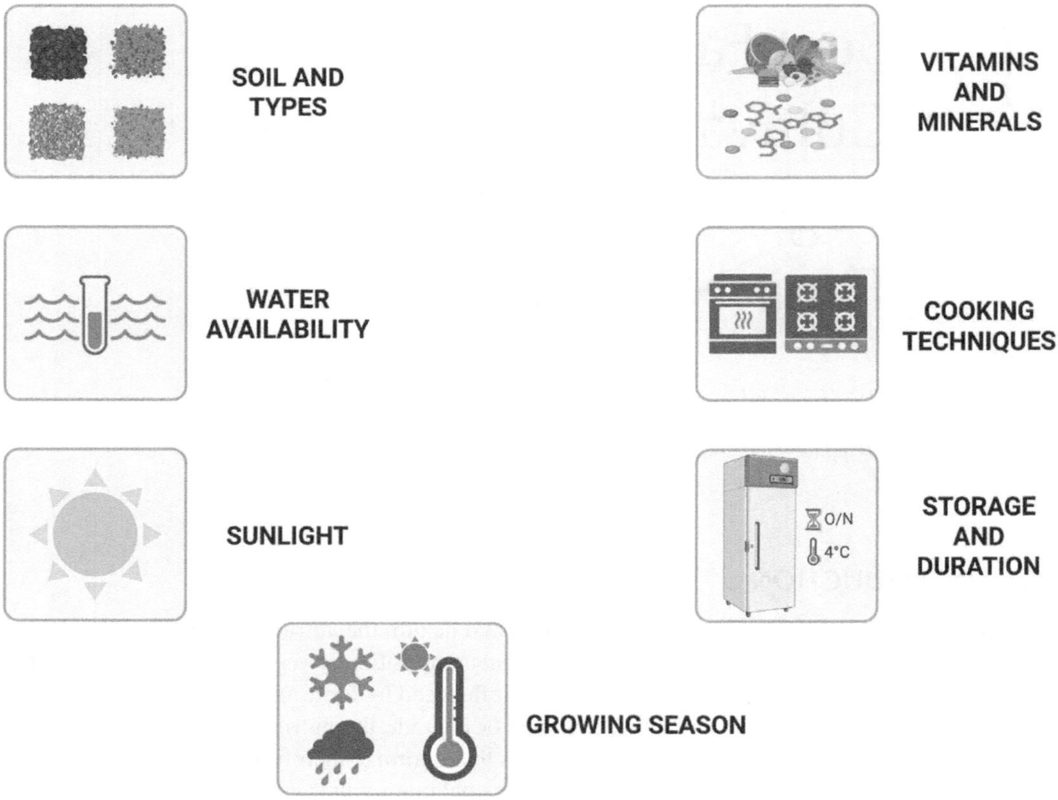

FIGURE 17.1 Factors impacting nutritional value of vegetables.

17.1.2.1 Genetic Factors

The genetic makeup of a vegetable species determines its inherent nutrient content. Different varieties of the same vegetable can have varying levels of vitamins, minerals, and bioactive compounds. For example red bell peppers contain higher amounts of vitamin C and carotenoids compared to green bell peppers. Similarly, heirloom varieties of tomatoes often have a higher concentration of antioxidants than commercially bred hybrids. Selective breeding and genetic improvement can help develop vegetable varieties with enhanced nutritional profiles (Rajasree & Pugalendhi, 2021).

17.1.2.2 Soil Quality and Nutrient Availability

Soil health is a critical determinant of vegetable nutrition. Poor soil fertility, excessive use of chemical fertilizers, and soil erosion can lead to nutrient depletion, resulting in lower-quality produce. Organic farming practices, such as composting, crop rotation, and the use of biofertilizers, can help maintain soil health and improve vegetable nutrition (Scherr, 2019). Moreover, seaweeds, due to their diversified biochemical profiles (Barot et al., 2015, 2019; JI et al., 2017), have been used to increase the soil fertility (Singh et al., 2025; Yasmeen et al., 2025).

17.1.2.3 Irrigation Practices

Water availability and irrigation methods also impact the nutrient density of vegetables. Over-irrigation can lead to nutrient leaching, while water scarcity can cause stress-induced reductions in certain vitamins and phytochemicals (Nyathi et al., 2018).

17.1.2.4 Use of Fertilizers and Pesticides

The type and amount of fertilizers used significantly influence vegetable nutrition. While synthetic fertilizers can increase yield, excessive use can degrade soil health and reduce nutrient density over time. Organic and bio-based fertilizers, such as manure, compost, and plant-based amendments, can improve nutrient availability without harmful environmental effects. Similarly, excessive pesticide application can lead to residue accumulation, which may negatively impact human health. Integrated pest management (IPM) and the use of natural biopesticides can provide a safer alternative while preserving vegetable quality (Priya et al., 2023).

17.1.2.5 Climatic Conditions

Temperature, sunlight, and atmospheric conditions also play a role in determining the nutrient composition of vegetables. Sunlight exposure enhances the synthesis of certain vitamins such as vitamin D in mushrooms and beta-carotene in carrots. On the other hand, extreme temperatures and environmental stress can lead to nutrient degradation. CEA, such as greenhouse farming and vertical farming, can help mitigate the effects of climate variability and improve nutrient retention in vegetables (Bian et al., 2020).

17.1.2.6 Harvesting and Post-Harvest Handling

The timing and method of harvesting significantly impact vegetable nutrition. Harvesting vegetables at their peak maturity ensures maximum nutrient content, while delayed harvesting can lead to nutrient loss (Parveen et al., 2023). Post-harvest handling practices, such as proper storage, refrigeration, and minimal processing, help retain the freshness and nutrient value of vegetables. Exposure to air, light, and high temperatures can degrade vitamins and antioxidants, emphasizing the need for proper preservation techniques (Elik et al., 2019).

17.1.3 Overview of Farming and Cultivation Methods

There are several farming and cultivation techniques that can enhance the nutritional quality of vegetables. These methods focus on soil health, water management, pest control, and sustainable practices to ensure that vegetables retain their maximum nutrient potential as seen in Figure 17.2.

17.1.3.1 Organic Farming

Organic farming method avoids synthetic chemicals, ensuring that vegetables are free from harmful residues. Studies have shown that organically grown vegetables often have higher antioxidant levels and better mineral content compared to conventionally grown produce (Singh et al., 2020).

17.1.3.2 Precision Agriculture

This method of irrigation involves the use of modern technologies such as Global Positioning System (GPS), soil sensors, and data analytics to optimize nutrient management and improve vegetable quality. By monitoring soil conditions and adjusting inputs accordingly, farmers can enhance nutrient uptake and minimize resource wastage (Yadav & Sidana, 2023).

17.1.3.3 Hydroponics and Aeroponics

Hydroponic and aeroponic systems involve growing vegetables in nutrient-rich water solutions without soil. These methods allow for precise control over nutrient delivery, resulting in vegetables with enhanced mineral content. Additionally, hydroponics and aeroponics use less water and reduce the risk of soil-borne diseases, making them efficient alternatives for urban and climate-controlled farming (Kumar et al., 2023).

17.1.3.4 Agroecology and Permaculture

Agroecological farming integrates ecological principles into agriculture, promoting biodiversity and sustainability. Practices such as intercropping, companion planting, and crop diversification

FIGURE 17.2 Farming and cultivation methods.

enhance soil health and nutrient availability. Permaculture, which mimics natural ecosystems, also supports the growth of nutrient-rich vegetables while reducing environmental impact (Yadav et al., 2021).

17.1.3.5 Biofortification

Biofortification is the process of increasing the nutritional value of vegetables through natural breeding techniques or genetic modification. Crops such as iron-rich spinach, vitamin A-enriched sweet potatoes, and zinc-fortified beans have been developed to address micronutrient deficiencies and improve public health outcomes (Ofori et al., 2022).

17.1.3.6 Controlled Environment Agriculture

Greenhouse farming, vertical farming, and aquaponics provide controlled growing conditions that optimize temperature, humidity, and nutrient supply. These methods ensure consistent production of high-quality vegetables with enhanced nutritional value (Lal et al., 2020).

17.2 SOIL MANAGEMENT FOR NUTRIENT-RICH VEGETABLES

17.2.1 Importance of Soil Health in Vegetable Cultivation

Soil health is a fundamental aspect of vegetable cultivation, directly influencing plant growth, nutrient availability, and overall crop yield. A well-balanced and nutrient-rich soil provides essential minerals and organic matter that support plant development and ensure high nutritional quality in vegetables (Xing et al., 2025). Soil acts as a reservoir for water, nutrients, and beneficial microorganisms that contribute to plant resilience and productivity. For instance nitrogen is crucial for leafy vegetable growth, phosphorus supports root development, and potassium enhances disease resistance (Bitew & Alemayehu, 2017). Moreover, soil with an appropriate balance of organic matter improves water retention, aeration, and microbial activity, leading to better plant health. Conversely, poor soil management can result in nutrient depletion, increased susceptibility to pests and diseases, and reduced vegetable quality. Intensive farming practices, excessive use of chemical fertilizers, and soil erosion contribute to declining soil fertility over time. Depleted soils lead to nutrient-poor vegetables, which in turn affect human health by providing lower levels of essential vitamins and minerals. Therefore, adopting sustainable soil management practices is crucial to maintaining high vegetable nutrition levels. Practices such as organic amendments, proper crop rotation, mulching, and minimal soil disturbance can help maintain soil health and enhance the nutrient uptake of vegetables (Montgomery & Biklé, 2021).

17.2.2 Organic versus Conventional Farming for Nutrient Retention

Farming practices play a crucial role in determining soil quality and, consequently, the nutritional value of vegetables (Khare et al., 2025a). Both organic and conventional farming approaches have distinct effects on soil health and nutrient retention. Understanding their differences can help in choosing the best methods to improve soil quality and enhance vegetable nutrition (Gamage et al., 2023).

17.2.2.1 Organic Farming and Soil Health

Organic farming relies on natural processes to maintain soil fertility and minimize environmental impact. It promotes the use of compost, manure, cover crops, and crop rotations to enhance soil structure, increase microbial diversity, and improve nutrient availability. Organic fertilizers release nutrients slowly, ensuring a sustained supply of minerals essential for plant growth (Verma et al., 2019). Additionally, organic farming avoids synthetic pesticides and herbicides, reducing soil contamination and preserving beneficial soil microbes that assist in nutrient absorption (Patani et al., 2023). This is because organic soils have higher organic matter content, which improves nutrient retention and reduces leaching. Furthermore, organic farming enhances soil biological activity, leading to better mineralization of nutrients and increased availability to plants (Reganold & Wachter, 2016).

17.2.2.2 Conventional Farming and Its Effects on Soil Quality

Conventional farming, on the other hand, relies heavily on synthetic fertilizers, pesticides, and herbicides to maximize crop yields. While these inputs provide rapid nutrient availability, they can degrade soil structure and microbial diversity over time. Excessive fertilizer application often leads to nutrient leaching, soil acidification, and water pollution, reducing the overall sustainability of the farming system. Additionally, heavy pesticide use can negatively impact beneficial soil organisms, further reducing the soil's ability to sustain plant health. Over time, conventional farming methods may result in nutrient imbalances, lower soil fertility, and diminished nutritional quality of vegetables (Średnicka-Tober et al., 2016).

17.2.3 Soil Enrichment Techniques

Maintaining soil fertility is essential for producing nutrient-rich vegetables. Several soil enrichment techniques can be employed to enhance soil quality, improve nutrient availability, and support sustainable vegetable farming as seen in Figure 17.3.

17.2.3.1 Composting: The Power of Organic Matter

Composting is one of the most effective methods to enrich soil and improve its structure. Compost enhances soil aeration, water retention, and microbial activity, leading to improved plant nutrient

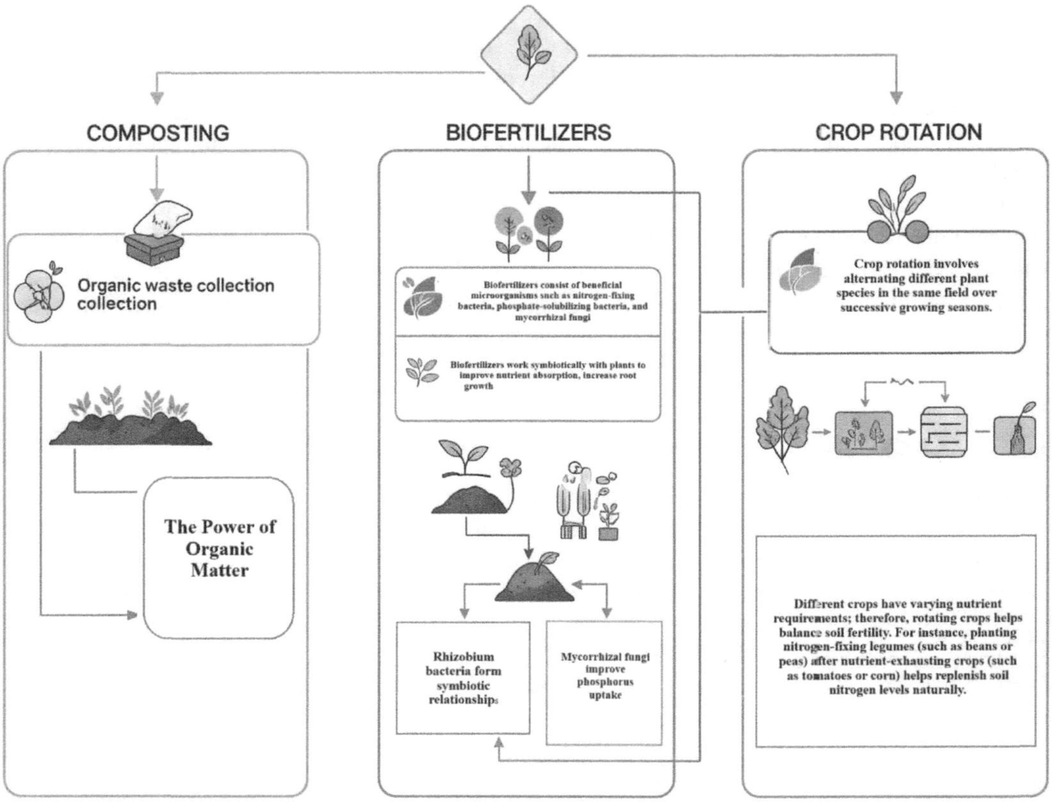

FIGURE 17.3 Soil enrichment techniques.

uptake. It also provides essential macronutrients and micronutrients, promoting balanced plant growth and development. Moreover, composting reduces dependency on chemical fertilizers, mitigates greenhouse gas (GHG) emissions, and recycles organic waste. Farmers using compost-enriched soil often observe improved vegetable quality, enhanced disease resistance, and increased productivity (Waqas et al., 2023).

17.2.3.2 Biofertilizers: Enhancing Soil Microbial Activity

Unlike chemical fertilizers, biofertilizers work symbiotically with plants to improve nutrient absorption, increase root growth, and enhance overall plant vigour (Khare et al., 2025b). Similarly, mycorrhizal fungi improve phosphorus uptake, while Azotobacter enhances soil fertility. Using biofertilizers reduces the need for synthetic inputs, minimizes soil degradation, and contributes to a healthier agroecosystem (Thomas & Singh, 2019).

17.2.3.3 Crop Rotation: Preventing Soil Nutrient Depletion

Crop rotation involves alternating different plant species in the same field over successive growing seasons. This practice prevents soil depletion, reduces pest and disease incidence, and enhances soil structure. Different crops have varying nutrient requirements; therefore, rotating crops helps balance soil fertility. For instance planting nitrogen-fixing legumes (such as beans or peas) after nutrient-exhausting crops (such as tomatoes or corn) helps replenish soil nitrogen levels naturally.

Similarly, deep-rooted crops can bring up nutrients from lower soil layers, making them available to shallow-rooted vegetables. Crop rotation also disrupts pest cycles, reducing the need for chemical pesticides and improving overall soil health (Boincean et al., 2019).

17.2.4 Impact of Soil pH and Microbial Activity on Nutrient Absorption

Soil pH and microbial activity play a significant role in determining the availability of essential nutrients for vegetable crops. Properly managed soil pH and microbial communities enhance plant nutrient absorption, leading to healthier and more nutrient-dense vegetables (Neina, 2019).

17.2.4.1 Understanding Soil pH and Its Role in Nutrient Availability

Soil pH is a measure of soil acidity or alkalinity and is crucial for plant nutrient uptake. Different nutrients are more readily available at specific pH levels. For instance:

- Acidic soils (pH < 6.0) can lead to deficiencies in calcium, magnesium, and phosphorus while increasing the availability of toxic elements such as aluminium and manganese.
- Neutral to slightly acidic soils (pH 6.0–7.0) provide the optimal range for nutrient availability and microbial activity.
- Alkaline soils (pH > 7.5) may lead to reduced availability of iron, zinc, and phosphorus, potentially limiting plant growth (Neina, 2019).

17.2.5 Microbial Activity and Its Impact on Soil Fertility

Beneficial microbes, such as bacteria, fungi, and actinomycetes, contribute to improved soil fertility and plant nutrition (Khatoon et al., 2017). For example mycorrhizal fungi form symbiotic relationships with plant roots, increasing nutrient uptake efficiency. Similarly, bacteria such as Bacillus and Pseudomonas produce enzymes that break down organic matter and release essential nutrients. Maintaining high microbial diversity through organic amendments, minimal soil disturbance, and avoiding excessive chemical inputs supports soil microbial health and enhances nutrient availability (Biswas & Kole, 2018).

17.3 BEST CULTIVATION PRACTICES TO ENHANCE NUTRITIONAL VALUE

17.3.1 Organic Farming and Its Benefits

Organic farming is a sustainable agricultural practice that enhances soil fertility, biodiversity, and the nutritional value of vegetables. It emphasizes the use of natural fertilizers, crop rotation, biological pest control, and minimal use of synthetic chemicals. Unlike conventional farming, which relies on chemical inputs for quick plant growth, organic farming nurtures the soil's microbial ecosystem, ensuring that essential nutrients are retained and made available to crops over time (Soni et al., 2022; Khare et al., 2025c). This is primarily due to the absence of synthetic fertilizers, which can lead to rapid but shallow nutrient absorption. Instead, organic farming practices promote slow nutrient release, allowing plants to absorb minerals such as calcium, magnesium, and iron more effectively (Benbrook et al., 2021). Another critical advantage of organic farming is its positive impact on soil health. By incorporating organic matter such as compost, manure, and plant residues, organic farmers enhance soil structure, moisture retention, and microbial activity. This creates an environment that supports healthy root growth and maximizes nutrient uptake. Additionally, the use of cover crops and crop rotation in organic farming prevents soil depletion, ensuring a consistent supply of essential nutrients for vegetable crops (Ramakrishnan et al., 2021). Organic farming also reduces exposure to harmful pesticides and chemical residues in vegetables. As organic farmers use biological pest control methods and natural repellents, consumers benefit from produce that is free from synthetic toxins. This contributes to better human health by reducing the risk of pesticide-related illnesses and enhancing the bioavailability of nutrients. From an environmental perspective, organic farming helps mitigate climate change by reducing GHG emissions and improving carbon sequestration in soils. It also promotes biodiversity by encouraging the growth of beneficial insects, birds, and microorganisms that play a vital role in maintaining ecosystem balance. These factors collectively contribute to a more sustainable food production system that prioritizes both human health and environmental well-being (Gamage et al., 2023).

17.3.2 Precision Agriculture for Nutrient Optimization

By using data-driven techniques such as satellite imaging, soil sensors, and GPS mapping, farmers can make informed decisions about nutrient management and cultivation practices. Unlike traditional farming methods that involve blanket fertilization, precision agriculture employs variable-rate technology (VRT) to adjust nutrient application based on soil conditions and crop needs. This prevents nutrient imbalances and enhances the mineral content of vegetables (Getahun et al., 2024).

Through the use of drip irrigation and automated soil moisture sensors, farmers can regulate water distribution, preventing both drought stress and overwatering. Proper hydration is essential for nutrient uptake, as water acts as a medium for transporting minerals from the soil to the plant roots. By optimizing irrigation, precision agriculture ensures that vegetables receive adequate hydration while conserving water resources. Precision agriculture also facilitates real-time monitoring of plant health and nutrient deficiencies. Advanced tools such as remote sensing and drones can detect early signs of nutrient stress, enabling farmers to take corrective measures before deficiencies affect crop yield and quality. For example leaf colour analysis can indicate nitrogen or phosphorus shortages, prompting farmers to apply the necessary organic or synthetic amendments. By implementing site-specific nutrient management strategies, farmers can maintain soil fertility while producing vegetables with enhanced nutritional profiles. This technology-driven approach aligns with the goals of sustainable agriculture, ensuring food security and improved dietary health (Monteiro et al., 2021).

17.3.3 Use of Cover Crops and Green Manure

Cover cropping and green manure are essential agronomic practices that improve soil fertility, prevent erosion, and enhance the nutrient density of vegetables. These practices involve planting specific crops during off-seasons or between vegetable rotations to replenish soil nutrients and organic matter. Cover crops, such as legumes, mustard, clover, and rye, play a vital role in soil enrichment. Leguminous cover crops, for instance, have nitrogen-fixing capabilities, meaning they convert atmospheric nitrogen into a form that is accessible to plants. When incorporated into the soil, cover crops release nutrients that benefit subsequent vegetable crops, resulting in higher mineral content in the harvested produce (Toungos & Bulus, 2019).

Green manure, on the other hand, involves growing and ploughing under crops specifically for soil enhancement. Green manure crops decompose and release organic matter, improving soil structure, moisture retention, and microbial diversity. This organic matter acts as a slow-release fertilizer, providing a steady supply of nutrients for vegetable crops. Additionally, green manure increases the availability of micronutrients such as zinc, iron, and boron, which are essential for plant metabolism and human nutrition. Another advantage of cover cropping is its ability to suppress weeds and improve soil aeration. By competing with weeds for space and nutrients, cover crops reduce the need for herbicides, minimizing chemical residues in vegetables. Improved soil aeration enhances root development, enabling better absorption of essential minerals and water. Furthermore, cover crops and green manure help combat soil erosion by stabilizing the soil structure. Erosion can lead to nutrient loss and reduced soil fertility, negatively impacting vegetable nutrition. By protecting the soil from wind and water erosion, these practices ensure that essential nutrients remain available for crop uptake, ultimately enhancing the nutritional value of vegetables (Brust, 2019).

17.3.4 Role of Biofortification in Enhancing Vegetable Nutrition

Biofortification is a scientific approach that enhances the nutritional content of vegetables through conventional breeding, genetic modification, or agronomic interventions. This method aims to increase the levels of essential micronutrients such as iron, zinc, and vitamin A in food crops, addressing global malnutrition and improving public health. One of the primary methods of biofortification is selective breeding, where plant varieties with higher nutrient content are crossbred to develop nutrient-rich vegetable strains. For example biofortified carrots with higher beta-carotene levels have been developed to combat vitamin A deficiency, while iron-rich beans help address anaemia in populations with iron-deficient diets (Ofori et al., 2022).

Agronomic biofortification involves applying micronutrient-rich fertilizers to enhance nutrient uptake by plants. Soil or foliar application of zinc, selenium, or iodine fertilizers can significantly increase the concentration of these elements in vegetables. For instance zinc-enriched soil produces spinach and kale with higher zinc levels, benefiting human health by preventing zinc deficiency-related disorders. Genetic biofortification, though more controversial, offers the potential to engineer crops with enhanced nutrient profiles. Genetically modified (GM) vegetables, such as Golden Rice enriched with vitamin A, demonstrate the potential of biotechnology in combating nutrient deficiencies. However, public perception and regulatory challenges often limit the widespread adoption of GM biofortified crops. Biofortification plays a crucial role in addressing micronutrient deficiencies, particularly in regions where malnutrition is prevalent. By integrating biofortified vegetable varieties into diets, populations can achieve better health outcomes without relying on artificial supplements. This approach complements traditional agricultural practices by ensuring that vegetables naturally provide essential nutrients, reducing the burden of nutrient deficiencies (Nagar et al., 2024) as seen in Figure 17.4.

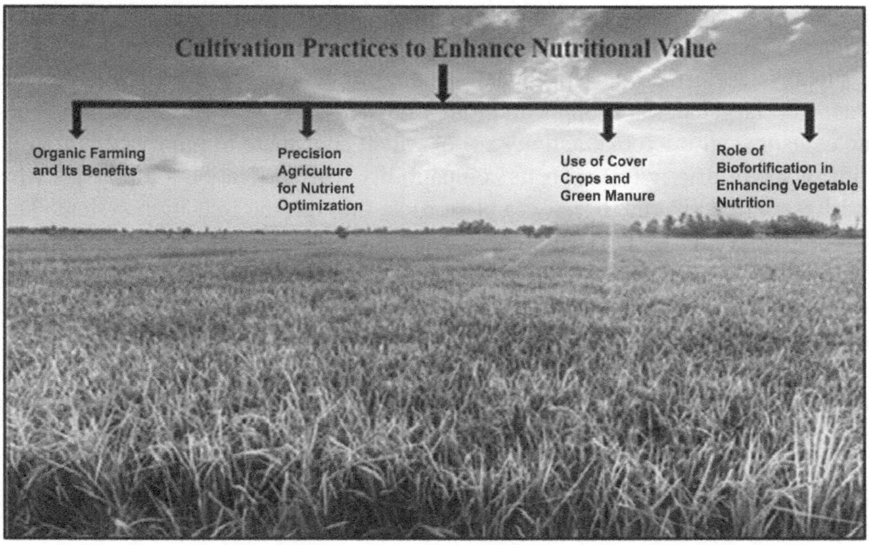

FIGURE 17.4 Cultivation practices to enhance nutritional value.

17.4 WATER MANAGEMENT FOR NUTRIENT RETENTION

17.4.1 Importance of Proper Irrigation Techniques

Water is an essential component in vegetable cultivation, directly influencing plant growth, nutrient uptake, and overall crop quality. Proper irrigation techniques ensure that plants receive adequate moisture for optimal physiological functions while maintaining soil structure and preventing nutrient leaching. Effective water management strategies contribute to higher yields, enhanced nutritional quality, and sustainable agricultural practices. One of the key benefits of proper irrigation is improved nutrient absorption. Water acts as a solvent, facilitating the transport of essential nutrients from the soil to plant roots. When irrigation is poorly managed, excessive or insufficient water can disrupt this process. Overwatering, for example, leads to nutrient leaching, where essential minerals such as nitrogen, potassium, and phosphorus are washed away from the root zone before plants can absorb them. This not only depletes soil fertility but also reduces the nutrient density of vegetables. Conversely, underwatering can result in drought stress, causing plants to limit nutrient uptake, leading to deficiencies that affect crop growth and quality (Rouphael et al., 2012).

Different irrigation methods impact nutrient retention differently. Traditional flood irrigation, while commonly used, is often inefficient as it results in water loss due to runoff and evaporation. This method also contributes to soil erosion, depleting organic matter and reducing microbial activity, both of which are crucial for nutrient availability. In contrast, modern irrigation methods such as drip and sprinkler irrigation allow for precise water application, minimizing waste and enhancing nutrient retention in the soil. Another critical aspect of irrigation management is timing. Watering crops at the right time of day, preferably early in the morning or late in the evening, helps reduce water loss through evaporation and ensures that plants can absorb moisture effectively. Frequent light irrigation may be beneficial in sandy soils that drain quickly, whereas deep but less frequent irrigation is better suited for clay soils that retain moisture longer. By adopting site-specific irrigation strategies, farmers can optimize nutrient availability and improve the nutritional profile of their vegetable crops (Njira & Nabwami, 2015).

17.4.2 Drip Irrigation versus Flood Irrigation for Nutrient Retention

Drip irrigation and flood irrigation represent two contrasting approaches to water management in vegetable farming. Each method has distinct advantages and challenges in terms of water efficiency, nutrient retention, and overall impact on soil health. By providing a steady supply of moisture, drip irrigation supports consistent nutrient uptake, leading to better crop growth and enhanced nutritional value. The targeted water application also prevents soil erosion, maintains soil structure, and promotes beneficial microbial activity, all of which contribute to improved nutrient retention. Another advantage of drip irrigation is its compatibility with fertigation – a practice where water-soluble fertilizers are delivered directly to plant roots through the irrigation system. This method ensures that nutrients such as nitrogen, phosphorus, and potassium are readily available to plants, reducing fertilizer wastage and enhancing nutrient absorption. Additionally, as drip irrigation reduces water contact with plant leaves, it lowers the risk of fungal diseases and pest infestations, leading to healthier crops with higher nutritional content (Ray & Majumder, 2024).

In contrast, flood irrigation is a traditional method where large volumes of water are applied to fields, allowing it to flow over the soil surface. While flood irrigation is relatively easy to implement and requires minimal equipment, it is highly inefficient in terms of water use and nutrient management. A significant portion of the water applied is lost through evaporation and runoff, reducing the amount of moisture available to plant roots. Furthermore, the excessive water flow can wash away essential soil nutrients, leading to nutrient imbalances and decreased vegetable quality. Flood irrigation can also contribute to soil degradation over time. Continuous waterlogging can lead to oxygen depletion in the root zone, negatively impacting root development and nutrient absorption.

In some cases, overuse of flood irrigation results in salinization, where dissolved salts accumulate in the soil, reducing fertility and limiting plant growth. This issue is particularly problematic in arid and semiarid regions where water scarcity is a concern. Despite these drawbacks, flood irrigation can still be improved through better management practices. Techniques such as laser land levelling help distribute water more evenly, reducing runoff and enhancing nutrient retention. Additionally, controlled flooding, where water is applied in short intervals rather than in a single, prolonged application, can help improve efficiency and reduce nutrient loss. When comparing drip irrigation and flood irrigation, it is evident that drip irrigation offers superior water efficiency and nutrient retention. However, for farmers with limited resources, modifications to flood irrigation techniques can help improve sustainability and nutrient management. The choice of irrigation method should be based on factors such as soil type, crop variety, climate conditions, and water availability to ensure optimal vegetable nutrition and farm productivity (Kapoor et al., 2022).

17.4.3 Rainwater Harvesting and Sustainable Water Use

Rainwater harvesting is an effective strategy for sustainable water management in vegetable farming. By collecting and storing rainwater for agricultural use, farmers can reduce their reliance on groundwater and municipal water supplies, ensuring a more sustainable and resilient farming system. This practice not only conserves water but also enhances nutrient retention by minimizing the risk of nutrient leaching associated with excessive irrigation. One of the primary benefits of rainwater harvesting is its ability to provide a clean and chemical-free water source for irrigation. Unlike groundwater, which may contain high levels of dissolved salts and contaminants, rainwater is naturally soft and free from pollutants. When used for irrigation, it supports healthy soil microbial activity and improves nutrient availability for crops. Additionally, rainwater can help dilute accumulated salts in the soil, reducing the risk of soil salinization and improving long-term soil fertility. There are several methods of rainwater harvesting, including rooftop collection, surface runoff harvesting, and the construction of farm ponds and reservoirs. Rooftop rainwater harvesting involves directing rainwater from building roofs into storage tanks, where it can be filtered and used for irrigation during dry periods. This method is particularly beneficial for small-scale vegetable farmers who need a cost-effective and sustainable water source. Surface runoff harvesting involves capturing rainwater from fields, roads, and other surfaces, directing it into storage structures such as check dams, percolation pits, or underground tanks. These systems help recharge groundwater, prevent soil erosion, and ensure that excess rainwater is utilized efficiently rather than being lost as runoff. By integrating rainwater harvesting with irrigation systems, farmers can maintain a consistent water supply for their crops, even during periods of low rainfall (Deo et al., 2022).

Farm ponds and reservoirs serve as long-term water storage solutions, allowing farmers to collect and store rainwater for use throughout the growing season. These structures help buffer against drought conditions, ensuring that vegetable crops receive adequate moisture for optimal growth and nutrient absorption. Additionally, farm ponds can support aquaculture, providing an additional source of income for farmers while enhancing farm biodiversity. To maximize the benefits of rainwater harvesting, it is essential to implement proper water conservation techniques. Mulching, for example, helps reduce evaporation and retain soil moisture, improving water efficiency. Drip irrigation systems can be integrated with harvested rainwater to optimize water distribution and nutrient retention. Furthermore, agroforestry practices, such as planting trees and shrubs around farm boundaries, help capture rainwater and reduce soil erosion, enhancing overall farm sustainability. Sustainable water management is critical for optimizing vegetable nutrition and ensuring long-term agricultural productivity. The integration of efficient water management strategies will be essential in addressing global challenges such as climate change, water scarcity, and soil degradation while promoting healthier and nutrient-rich vegetable production (Richards et al., 2021).

17.5 USE OF NATURAL AND BIO-BASED FERTILIZERS

17.5.1 Role of Organic Fertilizers in Improving Vegetable Nutrition

Organic fertilizers play a crucial role in enhancing the nutritional value of vegetables by enriching the soil with essential nutrients in a natural and sustainable manner. Unlike synthetic fertilizers, which often provide a quick but temporary nutrient boost, organic fertilizers release nutrients slowly, ensuring a steady supply for plant uptake. This gradual nutrient availability supports healthy plant growth, enhances soil fertility, and improves the overall quality of vegetable crops. The presence of organic matter also enhances soil aeration, reducing the risk of compaction and creating a favourable environment for root development and nutrient absorption (Singh et al., 2020).

17.5.2 Biofertilizers: Types and Benefits

Biofertilizers are an eco-friendly alternative to synthetic fertilizers that enhance soil fertility and plant growth by utilizing beneficial microorganisms. These microbial-based fertilizers improve nutrient availability, promote plant health, and contribute to sustainable agriculture by reducing dependence on chemical fertilizers. There are several types of biofertilizers, each serving a unique function in soil and plant nutrition. Nitrogen-fixing biofertilizers, such as Rhizobium, Azotobacter, and Azospirillum, help convert atmospheric nitrogen into a form that plants can readily absorb. These biofertilizers are particularly beneficial for legumes and nitrogen-demanding crops, reducing the need for synthetic nitrogen fertilizers while enhancing soil nitrogen levels.

Phosphate-solubilizing biofertilizers, including Bacillus and Pseudomonas species, play a crucial role in making phosphorus available to plants. These microorganisms produce organic acids and enzymes that break down phosphate compounds, increasing phosphorus availability and promoting root development and energy transfer in plants.

Potassium-mobilizing biofertilizers, such as Frateuria aurantia, help solubilize potassium, a vital nutrient for plant growth and disease resistance. Potassium enhances water uptake, improves enzyme activation, and strengthens plant cell walls, leading to healthier and more resilient vegetable crops (Manzoor et al., 2024).

17.6 PESTICIDE MANAGEMENT FOR HEALTHIER VEGETABLES

17.6.1 Organic Pest Control Methods

Organic pest control methods are essential for cultivating healthier vegetables while minimizing the negative impact of synthetic pesticides on human health and the environment. These methods involve natural and eco-friendly approaches to pest management, ensuring that vegetable crops remain free from harmful chemicals while maintaining their nutritional integrity. For example ladybugs are commonly introduced to control aphid infestations, while lacewings feed on soft-bodied insects that damage vegetable crops. Similarly, parasitic wasps can be used to target caterpillars and other pests, reducing the need for synthetic pesticides (Costa et al., 2023).

17.6.2 Use of Integrated Pest Management (IPM)

IPM is a sustainable approach to pest control that combines biological, cultural, physical, and chemical methods to minimize pest populations while reducing environmental and health risks. Unlike conventional pest control methods that rely heavily on synthetic pesticides, IPM focuses on prevention, monitoring, and targeted interventions to achieve effective pest management with minimal chemical use. The first step in IPM is pest identification and monitoring. Farmers need to regularly inspect their crops to detect early signs of pest infestations. This involves using pheromone traps, sticky traps, and field scouting to assess pest populations. Proper identification helps determine whether pest levels are high enough to cause economic damage, preventing unnecessary pesticide application (Deguine et al., 2021).

17.7 HARVESTING AND POST-HARVEST HANDLING FOR NUTRIENT PRESERVATION

17.7.1 Best Time for Harvesting to Maximize Nutrient Content

Harvesting vegetables at the right stage of maturity is crucial for ensuring maximum nutrient retention. The timing of harvest depends on several factors, including the vegetable type, environmental conditions, and the specific nutrient profile desired. Many vegetables reach peak nutritional value at a specific maturity stage, and delaying harvest beyond this point can result in nutrient degradation and reduced quality. Leafy vegetables, such as spinach, lettuce, and kale, should be harvested when they have reached full leaf expansion but before they become overly mature. As leaves age, they tend to lose moisture and essential nutrients like vitamin C, folate, and antioxidants. Harvesting early in the morning, when plants have the highest moisture content, can also help preserve freshness and nutrient levels (Quamruzzaman et al., 2022).

For fruiting vegetables like tomatoes, bell peppers, and cucumbers, the best time for harvest varies depending on the intended use. Vine-ripened tomatoes, for instance, contain higher levels of lycopene and vitamin C compared to those harvested prematurely and ripened off the vine. Similarly, bell peppers accumulate more beta-carotene and vitamin C as they transition from green to red. On the other hand, cucumbers are best harvested when they are still young and firm, as over-maturity can result in a decline in flavour and nutrient content. Root vegetables, such as carrots, beets, and radishes, should be harvested when they have reached full size but before they become woody. Over-mature roots tend to lose moisture and become fibrous, impacting both texture and nutrient availability. Additionally, exposure to prolonged sunlight after maturity can lead to a decline in key vitamins like vitamin C. The influence of environmental factors on nutrient retention should also be considered. Extreme heat, excessive rainfall, and prolonged exposure to direct sunlight can degrade nutrients in vegetables. Implementing proper harvesting schedules and techniques, such as handpicking delicate vegetables to avoid bruising, can help maintain nutrient density. By timing harvest correctly and employing careful handling techniques, farmers can ensure that vegetables reach consumers with optimal nutritional value (Elik et al., 2019).

17.7.2 Post-Harvest Storage Techniques (Cold Storage, Controlled Atmosphere)

Post-harvest storage plays a vital role in preserving the nutritional quality of vegetables. Without proper storage conditions, vegetables can lose essential vitamins, minerals, and antioxidants due to factors like oxidation, enzymatic activity, and microbial spoilage. Implementing effective storage techniques such as cold storage and controlled atmosphere storage can significantly extend shelf life while maintaining nutrient integrity. Cold storage is one of the most widely used methods for preserving vegetables after harvest. Most vegetables are highly perishable and require refrigeration to slow down respiration rates and prevent moisture loss. Leafy greens, for example, should be stored at temperatures close to 0°C with high humidity to retain their crispness and vitamin C content. Root vegetables like carrots and beets benefit from cool, moist conditions to prevent dehydration and nutrient depletion (El-Ramady et al., 2015).

17.7.3 Effects of Processing and Cooking on Nutritional Value

While fresh vegetables contain high levels of essential nutrients, various processing and cooking methods can impact their nutritional composition. Understanding how these methods affect nutrient retention can help consumers make informed choices about vegetable preparation. Blanching is a common pre-processing method used before freezing or canning vegetables. It involves briefly boiling or steaming vegetables to inactivate enzymes that cause spoilage. While blanching helps preserve colour and texture, it can lead to some nutrient loss, particularly water-soluble vitamins like vitamin C and B-complex vitamins. However, using minimal water and shorter blanching times can help reduce these losses. Freezing is one of the best preservation methods for retaining nutrients in vegetables. As freezing slows down enzymatic activity and microbial growth, frozen vegetables often retain higher levels of vitamins and minerals compared to fresh vegetables that have been stored for long periods. However, improper thawing methods, such as exposing vegetables to high heat, can lead to significant nutrient degradation. Canning is another widely used preservation method, but it often results in greater nutrient loss due to high-temperature processing. Water-soluble vitamins, such as vitamin C and folate, are particularly vulnerable to degradation during the canning process. However, canned vegetables still provide valuable nutrients and can be a convenient alternative when fresh produce is not available. Choosing low-sodium canned vegetables and consuming the liquid in which they are packed can help retain more nutrients (Richter Reis, 2017).

17.8 CONCLUSION

Throughout this study, we have explored various farming and cultivation methods aimed at enhancing the nutritional value of vegetables. Several best practices have emerged as effective strategies for ensuring nutrient-rich produce. Organic farming methods, including the use of compost, biofertilizers, and crop rotation, have been proven to improve soil health and increase the vitamin and mineral content of vegetables. Precision agriculture, which optimizes nutrient application and irrigation, has shown promise in enhancing plant nutrient uptake while minimizing environmental impact. Additionally, sustainable water management techniques such as drip irrigation and rainwater harvesting help preserve soil nutrients and improve vegetable quality. Another crucial aspect of nutrient-rich vegetable farming is the use of natural fertilizers and IPM systems. Organic and bio-based fertilizers enrich the soil with essential micronutrients, while reducing the harmful effects of synthetic fertilizers on soil and human health. Avoiding excessive pesticide use ensures that vegetables retain their full nutritional potential without contamination from harmful residues. Furthermore, post-harvest handling techniques, including proper storage and minimal processing, play a significant role in nutrient retention, ensuring that vegetables maintain their health benefits from farm to table. While significant advancements have been made in optimizing vegetable nutrition through farming techniques, further research is needed to refine these practices and develop innovative solutions. One promising avenue of research is biofortification, which focuses on enhancing the nutrient density of vegetables through selective breeding and genetic improvements. Future studies could explore the potential of biofortified vegetables in combating nutrient deficiencies in vulnerable populations. Another critical research area is the application of microbial inoculants and soil microbiome studies to enhance plant nutrient absorption. Investigating the role of beneficial microbes in increasing mineral uptake and plant resilience could lead to breakthroughs in sustainable agriculture. Moreover, advancements in CEA, such as vertical farming and hydroponics, warrant further exploration to optimize nutrient-rich vegetable production in urban settings. Additionally, future research should focus on the long-term impact of different farming practices on soil health and nutritional value. Comparative studies assessing the nutrient retention of vegetables grown under conventional, organic, and regenerative agricultural systems could provide valuable insights into the most effective and sustainable approaches. Enhancing the nutritional value of vegetables through improved farming methods is essential not only for public health but also for environmental sustainability. By implementing best practices, investing in future research, and adopting strong policy measures, we can ensure that future generations have access to high-quality, nutrient-dense vegetables. A holistic approach combining sustainable farming, technological innovations, and supportive policies will be key to achieving long-term nutritional security and agricultural resilience.

REFERENCES

Barot, M., Kumar, J., & Kumar, N. R. (2015). Seaweed species diversity in relation to hydrochemical characters of Okha Coast Western India. *International Journal of Recent Research and Review*, *8*(3), 16–28.

Barot, M., Nirmal Kumar, J. I., & Kumar, R. N. (2019). An evaluation of the nutritional composition of seaweeds as potential source of food and feed. *National Academy Science Letters*, *42*(6), 459–464.

Benbrook, C., Kegley, S., & Baker, B. (2021). Organic farming lessens reliance on pesticides and promotes public health by lowering dietary risks. *Agronomy*, *11*(7), 1266.

Bian, Z., Wang, Y., Zhang, X., Li, T., Grundy, S., Yang, Q., & Cheng, R. (2020). A review of environment effects on nitrate accumulation in leafy vegetables grown in controlled environments. *Foods*, *9*(6), 732.

Biswas, T., & Kole, S. C. (2018). Soil organic matter and microbial role in plant productivity and soil fertility. In *Advances in Soil Microbiology: Recent Trends and Future Prospects: Volume 2: Soil-Microbe-Plant Interaction* (pp. 219–238). Singapore: Springer Singapore.

Bitew, Y., & Alemayehu, M. (2017). Impact of crop production inputs on soil health: A review. *Asian Journal Plant Science*, *16*(3), 109–131.

Boincean, B., Dent, D., Boincean, B., & Dent, D. (2019). Crop rotation. In *Farming the Black Earth: Sustainable and Climate-Smart Management of Chernozem Soils* (pp. 89–124). Springer International Publishing.

Brust, G. E. (2019). Management strategies for organic vegetable fertility. In *Safety and Practice for Organic Food* (pp. 193–212). Academic Press.

Chaudhari, V. M., Singh, O. B., Gouthami, N. S., Thakur, N., Singh, R., Singh, S., . . . & Nagar, B. L. (2024). Unlocking the nutritional power of vegetables: A guide to vibrant health. *European Journal of Nutrition & Food Safety*, *16*, 247–261.

Costa, C. A., Guiné, R. P., Costa, D. V., Correia, H. E., & Nave, A. (2023). Pest control in organic farming. In *Organic Farming* (pp. 111–179). Woodhead Publishing.

Deguine, J. P., Aubertot, J. N., Flor, R. J., Lescourret, F., Wyckhuys, K. A., & Ratnadass, A. (2021). Integrated pest management: Good intentions, hard realities. A review. *Agronomy for Sustainable Development*, *41*(3), 38.

Deo, A., Karmakar, S., & Arora, A. (2022). Rainwater harvesting and water balance simulation-optimization scheme to plan sustainable second crop in small rain-fed systems. *Journal of Environmental Management*, *323*, 116135.

Elik, A., Yanik, D. K., Istanbullu, Y., Guzelsoy, N. A., Yavuz, A., & Gogus, F. (2019). Strategies to reduce post-harvest losses for fruits and vegetables. *Strategies*, *5*(3), 29–39.

El-Ramady, H. R., Domokos-Szabolcsy, É., Abdalla, N. A., Taha, H. S., & Fári, M. (2015). Postharvest management of fruits and vegetables storage. *Sustainable Agriculture Reviews*, *15*, 65–152.

Gamage, A., Gangahagedara, R., Gamage, J., Jayasinghe, N., Kodikara, N., Suraweera, P., & Merah, O. (2023). Role of organic farming for achieving sustainability in agriculture. *Farming System*, *1*(1), 100005.

Getahun, S., Kefale, H., & Gelaye, Y. (2024). Application of precision agriculture technologies for sustainable crop production and environmental sustainability: A systematic review. *The Scientific World Journal*, *2024*(1), 2126734.

Javed, I., Waseem, A., & Ammad, R. (2019). Vegetables as a source of important nutrients and bioactive compounds: their human health benefits. *MOJ Food Process Technol*, *7*(4), 136–146.

JI, N. K., Barot, M., & Kumar, R. N. (2017). Distribution and biochemical constituents of different seaweeds collected from Okha coast, Gujarat, India. *Indian Journal of Geo Marine Sciences*, *46*(2), 349–357.

Kapoor, R., Kumar, A., Sandal, S. K., Sharma, A., Raina, R., & Thakur, K. S. (2022). Water and nutrient economy in vegetable crops through drip fertigation and mulching techniques: A review. *Journal of Plant Nutrition*, *45*(15), 2389–2403.

Khare, N., Khare, P., & Singh, S. (2025a). Cancer drug development from the wild edible plants: Needs and current status. In *Exploring Traditional Wild Edible Plants* (pp. 144–159). CRC Press.

Khare, N., Khare, P., & Singh, S. (2025b). Molecular and physiological concepts: Macronutrients in crop plant growth and development. In *Agricultural Crop Improvement* (pp. 148–164). CRC Press.

Khare, N., Khare, P., Barot, M., Modi, A., Jain, T., Mitra, S., . . . & Singh, S. (2025c). Role of non-coding RNA in plant architecture. In *Non-Coding RNA in Plants* (pp. 99–119). Academic Press.

Khatoon, H., Solanki, P., Narayan, M., Tewari, L., Rai, J. P. N., & Hina Khatoon, C. (2017). Role of microbes in organic carbon decomposition and maintenance of soil ecosystem. *International Journal of Chemical Studies*, *5*(6), 1648–1656.

Kumar, D., Kumar, S., Bhadana, N. K., Singh, B., & Shekhar, C. (2020). Vegetables: Source of adequate health benefits. *Annals of Horticulture*, *13*(2), 124–130.

Kumar, S., Fandan, R., & Sachin, P. (2023). Hydroponics and aeroponics: Advancement in soilles cultivation. *Recent trends in Agriculture*, *10*, 99–112.

Lal, M. K., Kumar, A., Kardile, H. B., Raigond, P., Changan, S. S., Thakur, N., . . . & Singh, B. (2020). Biofortification of vegetables. In *Advances in Agri-Food Biotechnology* (pp. 105–129). Springer Singapore.
Manzoor, A., Naveed, M. S., Ali, R. M. A., Naseer, M. A., Ul-Hussan, M., Saqib, M., . . . & Farooq, M. (2024). Vermicompost: A potential organic fertilizer for sustainable vegetable cultivation. *Scientia Horticulturae*, *336*, 113443.
Monteiro, A., Santos, S., & Gonçalves, P. (2021). Precision agriculture for crop and livestock farming – Brief review. *Animals*, *11*(8), 2345.
Montgomery, D. R., & Biklé, A. (2021). Soil health and nutrient density: Beyond organic vs. conventional farming. *Frontiers in Sustainable Food Systems*, *5*, 699147.
Nagar, B. L., Thakur, S. S., Goutam, P. K., Prajapati, P. K., Kumar, R., Kumar, S., . . . & Kumar, A. (2024). The role of bio-fortification in enhancing the nutritional quality of vegetables: A review. *Advances in Research*, *25*(4), 64–77.
Neina, D. (2019). The role of soil pH in plant nutrition and soil remediation. *Applied and Environmental Soil Science*, *2019*(1), 5794869.
Njira, K. O., & Nabwami, J. (2015). A review of effects of nutrient elements on crop quality. *African Journal of Food, Agriculture, Nutrition and Development*, *15*(1), 9777–9793.
Nyathi, M. K., Van Halsema, G. E., Beletse, Y. G., Annandale, J. G., & Struik, P. C. (2018). Nutritional water productivity of selected leafy vegetables. *Agricultural Water Management*, *209*, 111–122.
Ofori, K. F., Antoniello, S., English, M. M., & Aryee, A. N. (2022). Improving nutrition through biofortification–A systematic review. *Frontiers in Nutrition*, *9*, 1043655.
Parveen, R., Khare, N., Singh, S., & Johri, P. (2023). An in silico outlook for the detection and surveillance of evolving and persistent plant pathogens. In *Detection, Diagnosis and Management of Soil-borne Phytopathogens* (pp. 17–40). Singapore: Springer Nature Singapore.
Patani, A., Patel, A., Prajapati, D., Khare, N., & Singh, S. (2023). Various agriculture crop plant-based bioactive compounds and their use in nanomaterial synthesis and applications. In *Nanomaterials from Agricultural and Horticultural Products* (pp. 223–241). Singapore: Springer Nature Singapore.
Priya, A. K., Alagumalai, A., Balaji, D., & Song, H. (2023). Bio-based agricultural products: A sustainable alternative to agrochemicals for promoting a circular economy. *RSC Sustainability*, *1*(4), 746–762.
Quamruzzaman, A. K. M., Islam, F., Akter, L., & Mallick, S. R. (2022). Effect of maturity indices on growth and quality of high value vegetables. *American Journal of Plant Sciences*, *13*(7), 1042–1062.
Rajasree, V., & Pugalendhi, L. (2021). Breeding vegetables for nutritional security. In *Veganism-a Fashion Trend or Food as a Medicine*. IntechOpen.
Ramakrishnan, B., Maddela, N. R., Venkateswarlu, K., & Megharaj, M. (2021). Organic farming: Does it contribute to contaminant-free produce and ensure food safety? *Science of the Total Environment*, *769*, 145079.
Ray, S., & Majumder, S. (2024). Water management in agriculture: Innovations for efficient irrigation. *Modern Agronomy*, 169–185.
Reganold, J. P., & Wachter, J. M. (2016). Organic agriculture in the twenty-first century. *Nature Plants*, *2*(2), 1–8.
Richards, S., Rao, L., Connelly, S., Raj, A., Raveendran, L., Shirin, S., . . . & Helliwell, R. (2021). Sustainable water resources through harvesting rainwater and the effectiveness of a low-cost water treatment. *Journal of Environmental Management*, *286*, 112223.
Richter Reis, F. (2017). Effect of blanching on food physical, chemical, and sensory quality. In *New Perspectives on Food Blanching* (pp. 7–48).
Rouphael, Y., Cardarelli, M., Schwarz, D., Franken, P., & Colla, G. (2012). Effects of drought on nutrient uptake and assimilation in vegetable crops. In *Plant Responses to Drought Stress: From Morphological to Molecular Features* (pp. 171–195).
Scherr, S. J. (2019). The future food security and economic consequences of soil degradation in the developing world. In *Response to Land Degradation* (pp. 155–170).
Sharma, S., Katoch, V., Kumar, S., & Chatterjee, S. (2021). Functional relationship of vegetable colors and bioactive compounds: Implications in human health. *The Journal of Nutritional Biochemistry*, *92*, 108615.
Singh, A., Sharma, K., Chahal, H. S., Kaur, H., & Hasanain, M. (2025). Seaweed-derived plant boosters: Revolutionizing sustainable farming and soil health. *Frontiers in Soil Science*, *5*, 1504045.
Singh, T. B., Ali, A., Prasad, M., Yadav, A., Shrivastav, P., Goyal, D., & Dantu, P. K. (2020). Role of organic fertilizers in improving soil fertility. In *Contaminants in Agriculture: Sources, Impacts and Management* (pp. 61–77). Springer International Publishing.

Soni, R., Gupta, R., Agarwal, P., & Mishra, R. (2022). Organic farming: A sustainable agricultural practice. *Vantage: Journal of Thematic Analysis*, *3*(1), 21–44.

Średnicka-Tober, D., Obiedzińska, A., Kazimierczak, R., & Rembiałkowska, E. (2016). Environmental impact of organic vs. conventional agriculture-a review. *Journal of Research and Applications in Agricultural Engineering*, *61*(4), 204–211.

Thomas, L., & Singh, I. (2019). Microbial biofertilizers: types and applications. In *Biofertilizers for Sustainable Agriculture and Environment* (pp. 1–19). Springer International Publishing.

Toungos, M. D., & Bulus, Z. W. (2019). Cover crops dual roles: Green manure and maintenance of soil fertility, a review. *International Journal of Innovative Agriculture and Biology Research*, *7*(1), 47–59.

Verma, B. C., Pramanik, P., & Bhaduri, D. (2019). Organic fertilizers for sustainable soil and environmental management. In *Nutrient Dynamics for Sustainable Crop Production* (pp. 289–313). Singapore: Springer.

Waqas, M., Hashim, S., Humphries, U. W., Ahmad, S., Noor, R., Shoaib, M., . . . & Lin, H. A. (2023). Composting processes for agricultural waste management: A comprehensive review. *Processes*, *11*(3), 731.

Xing, Y., Wang, X., & Mustafa, A. (2025). Exploring the link between soil health and crop productivity. *Ecotoxicology and Environmental Safety*, *289*, 117703.

Yadav, N., & Sidana, N. (2023). Precision agriculture technologies: Analysing the use of advanced technologies, such as drones, sensors, and GPS, in precision agriculture for optimizing resource management, crop monitoring, and yield prediction. *Journal of Advanced Zoology*, *44*.

Yadav, S. K., Banerjee, A., Jhariya, M. K., Raj, A., Khan, N., Meena, R. S., & Kumar, S. (2021). Agroecology towards environmental sustainability. In *Sustainable Intensification for Agroecosystem Services and Management* (pp. 323–352). Springer Singapore.

Yasmeen, A. R., Maharajan, T., Rameshkumar, R., Sindhamani, S., Banumathi, B., Prabakaran, M., Atchaya, S., & Rathinapriya, P. (2025). Role of seaweeds for improving soil fertility and crop development to address global food insecurity. *Crops*, *5*(3), 29.

18 Postharvest Management to Preserve the Nutritional Properties of Vegetables

Mostafa Abdelkader, Assiya Ansabayeva, Mahmoud A.A. Ali, and Karim M. Hassan

18.1 INTRODUCTION

To achieve food security, the food system must have the capacity to provide sufficient energy and nutrients for healthy diets while maintaining social, economic, and environmental balance (Mutungi et al., 2022). After harvest, it is essential to keep nutrients in food to ensure food security, public health, and economic stability (Khan et al., 2024). The time immediately after harvesting is one of the most critical periods in the food supply chain, as both quantity (e.g. weight and volume) and quality (e.g. nutrients, flavour, and appearance) can be quickly and significantly affected (Prusky, 2011). If a product isn't handled, stored, or moved correctly, it could lose important vitamins and minerals, making it less healthy and less appealing to customers (Yadav et al., 2023).

These losses not only degrade food quality but also have significant adverse effects on the economy and the environment, as the land, water, energy, and labor required to produce that food are wasted when it expires or goes uneaten (Sawaya, 2017). Preserving nutrients (PN) is vital for eliminating malnutrition and achieving food security. If postharvest procedures do not preserve essential vitamins and minerals in food, it may not have enough nutrition for a balanced diet when it reaches consumers, especially in impoverished areas (Adeyeye, 2017a; Khan et al., 2024). This might cause 'hidden hunger'.

Hidden hunger is a type of malnutrition that occurs when people have sufficient food quantity but inadequate food quality (Kennedy et al., 2003; Scrimshaw, 1994). PN is the appropriate method that helps food last longer, ensures its availability, and lessens the effects of seasonal shortages. Generally, vegetables must be stored for freshness and nutrition. Vegetables produces are kept for extended periods in stores, making them less likely to spoil (Lewis, 1957). It also makes long-distance travel and trade safer. This encourages more people to eat a wider variety of foods and gives communities access to more nutrient-rich foods, helping them stay healthy in the long term (El-Ramady et al., 2015).

Farmers become more financially unstable when their products are not sold directly after harvesting. Losing nutrients and spoiling food can cause farmers to lose their income, which could put their lives at risk (Johnson et al., 2019). The FAO notes that every year, nearly one-third of all food produced worldwide is lost or wasted (Boliko, 2019; Flanagan et al., 2019). Improving how nutrients are stored helps reduce waste, optimize resource use, and enhance the efficiency of food systems. Building cold storage facilities, drying systems, and processing units after harvest creates more jobs, supports local economies, and helps international trade by keeping food fresher for longer (Kitinoja, 2013). These changes not only strengthen farming communities but also make the country's food supply more stable (Adeyeye, 2017b).

Keeping nutrients in the soil is also very good for the environment. When food goes bad, it wastes resources, including water, energy, fertilizers, and land. By reducing these losses and maintaining

DOI: 10.1201/9781003606710-18

high food quality, we can feed the same number of people with less land dedicated to cropland (Wirsenius et al., 2010). This lowers GHG emissions and makes farming activities less harmful to the ecosystem. When inputs for food production are applied wisely, ecosystems are less stressed, which helps manage resources more sustainably (Rehman et al., 2022; Sahu & Arya, 2024).

Several factors can delay nutrient retention after harvest. High temperatures accelerate respiration and enzymatic activities. High temperatures also accelerate food degradation, leading to the loss of sensitive vitamins, such as A, C, and riboflavin (Ottaway, 2010; Paulus, 1989). Losing moisture causes wilting, shrivelling, and weight loss (Nunes & Emond, 2007). Food can break down vitamins and colours when it is exposed to light and oxygen. This process makes the food less healthy and less visually appealing (Choe et al., 2005). Also, pests and microbes cause decay and lose nutritional value, rendering the food unsuitable for consumption.

Various methods for PN after harvest can help alleviate the issues previously mentioned. Harvesting crops at the right time and in the cooler parts of the day helps keep the field cool and slows spoilage. One of the most effective ways to extend the shelf life of perishable goods is to control temperature, through either cold storage or the cold chain (Aung & Chang, 2014). Controlling the humidity is also very significant. High humidity prevents fresh vegetables from drying out, while low moisture levels in grains inhibit mould growth (Magan & Aldred, 2007).

Packaging in a modified environment may alter the surrounding gases, preventing oxygen from reaching the food and slowing down its ripening. Advanced storage, such as hermetic containers, makes it difficult for pests and bacteria to survive in low-oxygen conditions without chemical usage (Navarro, 2012, 2010). Training is just as vital, such as teaching farmers, dealers, and handlers the best practices for handling and treating food after it is harvested (Soon & Baines, 2012). This helps them retain the nutritious content of food and reduce losses throughout the supply chain.

In this chapter, we discuss the biological changes that may occur in postharvest vegetables. We will emphasize the key factors contributing to postharvest nutrient losses. This work focuses on the roles of temperature, relative humidity, handling, packaging, and storage practices. The chapter will also explore the effects of minimal processing on vegetable quality. Finally, we will investigate the socioeconomic and sustainability aspects involved in minimizing postharvest losses of vegetables.

18.2 PHYSIOLOGICAL AND BIOCHEMICAL CHANGES AFTER HARVESTING

Postharvest, vegetables remain biologically active and continue to undergo various physical and chemical transformations. These metabolites affect their freshness, shelf life, and nutritional content. Despite being detached from the plant, their tissues continue to respire, lose water, and undergo enzymatic and microbiological processes, resulting in deterioration (Song et al., 2002). Understanding the mechanics of these processes is important for effectively handling and storing vegetables. Hence, it reduces waste and maintains their freshness for a longer period.

Respiration is one of the primary processes that persist postharvest (Table 18.1). Vegetables continue to use oxygen and emit carbon dioxide, water, and heat as they metabolize stored carbohydrates, sugars, and organic acids for energy. This continuous breathing gradually depletes the plant's energy reserves, hastening ageing and tissue degradation. Temperature, gas levels, and vegetable type influence the respiration rate. Leafy greens, such as spinach and lettuce, exhibit rapid respiration, resulting in a short shelf life (Ferdousi et al., 2024b).

In contrast, carrots and potatoes respire more slowly, thereby extending their longevity. Excessive temperature or inadequate ventilation leads to the accumulation of heat and moisture, thereby accelerating degradation. Conversely, keeping vegetables in colder or controlled-atmosphere environments reduces respiration and helps maintain quality (Mishra & Gamage, 2007).

Water loss is a significant postharvest challenge that impacts both aesthetics and texture. Upon harvesting, vegetables experience moisture loss via their surfaces and any injuries, disturbing their internal water equilibrium. The rate of occurrence depends on temperature, humidity, air circulation, and the vegetable's physical attributes (Atanda et al., 2011; Kiaya, 2014). Leafy vegetables

TABLE 18.1
Physiological and Biochemical Changes After Harvesting Vegetables

Change Type	Description/Examples	Impact on Quality/Shelf Life	Citations
Respiration	Rapid increase postharvest, then decline; continues to consume nutrients and energy.	Accelerates senescence and nutrient loss	King et al. (1990), Umeohia and Olapade (2024)
Water Loss (Transpiration)	Loss of water due to continued transpiration after detachment	Leads to wilting, loss of texture, and freshness	Jiang et al. (2021), Shezi et al. (2024)
Ethylene Production	Increases after harvest, especially in wounded tissues	Promotes ripening and senescence	P. Ludford (1995), Shezi et al. (2024)
Enzymatic Activity	Increased activity of enzymes (PPO, POD, PAL) leading to browning and degradation	Causes discolouration, loss of nutritional value	Toivonen and Brummell (2008)
Chlorophyll Degradation	Breakdown of chlorophyll, especially in leafy greens	Yellowing, loss of visual appeal	Tonto et al. (2023)
Carotenoid Changes	Carotenoid content may increase or decrease depending on storage conditions	Alters nutritional value (provitamin A)	Hamieh et al. (2023)
Protein and Amino Acid Changes	Initial increase, then decline in protein; amino acid composition shifts (e.g. glutamine decreases)	Affects nutritional quality and flavour	King et al. (1990), Lill et al. (1990)
Antioxidant Compounds	Changes in phenolics, vitamins, and antioxidant activity	Impacts health benefits and shelf life	Nassarawa et al. (2020), Sivakumar et al. (2022)
Cell Wall Modification	Cell wall disassembly, loss of turgor, and increased lignification	Softening, loss of texture, increased toughness	Toivonen and Brummell (2008)
Hormonal Changes	Deprivation of hormones from parent plant, altered endogenous hormone levels	Affects senescence and stress responses	P. M. Ludford (1995)
Microbial Load	Increased susceptibility to microbial spoilage	Reduces shelf life, increases decay	Sonntag et al. (2023)

with extensive surface area dehydrate rapidly and wilt, whereas thicker-skinned varieties, such as cucumbers or peppers, retain moisture better. Even a slight dryness can cause vegetables to lose their firmness and become withered. Ensuring elevated humidity, reducing ventilation, and using packaging or wax coatings may help retain moisture and mitigate wilting (Devi et al., 2022).

Enzymes within vegetables remain active, facilitating chemical reactions that modify flavour, colour, and texture. PPO facilitates browning in cut potatoes and eggplants, whereas chlorophyllase degrades chlorophyll, resulting in the yellowing of green vegetables. Pectin methylesterase and polygalacturonase are enzymes that break down cell walls, leading to tissue softening. Lipoxygenase accelerates the oxidation of fats, leading to the formation of off-flavours, while ascorbic acid oxidase diminishes vitamin C concentrations. Storing vegetables at lower temperatures and minimizing damage can reduce enzymatic spoilage, as these reactions are accelerated by warmth (Dornenburg & Davies, 1999; Temiz & Ayhan, 2017).

Microbial activity constitutes a vital factor in postharvest decay. When a vegetable's inherent defences are diminished, microorganisms may infiltrate. Pathogens such as Erwinia carotovora cause soft rot in carrots and cucumbers. Fungi such as Botrytis cinerea and Rhizopus stolonifer cause mould formation and tissue degradation (Lalpekhlua et al., 2024). Microbes infiltrate the body through wounds or natural openings, releasing enzymes that degrade cell walls, leading to tissue softening, leakage, and subsequent decay. Warm and humid environments promote microbial proliferation. This can lead to undesirable odours or toxins. Preventing this requires strict sanitation, rapid postharvest cooling of produce, and controlled-atmosphere storage to alleviate microbial growth (Das et al., n.d.).

A complex interplay of respiration, moisture loss, enzymatic activity, and microbial spoilage influences the postharvest longevity of vegetables. The interrelation of these processes is evident; for instance accelerated respiration and dehydration establish optimal conditions for enzymatic reactions and microbial invasion. Effective postharvest management involves reducing metabolic rates, maintaining moisture equilibrium, minimizing oxidation, and preventing microbial contamination by precisely regulating temperature and humidity. When effectively managed, these measures can substantially prolong shelf life, maintain nutritional quality, and reduce waste across the entire supply chain, from production to market (Rolle, 2006).

18.2.1 Factors Contributing to Postharvest Nutrient Losses

Some vegetables deteriorate rapidly after harvest, while others remain fresh for extended periods. The key typically resides not in the events that follow selection, but in the critical decisions and agricultural methods implemented months before the harvest. Preharvest variables significantly influence the postharvest quality and shelf life of vegetables, rendering them either robust champions or susceptible victims along the supply chain (Agriculture Institute, 2023).

The foundation of postharvest success lies in the field. Preharvest management is analogous to constructing a home; a strong foundation is essential for the stability of the whole structure. The choices farmers make during the growing season directly affect the resilience of their food throughout storage, transit, and market handling. Studies indicate that up to 60% of postharvest losses may be attributed to preharvest factors and management strategies (Table 18.2). Comprehending these

TABLE 18.2
Factors Contributing to Postharvest Nutrient Losses

Factor	Description/Impact	Citations
Temperature	Accelerates nutrient degradation, especially vitamin C; improper cold storage increases losses	Ahmad and Siddiqui (2015), Akanbi (2021), Etefa et al. (2022), Lee and Kader (2000), Makule et al. (2022)
Mechanical Damage	Bruising, cuts, and rough handling increase decay and nutrient loss	Elik et al. (2019b), Etana (2019), Teshome and Leweye (2020)
Water Loss/ Dehydration	Leads to rapid vitamin C and quality loss, especially in leafy vegetables	Lee and Kader (2000), Singh et al. (2022)
Storage Duration	Extended storage increases nutrient loss, especially under non-optimal conditions.	Ahmad and Siddiqui (2015), Teshome and Leweye (2020)
Poor Packaging	Inadequate or inappropriate packaging increases exposure to air, moisture loss, and contamination.	Elik et al. (2019b), Etana (2019)
Inadequate Cooling/ Cold Chain	Lack of precooling and cold storage infrastructure accelerates spoilage and nutrient loss.	Akanbi (2021), Makule et al. (2022)
Pest and Disease Infestation	Insects, fungi, and rodents cause direct nutrient loss and spoilage.	Etefa et al. (2022), Hua et al. (2018), Workineh and Enyew (2021)
Improper Harvesting Techniques	Harvesting at the wrong maturity, rough handling, or delayed harvest increases losses.	Workineh and Enyew (2021), Yeshiwas and Tadele (2021)
Humidity	Promotes microbial growth and nutrient degradation.	Etefa et al. (2022), Lee and Kader (2000)
Socioeconomic	Lack of knowledge, poor market access, inadequate storage, and transport facilities exacerbate loss.	Qange et al. (2024)
Processing Methods	Blanching, dehydration, and improper cooking can cause significant loss of vitamins and minerals.	Khan (1989)

relationships enables farmers to make informed decisions that provide long-term benefits, successfully optimizing their crops for optimal success (P. M. Ludford, 1995).

Choosing the appropriate cultivar selection is one of the most important preharvest considerations. Not all cultivars exhibit uniform performance postharvest. Breeders have developed cultivars with different characteristics, including thicker skins, firmer meat, and less water content. For example Roma tomatoes are cultivated for processing and transportation. Their strong walls and little water content make it suitable for long-distance transportation (Smith et al., 2006). Certain kinds have genetic characteristics that inhibit ripening and senescence, hence extending shelf life. Furthermore, cultivars with inherent disease resistance reduce the likelihood of postharvest deterioration by reducing the amount of disease spores transported into storage (Agriculture Institute, 2023). Successful farmers choose varieties recognized for their superior storage capacity, implement meticulous irrigation and nutrition regimens that prioritize calcium, conduct careful pruning to enhance air circulation, and safeguard their orchards from temperature fluctuations and wind damage (John et al., 2023).

Managing actions during the growth season significantly impacts plants. It serves as a preparatory step for postharvest challenges. Pruning increases productivity, fruit quality, and shelf life. By eliminating surplus vegetative growth, farmers enhance air circulation and sunshine penetration. This is resulting in thicker skins and more robust cell walls of fruits. Pruning balances the plant's energy expenditure, directing resources toward fruit development instead of excessive vegetative growth (Asrey et al., 2013).

Crop cleanliness and field sanitation are fundamental aspects of preharvest quality. A pristine field functions similar to a sterile medical setting; it prevents issues before they arise. Consistently removing infected plant matter, controlling weeds, properly disposing of agricultural waste, and sanitizing equipment reduce pathogen spread in the field. Weeds compete for nutrients and harbour pests and diseases that can impact harvested crops (Elik et al., 2019a).

Irrigation is crucial for achieving preharvest success. Plants subjected to water stress during development usually yield fruits that subsequently exhibit cracking, compromised cell integrity, or reduced storage longevity. Conversely, excessive watering results in fruits with elevated water content, making them more susceptible to rot and reducing their shelf life. Adjusted moisture levels, especially during fruit growth, are crucial for maintaining quality. Optimal timing and regulated irrigation help develop firm texture, concentrated sugars, and resilient skins in fruits (Agriculture Institute, 2023).

Nutrient management is a crucial aspect influencing the postharvest longevity of vegetables (El-Ramady et al., 2015). Supplying the body with appropriate nutrients throughout development is akin to providing an athlete with a balanced diet; it enhances resilience and endurance. Calcium is critical for refreshing cell walls and membranes. It reduces susceptibility to mechanical injury and deterioration. Its deficiencies may lead to illnesses such as blossom end rot in tomatoes, which reduces marketability (Abdelkader et al., 2024a; Moghazee et al., 2025).

Potassium is essential because it affects fruit hardness and sugar concentration; however, excessive amounts may disrupt calcium absorption, underscoring the need for balance. Excessive nitrogen results in soft, decay-susceptible fruits, necessitating reduced treatments as harvest nears. Micronutrients are crucial for maintaining structural integrity and preventing underlying defects that may manifest postharvest (Weston & Barth, 1997).

Environmental factors have lasting impacts on postharvest performance (Galindo et al., 2004). Exposing to high temperatures during growth makes fruits more susceptible to rot. Cold stress can interfere with ripening and reduce storage longevity. Wind can generate micro-injuries that serve as portals for infections. Although many preharvest activities may require more labor or expense, they often yield returns through enhanced produce quality and reduced waste. Investing in effective preharvest management boosts profitability and promotes sustainability by reducing losses and ensuring that products reach customers in good condition.

18.2.1.1 Temperature and Relative Humidity

Vegetables experience postharvest changes, leading to accelerated ripening and senescence, increased vulnerability to postharvest infections, and reduced economic value and shelf life (Table 18.3). Physical stress and microbial infection lead to direct losses and physiological responses, including increased ethylene production, which accelerates ripening and senescence (Ansari & Tuteja, 2015). Prevalent postharvest treatments include low temperature, regulated or modified environment, and ethylene inhibitors. These therapies are often used with low-temperature protocols, yielding synergistic results (Zuo et al., 2025a).

Reducing the temperature of postharvest produce and preserving the cold chain throughout the postharvest phase are essential for proper management of horticultural products. Temperature regulation is the most effective method for prolonging the shelf life of fresh horticulture products. The ideal storage temperature depends on the origin of the produce, the plant organ, and its stage of development. Low temperatures impede metabolism and reduce the pace of compositional alterations, while temperatures below a certain threshold may cause chilling damage (Aghdam & Bodbodak, 2014).

Refrigeration profoundly influences the entire metabolism of plants and tissues. The temperature coefficient measures the changes in biological systems resulting from temperature variations. Reducing the temperature from 15°C to 5°C lowers respiration by a factor of 2, thereby enhancing shelf life. Low temperatures affect ethylene physiology by consistently reducing ethylene production rates, which is attributed to downregulation of ACS and ACO gene expression in both tropical and temperate fruit species. Reduced temperatures also influence enzyme activities, including ACS activity (Zuo et al., 2025a).

TABLE 18.3
Effects of Temperature and Relative Humidity on Postharvest Quality of Vegetables

Vegetable	Optimal Temperature (°C)	Optimal Relative Humidity (%)	Effects of Suboptimal Conditions	Shelf Life/ Quality Impact	Citations
Leafy vegetables	0–2	95–100	High temp./low RH: rapid wilting, weight loss, yellowing	Shelf life reduced by up to 75% at low RH	Agüero et al. (2011), Ambuko et al. (2017), Schmitz et al. (2019)
Root vegetables	0	98	Low RH: increased water loss, texture loss	Best quality at 0°C, 98% RH	Ilić et al. (2016), Zainalabidin et al. (2019)
Tomatoes	10 (short-term), 22 (room temp.)	95 (low temp.), 65 (room temp.)	High temp./low RH: increased chemical/ nutritional loss, weight loss	Quality loss faster at higher temp/RH	Al-Dairi et al. (2021), Ferdousi et al. (2024a)
Brassica (cabbage, broccoli, kale)	0–4	95–100	High temp.: increased senescence, water loss	Cold storage slows senescence	Ahlawat and Liu (2021), Zainalabidin et al. (2019)
Zucchini	Near 98	98	Low RH: increased chilling injury, weight loss	High RH reduces chilling injury	Wu et al. (2023)
Lettuce	0–2	95–100	Low RH: rapid weight loss, pigment loss, poor sensory quality	Shelf life 5 days at low RH, 20 at high	Hoffmann et al. (2021), Schmitz et al. (2019)
Amaranth	0–2	95–100	High temp./low RH: rapid vitamin C loss, wilting	Best preserved at low temp./high RH	Ambuko et al. (2017), Zainalabidin et al. (2019)

The cold storage of fruits may influence ethylene detection and signal transduction. This affects the expression of receptor genes and signalling cascades. In nonclimacteric fruit, the expression of ethylene receptor genes is influenced by low-temperature storage (Rugkong et al., 2011). Storing fruit in cold conditions may lead to chilling injury. Prolonged exposure to low temperatures in tomatoes results in visible damage, including loss of fragrance, blotchy ripening, excessive softening, pitting, increased susceptibility to rot, electrolyte leakage, and an inability to ripen (Tao et al., 2024).

The initiation and progression of chilling damage are contentious and appear to vary across species. Low temperatures induce ethylene synthesis in several cold-sensitive fruits. This initiates chilling injury. Implementing strategies to enhance ethylene production before or during cold storage can significantly reduce the incidence of chilling injury. Ethylene production is enhanced during extended cold exposure, and this effect is further enhanced when the fruits are transferred to room temperature (20°C). This physiological response may enhance quality, but might be mitigated by preserving fruits in a controlled climate (Zuo et al., 2025a).

Vegetables are nutritionally rich and offer several health benefits, but they are particularly susceptible to quality degradation and senescence due to biological processes and inadequate storage conditions (Tort et al., 2022). Water plays crucial biochemical roles in postharvest quality, mechanical characteristics, and physiological metabolism (Rodov et al., 2010).

Fresh vegetables, once separated from the plant, experience reduced water supply and are particularly susceptible to moisture loss through physiological processes, namely transpiration and respiration, leading to quality degradation and economic losses during storage. Transpiration is recognized as a cause of water loss in fresh produce and is influenced by relative humidity. Meanwhile, respiration generates heat, increasing the vapor pressure difference and leading to additional water loss (Galindo et al., 2004). Effective control of water loss throughout the storage life of vegetables is essential for maintaining optimal quality and marketability.

Relative humidity is the moisture content of the air, expressed as a percentage, defined by the ratio of the actual vapor pressure to the saturation vapor pressure at a specific temperature. It is a crucial environmental factor impacting the evaporation rate of vegetables during postharvest storage (Díaz-Pereira et al., 2020). The moisture content of most fresh vegetables ranges from 80 to 96%, and relative humidity below 80% leads to water loss, respiratory exhaustion, and physiological damage (Rodov et al., 2010). The relative humidity of ordinary ambient conditions, typical refrigerator-freezers, and storage cold rooms usually varies from 40% to 80%.

Storage of fresh vegetables requires a relative humidity exceeding 85%, in contrast to the recommended range of 40%–60% for dried fruits (Galindo et al., 2004). Inadequate relative humidity control at each postharvest step of vegetables, including storage, processing, marketing, and ultimate delivery to the customer, may result in both quantitative and qualitative food losses (Prusky, 2011).

The water content of most fresh vegetables is often regarded as saturated. Elevated relative humidity may diminish the water vapor pressure differential between the storage environment and the product, thereby impeding water loss from the produce through transpiration. A growing body of research indicates that elevated relative humidity during storage effectively mitigates moisture loss from vegetables, thereby delaying quality degradation caused by water loss during storage and distribution (Mukama et al., 2018). Moreover, elevated relative humidity enhances cell ultrastructural stability (Zan et al., 2020) and influences various physiological processes, including antioxidant responses (Xiao et al., 2019), degradation of bioactive compounds (Cheng et al., 2020; Ktenioudaki et al., 2019), ripening and senescence (Lufu et al., 2021), and stress tolerance (Henriod, 2006) in vegetables during postharvest storage.

Except for certain low-water-potential commodities, such as garlic and onions, the ideal relative humidity for storing most vegetables ranges from 85% to 98% RH, exceeding typical atmospheric conditions (Mahmood et al., 2019). Consequently, establishing an artificial storage environment with elevated relative humidity is crucial for reducing postharvest losses. The use of elevated relative humidity has lately emerged as a crucial method for the postharvest preservation of vegetables. Two primary methods are commonly employed: humidification within a storage facility and the creation of a localized microenvironment with elevated relative humidity through packaging. This has been validated to provide high relative humidity around the stored product (Rodov et al., 2010).

Specific mechanical humidification techniques have been documented to preserve postharvest product quality during storage, including ultrasonic, nano-mist, and dry-fog humidifier systems (Zuo et al., 2025b). Modified RH packaging (MHP) is an effective technique for preventing water loss and quality degradation during postharvest storage, including modified carton boxes (Harima et al., 2002), modified polyethylene bags, perforated PE bags, and shrink wrap.

Furthermore, various systems and equipment for the storage of fresh produce, including an evaporative cooling chamber (ECC) (Chopra et al., 2022), a pallet unit load-controlled atmosphere and relative humidity system, and a modified humidity clamshell (Olmedo et al., 2025), have been developed to maintain elevated relative humidity in the storage environment. Moreover, the characteristics of the produce and the prevention of condensation are crucial for the implementation of humidification technologies, presenting a challenge for their further application in the preservation of vegetables.

Despite growing research on water loss in postharvest horticultural products, there is a lack of systematic analysis of water loss, transpiration, and the implications and uses of high relative humidity storage for postharvest vegetables (Bisht & Singh, 2024).

18.2.1.2 Handling, Packing, and Storage

Annually, millions of tons of fresh vegetables are wasted between harvest and consumption owing to inadequate handling, packaging, storage, and transportation (Table 18.4). These postharvest losses signify wasted food and significant economic detriments for stakeholders. Understanding effective postharvest management of fresh produce is crucial for optimizing quality and shelf life. This will boost profitability and safeguard food security (Budhathoki et al., 2022).

Postharvest vegetable losses can reach 20–40%, primarily due to inadequate handling. Such losses arise from physical damage, physiological decline, and pathological degradation,

TABLE 18.4
Impact of Handling, Packing, and Storage on Postharvest Quality of Vegetables

Factor	Key Impacts on Postharvest Quality	Recommended Practices/Technologies	Citations
Handling	Poor handling causes bruising, mechanical damage, water loss, decay, and rapid deterioration in quality.	Gentle handling, cleaning, sorting, grading, precooling	Budhathoki et al. (2022), Pokhrel (2020), Ufitinema et al. (2024), Zainalabidin et al. (2019)
Packing	Inadequate packaging leads to physical damage, increased spoilage, and reduced shelf life.	Use of strong, suitable materials (e.g. LDPE, MAP, vacuum packs)	Batziakas et al. (2020), Nasrin et al. (2018), Yeboah et al. (2023)
Storage	Improper temperature/humidity accelerates senescence, decay, and nutrient loss; optimal storage extends shelf life.	Rapid cooling, cold/evaporative storage, controlled/modified atmospheres	Kaur et al. (2021), Kunwar et al. (2024), Yeshiwas and Tadele (2021)
Technological Advances	Innovative packaging, nano-coatings, AI-based quality monitoring, and advanced storage improve quality and reduce losses.	Nano-packaging, AI monitoring, cold chain, nondestructive sensors	Palumbo et al. (2022), Singh et al. (2022), Verma et al. (2025)
Socioeconomic Factors	Lack of infrastructure, training, and resources increases losses, especially in developing countries.	Training, infrastructure investment, supply chain management	Kaur and Watson (2024), Kunwar et al. (2024), Yeshiwas and Tadele (2021)

commencing immediately upon harvest. In contrast to grains, which can withstand harsh handling, fresh food remains viable and continues to respire, ripen, and deteriorate postharvest. The economic ramifications of these losses extend far beyond the apparent waste; when product deteriorates during storage or transit, all investments in seeds, fertilizers, labor, and irrigation are essentially forfeited. Moreover, decayed vegetables may contaminate adjacent healthy produce, initiating a chain reaction that exacerbates losses throughout the supply chain (Zainalabidin et al., 2019).

Mechanical injuries are one of the most apparent and avoidable causes of postharvest loss. These injuries arise when the product is exposed to impact during manipulation. Even minor lacerations may serve as entry points for bacteria, hastening deterioration and diminishing market value. Harvesting techniques may result in accidents. Using inappropriate instruments, plucking fruits rather than severing stems, or mishandling them may easily lead to punctures and bruises. Transportation significantly influences outcomes; overcrowded containers, insufficient padding, and vehicle vibrations may lead to surface abrasions or interior bruising. In storage facilities, inadequate stacking methods, inappropriate container dimensions, and negligent handling during loading or unloading exacerbate mechanical damage (Batziakas et al., 2020).

Mitigating these losses starts with careful handling techniques. People must be instructed to recognize the fragile nature of fresh food and to use proper harvesting practices. This includes carefully severing stems and supporting fruits during collection. The use of suitable containers, constructed from smooth materials with sufficient cushioning, mitigates mechanical damage during transportation and storage. Temperature regulation is perhaps the most critical element in prolonging the postharvest lifespan of vegetables (El-Ramady et al., 2015; Zuo et al., 2025b). Each product type has a specific ideal storage temperature, and slight fluctuations may accelerate spoilage. Conversely, temperate fruits require near-freezing temperatures to decelerate respiration and preserve firmness. This notion underpins the 'cold chain' – a continuous system of low-temperature management from harvest to retail presentation. A minor interruption in this network may significantly reduce shelf life and compromise quality (Maheshwar & Chanakwa, 2006; Mounika et al., 2025).

Effective temperature regulation encompasses numerous essential processes. Swiftly eliminating field heat postharvest using techniques such as hydrocooling, vacuum cooling, or forced-air chilling may prolong product shelf life by several days or even weeks. Controlled environment storage, characterized by reduced oxygen and elevated carbon dioxide levels, may effectively slow down respiration and delay senescence. Dependable temperature monitoring systems equipped with alerts are essential for maintaining consistent temperatures and promptly identifying variations that may compromise product quality (Roberts, 1972).

Humidity regulation works in conjunction with temperature control. Most vegetables require a humidity range of 85–95% to prevent water loss, which can result in withering, shrivelling, and diminished freshness. Fresh food consistently loses moisture through transpiration, and when humidity is very low, this process intensifies, diminishing turgor pressure and resulting in observable quality degradation (Mukama et al., 2018; Olmedo et al., 2025).

Conversely, excessively high humidity fosters condensation, facilitating the proliferation of fungi and bacteria. Leafy vegetables are highly sensitive and quickly lose freshness in arid conditions. Root crops, although more resilient, may still degrade in texture and weight when stored at inadequate humidity levels. Optimal humidity requires effective ventilation control to balance moisture and airflow, minimizing moisture reduction while maintaining a suitable humidity level. Packaging with perforated plastic sheets helps maintain an optimal microclimate for the produce. Misting or humidification systems are often used in extensive storage facilities to provide uniform humidity (Alferez et al., 2005).

Packaging is a vital factor in prolonging the postharvest shelf life of vegetables. Appropriate packaging protects goods from physical harm and helps regulate the ambient environment. Selecting packing materials depends on their gas permeability, moisture retention, and protective efficacy. Perforated plastic films are optimal for vegetables that require ventilation and moisture regulation.

At the same time, permeable paper wraps are suitable for produce that requires air circulation (Zhang et al., 2015). Modified atmosphere packaging (MAP) enhances preservation. It alters the gas composition within the package, decreasing oxygen levels and increasing carbon dioxide levels to inhibit respiration and delay ripening.

Active packaging solutions, including ethylene absorbers, moisture regulators, and antibacterial agents, provide enhanced protection against spoilage (Farber, 1991). The shape and dimensions of containers significantly impact postharvest quality. Oversized containers permit excessive product movement during transit, leading to bruising, while small containers may cause compression damage. Well-constructed containers have smooth interiors, air openings, and balanced weight distribution to prevent crushing (Kitinoja & Kader, 1995).

Transportation is also a crucial phase during which several postharvest losses occur. Vehicles used to transport produce must be sanitized, well ventilated, and equipped with temperature-control systems where required. Adhering to proper loading protocols, such as ensuring equitable weight distribution and minimizing handling, helps mitigate damage. Produce must be categorized on the basis of its ethylene production and sensitivity; for example ethylene-emitting fruits like apples should not be stored or transported with ethylene-sensitive vegetables such as lettuce or cucumbers, as this hastens ripening and deterioration. Optimized route design that minimizes trip duration and handling further mitigates the likelihood of degradation (Al-Dairi et al., 2021).

The design and management of storage facilities are equally essential for postharvest success. An efficient storage infrastructure must have appropriate insulation, temperature and humidity regulation systems, adequate ventilation, and sufficient drainage to avert moisture accumulation (Kaur et al., 2021). Facilities must be structured to facilitate efficient loading and unloading, thereby reducing product handling. Implementing a first-in, first-out (FIFO) inventory rotation ensures that older products are dispatched before fresh arrivals, thereby minimizing the risk of overstocking and deterioration. Consistent monitoring and inspection of stored produce enable the early identification of issues and prompt action to prevent substantial losses (Kunwar et al., 2024).

Implementing effective postharvest management measures yields significant economic benefits. Farmers and distributors who utilize efficient handling, packaging, storage, and transportation methods often experience a 15–30% increase in marketable product. Despite the initial expenditures for equipment, infrastructure, or training, the resultant benefits of reduced losses and enhanced product quality substantially outweigh these expenses. Premium fruits and vegetables that arrive at the market in optimal condition command higher prices, possess extended shelf lives, and have enhanced nutritional and sensory attributes. These traits enhance customer satisfaction, facilitate access to premium markets, and foster enduring commercial partnerships built on trustworthiness and quality (Yeshiwas & Tadele, 2021).

18.3 MINIMAL PROCESSING OF VEGETABLES PRODUCES

Minimal processing (Table 18.5) denotes the methods used to prepare vegetables while preserving their fresh-like flavour and nutritional integrity (Corbo et al., 2015). In contrast to conventional food-processing techniques, which often involve significant heating, drying, or chemical treatments, minimum processing aims to maintain the product's proximity to its original condition. These tasks include washing, peeling, chopping, and packing. The objective is to provide customers with goods that are convenient, safe, and healthy, while maintaining their flavour and nutritional benefits. Minimal processing methods aim to prolong the shelf life of fresh food while preserving its sensory and nutritional attributes (Ali et al., 2018; Temiz & Ayhan, 2017).

The increasing demand for minimally processed vegetables is driven by multiple factors. In the contemporary fast-paced environment, convenience has become a major determinant of consumer preference. Consumers consistently seek ways to save time in their daily activities, and less-processed items offer a solution. Pre-cut vegetables obviate the need for washing, peeling, and chopping, thereby facilitating quick, effortless meal preparation. This convenience is desirable to

TABLE 18.5
Minimal Processing of Vegetables and Its Impact on Postharvest Quality

Minimal Processing Technique	Impact on Postharvest Quality	Key Advantages/ Disadvantages	Citations
Cutting, peeling, slicing	Increases respiration, ethylene production, and susceptibility to microbial spoilage; reduces shelf life	Convenience, but accelerates senescence and quality loss.	Alexandre et al. (2022), Ali et al. (2018)
Washing and sanitization (e.g. chlorine, ozone, electrolysed water)	Reduces microbial load; may cause tissue stress and nutrient loss; chlorine can form harmful by-products	Microbial safety, but with potential chemical residues and tissue damage	Alexandre et al. (2022), Chiesa (2003), De Chiara et al. (2024)
Edible coatings (e.g. aloe vera, mucilage, chitosan)	Slows moisture loss, oxidation, and browning; maintains sensory and nutritional quality; extends shelf life	Eco-friendly, preserves taste and nutrients, but may not control all microbes	Jiang et al. (2021), Leon et al. (2023), Liguori et al. (2022), Ncama et al. (2018), Papoutsis and Edelenbos (2021)
Modified Atmosphere Packaging (MAP)	Reduces respiration and microbial growth; helps retain nutrients and texture	Extends shelf life, but improper use can cause off-flavours or spoilage.	Chiesa (2003), Ncama et al. (2018)
Physical treatments (e.g. UV, heat, high pressure, cold plasma)	Reduces spoilage and pathogens; can maintain or enhance phytonutrients; excessive treatment may damage tissue	Nonthermal, residue-free, but requires optimization to avoid quality loss	Duarte-Sierra et al. (2020), Palumbo et al. (2022), Perera et al. (2022)
Cold storage	Slows metabolic and microbial activity; preserves quality	Essential for shelf life extension, but not always accessible in all regions	Elolu et al. (2023), Gouda and Duarte-Sierra (2024)

busy people and families seeking to enjoy nutritious meals without dedicating excessive time to cooking (Shanky Bhat, 2013).

Consumers are getting more selective about the quality of the food they ingest. Minimally processed items are perceived as fresher and closer to their natural state than fully processed counterparts. The perception of superior quality is a crucial factor influencing demand for minimally processed vegetables (Siddiqui et al., 2011). Health awareness is increasing as more individuals pursue healthy diets. Minimal processing preserves the nutritional integrity of vegetables, making them appealing to health-conscious customers (Klein, 1987). These items are often free of chemicals and preservatives, reflecting the growing preference for clean, natural meals. The rise in popularity of minimally processed food may be ascribed to changing lifestyles and demographic changes. Minimal processing has various advantages as follows:

Prolonged shelf life: The primary advantage of minimal processing is the extended shelf life of fresh goods. Methods such as MAP and controlled atmosphere storage inhibit ripening and spoilage, enabling vegetables to remain fresh for extended periods. This minimizes food waste and provides customers with more opportunities to enjoy their purchases (Gómez-López et al., 2008).

Preservation of nutritional integrity: Minimal processing procedures aim to maintain the nutritional integrity of vegetables. In contrast to conventional processing techniques that sometimes use high temperatures or chemical treatments, minimum processing aims to preserve the inherent vitamins, minerals, and antioxidants in produce. This ensures that customers derive the utmost health benefits from their food (Pasha et al., 2014).

Improved food safety: Minimal processing includes procedures such as washing and sanitizing, which help eliminate dirt, pesticides, and pathogenic microbes from the food. This improves food safety, hence decreasing the likelihood of foodborne infections. Appropriate packaging and storage enhance the safety and quality of minimally processed items (Ohlsson & Bengtsson, 2002).

Accessible and pre-prepared: Minimally processed vegetables are often pre-cut, peeled, and packed, making them ready to use immediately upon removal from the packaging. This convenience offers a significant benefit for customers, particularly those with busy lives. It obviates the need for laborious preparatory processes, facilitating fast and effortless meal preparation (Monteiro, 2009). On the other hand, minimal processing has many disadvantages, such as:

Expense: Minimally processed items often command a greater premium than their raw equivalents. The supplementary procedures for processing, packing, and ensuring food safety increase the overall cost. This may provide an obstacle for some customers, especially those with constrained financial resources (Shewfelt, 1987).

Ensuring quality: Maintaining the quality and freshness of minimally processed food throughout the supply chain is a significant challenge. Appropriate management, storage, and conveyance are essential to avoid deterioration and preserve the intended sensory characteristics. Any deficiencies in these domains may lead to a decline in product quality and user satisfaction (De Corato, 2020).

Restricted longevity: Although minimum processing prolongs the shelf life of fresh produce, it remains comparatively shorter than that of fully processed or preserved items. Consumers must be aware of expiry dates and the proper storage conditions to maximize the shelf life of minimally processed items (Ohlsson & Bengtsson, 2002).

Final Assessment: The food business is increasingly adopting minimal processing, driven by customer demand for convenience, quality, and health-oriented choices. The shift in lifestyles, the prevalence of nuclear families, the increase in working women, and rising earnings are driving the appeal of minimally processed vegetables (De Corato, 2020). These products offer several advantages, including prolonged shelf life, preservation of nutritional integrity, enhanced food safety, and convenience. Nonetheless, issues such as expense, quality preservation, and limited shelf life must be addressed to ensure the ongoing success and expansion of this industry. As customers increasingly recognize the benefits of minimal processing, demand for such goods is anticipated to grow, shaping the future of the food sector (Pasha et al., 2014).

18.4 SOCIOECONOMIC AND SUSTAINABILITY ASPECTS FOR REDUCING POSTHARVEST LOSSES

Vegetable production represents one of the most dynamic components in agriculture. They are vital for human nutrition, providing essential bioactive compounds that promote health and reduce the risk of chronic diseases (Abdelkader et al., 2024b, 2022). However, postharvest losses, often reaching 30–50% in developing regions, pose a significant challenge for food security. These losses are primarily due to improper handling, inadequate storage, lack of cooling facilities, and poor market infrastructure. Sustainable solutions that integrate social, economic, and environmental dimensions are essential to ensure that vegetable production remains resilient under changing climatic and market conditions (Hailu & Derbew, 2015; Kiaya, 2014).

Postharvest vegetable losses have far-reaching economic consequences (Table 18.6). For farmers, these losses translate into reduced income, wasted labor, and lower returns on investment. For consumers, they result in higher prices and decreased availability of fresh and nutritious food. In some countries, limited access to cold storage, poor transportation systems, and insufficient packaging materials amplify these problems. Studies show that a 10% reduction in postharvest losses can increase farmers' income by 20–30%. This shows the critical importance of improving postharvest management. Investment in postharvest infrastructure such as cold rooms, refrigerated transport, and improved packing centres can create economic opportunities (Kiaya, 2014; Yahia et al., 2019).

Sustainability in vegetable production reduces environmental footprints while maintaining high productivity and profitability. Effective postharvest can significantly contribute to sustainability

TABLE 18.6
Cost-Effectiveness and Farmer Practices

Practice/Intervention	Impact on Losses/Costs	Food Security/Income Effect	Citations
Good Agricultural Practices (GAP)	Reduces agrochemical use by >40%	Improves soil health, safety, and image	Kharel et al. (2023)
Conservation Agriculture & IPM	Fewer pesticide sprays, less labor	Higher yields, better benefit-cost ratio	Paudel et al. (2020)
Optimized Fertilization	Cuts N losses by 29–61%, lowers social costs by 48–60%	Maintains yields, increases net income	Yang et al. (2016)
Postharvest Technology	Reduces losses by up to 25%	Boosts farmer income, food security	Nitikaroon and Petrat (2024), Shahbazi et al. (2025)

by minimizing waste, conserving natural resources, and lowering GHG emissions (Juroszek et al., 2008). These innovations align with the circular economy and achieve the SDGs, particularly Zero Hunger, Responsible Consumption and Production, and Climate Action.

Empowering farmers through knowledge and training is a cornerstone of reducing postharvest losses. Traditional practices, though valuable, are often insufficient under modern market and climatic pressures. Therefore, the introduction of improved, cost-effective technologies and better management skills is necessary. Training programmes and extension services play a dynamic role in transferring such knowledge and improving the efficiency of vegetable value chains (Mohammed & Tokala, 2018).

Reducing postharvest losses directly supports food security by increasing the suitable supply of vegetables without requiring additional land or inputs. Also, improved postharvest systems enhance year-round food availability. At the community level, better postharvest practices reduce food price volatility, encourage rural entrepreneurship (e.g. small-scale packaging or processing units), and create employment opportunities, particularly for women and youth. These benefits strengthen local economies and contribute to sustainable rural development (Mohammed & Tokala, 2018; Yumbya et al., 2025).

Strong institutional frameworks are essential to sustain progress in postharvest management. Policies encouraging public–private partnerships can accelerate innovation and improve access to the technologies in rural areas (Bisheko & Rejikumar, 2024).

18.5 CONCLUSION

Effective postharvest management is vital to sustaining the nutritional quality of vegetables throughout the supply chain. The rapid biological changes that occur after harvest can lead to significant nutrient losses, reduced shelf life, and economic waste. By implementing scientifically based handling, storage, and processing techniques, it is possible to minimize these losses. This could reserve the essential vitamins, minerals, and bioactive compounds that make vegetables vital for human health. Furthermore, adopting eco-friendly, low-cost preservation methods can ensure that nutritional quality is maintained without compromising sustainability or economy. Conserving the nutritional value of vegetables after harvest requires a comprehensive approach. This approach combines physiology, technology, and sustainability principles. Strengthening research, education, and innovation in postharvest management will be key to achieving food security. This leads to reduced losses and enhanced resilience of agri-food systems in the face of global climate challenges.

REFERENCES

Abdelkader, M., Elkhawaga, F.A., Suliman, A.A., Puchkov, M., Kuranova, K.N., Mahmoud, M.H., Abdelkader, M.F., 2024a. Understanding the regular biological mechanism of susceptibility of tomato plants to low incidences of blossom-end rot. *Horticulturae*, *10*, 648.

Abdelkader, M., Zargar, M., Bayat, M., Pakina, E., Shehata, A.S., Suliman, A.A., 2024b. Biogenic nano-fertilizers as a sustainable approach to alleviate nitrate accumulation and enrich quality traits of vegetable crops. *Horticulturae*, *10*, 789.

Abdelkader, M., Zargar, M., Murtazova, K.M.-S., Nakhaev, M.R., 2022. Life cycle assessment of the cultivation processes for the main vegetable crops in southern Egypt. *Agronomy*, *12*, 1527.

Adeyeye, S.A.O., 2017a. The role of food processing and appropriate storage technologies in ensuring food security and food availability in Africa. *Nutrition & Food Science*, *47*, 122–139.

Adeyeye, S.A.O., 2017b. The role of food processing and appropriate storage technologies in ensuring food security and food availability in Africa. *Nutrition & Food Science*, *47*, 122–139.

Aghdam, M.S., Bodbodak, S., 2014. Postharvest heat treatment for mitigation of chilling injury in fruits and vegetables. *Food bioprocess technol*, *7*, 37–53. https://doi.org/10.1007/s11947-013-1207-4

Agriculture Institute, 2023. *Understanding pre-harvest factors leading to postharvest losses of fruits and vegetables – agriculture notes by agriculture institute.* URL https://agriculture.institute/post-harvest-mgt-principles/pre-harvest-factors-post-harvest-losses-fruits-vegetables/ (accessed 11.5.25).

Agüero, M., Ponce, A., Moreira, M., Roura, S., 2011. Lettuce quality loss under conditions that favor the wilting phenomenon. *Postharvest biology and technology*, *59*, 124–131. https://doi.org/10.1016/j.postharvbio.2010.08.018

Ahlawat, Y., Liu, T., 2021. Varied expression of senescence-associated and ethylene-related genes during postharvest storage of brassica vegetables. *International journal of molecular sciences*, *22*. https://doi.org/10.3390/ijms22020839

Ahmad, M., Siddiqui, M., 2015. *Factors affecting postharvest quality of fresh fruits*, 7–32. https://doi.org/10.1007/978-3-319-21197-8_2

Akanbi, 2021. Common causes and prevention of post harvest losses in fruits and vegetables. *Journal of research in agriculture and animal science*, *8*(1), 21–24.

Al-Dairi, M., Pathare, P., Al-Yahyai, R., 2021. Chemical and nutritional quality changes of tomato during postharvest transportation and storage. *Journal of the Saudi society of agricultural sciences*. https://doi.org/10.1016/j.jssas.2021.05.001

Alexandre, A.C.S., Gomes, B.A.F., Duarte, G.N., Piva, S.F., Zauza, S.B., Boas, E.V.V., 2022. Recent advances in processing and preservation of minimally processed fruits and vegetables: a review – part 1: fundamentals and chemical methods. *Journal of food processing and preservation*. https://doi.org/10.1111/jfpp.16757

Alferez, F., Zacarias, L., Burns, J.K., 2005. Low relative humidity at harvest and before storage at high humidity influence the severity of postharvest peel pitting in citrus. *Jashs*, *130*, 225–231. https://doi.org/10.21273/JASHS.130.2.225

Ali, A., Yeoh, W., Forney, C., Siddiqui, M., 2018. Advances in postharvest technologies to extend the storage life of minimally processed fruits and vegetables. *Critical reviews in food science and nutrition*, *58*, 2632–2649. https://doi.org/10.1080/10408398.2017.1339180

Ambuko, J., Wanjiru, F., Chemining'wa, G., Owino, W., Mwachoni, E., 2017. Preservation of postharvest quality of leafy amaranth (amaranthus spp.) vegetables using evaporative cooling. *Journal of food quality*, 2017, 1–6. https://doi.org/10.1155/2017/5303156

Ansari, M.W., Tuteja, N., 2015. Postharvest quality risks by stress/ethylene: management to mitigate. *Protoplasma*, *252*, 21–32. https://doi.org/10.1007/s00709-014-0678-0

Asrey, R., Patel, V.B., Barman, K., Pal, R.K., 2013. Pruning affects fruit yield and postharvest quality in mango (Mangifera indica L.) cv. Amrapali. *Fruits*, *68*, 367–380.

Atanda, S., Agoda, S., Ikotun, I., 2011. The concepts and problems of postharvest food losses in perishable crops. *African journal of food science*, *5*(11), 603–613.

Aung, M.M., Chang, Y.S., 2014. Temperature management for the quality assurance of a perishable food supply chain. *Food control*, *40*, 198–207.

Batziakas, K., Singh, S., Ayub, K., Kang, Q., Brecht, J., Rivard, C., Pliakoni, E., 2020. Reducing postharvest losses of spinach stored at nonoptimum temperatures with the implementation of passive modified atmosphere packaging. *Hortscience*. https://doi.org/10.21273/hortsci14732-19

Bhat, Shanky, 2013. Minimal processing and preservation of fruits and vegetables by active packaging. *International journal of herbal medicine*, *1*(2), 131–138.
Bisheko, M.J., Rejikumar, G., 2024. Mitigating postharvest losses: a policy-driven approach to empowering smallholder farmers in Tanzania. *Scientific African*, *23*, e02119.
Bisht, A., Singh, S.P., 2024. Postharvest losses and management of horticultural produce: a review. *Journal of scientific research and reports*, *30*, 305–320.
Boliko, M.C., 2019. FAO and the situation of food security and nutrition in the world. *Journal of nutritional science and vitaminology*, *65*, S4–S8.
Budhathoki, P., Pandey, S., Magar, P., 2022. Review on postharvest handling of fruits and vegetables to minimize loss. *Food and agri economics review*. https://doi.org/10.26480/faer.01.2022.37.40
Cheng, G., Chang, P., Shen, Y., Wu, L., El-Sappah, A.H., Zhang, F., Liang, Y., 2020. Comparing the flavor characteristics of 71 tomato (*solanum lycopersicum*) accessions in central shaanxi. *Frontiers in plant science*, *11*, 58683.
Chiesa, A., 2003. Factors Determining Postharvest Quality of Leafy Vegetables 519–524. https://doi.org/10.17660/actahortic.2003.604.59
Choe, E., Huang, R., Min, D.B., 2005. Chemical reactions and stability of riboflavin in foods. *Journal of food science*, *70*, R28–R36. https://doi.org/10.1111/j.1365-2621.2005.tb09055.x
Chopra, S., Müller, N., Dhingra, D., Mani, I., Kaushik, T., Kumar, A., Beaudry, R., 2022. A mathematical description of evaporative cooling potential for perishables storage in India. *Postharvest biology and technology*, *183*, 111727. https://doi.org/10.1016/j.postharvbio.2021.111727
Corbo, M.R., Campaniello, D., Speranza, B., Bevilacqua, A., Sinigaglia, M., 2015. Non-conventional tools to preserve and prolong the quality of minimally-processed fruits and vegetables. *Coatings*, *5*, 931–961.
Das, K., Prajapati, J., Kumar, S., Yadav, S., Jangu, R., n.d. Postharvest diseases and disorders of vegetable and horticultural crops. *Postharvest management and value addition of fruits and vegetables*, *38*.
De Chiara, M., Castagnini, J., Capozzi, V., 2024. Cutting-edge physical techniques in postharvest for fruits and vegetables: unveiling their power, inclusion in 'hurdle' approach, and latest applications. *Trends in food science & technology*. https://doi.org/10.1016/j.tifs.2024.104619
De Corato, U., 2020. Improving the shelf-life and quality of fresh and minimally-processed fruits and vegetables for a modern food industry: a comprehensive critical review from the traditional technologies into the most promising advancements. *Critical reviews in food science and nutrition*, *60*, 940–975. https://doi.org/10.1080/10408398.2018.1553025
Devi, L.S., Kalita, S., Mukherjee, A., Kumar, S., 2022. Carnauba wax-based composite films and coatings: recent advancement in prolonging postharvest shelf-life of fruits and vegetables. *Trends in food Science & technology*, *129*, 296–305.
Díaz-Pereira, E., Romero-Díaz, A., de Vente, J., 2020. Sustainable grazing land management to protect ecosystem services. *Mitigation and adaptation strategies for global change*, *25*, 1461–1479.
Dornenburg, H., Davies, C., 1999. The relationship between lipid oxidation and antioxidant content in postharvest vegetables. *Food reviews international*, *15*, 435–453. https://doi.org/10.1080/87559129909541198
Duarte-Sierra, A., Tiznado-Hernández, M., Jha, D., Janmeja, N., Arul, J., 2020. Abiotic stress hormesis: an approach to maintain quality, extend storability, and enhance phytochemicals on fresh produce during postharvest. *Comprehensive reviews in food science and food safety*, *19*(6), 3659–3682. https://doi.org/10.1111/1541-4337.12628
Elik, A., Yanik, D.K., Istanbullu, Y., Guzelsoy, N.A., Yavuz, A., Gogus, F., 2019a. Strategies to reduce postharvest losses for fruits and vegetables. *Strategies*, *5*, 29–39.
Elik, A., Yanık, D.K., Istanbullu, Y., Guzelsoy, N.A., Yavuz, A., Göğüş, F., 2019b. Strategies to reduce postharvest losses for fruits and vegetables. *International journal of scientific and technological research*. https://doi.org/10.7176/jstr/5-3-04
Elolu, S., Byarugaba, R., Opiyo, A., Nakimbugwe, D., Mithöfer, D., Huyskens-Keil, S., 2023. *Improving nutrition-sensitive value chains of African Indigenous vegetables: current trends in postharvest management and processing*, *7*. https://doi.org/10.3389/fsufs.2023.1118021
El-Ramady, H.R., Domokos-Szabolcsy, É., Abdalla, N.A., Taha, H.S., Fári, M., 2015. Postharvest management of fruits and vegetables storage, in: Lichtfouse, E. (Ed.), *Sustainable agriculture reviews, sustainable agriculture reviews*. Springer International Publishing, Cham, pp. 65–152. https://doi.org/10.1007/978-3-319-09132-7_2
Etana, M.B., 2019. A detailed review on common causes of postharvest loss and quality deterioration of fruits and vegetables in Ethiopia. *Food science and quality management*. https://doi.org/10.7176/jbah/9-7-07

Etefa, O., Forsido, S., Kebede, M.T., 2022. Postharvest loss, causes, and handling practices of fruits and vegetables in Ethiopia: scoping review. *Journal of horticultural research, 30*, 1–10. https://doi.org/10.2478/johr-2022-0002

Farber, J.M., 1991. Microbiological aspects of modified-atmosphere packaging technology: a review. *Journal of food protection, 54*, 58–70.

Ferdousi, J., Hossain, M.I., Saha, S.R., Rob, M., Afroz, T., Pramanik, S., Islam, M., Nath, D., 2024a. *Postharvest physiology of fruits and vegetables and their management technology: a review*. April. https://doi.org/10.36899/japs.2024.2.0717

Ferdousi, J., Hossain, M.I., Saha, S.R., Rob, M., Pramanik, T.A., Islam, M.R., Nath, D.D., 2024b. Postharvest physiology of fruits and vegetables and their management technology: a review. *JAPS: journal of animal & plant sciences, 34*.

Flanagan, K., Robertson, K., Hanson, C., 2019. Reducing food loss and waste. *Setting the global action agenda*.

Galindo, F., Herppich, W., Gekas, V., Sjöholm, I., 2004. Factors affecting quality and postharvest properties of vegetables: integration of water relations and metabolism. *Critical reviews in food science and nutrition, 44*, 139–154. https://doi.org/10.1080/10408690490424649

Gómez-López, V.M., Ragaert, P., Debevere, J., Devlieghere, F., 2008. Decontamination methods to prolong the shelf-life of minimally processed vegetables, state-of-the-art. *Critical reviews in food science and nutrition, 48*, 487–495. https://doi.org/10.1080/10408390701638878

Gouda, M.H.B., Duarte-Sierra, A., 2024. An overview of low-cost approaches for the postharvest storage of fruits and vegetables for smallholders, retailers, and consumers. *Horticulturae*. https://doi.org/10.3390/horticulturae10080803

Hailu, G., Derbew, B., 2015. Extent, causes and reduction strategies of postharvest losses of fresh fruits and vegetables: a review. *Journal of biology, agriculture and healthcare, 5*, 49–64.

Hamieh, B., Borel, P., Raouche, S., Bruzzese, L., Adjriou, N., Halimi, C., Marconot, G., Gillet, G., Rostain, J.-C., Guieu, R., Desmarchelier, C., 2023. Postharvest atmospheric pressure and composition modify the concentration and bioaccessibility of α- and β-carotene in carrots and sweet potatoes. *Foods, 12*, 4262. https://doi.org/10.3390/foods12234262

Harima, S., Nakano, R., Yamauchi, S., Kubo, Y., Inaba, A., Kitano, Y., 2002. Suppression of postharvest softening of forcing-cultured 'tonewase' Japanese persimmon by packaging in improved carton boxes. *Journal of the Japanese society for horticultural science, 71*, 583–587.

Henriod, R.E., 2006. Postharvest characteristics of navel oranges following high humidity and low temperature storage and transport. *Postharvest biology and technology, 42*, 57–64. https://doi.org/10.1016/j.postharvbio.2006.05.012

Hoffmann, T., Ronzoni, A., Silva, D., Bertoli, S., Souza, C., 2021. Cooling kinetics and mass transfer in postharvest preservation of fresh fruits and vegetables under refrigerated conditions. *Chemical engineering transactions, 87*, 115–120. https://doi.org/10.3303/cet2187020

Hua, L., Yong, C., Zhang, Z., Li, B., Qin, G., Tian, S., 2018. Pathogenic mechanisms and control strategies of botrytis cinerea causing postharvest decay in fruits and vegetables. *Food quality and safety*. https://doi.org/10.1093/fqsafe/fyy016

Ilić, Z., Šunić, L., Mastilović, J., Kevrešan, Ž., Pestorić, M., Belović, M., Magazin, N., 2016. Quality of root vegetables during prolonged storage. *Agriculturae conspectus scientificus, 81*, 115–122.

Jiang, Y.-T., Yang, L.-H., Ferjani, A., Lin, W.-H., 2021. Multiple Functions of the Vacuole in Plant Growth and Fruit Quality. Mol Horticulture 1, 4. https://doi.org/10.1186/s43897-021-00008-7

John, A., Isaac, W.-A., Daley, O., Kahramanoğlu, İ., 2023. Biological and environmental factors affecting postharvest quality of fruits and vegetables, in: *postharvest physiology and handling of horticultural crops*. CRC Press, pp. 36–64.

Johnson, L.K., Bloom, J.D., Dunning, R.D., Gunter, C.C., Boyette, M.D., Creamer, N.G., 2019. Farmer harvest decisions and vegetable loss in primary production. *Agricultural systems, 176*, 102672.

Juroszek, P., Lumpkin, T.A., Palada, M.C., 2008. Sustainable vegetable production systems. *Acta horticulturae, 767*, 133.

Kaur, P., Devgan, K., Kumar, N., Kaur, A., Kumar, M., Sandhu, K., 2021. Quality retention and shelf-life prolongation of cucumbers (cucumis sativus l.) under different cool storage systems with passive modified atmosphere bulk packaging. *Packaging technology and science, 34*, 567–578. https://doi.org/10.1002/pts.2595

Kaur, R., Watson, J., 2024. A scoping review of postharvest losses, supply chain management, and technology: implications for produce quality in developing countries. *Journal of the ASABE*. https://doi.org/10.13031/ja.15660

Kennedy, G., Nantel, G., Shetty, P., 2003. The scourge of "Hidden hunger": global dimensions of micronutrient deficiencies. *Food nutrition and agriculture*, 8–16.

Khan, A.A., Siddiqui, Y., Siddique, K.H.M., Bobo, J., Ali, A., 2024. Minimizing postharvest food losses: a vital strategy to alleviate food insecurity and malnutrition in developing nations: a review. *Discover food*. https://doi.org/10.1007/s44187-024-00129-0

Khan, A.W., 1989. *Post-harvest losses during processing and preservation of fruits and vegetables* (doctoral dissertation, university of punjab).

Kharel, M., Raut, N., Dahal, B.M., 2023. An assessment of good agriculture practices for safe and sustainable vegetable production in mid-hills of Nepal. *Journal of agriculture and food research, 11*, 100518. https://doi.org/10.1016/j.jafr.2023.100518

Kiaya, V., 2014. Postharvest losses and strategies to reduce them. *Technical paper on postharvest losses, action contre la faim (ACF), 25*, 1–25.

King, G., Woollard, D., Irving, D., Borst, W., 1990. Physiological changes in asparagus spear tips after harvest. *Physiologia plantarum, 80*, 393–400. https://doi.org/10.1111/j.1399-3054.1990.tb00058.x

Kitinoja, L., 2013. Use of cold chains for reducing food losses in developing countries. *Population, 6*, 5–60.

Kitinoja, L., Kader, A.A., 1995. *Small-scale postharvest handling practices: a manual for horticultural crops*. Department of Pomology, University of California.

Klein, B.P., 1987. Nutritional consequences of minimal processing of fruits and vegetables. *Journal of food quality, 10*, 179–193. https://doi.org/10.1111/j.1745-4557.1987.tb00857.x

Ktenioudaki, A., O'Donnell, C.P., Do Nascimento Nunes, M.C., 2019. Modelling the Biochemical and Sensory Changes of Strawberries During Storage Under Diverse Relative Humidity Conditions. *Postharvest Biology and Technology, 154*, 148–158. https://doi.org/10.1016/j.postharvbio.2019.04.023

Kunwar, A., Bist, D.R., Khatri, L., Dhami, R., Joshi, G.R., 2024. Optimizing postharvest handling practices to reduce losses and enhance quality of fruits and vegetables. *Food and agri economics review*. https://doi.org/10.26480/faer.02.2024.78.82

Lalpekhlua, K., Tirkey, A., Saranya, S., Babu, P.J., 2024. Postharvest management strategies for quality preservation in crops. *International journal of vegetable science, 30*, 587–635. https://doi.org/10.1080/19315260.2024.2392551

Lee, S., Kader, A., 2000. Preharvest and postharvest factors influencing vitamin c content of horticultural crops. *Postharvest biology and technology, 20*, 207–220. https://doi.org/10.1016/s0925-5214(00)00133-2

Leon, A., Logegaray, V., Casajús, V., Perini, M., Lourenco, A.B., Salinas, C., Jara, A.R., Civello, P., Galelli, M., Martínez, G., Picallo, A., Frezza, D., 2023. Effect of aloe vera gel coating on organoleptic and nutritional quality of minimally processed carrot. *Food science and technology international, 31*, 237–247. https://doi.org/10.1177/10820132231199509

Lewis, W.E., 1957. *Maintaining produce quality in retail stores*. US Department of Agriculture.

Liguori, G., Greco, G., Gaglio, R., Settanni, L., Gentile, C., Inglese, P., 2022. Mucilage-based and calcium ascorbate edible coatings improve postharvest quality and storability of minimally processed cactus pear fruit stored under passive atmosphere. *Horticulturae*. https://doi.org/10.3390/horticulturae9010015

Lill, R., King, G., O'donoghue, E., 1990. Physiological changes in asparagus spears immediately after harvest. *Scientia horticulturae, 44*, 191–199. https://doi.org/10.1016/0304-4238(90)90119-y

Ludford, P.M., 1995. Postharvest hormone changes in vegetables and fruit, in: Davies, P.J. (Ed.), *Plant hormones*. Springer Netherlands, Dordrecht, pp. 725–750. https://doi.org/10.1007/978-94-011-0473-9_33

Lufu, R., Ambaw, A., Opara, U.L., 2021. The influence of internal packaging (liners) on moisture dynamics and physical and physiological quality of pomegranate fruit during cold storage. *Foods, 10*, 1388. https://doi.org/10.3390/foods10061388

Magan, N., Aldred, D., 2007. Postharvest control strategies: minimizing mycotoxins in the food chain. *International journal of food microbiology, 119*, 131–139.

Maheshwar, C., Chanakwa, T.S., 2006. Postharvest losses due to gaps in cold chain in India – a solution. *Acta hortic*, 777–784. https://doi.org/10.17660/ActaHortic.2006.712.100

Mahmood, M.H., Sultan, M., Miyazaki, T., 2019. Significance of temperature and humidity control for agricultural products storage: overview of conventional and advanced options. *International journal of food engineering, 15*, 20190063. https://doi.org/10.1515/ijfe-2019-0063

Makule, E., Dimoso, N., Tassou, S., 2022. Precooling and cold storage methods for fruits and vegetables in sub-saharan africa: a review. *Horticulturae*. https://doi.org/10.3390/horticulturae8090776

Mishra, V.K., Gamage, T.V., 2007. Postharvest physiology of fruit and vegetables, in: *Handbook of food preservation*. CRC Press, pp. 37–66.

Moghazee, M.M., Abdelaziz, S.A., Abdelkader, M., Suliman, A.A., Sayed, L.M.K., 2025. Evaluation of some tomato (Solanum Lycopersicum L.) genotypes for blossom-end rot incidences through physiological measurements and ISSR-markers. *Egyptian journal of chemistry, 68*, 233–240.

Mohammed, M., Tokala, V.Y., 2018. *Postharvest extension and capacity building for the developing world*. CRC Press.

Monteiro, C.A., 2009. Nutrition and health: the issue is not food, nor nutrients, so muchas processing. *Public health nutrition*, *12*, 729–731.

Mounika, K., Reddy, P.P., Singh, R.P., Pandey, A., Bharty, S.K., Kumar, A., Tiwari, K., Jain, S., 2025. Postharvest technologies and cold chain management: a review. *Plant archives*, *25*, 3275–3283.

Mukama, M., Ambaw, A., Berry, T.M., Opara, U.L., 2018. Effect of relative humidity on pomegranate quality under simulated ambient storage conditions. *Acta hortic*, 265–272. https://doi.org/10.17660/ActaHortic.2018.1201.36

Mutungi, C., Tungu, J., Amri, J., Gaspar, A., Abass, A., 2022. Nutritional benefits of improved postharvest handling practices for maize and common beans in northern Tanzania: a quantitative farm-level assessment. *Journal of stored products research*, *95*, 101918.

Nasrin, T.A.A., Islam, M.N., Rahman, M.A., Arfin, M.S., Ullah, M.A., 2018. *Evaluation of postharvest quality of edible coated Mandarin at ambient storage*. https://doi.org/10.22004/AG.ECON.305444

Nassarawa, S., Abdelshafy, A., Xu, Y., Li, L., Luo, Z., 2020. Effect of light-emitting diodes (LEDs) on the quality of fruits and vegetables during postharvest period: a review. *Food and bioprocess technology*, *14*, 388–414. https://doi.org/10.1007/s11947-020-02534-6

Navarro, S., 2010. Commercial applications of oxygen depleted atmospheres for the preservation of food commodities, in: *Case studies in novel food processing technologies*. Elsevier, pp. 321–350.

Navarro, S., 2012. The use of modified and controlled atmospheres for the disinfestation of stored products. *Journal of pest science*, *85*, 301–322. https://doi.org/10.1007/s10340-012-0424-3

Ncama, K., Magwaza, L., Mditshwa, A., Tesfay, S., 2018. Plant-based edible coatings for managing postharvest quality of fresh horticultural produce: a review. *Food packaging and shelf life*. https://doi.org/10.1016/j.fpsl.2018.03.011

Nitikaroon, K., Petrat, K., 2024. Economic evaluation of postharvest vegetable losses. *International journal of finance management economics*, *7*, 98–100. https://doi.org/10.33545/26179210.2024.v7.i1.265

Nunes, C.N., Emond, J.-P., 2007. Relationship between weight loss and visual quality of fruits and vegetables, in: *Proceedings of the florida state horticultural society*. Florida State Horticultural Society, USA, 120, pp. 235–245.

Ohlsson, T., Bengtsson, N., 2002. *Minimal processing technologies in the food industries*. Elsevier.

Olmedo, G.M., Plotto, A., Mattia, M., Zhao, W., Hunter, W., Rosskopf, E.N., Obenland, D., Bai, J., 2025. Evaluation of modified humidity and controlled-release thymol on the preservation of blueberries and blackberries. *Postharvest biology and technology*, *219*, 113237. https://doi.org/10.1016/j.postharvbio.2024.113237

Ottaway, P.B., 2010. Stability of vitamins during food processing and storage, in: *Chemical deterioration and physical instability of food and beverages*. Elsevier, pp. 539–560.

Palumbo, M., Attolico, G., Capozzi, V., Cozzolino, R., Corvino, A., De Chiara, M., Pace, B., Pelosi, S., Ricci, I., Romaniello, R., Cefola, M., 2022. Emerging postharvest technologies to enhance the shelf-life of fruit and vegetables: an overview. *Foods*, *11*. https://doi.org/10.3390/foods11233925

Papoutsis, K., Edelenbos, M., 2021. Postharvest environmentally and human-friendly pre-treatments to minimize carrot waste in the supply chain caused by physiological disorders and fungi. *Trends in food science and technology*, *112*, 88–98. https://doi.org/10.1016/j.tifs.2021.03.038

Pasha, I., Saeed, F., Sultan, M.T., Khan, M.R., Rohi, M., 2014. Recent developments in minimal processing: a tool to retain nutritional quality of food. *Critical reviews in food science and nutrition*, *54*, 340–351. https://doi.org/10.1080/10408398.2011.585254

Paudel, S., Sah, L.P., Devkota, M., Poudyal, V., Prasad, P.V.V., Reyes, M.R., 2020. Conservation agriculture and integrated pest management practices improve yield and income while reducing labor, pests, diseases and chemical pesticide use in smallholder vegetable farms in Nepal. *Sustainability*, *12*, 6418. https://doi.org/10.3390/su12166418

Paulus, K., 1989. Vitamin degradation during food processing and, in: *Nutritional impact of food processing: 25th symposium of the group of European nutritionists, Reykjavik*, September 1987. Karger Medical and Scientific Publishers, p. 173.

Perera, W.P.T.D., Navaratne, S., Wickramasinghe, I., 2022. Impact of spectral composition of light from light-emitting diodes (LEDs) on postharvest quality of vegetables: A review. *Postharvest biology and technology*. https://doi.org/10.1016/j.postharvbio.2022.111955

Pokhrel, B., 2020. Review on postharvest handling to reduce loss of fruits and vegetables. *International journal of horticulture and food science*. https://doi.org/10.33545/26631067.2020.v2.i2a.52

Prusky, D., 2011. Reduction of the incidence of postharvest quality losses, and future prospects. *Food security*, *3*, 463–474. https://doi.org/10.1007/s12571-011-0147-y

Qange, S., Mdoda, L., Mditshwa, A., 2024. Modeling the critical causal factors of postharvest losses in the vegetable supply chain in ethekwini Metropolitan municipality: the log-linear regression model. *Heliyon, 10*. https://doi.org/10.1016/j.heliyon.2024.e39565

Rehman, A., Farooq, M., Lee, D.-J., Siddique, K.H.M., 2022. Sustainable agricultural practices for food security and ecosystem services. *Environmental science and pollution research, 29*, 84076–84095. https://doi.org/10.1007/s11356-022-23635-z

Roberts, E.H., 1972. Storage environment and the control of viability, in: Roberts, E.H. (Ed.), *Viability of seeds*. Springer Netherlands, Dordrecht, pp. 14–58. https://doi.org/10.1007/978-94-009-5685-8_2

Rodov, V., Ben-Yehoshua, S., Aharoni, N., Cohen, S., 2010. Modified humidity packaging of fresh produce, in: Janick, J. (Ed.), *Horticultural reviews*, Volume 37. Wiley, pp. 281–329. https://doi.org/10.1002/9780470543672.ch5

Rolle, R.S., 2006. Improving postharvest management and marketing in the asia-pacific region: issues and challenges. *Postharvest management of fruit and vegetables in the asia-pacific region, 1*, 23–31.

Rugkong, A., McQuinn, R., Giovannoni, J.J., Rose, J.K., Watkins, C.B., 2011. Expression of ripening-related genes in cold-stored tomato fruit. *Postharvest biology and technology, 61*, 1–14.

Sahu, S., Arya, S.K., 2024. Mitigation of greenhouse gas emissions in agriculture and food processing through sustainable management practices for climate change, in: *Advances and technology development in greenhouse gases: emission, capture and conversion*. Elsevier, pp. 71–96.

Sawaya, W.N., 2017. Impact of food losses and waste on food security, in: Murad, S., Baydoun, E., Daghir, N. (Eds.), *Water, energy & food sustainability in the Middle East*. Springer International Publishing, Cham, pp. 361–388. https://doi.org/10.1007/978-3-319-48920-9_16

Schmitz, F.R.W., De Carvalho, L.F., Bertoli, S., De Souza, C., 2019. Effect of refrigerated storage conditions on leafy vegetables. *MOJ food processing & technology*. https://doi.org/10.15406/mojfpt.2019.07.00223

Scrimshaw, N.S., 1994. The consequences of hidden hunger for individuals and societies. *Food nutrition bulletin, 15*, 1–25. https://doi.org/10.1177/156482659401500117

Shahbazi, F., Shahbazi, S., Zare, D., 2025. Losses in agricultural produce: causes and effects on food security. *Food and energy security, 14*, e70086. https://doi.org/10.1002/fes3.70086

Shewfelt, R.L., 1987. Quality of minimally processed fruits and vegetables. *Journal of food quality, 10*, 143–156. https://doi.org/10.1111/j.1745-4557.1987.tb00855.x

Shezi, S., Ngcobo, M., Khanyile, N., Ncama, K., 2024. Physio-metabolic mechanisms behind postharvest quality deterioration in broccoli (Brassica oleracea var. Italica) and Swiss Chard (Beta vulgaris L. var. Cicla): A review. *Plants, 13*. https://doi.org/10.3390/plants13223174

Siddiqui, M.W., Chakraborty, I., Ayala-Zavala, J.F., Dhua, R.S., 2011. Advances in minimal processing of fruits and vegetables: a review. *Journal of scintific & industrial research, 70*, 823–834.

Singh, A., Vaidya, G., Jagota, V., Darko, D., Agarwal, R.K., Debnath, S., Potrich, E., 2022. Recent advancement in postharvest loss mitigation and quality management of fruits and vegetables using machine learning frameworks. *Journal of food quality*. https://doi.org/10.1155/2022/6447282

Sivakumar, D., Sultanbawa, Y., Cooperstone, J., Ziv, C., 2022. Editorial: phytochemical changes in vegetables during postharvest storage and processing, and implications for consumer benefits. *Frontiers in nutrition, 9*. https://doi.org/10.3389/fnut.2022.1025361

Smith, D.L., Stommel, J.R., Fung, R.W., Wang, C.Y., Whitaker, B.D., 2006. Influence of cultivar and harvest method on postharvest storage quality of pepper (Capsicum annuum L.) fruit. *Postharvest biology and technology, 42*, 243–247.

Song, Y., Vorsa, N., Yam, K.L., 2002. Modeling respiration–transpiration in a modified atmosphere packaging system containing blueberry. *Journal of food engineering, 53*, 103–109. https://doi.org/10.1016/S0260-8774(01)00146-7

Sonntag, F., Liu, H., Neugart, S., 2023. Nutritional and physiological effects of postharvest UV radiation on vegetables: a review. *Journal of agricultural and food chemistry, 71*, 9951–9972. https://doi.org/10.1021/acs.jafc.3c00481

Soon, J.M., Baines, R.N., 2012. Food safety training and evaluation of handwashing intention among fresh produce farm workers. *Food control, 23*, 437–448.

Tao, J., Zuo, J., Watkins, C.B., Bai, C., He, X., Liu, S., Han, L., Zhao, X., Liu, Y., Li, J., Zheng, Y., 2024. Low storage temperature affects quality and volatile compounds in fresh tomatoes. *Food chemistry, 460*, 140400.

Temiz, A., Ayhan, D.K., 2017. Enzymes in minimally processed fruits and vegetables, in: Yildiz, F., Wiley, R.C. (Eds.), *Minimally processed refrigerated fruits and vegetables, food engineering series*. Springer US, Boston, MA, pp. 93–151. https://doi.org/10.1007/978-1-4939-7018-6_4

Teshome, Y., Leweye, A., 2020. A review on the status of postharvest loss in vegetable crop and their management options. *International journal of agriculture and food science*. https://doi.org/10.33545/2664844x.2020.v2.i1a.32

Toivonen, P., Brummell, D., 2008. Biochemical bases of appearance and texture changes in fresh-cut fruit and vegetables. *Postharvest biology and technology*, *48*, 1–14. https://doi.org/10.1016/j.postharvbio.2007.09.004

Tonto, T.C., Cimini, S., Grasso, S., Zompanti, A., Santonico, M., De Gara, L., Locato, V., 2023. Methodological pipeline for monitoring postharvest quality of leafy vegetables. *Science reports*, *13*, 20568. https://doi.org/10.1038/s41598-023-47873-4

Tort, Ö.Ö., Vayvay, Ö., Çobanoğlu, E., 2022. A systematic review of sustainable fresh fruit and vegetable supply chains. *Sustainability*, *14*, 1573. https://doi.org/10.3390/su14031573

Ufitinema, B., De Dieu Niyomugabo, J., Hakizima, M., 2024. Relationship between damages and other postharvest stress factors affecting quality of fruit and vegetables. *Journal of food technology & nutrition sciences*. https://doi.org/10.47363/jftns/2024(6)178

Umeohia, U., Olapade, A., 2024. Physiological processes affecting postharvest quality of fresh fruits and vegetables. *Asian food science journal*. https://doi.org/10.9734/afsj/2024/v23i4706

Verma, S., Kumar, S., Kumar, V., Yadav, S., Singh, P., Saurabh, A., Kumar, A., Jain, M., 2025. Recent advancements and innovations in postharvest handling, storage, and technology for vegetables: a review. *Archives of current research international*. https://doi.org/10.9734/acri/2025/v25i21076

Weston, L.A., Barth, M.M., 1997. Preharvest factors affecting postharvest quality of vegetables. *Hortscience*, *32*, 812–816.

Wirsenius, S., Azar, C., Berndes, G., 2010. How much land is needed for global food production under scenarios of dietary changes and livestock productivity increases in 2030? *Agricultural systems*, *103*, 621–638.

Workineh, M., Enyew, M., 2021. Review the extent and cause of post harvest loss of fruits and vegetables in Ethiopia. *Journal of biology, agriculture and healthcare*. https://doi.org/10.7176/jbah/11-13-01

Wu, J., Tang, R., Fan, K., 2023. Recent advances in postharvest technologies for reducing chilling injury symptoms of fruits and vegetables: a review. *Food chemistry*: *X*, *21*. https://doi.org/10.1016/j.fochx.2023.101080

Xiao, L., Li, T., Jiang, G., John, A., Zhang, D., Jin, W., Duan, X., Jiang, Y., 2019. Effects of dry fog humidification on pericarp browning and quality of litchi fruit stored at low temperature. *International journal of agricultural and biological engineering*, *12*, 192–196. https://doi.org/10.25165/j.ijabe.20191204.4420

Yadav, S., Arora, S.K., Vats, S., 2023. Vitamins and minerals: a review on processing losses and strategies to control it. *Modern concepts developments in agronomy*, *12*, 1178–1182.

Yahia, E.M., Fonseca, J.M., Kitinoja, L., 2019. Postharvest losses and waste. *Postharvest technology of perishable horticultural commodities* 43–69.

Yang, L., Huang, B., Mao, M., Yao, L., Niedermann, S., Hu, W., Chen, Y., 2016. Sustainability assessment of greenhouse vegetable farming practices from environmental, economic, and socio-institutional perspectives in China. *Environmental science and pollution research*, *23*, 17287–17297. https://doi.org/10.1007/s11356-016-6937-1

Yeboah, S., Hong, S.-J., Park, Y., Choi, J.-H., Eum, H., 2023. Postharvest quality improvement of bell pepper (Capsicum annuum L. cv Nagano) with forced-air precooling and modified atmosphere packaging. *Foods*, *12*. https://doi.org/10.3390/foods12213961

Yeshiwas, Y., Tadele, E., 2021. An investigation into major causes for postharvest losses of horticultural crops and their handling practice in debre markos, north-western Ethiopia. *Advances in agriculture*. https://doi.org/10.1155/2021/1985303

Yumbya, P., Becker, K., Gregerson, K., Hayashi, H., Jarman, A., Kabirigi, M., McGuire, E., 2025. A call for locally Led postharvest innovation to urgently address global nutrition insecurity. *Frontiers in horticulture*, *4*, 1537852.

Zainalabidin, F.A., Sagrin, M., Azmi, W.N.W., Ghazali, A., 2019. Optimum postharvest handling-effect of temperature on quality and shelf life of tropical fruits and vegetables. *Journal of tropical resources and sustainable science (JTRSS)*. https://doi.org/10.47253/jtrss.v7i1.505

Zan, X., Jia, W., Zhuang, H.N., Cui, F.J., Li, N., Zhang, J.S., Sun, W.J., Zhao, X., 2020. Energy status and mitochondrial metabolism of volvariella volvacea with controlled ultrasound treatment and relative humidity. *Postharvest biology and technology*, *167*, 111250. https://doi.org/10.1016/j.postharvbio.2020.111250

Zhang, M., Meng, X., Bhandari, B., Fang, Z., Chen, H., 2015. Recent application of modified atmosphere packaging (MAP) in fresh and fresh-cut foods. *Food reviews international*, *31*, 172–193. https://doi.org/10.1080/87559129.2014.981826

Zuo, X., Wang, J., Li, Y., Zhang, J., Wu, Z., Jin, P., Cao, S., Zheng, Y., 2025a. Recent advances in high relative humidity strategy for preservation of postharvest fruits and vegetables: a comprehensive review. *Food chemistry*, 144130.

Zuo, X., Wang, J., Li, Y., Zhang, J., Wu, Z., Jin, P., Cao, S., Zheng, Y., 2025b. Recent advances in high relative humidity strategy for preservation of postharvest fruits and vegetables: a comprehensive review. *Food chemistry*, *481*, 144130. https://doi.org/10.1016/j.foodchem.2025.144130

19 Scope of Vegetables as Edible Vaccines

Swati Patel, Ashutosh Pathak, and Poonam Mishra

19.1 INTRODUCTION

Transmissible diseases claim the lives of more than 10 lakhs people annually, with nearly half of these cases attributed to pathogens that target the mucosal membranes of the mammalian host (Zeitlin et al., 1999). Vaccines are biological preparations designed to enhance the body's immune response, and the concept was first introduced by Edward Jenner in 1796 while treating smallpox. For over a century, vaccines have played a remarkable role in avoiding infectious diseases, dropping hospitalizations, and saving lives by offering targeted and effective immune protection (Saxena & Rawat, 2014; Jiji et al., 2024). Unlike traditional treatments, which are typically administered after the onset of disease, vaccination equips the immune system to recognize and combat infections before they occur. This proactive approach not only protects against future infections but often provides long-lasting immunity. The rapid development and deployment of COVID-19 vaccines during the 2020–2021 pandemic exemplified the potential of modern vaccine science to achieve global immunization in a remarkably short period and help in saving lots of lives (Le et al., 2020). Traditionally, the vaccines are produced using inactivated or attenuated pathogens and purified antigenic components, administered primarily through injections or oral routes. In certain situations, secondary immunity caused by antibodies is employed to confer immediate, short-term protection, particularly against diseases such as polio, tetanus, measles, mumps, and hepatitis. (Chen et al., 2012; Altindis et al., 2014). Despite these advancements, global immunization faces persistent challenges; as of 2023, approximately 14.5 million infants were unvaccinated, with 60% residing in developing countries which highlight significant disparities in access to vaccines and healthcare infrastructure (WHO, 2024). Conventional vaccines also face limitations including high production costs, complex storage, distribution requirements (often reliant on refrigeration), and insufficient delivery systems (Sohrab et al., 2017; Sohrab, 2020). Moreover, certain DNA-based vaccines are costly, sometimes ineffective, and may elicit adverse immune reactions (Faith et al., 2015). These barriers continue to restrict vaccine accessibility and effectiveness, particularly in low-resource settings (Hefferon, 2014; Patel et al., 2022). In addition, they are manufactured through complex industrial processes, making them costly and difficult to distribute, particularly in developing countries (Wurm, 2004; Ulmer et al., 2006).

To overcome these limitations, there is a urgent need for alternative vaccination strategies which are safe, affordable, and easy to store and administer. This chapter shows that edible vaccines are being explored as a promising substitute, offering a more affordable, accessible, and scalable solution for global immunization. It falls within the concept of vaccines, that is 'biological agents designed to stimulate the production of antibodies in humans and animals, and thereby enhancing immunity against specific diseases' (Sahoo et al., 2020; Mamedov et al., 2021). The theory of edible vaccines (EVs) emerged in initial stage of 1990s when scientists discovered that plants can be genetically engineered to express foreign proteins, including antigens derived from viruses and bacteria. Initial studies have demonstrated that both humans and animals could mount immune responses after consuming these modified plants. One of the first experimental edible vaccines was developed in potatoes to combat the Norwalk virus (Tacket et al., 1998). Since then, research has

DOI: 10.1201/9781003606710-19

escalated to include various plant hosts such as tomatoes, bananas, and lettuce, targeting a broad spectrum of diseases including cholera, hepatitis B, and HIV. A major milestone in this field was achieved by Dr Charles Arntzen, a leading figure in edible vaccine research, who demonstrated the feasibility of producing hepatitis B antigens in tobacco plants. His pioneering work significantly advanced the expansion and improvement of plant-based vaccines and laid the foundation for ongoing innovations in this area.

Progressive developments in nature-based sciences, biotechnology, and human genetics have significantly enhanced our perception and opinion of transmittable diseases and paved the way for the development of edible vaccines. These vaccines, including food-based, oral, subunit, and green vaccines, represent an innovative and promising method to immunization (Saxena & Rawat, 2014). As subunit vaccines, they involve the established integration of genes encoding specific antigens into the nuclear or chloroplast genomes of suitable plants such as potatoes, bananas, and tomatoes (Esmael & Hirpa, 2015; Gupta et al., 2022). This technology enables the formation of complex, well-folded, and bioactive recombinant proteins, facilitating efficient conversion, manifestation, and practical stability (Khalid et al., 2022). A key advantage of edible vaccines is their elimination of costly refinement and subsequent stages, making them major cost enhancers in traditional vaccine manufacturing (Rice et al., 2005; Gupta et al., 2022). These plant-based vaccines are not only safe and environmentally sustainable but also offer great scalability, and it would be possible to achieve mass production at low costs and within shorter time duration. It has numerous advantages such as oral delivery, which allows for self-administration, improved patient compliance, and simplified distribution logistics (Aboul-Ata et al., 2014; Faith et al., 2015; Kurup & Thomas, 2020). Currently, the edible vaccines are being developed to target different human infections such as measles, cholera, hepatitis B, C, and E, and foot-and-mouth disease. Furthermore, an emerging study is investigating their role in combating SARS-CoV-2 infections (Kumar et al., 2021; Singh et al., 2023).

19.2 HISTORY OF EDIBLE VACCINES

The development of edible vaccines began with a major milestone achieved in 1990, when an immune response was successfully detected in rodents against the surface protein antigen (SpaA) of *Streptococcus* that was cloned and expressed in tobacco species (Curtiss & Cardineau, 1997; Mason et al., 1998; Tacket et al., 1998). Further progress was made in 1996 with the heterologous expression of hepatitis B virus surface antigen (HBsAg) in potato tubers, indicating that plant-derived HBsAg could elicit mucosal immune responses when administered orally (Mason et al., 1996; Tacket et al., 2000). Later on, a pivotal moment happened in 1998 with the first human trial which involved GM potatoes expressing an altered fragment of heat-labile toxin from *Escherichia coli*, a common cause of diarrhoea. The trial demonstrated a four-fold increase in intestinal antibodies post-immunization, with no significant side effects, demonstrating the substantial edible vaccine-based human immune response (Arntzen, 1998; Tacket et al., 1998; Mason et al., 2002; Gupta et al., 2022; Saleem et al., 2024). Subsequent studies have confirmed that recombinant proteins synthesized in transgenic plants could activate immune responses in mucosal tissues (Sala et al., 2003; Patel et al., 2022). Following these initial breakthroughs, researchers achieved remarkable expression of hepatitis B antigen and the β subunit of heat-labile toxin in both tobacco and potato plants (Richter et al., 2000). This led to the emergence of using transgenic edible plant organs, such as fruits and tubers, as a novel means for delivering subunit vaccines, offering a promising strategy to overcome many limitations of traditional vaccination methods. For example the expression of HBsAg was later extended to numerous edible crops like cherry tomato (Gao et al., 2003), banana (Kumar et al., 2005), tomato (Lou et al., 2007; Srinivas et al., 2008), rice (Oszvald et al., 2008; Qian et al., 2008), and algae (Specht & Mayfield, 2014). These advancements set the stage for the development of edible vaccines targeting other diseases, which are discussed in subsequent sections.

19.3 PLANT-BASED EDIBLE VACCINES (PBEVS)

Plants have recently been extensively exploited to build innovative biopharmaceutical mechanisms that promote the appropriate conformation of expressed proteins while remaining economically feasible (Hefferon, 2014; Hernández et al., 2014). Researchers have attempted to address the shortcomings of traditional vaccinations by developing edible formulations (Landry et al., 2010). This phenomenon is also called as 'molecular farming' in which biomolecules with higher economic significance from GM plants are formed, e.g. banana, potato, lettuce for hepatitis B; tomato for diphtheria tetanus; rice for cholera; tobacco for Avian flu; maize for diarrhoea; spinach and carrot for HIV, etc. (Figure 19.1). Various clinical experiments are currently underway, with purified antigens briefly synthesized in injectable vaccine derived from tobacco. For example a phase II clinical trial using a plant-based flu vaccine (VLP) has been conducted, followed by a phase III clinical investigation (Rybicki, 2014). The success of the ZMapp, an experimental Ebola medicine, produced in tobacco plants, raised the profile of molecular farming (Chan & Daniell, 2015; Rosales-Mendoza et al., 2016). Unlike biomolecule synthesis, edible vaccine preparations don't require pre-administration treatment or purification, and it has significantly lower production costs.

19.4 WHAT MAKES A CANDIDATE PLANT?

Selecting the suitable plant species is a critical step for generating an effective edible vaccine, as the chosen plant should express high levels of vaccine, be cultivable on a wide scale, and be consumed orally (Esmael & Hirpa, 2015). According to Tiwari et al. (2009), effective animal immunization requires continuous incorporation of the plant into the animal's diet. Additionally, it is crucial to choose an edible plant that can show characteristics of suitable antigens while remaining

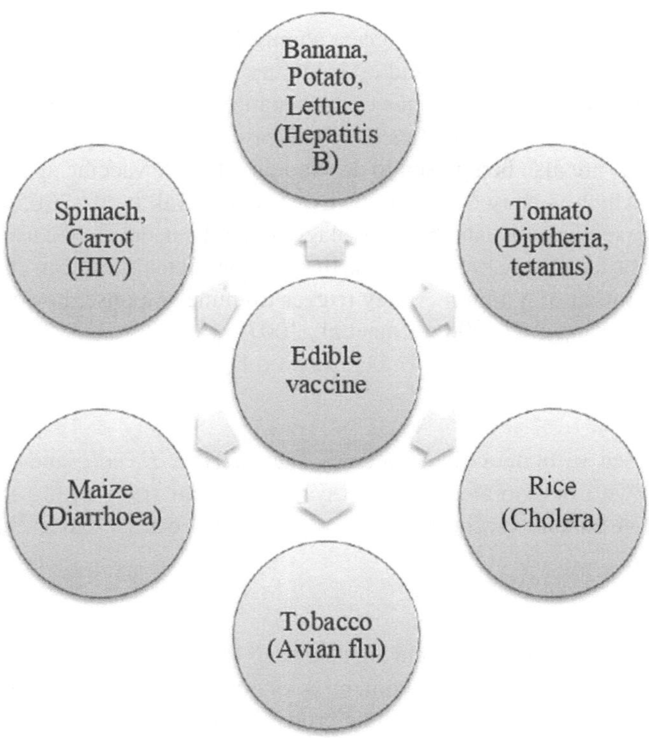

FIGURE 19.1 Different types of plant-based edible vaccines against different diseases.

tasty. The candidate plants are those that are best suited to edible vaccine manufacturing, and several elements contribute to a plant's suitability as an edible vaccination candidate such as: (i) the plant or edible component of the plant must have an extended shelf life, i.e. it must be kept fresh for an extended period of time, and rice, maize, and wheat are excellent examples of such plants; (ii) it must grow rapidly, i.e. 'tree fruits' are unsuitable because they take much time to develop and mature, whereas fast-growing plants such as tobacco and tomato are more appropriate; and (iii) it should be easily transported, i.e. the preparation of edible vaccine should be easy. Previous studies examined the potential of several food crops to produce PBEVs (Figure 19.1) (PBEVs) and are summarized below

19.4.1 Potato

The usage of potatoes as an edible vaccine has considerable benefits because of their ease of modification and distribution, and it eliminates the requirement for refrigeration. It offers a diverse foundation for creating vaccines against diseases like tetanus, hepatitis B, and diphtheria. (Gupta et al., 2022). This began when Cornell University's Arntzen and Mason's research team first successfully created and produced transgenic potatoes (Haq et al., 1995; Mishra et al., 2008). A subsequent experiment assessed the effectiveness of antigens derived from potatoes against the Norwalk virus capsid pathogen and the non-toxic *Vibrio cholera* endotoxin, and the study included testing on both rats and human volunteers (Mason et al., 1996; Tacket et al., 1998). This was the first time a potato-based consumable vaccination provided immunization in mice against the *E. coli* LT-B strain (Tacket et al., 2004). However, one of the major limitations on the usage of potato plants for edible vaccine is that the vegetable cannot be edible in its raw form, and the cooking will denature the antigens, reducing its efficacy (Waheed et al., 2016; Concha et al., 2017).

19.4.2 Tomatoes

Tomatoes are reported to have more effectiveness against Norwalk virus than potato-based vaccinations. In addition, leaves, stems, fruits, and other tissues can express *V. cholera* B toxin known as CT-B proteins and HBsAg, a surface antigen of the hepatitis B virus (Kumar et al., 2005). This plant produced a potent vaccination against Alzheimer's disease by expressing β-amyloid proteins (Youm et al., 2008). Tomatoes are also being used to develop an effective vaccine against acute respiratory syndrome (SARS), which is caused by coronavirus (Pogrebnyak et al., 2005). Furthermore, vaccines for pneumonia, septicaemia, and bubonic plagues have been developed using tomatoes (Kurup & Thomas, 2020). The advantage of this plant is that it grows quickly and can cultivate broadly, also the elevated concentration of Vitamin A may trigger immune response; however, the drawback is that it readily spoils (Zhang et al., 2006; Lou et al., 2007).

19.4.3 Carrots

Carrots are being used to make edible vaccines against HIV, *E. coli*, and *Helicobacter pylori*. In addition, people with deteriorated immune systems benefit from taking this sort of antigen-comprising carrot consumable vaccine (Karasev et al., 2005; Ye & Li, 2010; Zhang et al., 2010).

19.4.4 Soybean

Soybeans' high protein content makes them an ideal basis for edible vaccine production. The cooked soybeans have 8.5-times higher protein content as compared to potatoes, which can potentially result into elevated concentration of therapeutic protein (Sahoo et al., 2020). Successful expression of *E. coli* β-subunit thermolabile toxin, a key vaccine candidate, in the endoplasmic reticulum of soybean seeds has been reported recently (Patel et al., 2022).

19.4.5 Banana

Banana is a widely cultivated tropical plant, valued for their consumption in both raw and cooked forms. Their ability to thrive in tropical climates, i.e. regions often characterized by limited access to conventional healthcare, further enhances their potential, as many countries most in need of vaccination solutions are located in these areas. The main advantage is that its proteins remain stable after cooking, which makes banana a promising candidate for a source of edible vaccine (Qian et al., 2008; Esmael & Hirpa, 2015). Experimental studies have reported successful expression of HBsAg in banana using various genetic constructs, and the expressed antigen is predominantly localized in banana leaves (Kumar et al., 2005; Sahai et al., 2013; Sahoo et al., 2020). Despite these promising results, the extended cultivation period, limiting scalability and timely production, and quickly spoiling nature once matured continue to pose a barrier to the widespread application of banana-based vaccines (Qian et al., 2008; Aryamvally et al., 2017).

19.4.6 Maize

Maize is a front runner crop for producing edible vaccines for human and animal applications (Shah et al., 2022). Furthermore, its widespread consumption in processed forms such as cornflakes, corn chips, and tortillas offers a practical route for oral vaccine delivery. Research at Iowa State University has focused on developing flu vaccines for humans and pigs through the consumption of GM corn, potentially benefiting millions of people affected by diseases like hepatitis B (Shah et al., 2022). In addition to human applications, maize has shown promise in veterinary immunization. For instance researchers managed by Octavio Guerrero-Andrade at the Institute for Research and Advanced Studies in Central Mexico successfully integrated a gene from the Newcastle disease virus (NDV) into maize DNA. Poultry that used up the transgenic corn formed antibodies against NDV, highlighting the viability of maize-based oral vaccines for livestock (Faith et al., 2015). ProdiGene, a U.S.-based biotechnology company, explored the usage of genetically constructed maize to express a surface protein from a monkey strain of HIV, with the goal of better safety and easily digestible human HIV vaccines (Meyers et al., 2008). These developments underscore maize's potential as a scalable and effective candidate for plant-derived vaccines across species.

19.4.7 Rice

Rice holds considerable promise as a platform for vaccine manufacture due to the global dietary prevalence, particularly in infant nutrition, and its capacity for high-level antigen expression, especially in regions where rice is a dietary staple (Mason et al., 1996; Rice et al., 2005; Oszvald et al., 2007). For example major T-cell epitope peptides resulting from Japanese cedar pollen allergens have been successfully expressed in rice seeds and targeted to the mucosal immune system (MIS) to counteract allergic immune responses, particularly those mediated by allergen-specific Th2 cells (Wu et al., 2007). This strategy effectively inhibited the formation of allergen-specific IgE, inhibited histamine discharge from mast cells, and alleviated warning sign of pollinosis such as sneezing. These discoveries provide groundwork for the development of oral immunotherapy preparations using transgenic plants which express allergen-specific T-cell epitopes. Additionally, rice has been genetically engineered to express HBsAg and bacterial genes encoding cholera toxin (Oszvald et al., 2007). A report suggested that transgenic rice grains expressing this toxin conferred protection in mice against cholera-causing bacteria upon ingestion, demonstrating the feasibility of rice-based vaccines against infectious diseases such as cholera (Nochi et al., 2009).

19.4.8 Spinach

Spinach is consumed worldwide as leafy vegetable, and hence GM spinach is potential source for edible vaccinations (Sussman, 2003). Exploration of spinach is possible for anthrax immunization due to its ability to transport HIV-1 Tat protein, a potential vaccine candidate (Oszvald et al., 2007). The recombinant tobacco mosaic virus (TMV) carrying a protective antigen (PA) was expressed in spinach, which was highly immunogenic in animals (Karasev et al., 2005). Antigens accumulated to up to 2.4% of total seed protein after drying (Patel et al., 2022).

19.4.9 Lettuce

Lettuce is another leafy vegetable popular worldwide, and hence it emerged as a promising platform for the production of edible vaccines, notably through its ability to express antigenic proteins relevant to both human and animal health. This plant is primarily used uncooked and has favourable effects against the hepatitis B virus (Kim et al., 2007; Srinivas et al., 2008). Lettuce was plotted to direct the E2 glycoprotein of the classical swine fever virus, further demonstrating its potential in veterinary vaccine development (Sahoo et al., 2020). More recently, researchers expressed the native spike protein of SARS-CoV-2 in lettuce chloroplasts, which was an innovative advancement aimed at developing oral booster vaccines to enhance immunity against COVID-19 (Singh et al., 2023).

19.4.10 Tobacco

Although tobacco is a non-edible plant, it has emerged as a valuable platform for the production of vaccines, demonstrating notable versatility in supporting a wide range of vaccine types and targets. Since the successful expression of a *Streptococcus* surface protein in tobacco in 1990, the plant has been further engineered to produce edible vaccines such as those targeting hepatitis B by expressing the viral s-gene as an antigen (Mason et al., 1996; Tacket, 2009; Jeshvaghani et al., 2019). Transgenic tobacco plants have shown significant accumulation of the target proteins in both seeds and leaves. The reports suggested that seed-dependent production proposed several benefits over conventional leaf-based systems, including higher yield, simplified purification, and room-temperature stability, making it a promising approach for scalable vaccine production. Moreover, tobacco has been utilized to manufacture additional hepatitis B-related polypeptides, further expanding its utility in vaccine formation (Mason et al., 2002; Tiwari et al., 2009). In addition, a distinguished advancement includes the development of a rabies vaccine incorporating the rabies G protein fused to cholera toxin B (CtxB), successfully expressed in tobacco seeds (Tiwari et al., 2009). Other studies have reported that transgenic tobacco expresses VP1 protein, which protects against chicken infectious anaemia, a polypeptide linked to hepatitis B, as well as vaccine against coccidiosis (Wigdorovitz et al., 1999; Hahn et al., 2007; Gómez et al., 2009).

19.4.11 Alfalfa

Alfalfa is utilized to create consumable vaccinations, primarily for veterinary use. Transgenic alfalfa with hog pest virus glycoprotein E2 was formed in 2005, and it was also engineered to express Eeg95-EgA31 from *Echinococcus ganulosus* (Wigdorovitz et al., 1999).

Beyond these, other edible plants, e.g. sweet potatoes, peanuts, and watermelons, are also being explored for vaccine production. For that, specific mechanism and procedure can be followed (Figure 19.2). These platforms are being used to target a broad range of ailments, including hepatitis B, non-Hodgkin's lymphoma, HIV-1, and rabies (Gupta et al., 2022).

19.5 MECHANISM OF ACTION OF PBEVS

After oral intake, the antigens which are present in edible vaccines influence the intestinal mucosal surfaces, permitting the immune system to absorb and process them. The mucosal epithelial tissue that connects to the respiratory, urinary/reproductive, and gastrointestinal systems is the

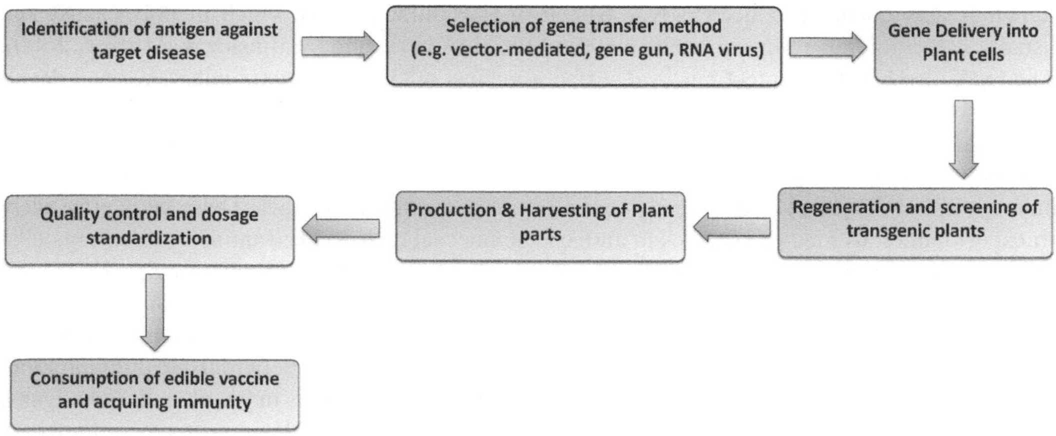

FIGURE 19.2 Schematic representation of steps involved in the development and immunization process of plant-based edible vaccines.

body's biggest immunologically energetic tissue, serving as the principal entrance site for infections (Saxena & Rawat, 2014). Edible vaccines stimulate integrated mucosal and systemic immune responses, which effectively battle a variety of mucosa-invading infections. The MIS serves as primary safeguard and provides an optimal location for targeted immunization against certain illnesses. Oral antigens target to boost mucosal and humoral immunity in order to resist toxicities (Tregoning, 2020). Bio-encapsulation (a surface coating of plant cells) protects the antigens from gastric acids till they reach the intestines, enabling antigen survival and successful delivery to the MIS (Sohrab, 2020). These antigens bind to M cells in the intestinal lining, especially gut-associated lymphoid tissue (GALT) and Peyer's patches (PP), which are actually crucial components of the MIS (Patel et al., 2022). Following acceptance, antigens are transferred to local lymphocytes, macrophages, and other antigen-dependent cells, causing serum IgG, IgA, and IgE reactions. Production of memory cells plays an important role in quickly deactivating subsequent attacks occurred by pathogenic pathogens. Microfold (M) cells (M-cells) play a significant role in antigen uptake and are found in the intestinal lining, above PP in the ileum, and in gut-assisted lymphoid tissue. Mucosal tissues are rich in plasma cells that generate IgA from PP and play a key role in mucosal immune effectors. Lymphoid nodules on the colon surface contain follicles that generate germinal centres upon antigenic stimulation. The colon is also the site where most plant cell collapses and is affected by bacterial activity and metabolic enzymes. Edible vaccine degrade in the PP, after the antigen penetrates the intestinal epithelium via the follicles, it eventually concentrates around the organized lymphoid tissue. M-cells produce class II MHC molecules, bind to antigens, facilitate mucosal membrane transit, and stimulate B cells in lymphoid follicles. Diffuse mucosal-associated lymphoid tissue-derived activated B cells (MALT) leave lymphoid follicles to grow within the plasma cells which release IgA types of antibodies (Saxena & Rawat, 2014). Epithelial cells then transport IgA antibodies to the lumen, where they bind to antigens on pathogenic agents and neutralizing them.

The antigen penetrates gut mucosal layer via number of mechanisms, like M-cells and macrophages. Interferon gamma specifically triggers macrophages, and this stimulation causes macrophages to transmit fragmented peptides to helper T-cells and then generate antibodies (Johansen et al., 1999). M-cells are kind of cells which can deliver antigens to T-cells. Following the display of antigenic epitopes on the APC surface by helper T-cells, B cells are activated. These cells then proceed to mesenteric lymph nodes, where they mature into plasma cells, move to mucosal membranes, and release immunoglobulin A (IgA). Secretory IgA (sIgA), which is transported into the lumen, is subsequently produced by IgA. B1 cells in the lamina propria, which are polyreactive and usually detect foreign antigens, produce 50% of the sIgA in the gut (Walmsley & Arntzen, 2000). The sIgA also helps to protect mucosal surfaces against microbe and toxin adherence and development of new platforms for the delivery of pathogens or toxin-specific sIgA, and systemic IgG is critical to improving vaccine efficacy (Dietrich et al., 2003; Kunisawa et al., 2012). M-cells are type of primary pathways of antigen uptake in the intestine. A small subset of follicular-associated enterocytes (FAE), M-cells are mostly found in the gastrointestinal tract. These cells effectively transport numerous macromolecules to antigen submucosal cells (APCs, antigen-presenting cells) on Peyer's patches from small intestinal lumens (Mabbott et al., 2013). Dendritic cells (DCs) refer to the most influential antigenic cells capable of inducing an adaptive immune response in priming naive T-cells (Mildner & Jung, 2014). In the immediate phase, DC is steady, with high endocytic activity and poor capacity for producing naive T-cells. DCs, conversely, during inflammation, grow, create more co-stimulatory chemicals, and migrate to T-cell areas in lymph nodes. Antigens, as well as cytokine release, assist convert naive antigen-specific T-cells into effector cells which move to a specific inflammatory location (Dalod et al., 2014). By directly stimulating or promoting transformed T-17 cells into Tfh, intestinal DCs can increase naïve T-cell activation and follicular T-helper differentiation (Tfh) (Milpied & McHeyzer-Williams, 2013; Shin et al., 2015). Such active B cells leave the follicles and go to mucosa-associated lymphoid tissue (MALT) where plasma cells release anti-IgA antibodies (Mishra et al., 2008). The same IgA antibodies are transferred to the lumen in secretions against epithelial cells to interact with antibodies (Milpied & McHeyzer-Williams, 2013). Intestinal DCs also sample luminous antigens by extending dendritic projection through the epithelial cell layer into the lumen (Rescigno et al., 2001). Intravital microscopy shows that goblet cells, a type of mucin-producing cell, can directly acquire and supply intestinal antigens. Goblet cells were a relatively new technique for antigen capture in the small intestine (McDole et al., 2012).

Although the term 'oral tolerance' refers to the T-cell-mediated paradox in which the immune response to a previously encountered antigen via the oral route is reduced, it remains one of the challenges associated with oral vaccination (Hernández et al., 2014; Chan & Daniell, 2015). Antigens are released in the intestinal immune system due to a lack of inflammation, and immature DCs introduce T-cells, resulting in tolerance (Lamichhane et al., 2014). Regulatory T-cells release cytokines like IL-10 and establish close contact with DCs when they limit their growth and development to alter their resistance mechanism (Richman et al., 1978). Recurrent delivery of mucosal antigen may similarly reduce the humoral immune response, making it difficult to develop vaccines with consistent efficacy (Kesik-Brodacka et al., 2017). Some approaches can be used to resolve this issue, such as triamcinolone can suppress the immune system. However, this must be done in tiny amounts to avoid significant health issues or even death. Increasing the vaccine dosage sufficiently can often result in a jump start in the immunological response. According to Silin and Lyubomska (2002), multiple dosages over a certain time period are recommended.

19.6 APPLICATIONS OF PBEVS AGAINST DIFFERENT DISEASES

Innovative vaccinations delivered using edible plants provide various advantages over standard methods, and thus edible vaccinations have the potential to revolutionize disease prevention and management. However, to develop effective and extensively accepted edible vaccines, several elements must be carefully considered. These include the ideal antigen dose, the immunogenic part which activates the immune response, and the plant's availability for consumption. To guarantee constant dosage, the antigen concentration must be uniform throughout the plant material. Another aspect is antigen thermostability, which means the vaccine antigen can be steady at particular temperatures for lengthy time durations, which enable proficient storage and distribution. Till date, several vaccines have been developed to target different diseases.

19.6.1 Malaria

This is transmitted via female Anopheles mosquitoes infected by the most lethal parasites *Plasmodium vivax* and *P. falciparum* (Mishra et al., 2008). It has a significant impact on world health as it affects around 300–500 million individuals every year, causing 1.5–2.7 million losses. Transgenic tomato plants were developed to create an edible antimalarial vaccine in order to overcome the challenges of administering and delivering the malaria vaccine (Soria-Guerra et al., 2007). Over 249 million malaria cases and over 6,08,000 deaths are reported globally, according to the WHO's 2023 World Malaria Report (Venkatesan, 2024). *P. falciparum* (MSP4 and MSP5) and *P. yoelii* (more especially, the merozoite surface protein MSP4/5) antigens were employed. Recombinant MSP 4/5, MSP 4, and MSP1 combined with cholera toxin B (CTB) and administered orally produced an antibody response against blood-level parasites (Vasudevan et al., 2024).

19.6.2 Measles

Measles is a greatly transmissible viral respiratory disease which poses an effective public health challenge, particularly in developing countries. The virus is primarily transmitted through respiratory droplets released via infected individual body parts like nose, mouth, and throat. If you follow an incubation period of 10 to 12 days, symptoms like high fever, runny nose, bloodshot eyes, and distinctive small white spots inside the mouth typically appear (Huang et al., 2001). In 2022, it affected approximately 8,00,000 individuals globally, with the majority of cases occurring in children under five years of age (Patel et al., 2022). In a notable experimental study, transgenic tobacco plants expressing the hemagglutinin (MV-H) protein from the Edmonston strain of the measles virus were used to immunize mice via oral administration. The immunized mice produced antibody contents roughly 5-times higher as compared to the protective threshold for humans, with detectable IgA antibodies also found in faecal samples. Moreover, combining oral and parenteral delivery of MV-H has the potential to boost human antibody titres to levels up to 20-times above the protective threshold (Zhang et al., 2010). In addition to tobacco, GM carrot plants have shown promise for measles immunization. Ongoing research also explores the incorporation of measles antigens into rice, cabbage, and baby food products, highlighting the potential of PBEVs to enhance measles prevention strategies, particularly in resource-limited settings.

19.6.3 Rabies

The lyssavirus, a single-stranded RNA virus with a tubular shape and bullet-shaped particles, is responsible for rabies (Sahoo et al., 2020). It can be prevented by using tomato plants which express rabies proteins to produce anti-rabies antibodies in mice. The Tobacco Mosaic Virus (TMV) can also be treated using this technique. Experiments in mammals have shown that genetically engineered plants of tomato, changed by CaMV and harbouring the rabies virus glycoprotein (G-protein) gene from the ERA strain, have physiological activity. Modified tomato plants expressing the rabies virus glycoprotein (G-protein) gene (ERA strain) have been shown to provide protection against rabies in animals (Lal et al., 2007; Shah et al., 2022).

19.6.4 Hepatitis B

Hepatitis B is a widespread virus-related infection, impacting approximately 400 million individuals globally and ranking among the most prevalent hominoid diseases (Sahoo et al., 2020). Roughly 360 million individuals suffer from prolonged hepatitis B virus (HBV) infection, contributing to over 600,000 deaths annually due to many reasons such as hepatocellular carcinoma and liver cirrhosis (Lal et al., 2007). One of the principal challenges in global hepatitis B immunization efforts is the high cost of vaccines and their administration. In this context, PBEVs offer a promising alternative, with the potential to reduce costs and increase accessibility (Sobrino et al., 2001; Hefferon, 2014). Notably, early breakthroughs confirmed the successful genetic modification of tobacco plants to express the hepatitis B surface antigen (HBsAg) under the control of a constitutive promoter (Mason et al., 2002). However, due to the high alkaloid content in tobacco, other crops such as potatoes have been explored as more suitable platforms for oral vaccine delivery (Saxena & Rawat, 2014). Potatoes are particularly favoured for edible hepatitis B vaccines, with transgenic roots showing higher levels of HBsAg expression as compared to leaf tissues (Thanavala et al., 2005). Despite these advances, current expression levels in potatoes remain insufficient for effective oral immunization (Sohrab et al., 2017; Sohrab, 2020). To overcome the limitation, researchers are examining alternative promoters, including the patatin promoter, and transcriptional regulatory elements to enhance antigen production. Studies on thermal degradation of the antigen revealed that cooked transgenic potatoes induce a lower antibody response compared to uncooked ones. Nevertheless, the natural encapsulation of HBsAg within potato tissues helps protect the antigen from heat-induced degradation, contributing to its stability and immunogenicity (Sahoo et al., 2020).

19.6.5 Cholera

It is a severe diarrhoeal disease caused by cholera toxin (CT), a protein complex formed by *V. cholerae* (Tiwari et al., 2009). It affects roughly 10 million people annually, primarily in emerging countries, with children being particularly vulnerable. Research has shown that transgenic potatoes expressing the *V. cholerae* CT-B gene can enhance immune responses in mice. For a month, mice were given these transgenic potatoes every week in addition to sporadic booster doses, and they exhibited increased immunity. Additionally, the nasal administration of LT-B and mCT-A antigens using plant-based delivery systems has proven successful. In one approach, the gene encoding LT-B, an enterotoxin from *E. coli*, was injected into plants. Mice that consumed these transgenic potatoes developed both serum and mucosal antibodies, offering protection against bacterial toxins *in vitro* (Yuki & Kiyono, 2003). In a separate study, rice was engineered to produce CT-B, and mice immunized with the resulting vaccine developed the neutralizing antibodies and were also secured after oral challenge (Nochi et al., 2009).

19.6.6 Diabetes

Diabetes is a promptly growing global health concern. The mice treated with PBEVs exhibited elevated levels of IgG1 antibodies, which are associated with cytokines that help modulate harmful immune responses. Antigens derived from GM plants were shown to retain effective and conferred defence beside diabetes in animal models. This research is considered the first document that supports the accessibility of edible vaccines for the treatment of autoimmune diseases (Carter & Langridge, 2002). A Canadian research team planned an experiment using genetically engineered potato and tobacco plants expressing the GAD67 gene, delivered into non-obese diabetic mice (Avesani et al., 2003). They typically develop diabetes spontaneously which was insulin-dependent. The study found that only 20% of prediabetic mice developed diabetes after the consumption of transgenic plants, compared to 70% in the untreated control group (Avesani et al., 2003, 2010).

19.6.7 HIV

Efforts have also focused on developing edible vaccines for infectious diseases such as HIV. In this, the virus targets the immune system, specifically CD4+ T cells, leading to their depletion and the eventual onset of Acquired Immunodeficiency Syndrome (AIDS), which increases vulnerability to life-threatening infections and cancer (Karasev et al., 2005). Preliminary progress has been made in incorporating HIV proteins into plant-based systems. For instance the cowpea mosaic virus (CPMV) has been engineered to express HIV antigens, resulting in protein expression across various plant tissues, including mature fruits and their progeny (Yusibov et al., 2002; Komarova et al., 2010). In another report, the Tat protein of HIV was delivered via tobacco mosaic virus (TMV) and expressed in spinach plants, and mice fed the modified spinach followed by DNA injection exhibited significantly higher antibody titres compared to controls four weeks post-inoculation (Karasev et al., 2005).

19.7 CLINICAL TRIALS OF PBEVS

Numerous biomedical and behavioural studies have been piloted to evaluate the safety, efficacy, purity, and effectiveness of edible vaccines. Still, clinical trials cannot initiate until enough preclinical data establishes the vaccine's safety exterior of human testing (Garduño-González et al., 2022). This highlights the importance of progressing through the established phases of clinical development. Successful antigen expression has been achieved in several types of plants, including rabies virus-G protein in tomatoes, LT-B in tobacco and potatoes, Norwalk virus in tobacco and potatoes, *V. cholerae* CT-B in potatoes, and HBsAg in tobacco and potatoes (Tacket et al., 1998; Sahoo et al., 2020). A notable clinical test conducted at the Center for Vaccine Development at the University of Maryland involved human participants consuming raw potatoes GM to express a vaccine for traveller's diarrhoea. The trial, approved by the U.S. Food and Drug Administration (FDA), marked a milestone in 1998 when its results became the first human clinical analysis of an edible vaccine (Kumar et al., 2005). However, due to ethical limitations surrounding direct human drug testing, initial edible vaccine studies have focused on proof-of-concept trials. In 2005, Avala's team of research have developed a potato-based booster vaccine to complement injectable hepatitis B immunizations. Similarly, Tacket et al. (2000) evaluated the immune response to potatoes expressing Norwalk virus capsid protein; 95% of participants (19 out of 20) showed an immune response, with some experiencing modest increases in antibody levels. That same year, Pogrebnyak et al. (2005) developed an edible vaccine against using tomato and tobacco plants to express the SARS-CoV spike protein (S1), and mice fed lyophilized tomato fruit followed by an S1 protein booster showed strong IgG1, IgG2a, and IgGb responses in their blood. In another study, 14 healthy adult contributors were roughly allotted to receive 100 g or 50 g of transgenic potato or 50 g of non-genetically modified organism (GMO) potato. The transgenic potatoes expressed LT-B concentrations ranging from 3.71 to 15.7 mg/g, depending on the tissue-specific promoter used (Pogrebnyak et al., 2005). The CT-B antigen remained stable even after brief boiling, maintaining its pentameric GM1-binding form for up to 3 minutes, until tissue breakdown occurred. Serological analysis showed that 10 out of 11 individuals developed IgG anti-LT antibodies after a single dose, and 6 of the 11 (55%) showed a fourfold increase in IgA anti-LT antibodies (Sahoo et al., 2020). The aim of trial like proof-of-concept was used to demonstrate that oral administration of GM plants could induce a defensive immune response in humans (Saxena & Rawat, 2014). In this study, 11 participants received GM potatoes while 3 received conventional potatoes. Blood and stool samples were regularly collected to monitor systemic and mucosal immune responses, and no serious adversarial events were documented during the trial. Research continues on edible vaccines targeting diseases such as Norwalk virus via transgenic potatoes and tomatoes and hepatitis B virus using modified bananas and potatoes. While plant-derived proteins have been accepted for certain human treatments and veterinary vaccines have reached the market, no plant-based vaccines have yet been approved for widespread human use. The WHO has not endorsed any plant-derived vaccines to date, although candidates for e.g. the Ebola virus vaccine have entered advanced clinical analysis (Tregoning, 2020).

19.8 RECENT DEVELOPMENTS IN THE PBEVS

Current discoveries in edible vaccines have been greatly driven by advances in genetic engineering technologies for creating suitable plants which carry desired gene expression. Edible vaccines are used in the evolution of human disease as well as in the veterinary field which is described in many documents – for example tobacco-seed-based oral vaccinations against verocytotoxic *E. coli* strains in piglets (Jabif et al., 2021; Rossi et al., 2023). This strategy addresses public health concerns about food-borne bacteria and offers a substitute of antibiotics in animal husbandry, reducing the possibility of antibiotic resistance (Debnath et al., 2022). In addition, *Wolffia globosa* (duckweed) has been studied for its potential as an edible vaccine for fish and aquatic diseases, highlighting the adaptability of edible vaccine applications across several species (Heenatigala et al., 2020). Recent advancement in the nanoparticles (NPs) for edible vaccine preparations reveals their tremendous prospective for increasing immunogenicity, serving as antigen transporters, and elevating vaccine delivery (Zafar et al., 2023). The distinctive features of NPs, such as form, size, and surface charge, enhance connection with APCs. Azuar et al. (2021) documented that NP-peptide presentation increased transport to APCs, resulting in greater immune responses as compared to non-covalent peptide vaccines. This strategy can be employed in consumable vaccines seeking to trigger better immune response. NPs are advantageous for edible vaccines because they can encapsulate vaccine antigens and protect them from proteolytic breakdown in the gastrointestinal tract. Francis et al. (2020) used a novel adjuvant solid lipid nanoparticle (SLN-A) platform to create a non-viral DNA vaccine delivery system. The vaccine candidate encodes the Urease alpha (UreA) antigen from Helicobacter pylori. In-vitro studies demonstrated that SLN-A particles upregulated the expression of TNF-α, a pro-inflammatory cytokine, in human macrophage-like Tohoku Hospital Pediatrics-1 cells. Lipoplexes demonstrated biocompatibility and effectively transfected murine immune cells to express recombinant *H. pylori* antigen UreA, suggesting their potential as a DNA vaccine delivery system (Francis et al., 2020). Using layered protein NPs has the potential to improve antigen identification and processing in edible vaccines, resulting in longer protection and less off-target immune reactions (Deng & Wang, 2018). Genetically altered bacteria such as *E. coli*, *Salmonella*, and *Lactobacillus* have been used as living transporters and vectors for antigen vaccinations. They have the capability to cling and proliferate within the intestinal mucosa, which is required for successful antigen delivery by oral administration, providing a significant advantage in triggering immunological responses. WhileWang et al. (2018) found that the *Lactobacillus casei*-based vaccination efficiently generated an immune response in contrast to the porcupine epidemic diarrhoea virus (PEDV). A comparable investigation of recombinant *L. casei* with antigen Ah1 using oral administration beside *Aeromonas hydrophila* toxicities in carp fishes found high effectiveness and increased production of antibody in serum (Zhao et al., 2022).

In a similar vein, GM yeasts – *Saccharomyces cerevisiae* in particular – display remarkable versatility, serving both conventional fermentation applications and cutting-edge roles in medicine as probiotics and edible vaccine vectors. Traditionally, *S. cerevisiae* or brewery yeast is used in bakeries to convert carbohydrates to alcohol under anaerobic conditions. Additionally, it is used to create significant by-products like dietary additives, biofuels, and biopharmaceuticals (Ardiani et al., 2010; Palma et al., 2015). Genetic engineering techniques ranging from transformations (mutagenesis, hybridization) to advanced CRISPR/Cas9 editing have resulted in the development of recombinant *S. cerevisiae* to create medicinal molecules, demonstrating their commercial adaptability and relevance (Parapouli et al., 2020). This adaptability enables marker-free genetic changes, permitting tailored antigen synthesis that can provoke an immunological response at the time of consumption (Grissom et al., 2024). Strains like *S. cerevisiae* and *S. boulardii* help to balance the gut microbiota, have immunomodulatory effects, and improve mucosal health, making them useful for treating gastrointestinal illnesses (Stubbs et al., 2001; Palma et al., 2015, 2019). Yeast probiotics regulate the immune response and give therapeutic effects by releasing active chemicals which boost the body's defences against infections (Stubbs et al., 2001). Because *S. cerevisiae* can express a variety of antigens and effectively deliver them to the immune system, it may be utilized as an oral vaccination vector. Heat-killed *S. cerevisiae* formulations stimulate dendritic cell maturation and cross-presentation via MHC pathways, making them more immunogenic than live yeasts. This lowers the dangers of live vaccinations, especially for people with weakened immune systems (Ardiani et al., 2010). Current research focuses on synthetic yeast genomes, which offer superior control over genomic structures, in addition to antigen synthesis. In order to facilitate targeted genetic engineering, the International Synthetic Yeast Genome Project (Sc2.0) seeks to virtually create entire yeast genomes (Fachinetti et al., 2020; Kutyna et al., 2022). This precision engineering makes it possible to incorporate numerous vaccine genes into a single strain of yeast, increasing the efficiency of producing multi-antigen vaccines. *S. cerevisiae*'s high lipid content improves the durability and efficacy of orally delivered vaccinations, opening up new possibilities for vaccine delivery. Lipids can improve immune response by enhancing antigen absorption and vaccination particle retention in the gastrointestinal system (Zamith-Miranda et al., 2016). *S. cerevisiae* strains with optimized lipid profiles may improve the efficiency of oral vaccinations as suggested by Zamith-Miranda et al. (2016). Niosomes are biocompatible, non-ionic surfactant vesicles which can encapsulate lipophilic and hydrophilic vaccine antigens to improve immunological response and transport (Bishnoi et al., 2021; García-Díaz et al., 2019). Fallarini et al. (2023) discovered that niosomes can successfully encapsulate tumour-associated antigens, allowing them to travel safely across the gastric environment to the intestinal mucosa, where elicitation is performed for an immune response. Additionally, they stated that niosomes function as adjuvants, boosting a Th1-mediated immune response, which is essential for the efficacy of vaccinations. Another study found that niosome formulation affects the Th1/Th2 immune response in mice, protecting them from anthrax spore attacks (Gogoi et al., 2018). Future oral vaccines which call for targeted immunity will particularly benefit from the capacity to modify the immune response (Riccardi et al., 2024). Lipidic characteristics are used in oral lipid formulations (OLFs) for edible vaccines to protect vaccination antigens from proteolytic degradation and to stimulate immune responses (Aibani et al., 2022). According to current studies, combining OLFs with vitamins A, D, and E can considerably boost immunological responses by promoting the formation and activity of T-cells (Yu et al., 2023). Eskandari et al. (2018) suggested that lipid preparations with lipopeptide adjuvants have the potency to boost T-cell responses by activating DCs which play a key participation in the start of adaptive immunity.

19.9 REGULATORY, ETHICAL ASPECTS, AND CHALLENGES

Although PBEVs proposed a appropriate and potentially cost-effective immunization strategy, several regulatory and public perception challenges persist. In January 2005, WHO convened a gathering to discuss the monitoring examination of plant-based vaccines. The gathering finalized that the established criteria for the creation, evaluation, and use of vaccines manufactured using traditional methods can also be applied to the manufacturing of edible vaccines. There were special concerns with edible vaccines (Mason et al., 1998; Milpied & McHeyzer-Williams, 2013), and uncertainty remains over how these vaccines should be classified, whether as food, drugs, or agricultural products (Khalid et al., 2022; Koul et al., 2024). As GMOs, for stringent regulatory oversight to prevent accidental entry into the food chain, edible vaccines are subjected. Consequently, cultivation is restricted to specialized greenhouses, with strict segregation during storage and processing. Patent laws now cover GMOs, including edible vaccines, though they exclude naturally arising organisms (Koul et al., 2024). Clinical trials must comply with GMP and meet the requirements of the U.S. Investigational New Drug (IND) application (Kim et al., 2004; Swapna, 2013). However, public acceptance remains a significant hurdle due to persistent misinformation and concerns about GMOs. Addressing these through education and transparency is crucial. With ongoing research, regulatory alignment, and improved public understanding, edible vaccines have significant potential to enhance global health outcomes (Garduño-González et al., 2022; Singh et al., 2023). However, in some areas, GM plants and goods are viewed as a threat to the planet, similar to demonic spirits, so the authorities must play a key role in awakening the people from this delusion. The second crucial criterion to examine is the steadiness of GM plants, and adequate plant isolation is required (Zapanta & Ghorab, 2014; Kim et al., 2016). Sometimes, the transgene causes allergies, and these allergic reactions occur at the time of post-translational alterations. Oral tolerance, when combined with oral adjuvants to primarily engage the MIS, may generate hypersensitivity responses to other proteins included in daily meals. Recurrent administration of PBEVs can increase regulatory T-cell activation, but vaccine antigens trigger hypersensitive reactions in cases of pollen or food allergy. It is mandatory to prevent environmental contamination as well as major adverse effects. Edible vaccines, when properly created, have the potential to be a superior disease prevention tool than other standard vaccines. This vaccination model is cost-effective, efficient, and safe (van der Laan et al., 2006; Lal et al., 2007; Maxwell, 2014). It should be noted that growing plants for edible vaccine manufacturing necessitates careful supervision, but even though the production of GM plants is organized, confirming their security and quality will be tough. During pollination, there may be cross-contamination between GM and non-GM plants, and the GM plants may become harmful. Few times, DNA or antigens are released into water sources as a result of insect or bird contact with plants, contaminating water bodies. Accidentally medicine may penetrate the human food chain and can damage wildlife populations (Twyman et al., 2005; Colson et al., 2010; Hirlekar & Bhairy, 2017).

Despite having numerous advantages, PBEVs have certain downsides. Vaccine-producing genetically engineered plants are not consumed in uncooked form, and heating reduces their efficiency. Furthermore, the immune system and metabolic enzymes can impact vaccine stability in fruits and vegetables. In addition, each and every one requires a different dosage, and in rare cases, they may induce allergic responses. Moreover, selecting the appropriate plants is a tough task as at the time of pollination, there is a possibility of cross-contamination between vaccine-producing GM and non-GM plants. Getting the authorities to authorize the edible immunization is another huge challenge. As a result, further research in this subject is required to solve these problems.

19.10 FUTURE PERSPECTIVE

Traditional vaccinations have an important role in improvising world health, but large-scale manufacture is costly and laborious (Garduño-González et al., 2022). The development and search for edible vaccines can address this issue. Since the early human clinical trials for PBEVs, there have been multiple challenges, like having a sufficiently good expression and its stability during post-harvest (Mamedov et al., 2021). Understanding their long-term impact, such as the probability of hitherto unknown late responses, has to be done. Although edible immunizations are not available now, advances in biotechnology and agriculture point to a future in which vaccination can be provided via the intake of tomatoes. This innovative approach has the potential to improve immunity against various pathogen-caused infections such as HIV, malaria, cardiovascular, respiratory, and coronary infections, impacting a large portion of the global population. Furthermore, in the future, crossing between two different plant generations and expressing therapeutically important antigens encoded by different genes may produce a novel vaccine with adjuvant qualities for those components (Yusibov et al., 2002). Furthermore, the introduction of a new adjuvant, in combination with previously known genes or individually, can improve its immunogenicity (Tiwari et al., 2009). The worldwide effect of the recent COVID-19 pandemic has been enormous, and tobacco has played an essential role in developing SARS-CoV-2 Virus-Like Particles (VLPs). Still, no licensed PBEVs are available which target the newly found corona virus. However, cloning the Spike (S) protein into a transgenic plant such as cabbage, tomato, or cucumber has the potential to produce a vaccine against this CoV (Khalid et al., 2022). These GMOs, when used as salad, might be used to immunize people against a recently discovered virus. Identification of such modifiable plants is an essential step in determining the continuing survival of GM Crops (Gupta et al., 2022). Research is ongoing to determine if edible vaccines meet WHO quality standards for potency, effectiveness, purity, and safety (Charland et al., 2022; Garduño-González et al., 2022). If produced effectively, these vaccinations can provide a one-time solution to prevent pandemic diseases worldwide. Therefore, few examples for plant-based edible vaccines are shown in Table 19.1.

TABLE 19.1
Plant-Based Edible Vaccines

Plant Source	Antigen	Diseases	Advantages	Applications	References
Tobacco	Hepatitis B antigen (HBsAg)	Hepatitis, cervical cancer	• Low cost, preservative system, • Easy purification, • Large harvests • Protective antibodies produced following intranasal immunization	• Good model for evaluating recombinant proteins • It carries immunologically active cost-effective vaccine against many diseases. • Systemic injection showing therapeutic activity • Indicative of humoral and mucosal immune responses by oral delivery	Mason et al. (1996)
	Tetanus vaccine antigen (TetC)	Tetanus			Tregoning (2020)
	IgG (rabies virus)	Rabies			Ko et al. (2004)
	MV-H (measles virus hemagglutinin)	Measles			Webster et al. (2005)
	L1	HPV			Sahoo et al. (2020),
	BV Protein	Multiple sclerosis			Arevalo-Villalobos et al. (2020)
Potato	Heat labile enterotoxin	Enteritis	• Dominated clinical trails • Easily transformed • Stored for long period without refrigerated	• Induced significantly greater immune responses in mice	Mason et al. (1998)
	HBsAg	Hepatitis B			Richter et al. (2000)
	HBsAg and HBsAg-VSPαS/VSPαL	Hepatitis			
	HBsAgS and preS2 antigens	Hepatitis			Joung et al. (2004)
	Norwalk virus capsid protein (NVCP)	Gastroenteritis			Arakawa et al. (1998)
	LT-B	Diarrhoea			Zhou et al. (2015)
	CTB–rota virus enterotoxin protein (NSP4)	*Cholera* and gastroenteritis by *V. cholerae* and rotavirus			Arakawa et al. (1997)
	CTAB-NSP4	Cholera and gastroenteritis			Kim and Langridge (2003)
	HPV11L1 major capsid protein	Cervical cancer			Warzecha et al. (2003)

(Continued)

TABLE 19.1 *(Continued)*
Plant-Based Edible Vaccines

Plant Source	Antigen	Diseases	Advantages	Applications	References
	Human β-amyloid (Aβ)	Alzheimer's			Kim et al. (2003)
	Structural protein VP1	Foot and mouth disease			Castañón et al. (2002)
	Structural protein VP60	Rabbit haemorrhagic syndrome			Castañón et al. (2002)
Tomato	CTB	Cholera	• Grow quickly • Cultivated broadly • High content of vitamin A may boost immune response • Overcome the spoilage problem by freeze-drying technology • Heat-stable, antigen-containing powders, made into capsules	• Immunogenic by oral delivery freeze-drying	Jiang et al. (2007)
	CTB-P4/CTB-P6/TCPA	Cholera			Sharma et al. (2008)
	LTB – Mouse zona pellucida3 epitope (ZP3)	Diarrhoea			Walmsley et al. (2003)
	HBsAg-ENV, GAG epitopes of HIV-1	Hepatitis			Shchelkunov et al. (2006)
	Epitopes of *C. diphtheriae, B. pertussis* and *C. tetani* exotoxins	Diphtheria, Pertussis and Tetanus (DPT)			Soria-Guerra et al. (2007)
	Human β-amyloid (Aβ)	Alzheimer's			Youm et al. (2008)
	RSVF protein	Serious respiratory tract disease			Sandhu et al. (2000)
	Antiphagocytic capsular envelope glycoprotein (F1) and low calcium response virulent antigen (V) fusion protein	Pneumonic/bubonic plague			Alvarez et al. (2006)
	CT-B protein	Vibrio cholerae B toxin			Jiang et al. (2007)
	HBsAg (surface protein of hepatitis B)	Hepatitis B			Guan et al. (2010)
	NVCP (*Norwalk virus* capsid protein)	Norwalk virus			Daniell et al. (2005), Concha et al. (2017),
	Antiphagocytic capsular envelope glycoprotein (F1) and low calcium response virulent antigen (V) fusion protein	Pneumonic/bubonic plague			Alvarez et al. (2006)

Rice	VP2 protein	Infectious bursitis	• Commonly used in baby food because of low allergenic potential • High expression of proteins/antigens • Easy storage/transportation • Expressed protein is heat-stable • Vaccine does not dissolve when exposed to stomach acids • Less risk of contaminating than normal crop and has a broader utility	-	Wu et al. (2007)
	HBsAg (surface protein of Hepatitis B)	Hepatitis B			
Algae *(Dunaliella salina)*	HBsAg (surface protein of hepatitis B)	Hepatitis B	-	-	Geng et al. (2003), Specht and Mayfield (2014)
	Glutamic acid decarboxylase	Diabetes			Specht and Mayfield (2014)
	E7 protein	Human papillomavirus			Specht and Mayfield (2014)
Lettuce	CT-B protein	SARS-CoV-2	• Fast growing		
	sCTB-KDEL	Cholera			Kim et al. (2006)
	β-subunit	Animal enteric			Singh et al. (2023)
	SARS-Cov-2	Oral booster			
	E2 glycoprotein	Swine fever			Sahoo et al. (2020)
Maize	LTB and TGEV glycoprotein S	Diarrhoea and swine transmissible	-	• Antibodies produced following feeding of	Lamphear et al. (2002)
	LTB (heat-labile toxin B)	Diarrhoea		potatoes to mice	Chikwamba et al. (2002)
Wild Tobacco *(Nicotiana benthamiana)*	Zmapp	Ebola			Venkataraman (2021)
	S protein	SARS-CoV			Charland et al. (2022)
	S1 protein	SARS-CoV			Mamedov et al. (2021)

19.11 CONCLUSION

Conventional vaccines have significantly reduced the global mortality rate from infectious illnesses. However, these accomplishments obscure the terrible vacuum in their administration and limitation in treatment worthiness, especially in undeveloped and distant areas with a low vaccine coverage. PBEVs, in contrast, provide a dual approach, offering protection against illnesses while also addressing malnutrition. Although still in their developmental phase, PBEVs have laid a foundational framework for future advancements. They hold a transformative potential for global immunization strategies by leveraging GM plants and innovative delivery systems. With continued research, investment, and collaboration, edible vaccines could significantly improve accessibility and strengthen disease prevention efforts worldwide. However, to make PBEVs accessible to people, further research is needed to assess the risk to the environment, public alertness, safety issues, industrial production, cost-effectiveness, and regulatory concerns before they can be integrated into mainstream vaccination programmes on a global scale.

ACKNOWLEDGMENT

Ashutosh Pathak would like to thank the Ministry of Human Resource Development (MHRD), Department of Higher Education, Government of India, New Delhi, India, for providing a Post-Doctoral Fellowship under RUSA 2.0 (Thematic Project III).

REFERENCES

Aboul-Ata, A.A., Vitti. A., Nuzzaci. M., El-Attar, A.K, Piazzolla, G., Tortorella, C., Harandi, A.M., Olson, O., Wright, S.A., Piazzolla, P. (2014). Plant-based vaccines: Novel and low-cost possible route for Mediterranean innovative vaccination strategies. *Adv Virus Res*, 89, 1–37.

Aibani, N., Patel, P., Buchanan, R., Strom, S., Wasan, K.M., Hancock, R.E.W., Gerdts, V., Wasan, E.K. (2022). Assessing the *in vivo* effectiveness of cationic lipid nanoparticles with a triple adjuvant for intranasal vaccination against the respiratory pathogen *Bordetella pertussis*. *Mol Pharm*, 19(6), 1814–1824.

Altindis, E., Iz, S.G., Ozen, M.O., Nartop, P., Gurhan, I.D., Gurel, A. (2014). Plant derived edible vaccines and therapeutics. In: Atta-ur-Rahman (Ed.), Frontiers in Clinical Drug Research: Anti-Infectives, Bentham Books, UAE, pp. 200–236.

Alvarez, M.L., Pinyerd, H.L., Crisantes, J.D., Rigano, M.M., Pinkhasov, J., Walmsley, A.M., Mason, H.S., Cardineau, G.A. (2006). Plant-made subunit vaccine against pneumonic and bubonic plague is orally immunogenic in mice. *Vaccine*, 24(14), 2477–2490.

Arakawa, T., Chong, D.K., Merritt, J.L., Langridge, W.H. (1997). Expression of cholera toxin B subunit ologomers in transgenic potato plants. *Transgenic Res*, 6, 403–413.

Arakawa, T., Yu, J., Chong, D.K., Hough, J., Engen, P.C., Langridge, W.H. (1998). A plant-based cholera toxin B subunit-insulin fusion protein protects against the development of autoimmune diabetes. *Nat Biotechnol*, 16, 934–938.

Ardiani, A., Higgins, J.P., Hodge, J.W. (2010). Vaccines based on whole recombinant *Saccharomyces cerevisiae* cells. *FEMS Yeast Res*, 10(8), 1060–1069.

Arevalo-Villalobos, J. I., Govea-Alonso, D. O., Bañuelos-Hernández, B., González-Ortega, O., Zarazúa, S., Rosales-Mendoza, S. (2020). Inducible expression of antigens in plants: A study focused on peptides related to multiple sclerosis immunotherapy. *J Biotechnol*, 318, 51–56. https://doi.org/10.1016/j.jbiotec.2020.05.003

Arntzen, C.J. (1998). Pharmaceutical foodstuffs-oral immunization with transgenic plants. *Nat Med*, 4, 502–503.

Aryamvally, A., Gunasekaran, V., Narenthiran, K.R., Pasupathi, R. (2017). New strategies toward edible vaccines: An overview. *J Diet Suppl*, 14, 101–116.

Avesani, L., Falorni, A., Tornielli, G.B., Marusic, C., Porceddu, A., Polverari, A., Faleri, C., Calcinaro, F., Pezzotti, M. (2003). Improved in planta expression of the human islet autoantigen glutamic acid decarboxylase (GAD65). *Transgenic Res*, 12(2), 203–212.

Avesani, L., Vitale, A., Pedrazzini, E., Devirgilio, M., Pompa, A., Barbante, A., Gecchele, E., Dominici, P., Morandini, F., Brozzetti, A., Falorni, A., Pezzotti, M. (2010). Recombinant human GAD65 accumulates to high levels in transgenic tobacco plants when expressed as an enzymatically inactive mutant. *Plant Biotechnol J*, 8(8), 862–872.

Azuar, A., Shibu, M.A., Adilbish, N., Marasini, N., Hung, H., Yang, J., Luo, Y., Khalil, Z.G., Capon, R.J., Hussein, W.M., Toth, I., Skwarczynski, M. (2021). Poly (hydrophobic amino acid) conjugates for the delivery of multiepitope vaccine against group A Streptococcus. *Bioconjug Chem*, 32(11), 2307–2317.

Bishnoi, S., Rehman, S., Dutta, S.B., De, S.K., Chakraborty, A., Nayak, D., Gupta, S. (2021). Optical-property-enhancing novel near-infrared active niosome nanoformulation for deep-tissue bioimaging. *ACS Omega*, 6, 22616–22624.

Carter, J.E., Langridge, W.H.R. (2002). Plant-based vaccines for protection against infectious and autoimmune diseases. *Crit Rev Plant Sci*, 21(2), 3–109.

Castañón, S., Martín-Alonso, J.M., Marín, M.S., Boga, J.A., Alonso, P., Parra, F., Ordás, R.J. (2002). The effect of the promoter on expression of VP60 gene from rabbit hemorrhagic disease virus in potato plants. *Plant Sci*, 162(1), 87–95.

Chan, H.T., Daniell, H. (2015). Plant-made oral vaccines against human infectious diseases-Are we there yet? *Plant Biotechnol J*, 13(8), 1056–1070.

Charland, N., Gobeil, P., Pillet, S., Boulay, I., Séguin, A., Makarkov, A., Heizer, G., Bhutada, K., Mahmood, A., Trépanier, S. (2022). Safety and immunogenicity of an AS03-adjuvanted plant-based SARS-CoV-2 vaccine in adults with and without comorbidities. *NPJ Vaccines*, 7(1), 142.

Chen, T.H., Chen, T.H., Hu, C.C., Liao, J.T., Lee, C.W., Liao, J.W., Lin, M.Y., Liu, H.J., Wang, M.Y., Lin, N.S. (2012). Induction of protective immunity in chickens immunized with plant-made chimeric Bamboo mosaic virus particles expressing very virulent infectious bursal disease virus antigen. *Virus Res*, 166(1–2), 109–115.

Chikwamba, R., Cunnick, J., Hathaway, D., McMurray, J., Mason, H., Wang, K. (2002). A functional antigen in a practical crop: LT-B producing maize protects mice against *Escherichia coli* heat labile enterotoxin (LT) and cholera toxin (CT). *Transgenic Res*, 11(5), 479–493.

Colson, P., Richet, H., Desnues, C., Balique, F., Moal, V., Grob, J.J., Berbis, P., Lecoq, H., Harlé, J.R., Berland, Y., Raoult, D. (2010). Pepper mild mottle virus, a plant virus associated with specific immune responses, fever, abdominal pains, and pruritus in humans. *PLoS One*, 5(4), 10041.

Concha, C., Cañas, R., Macuer, J., Torres, M.J., Herrada, A.A., Jamett, F., Ibáñez, C. (2017). Disease prevention: An opportunity to expand edible plant-based vaccines? *Vaccines*, 5(2), 14.

Curtiss, R., Cardineau, G.A. (1997). Oral immunization by transgenic plants. United States Patent US5679880A.

Dalod, M., Chelbi, R., Malissen, B., Lawrence, T. (2014). Dendritic cell maturation: Functional specialization through signaling specificity and transcriptional programming. *EMBO J*, 33(10), 1104–1116.

Daniell, H., Chebolu, S., Kumar, S., Singleton, M., Falconer, R. (2005). Chloroplast-derived vaccine antigens and other therapeutic proteins. *Vaccine*, 23(15), 1779–1783.

Debnath, N., Thakur, M., Negi, N.P., Gautam, V., Yadav, A.K., Kumar, D. (2022). Insight of oral vaccines as an alternative approach to health and disease management: An innovative intuition and challenges. *Biotechnol Bioeng*, 119(2), 327–346.

Deng, L., Wang, B.Z. (2018). A perspective on nanoparticle universal influenza vaccines. *ACS Infect Dis*, 4(12), 1656–1665.

Dietrich, G., Griot-Wenk, M., Metcalfe, I.C., Langm, A.B., Viret, J.F. (2003). Experience with registered mucosal vaccines. *Vaccine*, 21(7–8), 678–683.

Eskandari, S., Pattinson, D.J., Stephenson, R.J., Groves, P.L., Apte, S.H., Sedaghat, B., Chandurudu, S., Doolan, D.L., Toth, I. (2018). Influence of physicochemical properties of lipopeptide adjuvants on the immune response: A rationale for engineering a potent vaccine. *Chem Eur J*, 24(3), 9892–9902.

Esmael, H., Hirpa, E. (2015). Review on edible vaccine. *Acad J Nutr*, 4, 40–49.

Fachinetti, D., Masumoto, H., Kouprina, N. (2020). Artificial chromosomes. *Exp Cell Res*, 396(1), 112302.

Faith, V.H., Debnath, S., Lavanya, D., Jerusha, V., Madhuri, M. (2015). Food as a vaccine. *Res J Pharm Dos Forms Technol*, 7, 161–165.

Fallarini, S., Papi, F., Licciardi, F., Natali, F., Lombardi, G., Maestrelli, F., Nativi, C. (2023). Niosomes as biocompatible scaffolds for the multivalent presentation of tumor-associated antigens (TACAs) to the immune system. *Bioconjug Chem*, 34(1), 181–192.

Francis, J.E., Skakic, I., Dekiwadia, C., Shukla, R., Taki, A.C., Walduck, A., Smooker, P.M. (2020). Solid lipid nanoparticle carrier platform containing synthetic TLR4 agonist mediates non-viral DNA vaccine delivery. *Vaccines*, 8(3), 551.

Gao, Y., Ma, Y., Li, M., Cheng, T., Li, S.W., Zhang, J., Xia, N.S. (2003). Oral immunization of animals with transgenic cherry tomatillo expressing HBsAg. *World J Gastroenterol*, 9(5), 996–1002.

García-Díaz, M., Patiño, B., Vázquez, C., Gil-Serna, J. (2019). A novel niosome-encapsulated essential oil formulation to prevent *Aspergillus flavus* growth and aflatoxin contamination of maize grains during storage. *Toxins*, 11(11), 646.

Garduño-González, K.A., Peña-Benavides, S.A., Araujo, R.G., Castillo-Zacarias, C., Melchor-Martínez, E.M., Oyervides-Muñoz, M.A., Sosa-Hernández, J.E., Purton, S., Iqbal, H.M.N., Parra-Saldivar, R. (2022). Current challenges for modern vaccines and perspectives for novel treatment alternatives. *J Drug Deliv Sci Techno*, 70, 103222.

Geng, D., Wang, Y., Wang, P., Li, W., Sun, Y. (2003). Stable expression of hepatitis B surface antigen gene in *Dunaliella salina* (Chlorophyta). *J App Phycol*, 15(6), 451–456.

Gogoi, H., Mani, R., Bhatnagar, R. (2018). A niosome formulation modulates the Th1/Th2 bias immune response in mice and also provides protection against anthrax spore challenge. *Int J Nanomedicine*, 13, 7427–7440.

Gómez, E., Zoth, S.C., Asurmendi, S., Vázquez Rovere, C., Berinstein, A. (2009). Expression of hemagglutinin neuraminidase glycoprotein of Newcastle disease virus in agro infiltrated *Nicotiana benthamiana* plants. *J Biotechnol*, 144(4), 337–340.

Grissom, J.H., Moody, S.E., Chi, R.J. (2024). Marker-free genomic editing in *Saccharomyces cerevisiae* using universal donor templates and multiplexing crispr-cas9. *Yeast*, 41(9), 568–579.

Guan, Z.J., Guo, B., Huo, Y.L., Guan, Z.P., Wei, Y.H. (2010). Overview of expression of hepatitis B surface antigen in transgenic plants. *Vaccine*, 28(46), 7351–7362.

Gupta, P., Andankar, I., Gunasekaran, B., Easwaran, N., Muthukaliannan, G.K. (2022). Genetically modified potato and rice based edible vaccines: An overview. *Biocatal Agric Biotechnol*, 43, 102405.

Hahn, B.S., Jeon, I.S., Jung, Y.J., Kim, J.B., Park, J.S., Ha, S.H., Kim, K.H., Kim, H.M., Yang, J.S., Kim, Y.H. (2007). Expression of hemagglutinin-neuraminidase protein of Newcastle disease virus in transgenic tobacco. *Plant Biotechnol Rep*, 1, 85–92.

Haq, T.A., Mason, H.S., Clements, J.D., Arntzen, C.J. (1995). Oral immunization with a recombinant bacterial antigen produced in transgenic plants. *Science*, 268(5211), 714–716.

Heenatigala, P.P.M., Sun, Z., Yang, J., Zhao, X., Hou, H. (2020). Expression of LamB vaccine antigen in *Wolffia globosa* (duck weed) against fish vibriosis. *Front Immunol*, 11, 1857.

Hefferon, K.L. (2014). DNA virus vectors for vaccine production in plants: Spotlight on Gemini viruses. *Vaccines*, 2(3), 642–653.

Hernández, M., Rosas, G., Cervantes, J., Fragoso, G., Rosales-Mendoza, S., Sciutto, E. (2014). Transgenic plants: A 5-year update on oral antipathogen vaccine development. *Expert Rev Vaccines*, 13(12), 1523–1536.

Hirlekar, R., Bhairy, S. (2017). Edible vaccines: An advancement in oral immunization. *Asian J Pharma Clin Res*, 10(2), 71–77.

Huang, Z., Dry, I., Webster, D., Strugnell, R., Wesselingh, S. (2001). Plant-derived measles virus hemagglutinin protein induces neutralizing antibodies in mice. *Vaccine*, 19(15–16), 2163–2171.

Jabif, M.F., Gumina, E., Hall, J.W., Hernandez-Velasco, X., Layton, S. (2021). Evaluation of a novel mucosal administered subunit vaccine on colostrum IgA and serum IgG in sows and control of enterotoxigenic *Escherichia coli* in neonatal and weanling piglets: Proof of concept. *Front Vet Sci*, 8, 640228.

Jeshvaghani, F.S., Amani, J., Kazemi, R., Karimi Rahjerdi, A., Jafari, M., Abbasi, S., Salmanian, A.H. (2019). Oral immunization with a plant-derived chimeric protein in mice: Toward the development of a multipotent edible vaccine against E. coli O157: H7 and ETEC. *Immunobiology*, 224(2), 262–269.

Jiang, X.L., He, Z.M., Peng, Z.Q., Qi, Y., Chen, Q., Yu, S.Y. (2007). Cholera toxin B protein in transgenic tomato fruit induces systemic immune response in mice. *Transgenic Res*, 16(2), 169–175.

Jiji, M.G., Ninan, M.A., Thomas, V.P., Thomas, B.T. (2024). Edible microalgae: Potential candidate for developing edible vaccines. *Vegetos*, 37, 788–793.

Johansen, F.E., Pekna, M., Norderhaug, I.N., Haneberg, B., Hietala, M.A., Krajci, P., Betsholtz, C., Brandtzaeg, P. (1999). Absence of epithelial immunoglobulin a transport, with increased mucosal leakiness, in polymeric immunoglobulin receptor/secretory component–deficient mice. *J Exp Med*, 190, 915–922.

Joung, Y.H., Youm, J.W., Jeon, J.H., Lee, B.C., Ryu, C.J., Hong, H.J., Kim, H.C., Joung, H., Kim, H.S. (2004). Expression of the hepatitis B surface S and preS2 antigens in tubers of *Solanum tuberosum*. *Plant Cell Rep*, 22(12), 925–930.

Karasev, A.V., Foulke, S., Wellens, C., Rich, A., Shon, K.J., Zwierzynski, I., Hone, D., Koprowski, H., Reitz, M. (2005). Plant based HIV-1 vaccine candidate: Tat protein produced in spinach. *Vaccine*, 23(15), 1875–1880.

Kesik-Brodacka, M., Lipiec, A., Kozak, Ljunggren, M., Jedlina, L., Miedzinska, K., Mikolajczak, M., Plucienniczak, A., Legocki, A.B., Wedrychowicz, H. (2017). Immune response of rats vaccinated orally with various plant-expressed recombinant cysteine proteinase constructs when challenged with *Fasciola hepatica* metacercariae. *PLoS Negl Trop Dis*, 11(3), e0005451.

Khalid., F, Tahir, R., Ellahi, M., Amir, N., Rizvi, S.F.A., Hasnain, A. (2022). Emerging trends of edible vaccine therapy for combating human diseases especially COVID-19: Pros, cons, and future challenges. *Phyther Res*, 36(7), 2746–2766.

Kim, H.S., Euym, J.W., Kim, M.S., Lee, B.C., Mook-Jung, I., Jeon, J.H., Joung, H. (2003). Expression of human β-amyloid peptide in transgenic potato. *Plant Sci*, 165(6), 1445–1451.

Kim, N.S., Mbongue, J.C., Nicholas, D.A., Esebanmen, G.E., Unternaehrer, J.J., Firek, A.F., Langridge, W.H. (2016). Chimeric vaccine stimulation of human dendritic cell indoleamine 2,3-dioxygenase occurs via the noncanonical NF-kB pathway. *PLoS One*, 11(2), 1–16.

Kim, T.G., Galloway, D.R., Langridge, W.H. (2004). Synthesis and assembly of anthrax lethal factor-cholera toxin B-subunit fusion protein in transgenic potato. *Mol Biotechnol*, 28(3), 175–183.

Kim, T.G., Kim, M.Y., Kim, B.G., Kang, T.J., Kim, Y.S., Jang, Y.S., Arntzen, C.J., Yang, M.S. (2007). Synthesis and assembly of *Escherichia coli* heat-labile enterotoxin B subunit in transgenic lettuce (*Lactuca sativa*). *Protein Expr Purif*, 51(1), 22–27.

Kim, T.G., Langridge, W.H.R. (2003). Assembly of cholera toxin B subunit full-length rotavirus NSP4 fusion protein oligomers in transgenic potato. *Plant Cell Rep*, 21(9), 884–890.

Kim, Y.S., Kim, B.G., Kim, T.G., Kang, T.J., Yang, M.S. (2006). Expression of a cholera toxin B subunit in transgenic lettuce (*Lactuca sativa* L.) using *Agrobacterium*-mediated transformation system. *Plant Cell Tissue Organ Cult*, 87(2), 203–210.

Ko, K., Wei, X., Crooks, P.A., Koprowski, H.P. (2004). Elimination of alkaloids from plant-derived human monoclonal antibody. *J Immunol Method*, 286(1–2), 79–85.

Komarova, T.V., Baschieri, S., Donini, M., Marusic, C., Benvenuto, E., Dorokhov, Y.L. (2010). Transient expression systems for plant-derived biopharmaceuticals. *Expert Rev Vaccine*, 9(8), 859–876.

Koul, B., Pudhuvai, B., Bhanot, M., Tiwari, S. (2024). Updates on global status of transgenic and genome-edited crops. In: Tiwari, S., Koul, B. (Eds.), Genetic Engineering of Crop Plants for Food and Health Security, Springer, Singapore, pp. 469–510.

Kumar, G.B., Ganapathi, T.R., Revathi, C.J., Srinivas, L., Bapat, V.A. (2005). Expression of hepatitis B surface antigen in transgenic banana plants. *Plant*, 222(3), 484–493.

Kumar, M., Kumari, N., Thakur, N., Bhatia, S.K., Saratale, G.D., Ghodake, G., Mistry, B.M., Alavilli, H., Kishor, D.S., Du, X. (2021). A Comprehensive overview on the production of vaccines in plant-based expression systems and the scope of plant biotechnology to combat against SARS-CoV-2 virus pandemics. *Plants*, 10(6), 1213.

Kunisawa, J., Kurashima, Y., Kiyono, H. (2012). Gut-associated lymphoid tissues for the development of oral vaccines. *Adv Drug Deliv Rev*, 64(6), 523–530.

Kurup, V.M., Thomas, J. (2020). Edible vaccines: Promises and challenges. *Mol Biotechnol*, 62(2), 79–90.

Kutyna, D.R., Onetto, C.A., Williams, T.C., Goold, H.D., Paulsen, I.T., Pretorius, I.S., Johnson, D.L., Borneman, A.R. (2022). Construction of a synthetic *Saccharomyces cerevisiae* pan-genome neo-chromosome. *Nat Commun*, 13(1), 3628.

Lal, P., Ramachandran, V.G., Goyal, R., Sharma, R. (2007). Edible vaccines: Current status and future. *Ind J Med Microbiol*, 25(2), 93–102.

Lamichhane, A., Azegamia, T., Kiyonoa, H. (2014). The mucosal immune system for vaccine development. *Vaccine*, 32(49), 6711–6723.

Lamphear, B.J., Streatfield, S.J., Jilka, J.M., Brooks, C.A., Barker, D.K., Turner, D.D., Delaney, D.E., Garcia, M., Wiggins, B., Woodard, S.L., Hood, E.E. (2002). Delivery of subunit vaccines in maize seed. *J Control Release*, 85(1–3), 169–180.

Landry, N., Ward, B.J., Trépanier, S., Montomoli, E., Dargis, M., Lapini, G., Vézina, L.P. (2010). Preclinical and clinical development of plant-made virus-like particle vaccine against avian H5N1 influenza. *PLoS One*, 5(12), e15559.

Le, T.T., Cramer, J.P., Chen, R., Mayhew, S. (2020). Evolution of the COVID-19 vaccine development landscape. *Nat Rev Drug Discov*, 19(10), 667–668.

Lou, X.M., Yao, Q.H., Zhang, Z., Peng, R.H., Xiong, A.S., Wang, H.K. (2007). Expression of the human hepatitis B virus large surface antigen gene in transgenic tomato plants. *Clin Vaccine Immunol*, 14(4), 464–469.

Mabbott, N.A., Donaldson, D.S., Ohno, H., Williams, I.R., Mahajan, A. (2013). Microfold (M) cells: Important immunosurveillance posts in the intestinal epithelium. *Mucosal Immunol*, 6(4), 666–667.

Mamedov, T., Yuksel, D., Ilgın, M., Gürbüzaslan, I., Gulec, B., Mammadova, G., Ozdarendeli, A., Yetiskin, H., Kaplan, B., Islam Pavel, S.T., et al. (2021). Production and characterization of nucleocapsid and RBD cocktail antigens of SARS-CoV-2 in *Nicotiana benthamiana* plant as a vaccine candidate against COVID-19. *Vaccines*, 9(11), 1337.

Mason, H.S., Ball, J.M., Shi, J.J., Jiang, X., Estes, M.K., Arntzen, C.J. (1996). Expression of Norwalk virus capsid protein in transgenic tobacco and potato and its oral immunogenicity in mice. *Proc Natl Acad Sci USA*, 93(11), 5335–5340.

Mason, H.S., Haq, T.A., Clements, J.D., Arntzen, C.J. (1998). Edible vaccine protects mice against *Escherichia coli* heat-labile enterotoxin (LT): Potatoes expressing a synthetic LTB gene. *Vaccine*, 16(13), 1336–1343.

Mason, H.S., Warzecha, H., Mor, T., Arntzen, C.J. (2002). Edible plant vaccines: Applications for prophylactic and therapeutic molecular medicine. *Trends Mol Med*, 8(7), 324–329.

Maxwell, S. (2014). Analysis of laws governing combination products, transgenic food, pharmaceutical products and their applicability to edible vaccines. *BYU Prelaw Rev*, 28, 65–82.

McDole, J.R., Wheeler, L.W., McDonald, K.G., Wang, B., Konjufca, V., Knoop, K.A., Newberry, R.D., Miller, M.J. (2012). Goblet cells deliver luminal antigen to CD103+ DCs in the small intestine. *Nature*, 483(7389), 345–349.

Meyers, A., Chakauya, E., Shephard, E., Tanzer, F.L., Maclean, J., Lynch, A., Williamson, A.L., Rybicki, E.P. (2008). Expression of HIV-1 antigens in plants as potential subunit vaccines. *BMC Biotechnol*, 8, 53.

Mildner, A., Jung, S. (2014). Development and function of dendritic cells subsets. *Inmmunity*, 40(5), 642–646.

Milpied, P.J., McHeyzer-Williams, M.G. (2013). High-affinity IgA needs TH17 cell functional plasticity. *Nature Immunol*, 14(4), 313–315.

Mishra, N., Gupta, P.N., Khatri, K., Goyal, A.K., Vyas, S.P. (2008). Edible vaccines: A new approach to oral immunization. *Ind J Biotechnol*, 7, 283–294.

Nochi, T., Yuki, Y., Katakai, Y., Shibata, H., Tokuhara, D., Mejima, M., Kurokawa, S., Takahashi, Y., Nakanishi, U., Ono, F., et al. (2009). A rice-based oral cholera vaccine induces macaque-specific systemic neutralizing antibodies but does not influence pre-existing intestinal immunity. *J Immunol*, 183(10), 6538–6544.

Oszvald, M., Kang, T.J., Tomoskozi, S., Jenes, B., Kim, T.G., Cha, Y.S., Tamas, L., Yang, M.S. (2008). Expression of cholera toxin B subunit in transgenic rice endosperm. *Mol Biotechnol*, 40(3), 261–268.

Oszvald, M., Kang, T.J., Tomoskozi, S., Tamas, C., Tamas, L., Kim, T.G., Yang, M.S. (2007). Expression of a synthetic neutralizing epitope of porcine epidemic diarrhea virus fused with synthetic B subunit of Escherichia coli heat labile enterotoxin in rice endosperm. *Mol Biotechnol*, 35(3), 215–224.

Palma, M.L., Garcia-Bates, T.M., Martins, F.S., Douradinha, B. (2019) Genetically engineered probiotic *Saccharomyces cerevisiae* strains mature human dendritic cells and stimulate gag-specific memory CD8+ T cells ex vivo. *Appl Microbiol Biotechnol*, 103(13), 5183–5192.

Palma, M.L., Zamith-Miranda, D., Martins, F.S., Bozza, F.A., Nimrichter, L., Montero-Lomeli, M., Marques, E.T. Jr., Douradinha, B. (2015). Probiotic Saccharomyces cerevisiae strains as biotherapeutic tools: Is there room for improvement? *App Microbiol Biotechnol*, 99(16), 6563–6570.

Parapouli, M., Vasileiadis, A., Afendra, A.S., Hatziloukas, E. (2020). *Saccharomyces cerevisiae* and its industrial applications. *AIMS Microbiol*, 6(1), 1–32.

Patel, P., Patel, R., Patel, S., Patel, Y., Patel, M., Trivedi, R. (2022). Edible vaccines: A nutritional substitute for traditional immunization. *Pharmacogn Rev*, 16(32), 62–69.

Pogrebnyak, N., Golovkin, M., Andrianov, V., Spitsin, S., Smirnov, Y., Egolf, R., Koprowski, H. (2005). Severe acute respiratory syndrome (SARS) S protein production in plants: Development of recombinant vaccine. *Proc Natl Acad Sci USA*, 102(25), 9062–9067.

Qian, B., Shen, H., Liang, W., Guo, X., Zhang, C., Wang, Y., Li, G., Wu, A., Cao, K., Zhang, D. (2008). Immunogenicity of recombinant hepatitis B virus surface antigen fused with preS1 epitopes expressed in rice seeds. *Transgenic Res*, 17(4), 621–631.

Rescigno, M., Urbano, M., Valzasina, B., Francolini, M., Rotta, G., Bonasio, R., Granucci, F., Kraehenbuhl, J.P., Ricciardi-Castagnoli, P. (2001). Dendritic cells express tight junction proteins and penetrate gut epithelial monolayers to sample bacteria. *Nature Immunol*, 2(4), 361–367.

Riccardi, D., Baldino, L., Reverchon, E. (2024). Liposomes, transfersomes and niosomes: Production methods and their applications in the vaccinal field. *J Transl Med*, 22(1), 339.

Rice, J., Ainley, W.M., Shewen, P. (2005). Plant-made vaccines: Biotechnology and immunology in animal health. *Anim Heal Res Rev*, 6(2), 199–209.

Richman, L.K., Chiller, J.M., Brown, W.R., Hanson, D.G., Vaz, N.M. (1978). Enterically induced immunologic tolerance: I. Induction of suppressor T lymphoyctes by intragastric administration of soluble proteins. *J Immunol*, 121(6), 2429–2434.

Richter, L.J., Thanavala, Y., Arntzen, C.J., Mason, H.S. (2000). Production of hepatitis B surface antigen in transgenic plants for oral immunization. *Nat Biotechnol*, 18(11), 1167–1171.

Rosales-Mendoza, S., Angulo, C., Meza, B. (2016). Foodgrade organisms as vaccine biofactories and oral delivery vehicles. *Trends Biotechnol*, 34(2), 124–136.

Rossi, L., Dell'Anno, M., Turin, L., Reggi, S., Lombardi, A., Alborali, G.L., Filipe, J., Riva, F., Riccaboni, P., Scanziani, E., et al. (2023). Tobacco seed-based oral vaccination against verocytotoxic O138 Escherichia coli as alternative approach to antibiotics in weaned piglets. *Antibiotics*, 12(4), 715.

Rybicki, E.P. (2014). Plant-based vaccines against viruses. *Virology J*, 11(1), 205–220.

Sahai, A., Shahzad, A., Shahid, M. (2013). Plant edible vaccines: A revolution in vaccination. In: Shahid, M., Shahzad, A., Malik, A., Sahai, A. (Eds.), Recent Trends in Biotechnology and Therapeutic Applications of Medicinal Plants, Springer, Netherlands, pp. 225–252.

Sahoo, A., Mandal, A.K., Dwivedi, K., Kumar, V. (2020). A cross talk between the immunization and edible vaccine: Current challenges and future prospects. *Life Sciences*, 261, 118343.

Sala, F., Manuela Rigano, M., Barbante, A., Basso, B., Walmsley, A.M., Castiglione, S. (2003). Vaccine antigen production in transgenic plants: Strategies, gene constructs and perspectives. *Vaccine*, 21(78), 803–808.

Saleem, A., Saeed, M.A., Shah, N.A., Kaleem, I., Ahmed, H., Khattak, S.H. (2024). Using plants as vaccines. In: Hashmi, M.Z., Saeed, A., Musharraf, S.G., Shuhong, W. (Eds.), *Recent* Advances in Industrial Biochemistry, Springer International Publishing, Cham, pp. 49–76.

Sandhu, J.S., Krasnyanski, S.F., Domier, L.L., Korban, S.S., Osadjan, M.D., Buetow, D.E. (2000). Oral immunization of mice with transgenic tomato fruit expressing respiratory syncytial virus-F protein induces a systemic immune response. *Transgenic Res*, 9(2), 127–135.

Saxena, J., Rawat, S. (2014). Edible vaccines. In: Ravi, I., Baunthiyal, M., Saxena, J. (Eds.), Advances in Biotechnology, Springer, India, pp. 207–226.

Shah, V.V., Prajapati, R.A., Shah, S.P., Patel, S.R., Patel, H.P. (2022). A comprehensive review on edible vaccine. *J Drug Del Ther*, 12(2S), 192–201.

Sharma, M.K., Singh, N.K., Jani, D., Sisodia, R., Thungapathra, M., Gautam, J.K., Meena, L.S., Singh, Y., Ghosh, A., Tyagi, A.K., Sharma, A.K. (2008). Expression of toxin co-regulated pilus subunit A (TCPA) of *Vibrio cholerae* and its immunogenic epitopes fused to cholera toxin B subunit in transgenic tomato (*Solanum lycopersicum*). *Plant Cell Rep*, 27(2), 307–318.

Shchelkunov, S.N., Salyaev, R.K., Pozdnyakov, S.G., Rekoslavskaya, N.I., Nesterov, A.E., Ryzhova, T.S., Sumtsova, V.M., Pakova, N.V., Mishutina, U.O., Kopytina, T.V., Hammond, R.W. (2006). Immunogenicity of a novel, bivalent, plant-based oral vaccine against hepatitis B and human immunodeficiency viruses. *Biotechnol Lett*, 28(13), 959–967.

Shin, C., Han, J.A., Koh, H., Choi, B., Cho, Y., Jeong, H., Ra, J.S., Sung, P.S., Shin, E.C., Ryu, S., et al. (2015). CD8α – Dendritic cells induce antigenspecific T follicular helper cells generating efficient humoral immune responses. *Cell Rep*, 11(12), 1929–1940.

Silin, D.S., Lyubomska, V. (2002). Overcoming immune tolerance during oral vaccination against *Actinobacillus pleuropneumoniae*. *J Vet Med B Infect Dis Vet Public Health*, 49(4), 169–175.

Singh, R., Lin, S., Nair, S.K., Shi, Y., Daniell, H. (2023). Oral booster vaccine antigen – Expression of full-length native SARS-CoV-2 spike protein in lettuce chloroplasts. *Plant Biotechnol J*, 21, 887.

Sobrino, F., Sáiz, M., Jiménez-Clavero, M.A., Núñez, J.I., Rosas, M.F., Baranowski, E., Ley, V. (2001). Foot-and-mouth disease virus: A long known virus, but a current threat. *Vet Res*, 32(1), 1–30.

Sohrab, S.S. (2020). An edible vaccine development for coronavirus disease 2019: The concept. *Clin Exp Vaccine Res*, 9(2), 164–168.

Sohrab, S.S., Suhail, M., Ali, A., Kamal, M.A., Husen, A., Azhar, E.I. (2017). Edible vaccine: Current status and future perspectives. *Current Drug Metabolism*, 18, 831–841.

Soria-Guerra, R.E., Rosales-Mendoza, S., Márquez-Mercado, C., López-Revilla, R., Castillo-Collazo, R., Alpuche-Solís, A.G. (2007). Transgenic tomatoes express an antigenic polypeptide containing epitopes of the diphtheria, pertussis and tetanus exotoxins, encoded by a synthetic gene. *Plant Cell Rep*, 26(7), 961–968.

Specht, E.A., Mayfield, S.P. (2014). Algae-based oral recombinant vaccines. *Front Microbiol*, 5, 60.

Srinivas, L., Sunil Kumar, G.B., Ganapathi, T.R., Revathi, C.J., Bapat, V.A. (2008). Transient and stable expression of hepatitis B surface antigen in tomato (*Lycopersicon esculentum* L.). *Plant Biotechnol Rep*, 2, 1–6.

Stubbs, A.C., Martin, K.S., Coeshott, C., Skaates, S.V., Kuritzkes, D.R., Bellgrau, D., Franzusoff, A., Duke, R.C., Wilson, C.C. (2001). Whole recombinant yeast vaccine activates dendritic cells and elicits protective cellmediated immunity. *Nat Med*, 7(5), 625–629.

Sussman, H.E. (2003). Spinach makes a safer anthrax vaccine. *Drug Discov Today*, 8(10), 428–430.

Swapna, L.A. (2013). Edible vaccines: A new approach for immunization in plant biotechnology. *Sch Acad J Pharm*, 2(3), 227–232.

Tacket, C.O. (2009). Plant-based oral vaccines: Results of human trials. In: Karasev, A.V. (Ed.), Plant-Produced Microbial Vaccines, Springer, Berlin, pp. 103–117.

Tacket, C.O., Mason, H.S., Losonsky, G., Clements, J.D., Levine, M.M., Arntzen, C.J. (1998). Immunogenicity in humans of a recombinant bacterial antigen delivered in a transgenic potato. *Nat Med*, 4(5), 607–609.

Tacket, C.O., Mason, H.S., Losonsky, G., Estes, M.K., Levine, M.M., Arntzen, C.J. (2000). Human immune responses to a novel Norwalk virus vaccine delivered in transgenic potatoes. *J Infect Dis*, 182(1), 302–305.

Tacket, C.O., Pasetti, M.F., Edelman, R., Howard, J.A., Streatfield, S. (2004). Immunogenicity of recombinant LT-B delivered orally to humans in transgenic corn. *Vaccine*, 22(31–32), 4385–4389.

Thanavala, Y., Mahoney, M., Pal, S., Scott, A., Richter, L., Natarajan, N., Goodwin, P., Arntzen, C.J., Mason, H.S. (2005). Immunogenicity in humans of an edible vaccine for hepatitis B. *Proc Natl Acad Sci USA*, 102(9), 3378–3382.

Tiwari, S., Verma, P.C., Singh, P.K., Tuli, R. (2009). Plants as bioreactors for the production of vaccine antigens. *Biotechnol Adv*, 27(4), 449–467.

Tregoning, J.S. (2020). First human efficacy study of a plant-derived influenza vaccine. *Lancet*, 396(10261), 1464–1465.

Twyman, R.M., Schillberg, S., Fischer, R. (2005). Transgenic plants in the biopharmaceutical market. *Expert Opin Emerg Drugs*, 10(1), 185–218.

Ulmer, J.B., Valley, U., Rappuoli, R. (2006). Vaccine manufacturing: Challenges and solutions. *Nat Biotechnol*, 24(11), 1377–1383.

van der Laan, J.W., Minor, P., Mahoney, R., Arntzen, C., Shin, J., Wood, D. (2006). WHO informal consultation on scientific basis for regulatory evaluation of candidate human vaccines from plants, Geneva, Switzerland, 24–25 January 2005. *Vaccine*, 24(20), 4271–4278.

Vasudevan, S.S., Kandrikar, T.Y., Sayyed, A.A., Sarker, P., Nasir, N.S., Venugopalan, S., Mariajohn, R., Chavda, V.P., Gondaliya, P. (2024). Nanoparticle-based vaccines and future vaccine technologies. In: Chavda, V.P., Vora, L.K., Apostolopoulos, V. (Eds.), Advanced Vaccination Technologies for Infectious and Chronic Diseases, Academic Press, USA, pp. 477–495.

Venkataraman, S. (2021). Plant-based vaccines against infectious human viral diseases: Spotlight on HIV vaccines. *Journal of AIDS & HIV Treatment*, 3, 54–68.

Venkatesan, P. (2024). The 2023 WHO world malaria report. *Lancet Microbe*, 5(3), e214.

Waheed. M.T., Sameeullah, M., Khan, F.A., Syed, T., Ilahi, M., Gottschamel, J., Lössl, A.G. (2016). Need of cost-effective vaccines in developing countries: What plant biotechnology can offer? *SpringerPlus*, 5(1), 65.

Walmsley, A.M., Arntzen, C.J. (2000). Plants for delivery of edible vaccines. *Curr Opin Biotechnol*, 11(2), 126–129.

Walmsley, A.M., Kirk, D.D., Mason, H.S. (2003). Passive immunization of mice pups through oral immunization of dams with a plant-derived vaccine. *Immunol Lett*, 86(1), 71–76.

Wang, X.N., Wang, L., Zheng, D.Z., Chen, S., Shi, W., Qiao, X.Y., Jiang, Y.P., Tang, L.J., Xu, Y.G., Li, Y.J. (2018). Oral immunization with a Lactobacillus casei-based anti-porcine epidemic diarrhoea virus (PEDV) vaccine expressing microfold cell-targeting peptide Co1 fused with the COE antigen of PEDV. *J Appl Microbiol*, 124(2), 368–378.

Warzecha, H., Mason, H.S., Lane, C., Tryggvesson, A., Rybicki, E., Williamson, A.L., Clements, J.D., Rose, R.C. (2003). Oral immunogenicity of human papillomavirus-like particles expressed in potato. *J Virol*, 77(16), 8702–8711.

Webster, D.E., Thomas, M.C., Huang, Z., Wesselingh, S.L. (2005). The development of a plant-based vaccine for measles. *Vaccine*, 23(15), 1859–1865.

WHO (World Health Organization). 2024. Immunization Coverage. World Health Organization: 15 July 2024, URL: https://www.who.int/newsroom/fact-sheets/detail/immunization-coverage (accessed 15/2/2025).

Wigdorovitz, A., Pérez Filgueira, D.M., Robertson, N., Carrillo, C., Sadir, A.M., Morris, T.J., Borca, M.V. (1999). Protection of mice against challenge with foot and mouth disease virus (FMDV) by immunization with foliar extracts from plants infected with recombinant tobacco mosaic virus expressing the FMDV structural protein VP1. *Virology*, 264(1), 85–91.

Wu, J., Yu, L., Li, L., Hu, J., Zhou, J., Zhou, X. (2007). Oral immunization with transgenic rice seeds expressing VP2 protein of infectious bursal disease virus induces protective immune responses in chickens. *Plant Biotechnol J*, 5(5), 570–578.

Wurm, F.M. (2004). Production of recombinant protein therapeutics in cultivated mammalian cells. *Nat Biotechnol*, 22, 1393–1398.

Ye, Y.J., Li, W.G. (2010). The study on immune responses in BALB/c mice by immunization with the transgenic Alfalfa (*Medicago sativa*) containing Eg95-EgA31 fusion gene of *Echinococcus granulosus*. *Xi Bao Yu Fen Zi Mian Yi Xue Za Zhi*, 26(8), 750–753.

Youm, J.W., Jeon, J.H., Kim, H., Kim, Y.H., Ko, K., Joung, H., Kim, H. (2008). Transgenic tomatoes expressing human beta-amyloid for use as a vaccine against Alzheimer's disease. *Biotechnol Lett*, 30(10), 1839–1845.

Yu, X., Yu, C., Wu, X., Cui, Y., Liu, X., Jin, Y., Li, Y., Wang, L. (2023). Validation of an HPLC-CAD method for determination of lipid content in LNP-encapsulated COVID-19 mRNA vaccines. *Vaccines*, 11(5), 937.

Yuki, Y., Kiyono, H. (2003). New generation of mucosal adjuvants for the induction of protective immunity. *Rev Med Virol*, 13(5), 293–310.

Yusibov, V., Hooper, D.C., Spitsin, S.V., Fleysh, N., Kean, R.B., Mikheeva, T., Deka, D., Karasev, A., Cox, S., Randall, J., Koprowski, H. (2002). Expression in plants and immunogenicity of plant virus-based experimental rabies vaccine. *Vaccine*, 20(25–26), 3155–3164.

Zafar, A., Arshad, R., Ur Rehman, A., Ahmed, N., Akhtar, H. (2023). Recent developments in oral delivery of vaccines using nanocarriers. *Vaccines*, 11(2), 490.

Zamith-Miranda, D., Palma, M., Matos, G., Schiebel, J., Maya-Monteiro, C., Aronovich, M., Bozza, P.T., Bozza, F.A., Nimrichter, L., Montero-Lomeli. M. (2016). Lipid droplet levels vary heterogeneously in response to simulated gastrointestinal stresses in different probiotic *Saccharomyces cerevisiae* strains. *J Funct Food*, 21, 193–200.

Zapanta, P.E., Ghorab, S. (2014). Age of bioterrorism: Are you prepared? Review of bioweapons and their clinical presentation for otolaryngologists. *Otolaryngol Head Neck Surg*, 151(2), 208–214.

Zeitlin, L., Cone, R.A., Whaley, K.J. (1999). Using monoclonal antibodies to prevent mucosal transmission of epidemic infectious diseases. *Emerg Infect Dis*, 5(1), 54–64.

Zhang, H., Liu, M., Li, Y., Zhao, Y., He, H., Yang, G., Zheng, C. (2010). Oral immunogenicity and protective efficacy in mice of a carrot-derived vaccine candidate expressing UreB subunit against *Helicobacter pylori. Protein Expr Purif*, 69(2), 127–131.

Zhang, X., Buehner, N.A., Hutson, A.M., Estes, M.K., Mason, H.S. (2006). Tomato is a highly effective vehicle for expression and oral immunization with Norwalk virus capsid protein. *Plant Biotechnol J*, 4(4), 419–432.

Zhao, Z., Wang, H., Zhang, D., Guan, Y., Siddiqui, S.A., Feng-Shan, X., Cong, B. (2022). Oral vaccination with recombinant *Lactobacillus casei* expressing *Aeromonas hydrophila* Aha1 against A. hydrophila infections in common carps. *Virulence*, 13(1), 794–807.

Zhou, Z., Gong, S., Li, X.M., Yang, Y., Guan, R., Zhou, S., Yao, S., Xie, Y., Ou, Z., Zhao, J., Liu, Z. (2015). Expression of Helicobacter pylori urease B on the surface of *Bacillus subtilis* spores. *J Med Microbiol*, 64(1), 104–110.

20 Medicinally Important Vegetable Seeds

Rita Mahapatra, Disha Vainsh, and Poonam Mishra

20.1 INTRODUCTION

In a balanced diet for humans, vegetables are the key component to achieve global nutritional security by providing nutrients, vitamins, and minerals. Various vegetable seeds are also a rich source of nutrients, including carbohydrates, proteins, essential fatty acids, vitamins, and minerals. These seeds are often considered by-products of agriculture, but they are also rich in phytochemicals, including phenolic acids, glucosinolates, and flavonoids, making them valuable candidates in the treatment of various ailments. These bioactive compounds are shown to have various health benefits, including antidiabetic, antioxidant, anti-inflammatory, antimicrobial, and anticancer properties (Shahrajabian & Sun, 2023; Barot et al., 2016). As interest in holistic and natural health grows, vegetable seeds are receiving renewed attention for their potential in both preventive and therapeutic medicines, which is a practice that spans centuries, rooted in the traditions of various cultures worldwide. They provide a sustainable and accessible source of natural remedies, particularly in the context of growing interest in plant-based diets and natural medicine. While much of modern medicine focuses on the application of pharmaceutical drugs, there is a growing interest in natural therapies, including the therapeutic properties of vegetable seeds (Roy et al., 2022). This chapter examines the role of vegetable seeds in traditional and modern medicine, focusing on their medicinal properties, mechanisms of action, and potential health benefits.

20.2 PHYTOCHEMICAL COMPOSITION OF VEGETABLE SEED

Vegetable seeds are rich source of micro and macronutrients, making them highly nutritious and useful in medicinal applications (Petrovska, 2012). The basic composition of vegetable seeds includes:

a. **Proteins**: Seeds are an good source of plant-based proteins, which are essential for cellular repair and immune function. Bioactive peptides derived from seeds of soybean, black soybean, chia seeds, and peas have peptides as food sources having less than 20 amino acids and molecular weight below 8 kDa and possessing potent ROS and reactive-nitrogen-species-scavenging abilities, which can be attributed to the presence of aromatic and polar amino acids within them, thus qualifying them as potential dietary supplements or therapeutic agents for combating the ageing process (Li et al., 2023).
b. **Essential Fatty Acids**: Many vegetable seeds, such as flaxseed and pumpkin seeds, have omega-3 and omega-6 fatty acids, crucial for healthy heart and inflammation regulation (Peter & Jacob, 2016).
c. **Antioxidants**: Seeds contain high levels of antioxidants, including phenolic compounds, flavonoids, and vitamin E, which protect cells from damage caused due to oxidative burst (Cevallos-Casals & Cisneros-Zevallos, 2010).
d. **Vitamins and Minerals**: Vegetable seeds have vitamins (such as A, B, and C complex) and minerals (including magnesium, zinc, and iron), supporting various physiological functions (Matthäus & Musazcan Özcan, 2015).

 DOI: 10.1201/9781003606710-20

e. **Bioactive compounds**: Vegetable seeds contain a range of bioactive compounds, including phenolic acids, glucosinolates, and flavonoids. Glucosinolates are group of sulphur-containing compounds that are found in Brassica vegetables such as broccoli, cauliflower, and cabbage (Vig et al., 2009). Flavonoids are group of plant compounds which are found in a range of vegetables including tomatoes, peppers, and onions (Roy et al., 2022). Phenolic acids are a group of plant compounds which are found in a range of vegetables, including beans, lentils, and peas (for details, refer to Table 20.1).

TABLE 20.1
Bioactive Compounds of Vegetable Seeds with Their Medicinal Importance

Vegetable Seeds	Bioactive Compound	Medicinal Property	Key References
Anise seeds	Flavonoids, phenolic compounds, sulphur-containing compounds, tannins, alkaloids, phenolic diterpenes, vitamins, polyunsaturated fatty acids	Antioxidant antiseptic, antimicrobial, aperitif, digestive, antispasmodic, expectorant, galactagogue, estrogenic, anti-inflammatory, diuretic agents	Bettaieb Rebey et al. (2019b), Sun et al. (2019)
Amaranth seeds	Terpenoids, flavonoids, saponins, phenolics, minerals, amaranth oil, pigments	Antidiabetic, anti-inflammatory, antiulcer, hepatoprotective, antioxidant, antimicrobial, anticancer, antiviral, anti-depressant, bronchodilator, spasmolytic, laxative	Gandhi et al. (2021), M. P. Singh et al. (2024)
Basil seeds	Phenolic acids, flavanol, glycosides, fatty acids (stearic acid, oleic acid, palmitic acid, linoleic acid, myristic acid, α-linolenic acid, capric acid, lauric acid, and arachidonic-acid dominated basil species, major antioxidants (caffeic, vanillic, rosmarinic, quercetin, rutin, apigenin, chlorogenic, and p-hydroxybenzoic acids); basil's essential oils (eugenol, chavicol, and terpenoids); galacturonic acid, galactose, rhamnose, mannose, rabinose, and glucuronic acid	Anticancer, radioprotective, antimicrobial, anti-inflammatory, immunomodulatory, anti-stress, antidiabetic, anti-pyretic, anti-arthritic, antioxidant, prophylactic agent, insecticidal, antiparasitic, hepatoprotective, anti-osteoporotic, cardioprotective, neuroprotective	Dhama et al. (2019), Shahrajabian et al. (2020b), Vassilina et al. (2025b)

(Continued)

TABLE 20.1 *(Continued)*
Bioactive Compounds of Vegetable Seeds with Their Medicinal Importance

Vegetable Seeds	Bioactive Compound	Medicinal Property	Key References
Beetroot seeds	Highly pigmented betalains, ascorbic acid, carotenoids, polyphenols, flavonoids, saponins, high level of nitrates, glycine, folate	Antianaemic, anti-inflammatory, anti-hypertensive, antioxidant, anti-carcinogenic, antipyretic, antibacterial, detoxicant, diuretic properties, hepato-protective	Chhikara et al. (2019b), Rehman et al. (2024)
Broccoli seeds	Phenolics, glucosinolates, isothiocyanates, ascorbic acid, sulforaphane	Antioxidant, cardioprotective, anti-carcinogenic, anti-inflammatory, antioxidant	Torres-Contreras et al. (2017)
Borage seeds	Gamma-linolenic acid (GLA), linoleic acid, oleic acid, palmitic acid	Anti-inflammatory, skin health, hormonal balance, rheumatoid arthritis	Pieszak et al. (2012)
Baobab seeds	Vitamin C, fibre, polyphenols, and flavonoids	Antioxidant, supports the immune system, anti-inflammatory, aids in digestion	Gautam et al. (2012)
Chia seeds	Polyphenols, essential fatty acids, tocopherols, phytosterols, carotenoids	Antioxidant, supports digestive health, reduces inflammation, and enhances skin health.	W. Khalid et al. (2023), Motyka et al. (2023)
Celery Seeds	Flavonoids (luteolin), phthalides, coumarins	Anti-inflammatory, antioxidant	Yan et al. (2022)
Coriander seeds/Cilantro seeds	Essential oils, fatty acids, sterols, phenolic compound	Antimicrobial antioxidant, hypoglycaemic, hypolipidemic, anxiolytic, analgesic, anti-inflammatory, anti-convulsant, anticancer activities	Laribi et al. (2015a)
Cauliflower seeds	Polyamines, phenolic compounds, flavonoids, vitamins, minerals	Glucosinolates, Vitamins, phenolic compounds, carotenoids	Ayadi et al. (2022), Laribi et al. (2015b),
Cabbage seeds	Phytochemically, polyphenols, glucosinolates	Antioxidant, anticancer, antimicrobial, anti-inflammatory, antidiabetic, neuroprotective activities	Ayadi et al. (2022)

TABLE 20.1 *(Continued)*
Bioactive Compounds of Vegetable Seeds with Their Medicinal Importance

Vegetable Seeds	Bioactive Compound	Medicinal Property	Key References
Carrom seeds	Phenols, carvacrol, and mainly thymol, α-pinene, α-thujene, β-myrcene, β-pinene, β-phellandrene, o-cymene, γ-terpinene, limonene, 4-terpineol, dodecane, cis limonene oxide, βfenchyl alcohol, tetradecan, thymol, ethylene methacrylate, heptadecane, diethyl phthalate	Antifungal, antibacterial, anthelmintic, antioxidant and also used as a household remedy for diarrhoea, asthma, colic, and dyspepsia	Shahida et al. (2019)
Cumin seeds	Cumin aldehyde, Thymoquinone, cymene	Neuroprotective, cardioprotective, chemoprotective, antidiabetic, anti-inflammatory, antioxidant, dyspepsia, digestive stimulant	Srinivasan (2018)
Chamomile seeds	Flavonoids, coumarins, volatile oils, terpenes, organic acids, polysaccharides	Anticancer, anti-infective, anti-inflammatory, antithrombotic, antioxidant, hypolipidemic, hypoglycaemic, antihypertensive, antidepressant, neuroprotective activities	Dai et al. (2022)
Chestnut seeds	Saponins, flavonoids, tannins, alkaloids, glycosides, triterpenoids, steroids and phenolic substances	Antioxidative, anti-inflammatory, antiviral, immune-modulating, spasmolytic, anti-carcinogenic, neurodepressive	Allai et al. (2025), Joshi and Singh (2024)
Chicory seeds	Flavonoids, saponins, tannins, alkaloids, anthraquinones	Antioxidant, antimicrobial, supports liver health, digestive aid	Duda et al. (2024)
Cardamon seeds	Essential oil comprising cineole, terpineol, borneol, terpinene, fixed oil, protein	Aromatic, carminative stimulant, flavouring agent, condiment	Bhattacharya (2020)
Coffee seeds	Caffeine (alkaloid), tannin, fixed oil, protein, chlorogenic acid, sugars	Stimulant, diuretic, to antagonize the toxic effects of CNS-depressant drugs	Bhattacharya (2020)
Cocoa seeds	Alkaloids mainly theobromine, caffeine, polyphenols	Stimulant, diuretic, cutritive	Bhattacharya (2020)

(Continued)

TABLE 20.1 ***(Continued)***
Bioactive Compounds of Vegetable Seeds with Their Medicinal Importance

Vegetable Seeds	Bioactive Compound	Medicinal Property	Key References
Caraway seeds	Carvone, carway oil, limonene, flavonoids, terpenoids, alkaloids, tannins	Antibacterial, antifungal, antiparasitic, antineuraminidase, antioxidant, anti-inflammatory, anticancer activities	Noreen et al. (2023b), Rasooli and Allameh (2016)
Flax seeds	Omega-3 fatty acids (ALA), Lignans, fibre	Anti-inflammatory, heart health, lowers cholesterol, cancer prevention	Shim et al. (2024)
Fenugreek seeds	Diosgenin, 4-hydroxyisoleucine (4-HIL), trigonelline, galactomannan (GM), and polyphenols such as quercetin, lysine and L-tryptophan in their proteins, mucilaginous and gummy fibres and saponins, coumarins, sapogenins, and phytic acids	Antioxidant, anti-inflammatory, maintaining cardiovascular health	Srinivasa and Naidu (2021)
Fennel seeds	Volatile oils like anethole, estragole, fenchone, and limonene, as well as phenolic compounds such as rutin and quercetin	Hepatoprotective, antioxidant, anti-thrombotic, anti-inflammatory, antibacterial, and antifungal qualities	Barakat et al. (2022), Noreen et al. (2023b)
Garden Crase seeds/Halin seeds	Gallic acid, protocatechuic acid, ellagic, iso-ferulic, coumarin, hesperidin, rutin, quercitrin	Anti-asthmatic, anti-diarrhoea, hypocholesterolemia antioxidant, anti-inflammatory, antimicrobial, antidiabetic, hepatoprotective, cardioprotective mechanisms	Khalid et al. (2025)
Guava seeds	Linoleic acid, α-tocopherol, B-sitosterol, carotenoids, vanillin, gallic acid, catechins	Antioxidant, anti-inflammatory, supports heart health, weight management	Allaqaband et al. (2022)
Jackfruit seeds	Polyphenols, tannins, flavonoids, ferulic acid, gallic acid, caffeic acid	Anti-histamine, antioxidant, anti-tumour, cardioprotective, antimicrobial, anti-inflammatory	Dhani et al. (2025)

TABLE 20.1 *(Continued)*
Bioactive Compounds of Vegetable Seeds with Their Medicinal Importance

Vegetable Seeds	Bioactive Compound	Medicinal Property	Key References
Mustard seeds	Glucosinolate vitamins, protein, phenolic compound, flavonoid	Antimicrobial, antidiabetic, diuretic, expectorant, anti-catarrhal activity, anti-asthmatic, anticancer, galactagogue, anti-spasmodic activity, use in neuralgia, anti-inflammatory	Rahman et al. (2024)
Moringa Seeds	Vitamins carotenoids, polyphenols, phenolic acids, flavonoids, alkaloids, gluconolates, isothiocyanates, tannins, saponins, functional peptides; phenolic compounds essential amino acids	Diuretic cholesterol-lowering activities, antidiabetic activity, antiesthetic activity antimalarial protection against gastric ulcers, antidiabetic, hypotensive and anti-inflammatory thyroid hormone regulator	Abd Rani et al. (2018b), N. Kumar et al. (2021), Vergara-Jimenez et al. (2017)
Nigella seeds	Fixed oils, proteins, nigellone, thymoquinone, essential oil, saponins, alkaloids, ascorbic acid	Anti-inflammatory, antimicrobial, anthelmintic, in respiratory diseases (cough, bronchitis, asthma, flu)	Bhattacharya (2020)
Onion seeds	Flavonol, quercetin, flavanosulfer, cysteine sulphoxides, quercetin glucosides	Antioxidant, anticancer, anticoagulant, antivirus, immune-modulating, anti-inflammatory, anti-complementary, antidiabetic, antimicrobial, antiobesity, antidiabetic, cardiovascularprotective, neuroprotective, hepatorenal protective, respiratory protective, digestive system protective, reproductiveprotective, immunomodulatory properties	Ben Yakoub et al. (2018), Zhao et al. (2021)

(Continued)

TABLE 20.1 *(Continued)*
Bioactive Compounds of Vegetable Seeds with Their Medicinal Importance

Vegetable Seeds	Bioactive Compound	Medicinal Property	Key References
Olive seeds	Oleuropein, hydroxytyrosol, 62% oleic acid, 24% linoleic acid	Anticancer properties, neuroprotective potential, antidiabetic, antihyperlipidemic, antioxidant, anti-inflammatory, antimicrobial	Ghanbari et al. (2012); Maestri et al. (2019)
Pea seeds	Quercetin, kaempferol, isorhamnetin, apigenin, luteolin, quercetin, glycosides, alkaloids	Antioxidant, antimicrobial, anticancer, anti-inflammation, antidiabetic	Wu et al. (2023)
Poppy seeds	Omega-6 fatty acid, omega-9 fatty acid, palmitic acid, gamma tocopherol, caffeic and ferulic acids	Antioxidant, analgesic, sedative and calming, cardioprotective, bone strengthener,	Muhammad et al. (2021)
Pumpkin seeds	β-carotene, phenolic compound (kaempferol, quercetin, and ferulic acid); phytopherol, tocopherol, essential fatty acids	Antioxidant and anti-inflammatory effects, anticancer activity, metabolic health, benign prostatic hyperplasia (BPH), neuroprotection, antimicrobial, anthelmintic properties	Dotto and Chacha (2020b), A. Singh and Kumar (2024)
Purslane Seeds	Flavonoids, Alkaloids, omega-3 fatty acid, vitamins and minerals, polyphenols, polysaccharides, terpenoids, cinnamic acid	Anti-diabetes, anticancer, promotes eye and liver health, antioxidant, anti-inflammatory	Truc Thanh Tran et al. (2023) Hou et al. (2021)
Quinoa seeds	Bioactive peptides (chenopodin and lunasin); phytosterols (β-sitosterol, C campesterol, and stigmasterol); tocopherols (particularly γ-tocopherol); phenolic compounds (quercetin and kaempferol derivatives); terpenoid saponins (including oleanolic acid, hederagenin, Phytolacca genic acid, and serjanic acid)	Anticancer potential, cholesterol-lowering effects, antimicrobial properties, anti-inflammatory activity, antioxidant effects	Casalvara et al. (2024); Xi et al. (2024)

TABLE 20.1 ***(Continued)***

Bioactive Compounds of Vegetable Seeds with Their Medicinal Importance

Vegetable Seeds	Bioactive Compound	Medicinal Property	Key References
Rapeseeds	Brassinosteroids, tocopherols, carotenoids, flavonoids, glucosinolates, vitamin C, minerals, fatty acids	Antidiabetic, anti-inflammation, anticancer, obesity, atherosclerosis, hypertension, anti-viral activity	
Sunflower seeds	Phenolic acid (5-O-caffeoylquinic acid (5-CQA)), flavonoids (kaempferol and quercetin), alpha-tocopherol, essential amino acids (glutamic acid, aspartic acid, arginine, phenylalanine, tyrosine, leucine, methionine, and cysteine)	Anti-inflammation, antimicrobial, antioxidant, cardioprotective	Alexandrino et al. (2021), X.-S. Liu et al. (2020), Tileuberdi et al. (2022)
Sweet corn seeds	Phenolic acids (ferulic acid and vanillic acid), flavonoids (anthocyanins), carotenoids (lutein and zeaxanthin), bioactive peptides, polysaccharides and dietary fibre	Antioxidant activity, antidiabetic effects, anti-inflammatory, anticancer, cardiovascular health, hepatoprotective effects Bottom of Form	Cagnin et al. (2024b), Nile and Park (2014a, 2014b), Siyuan et al. (2018)
Tamarind seeds	Polyphenols (catechins, procyanidins, and flavonoids like taxifolin, apigenin, luteolin, and naringenin) Fatty Acids (Palmitic acid, oleic acid, linoleic acid, and eicosanoid acid) Camp sterol β-Sitosterol Glycosides (scilliroside and uzarigenin derivatives)	Anti-inflammatory, analgesic effects, antioxidant activity, antimicrobial properties, antidiabetic potential, wound healing, tissue engineering	Bonin et al. (2023), Chinta et al. (2025), Wandee et al. (2022)
Tomato seeds	Phenolic acids (gallic acid, ferulic acid, and caffeic acid) Flavonoids (quercetin, kaempferol, and rutin) Carotenoids (lycopene and β-carotene)	Gut microbiota modulation, cardiovascular health, antimicrobial, anticancer, anti-inflammatory, antioxidant	M. Kumar et al. (2022b)
Watermelon seeds	Phenolic acid and flavonoid (4-methylcatechol, pyrogallol, and 1,2,4-benzenetriol) Alkaloids (thiirane)	Antioxidant, antimicrobial, anti-inflammatory, hypogyxamic, cardiovascular	Anang et al. (2018b), Nissar et al. (2025b), Rezig et al. (2019b)
Chickpea	Saponins, flavonoids (e.g. quercetin, kaempferol), phenolic acids (e.g. gallic acid, caffeic acid); phytosterols (β-sitosterol, campesterol, stigmasterol)	Antioxidant, anti-inflammatory, cholesterol-lowering, anticancer (colon cancer), liver protection	Iriti and Varoni (2017b), Maria De Jesus Benevides et al. (2018a), B. Singh et al. (2017)

(Continued)

TABLE 20.1 *(Continued)*
Bioactive Compounds of Vegetable Seeds with Their Medicinal Importance

Vegetable Seeds	Bioactive Compound	Medicinal Property	Key References
Lentil	Phenolic acids (e.g. ferulic acid, p-coumaric acid) Flavonoids (e.g. catechin, epicatechin), tannins, phytosterols	Antioxidant, anti-inflammatory, cardiovascular health, blood sugar regulation	Iriti and Varoni (2017b), Maria De Jesus Benevides et al. (2018b), B. Singh et al. (2017)
Common bean	Phenolic acids (e.g. ferulic acid, p-coumaric acid) Aanthocyanins Flavonoids (e.g. quercetin, kaempferol), Phytosterols	Antioxidant, anti-inflammatory, cholesterol-lowering, cardiovascular protection	Iriti and Varoni (2017b), Maria De Jesus Benevides et al. (2018b), B. Singh et al. (2017)
Black-eyed peas	Phenolic acids (e.g. ferulic acid), flavonoids (e.g. quercetin) Phytosterols	Antioxidant, anti-inflammatory, cardiovascular health, cholesterol-lowering	Iriti and Varoni (2017a), B. Singh et al. (2017)
Pigeon pea	Saponins, flavonoids, phenolic acids, phytosterols	Antioxidant, anti-inflammatory, cholesterol-lowering, anticancer (colon cancer)	Iriti and Varoni (2017b), Maria De Jesus Benevides et al. (2018b), B. Singh et al. (2017)
Cowpea	Phenolic acids, flavonoids, phytosterols	Antioxidant, anti-inflammatory, cardiovascular health, cholesterol-lowering	Iriti and Varoni (2017b), Maria De Jesus Benevides et al. (2018b), B. Singh et al. (2017)
Lupin beans	Phenolic acids, flavonoids, phytosterols	Antioxidant, anti-inflammatory, cholesterol-lowering, cardiovascular protection	Iriti and Varoni (2017b), Maria De Jesus Benevides et al. (2018b), B. Singh et al. (2017)

20.3 TRADITIONAL USES OF VEGETABLE SEEDS AS MEDICINE

Historically, vegetable seeds have been used in various cultures for their healing properties. Ancient civilizations such as the Egyptians, Greeks, and Chinese employed seeds as medicinal remedies for various ailments. In India, where a large portion of the population relies on vegetarian diets, the importance of vegetable seeds in addressing nutritional needs cannot be overstated. Vegetable seeds also play a vital role in preserving and enhancing India's rich culinary heritage, from traditional regional vegetables to globally acclaimed spices. The scope of vegetable seeds extends beyond sustenance to cultural preservation and culinary innovation (Table 20.1). For example, seeds of the Apiaceae family are mostly cultivated for their use as spices, including fennel (*Foeniculum vulgare*), cumin (*Cuminum cyminum*), coriander (*Coriandrum sativum*), anise (*Pimpinella anisum*), and caraway (*Carum carvi*). From ancient times, these spices have been essential to the culinary art of flavouring and seasoning food (Otunola, 2022). Moreover, they are also used in indigenous medicines, pharmaceuticals, nutraceuticals, aroma therapy, preservatives, natural colours, perfumes, cosmetics, and biopesticides. Several dietary vegetable seeds and spices with medicinal importance are reviewed by Shahrajabian and Sun (2023).

Several vegetable seeds have been shown to have medicinal importance and have various types of bioactive compounds as shown in Figure 20.1. This section of chapter explores various seeds with medicinal properties.

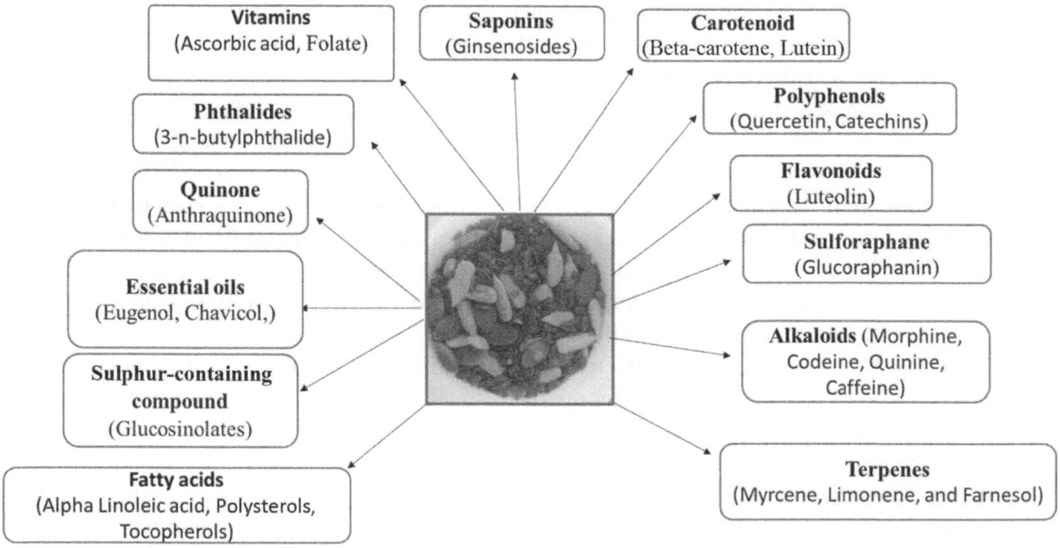

FIGURE 20.1 Different types of bioactive compounds present in the vegetable seeds.

1. **Aniseed (*Pimpinella anisum* L.)**: Aniseeds or the seeds of anise accumulate a phenolic compound, mainly naringin (17.55–32.49%) during the ripening of aniseed, followed by the reduction of syringic acids and rosmarinic, ellagic, and gallic acids. Concerning the chelating ability, antioxidant activity, 1, 1-diphenyl-2-picrylhydrazyl (DPPH) scavenging activity, and reducing power were maximal at full maturity in all aniseeds (Bettaieb Rebey et al., 2019a). Determination of optimal periods and attributions for antioxidant accumulation can be used to evaluate the quality of aniseeds. They have also shown that aniseeds and essential oils have antibacterial, antioxidant, antifungal, anti-inflammatory, anticonvulsant, analgesic, antidiabetic, gastro-protective, and antiviral activities. Other important benefits of aniseeds are that they are digestive, antispasmodic, being one of the best gas-releasing agents, antiseptic, antiepileptic, antihysteric, and antirheumatic and also have significant culinary value, contributing to cardiovascular health by helping in regulation of blood pressure, and play a role in maintaining hormonal balance, particularly in females. Additionally, they help in reducing the symptoms of depression, with hair as well as skin benefits. Aniseeds are used as an analgesic in migraine and as a carminative, aromatic, disinfectant, and diuretic in traditional medicine (Sun et al., 2019).
2. **Amaranth seeds**: Amaranthus species provide a good source of polyphenols and a greater nutritional value in their seeds and leaves (Gandhi et al., 2021). The seeds of amaranthus are famous as *ramdana*; they can be eaten raw or roasted. These seeds can also be ground into flour, which is then used to prepare bread, consumed as an alternative to cereals, and also as an additive for salads (Fiorito et al., 2017). It is found that amaranth seeds are rich in phytonutrients which play a significant role in the inhibition of both oxidative stress and free radical reactions with tissue and membranes (M. P. Singh et al., 2024; Yawadio Nsimba et al., 2008).
3. **Baobab (*Adansonia digitata* L.) Seeds**: The seeds of baobab are eaten either fresh or dry, or even ground and used in cooking, as they are rich in phosphorus, calcium, and magnesium and are also used to cure fever, diarrhoea, and cough. Seed oils are important sources of nutritional oils, with industrial and medicinal importance (Rahul et al., 2015).
4. **Basil (*Ocimum basilicum* L.)**: Also known as sweet basil, it has been known for its potent effects, such as antimicrobial, antifungal, insecticidal, antiparasitic, antioxidant,

immunomodulatory, anti-inflammatory, hepatoprotective, anti-osteoporotic, cardioprotective, neuroprotective, anticancer, and other beneficial health effects (Vassilina et al., 2025a). Thus, this herb has the potential for treating a myriad of ailments and disorders in both humans and animals. Different bioactive chemical compounds have been identified from basil, such as linalool, geraniol, methyl eugenol, methyl chavicol, *p*-allylanisole, 1,8–cineole, trans–α–bergamotene, and neryl acetate (Shahrajabian et al., 2020b)

5. **Beet root (*Beta vulgaris* L.) seeds**: The seeds of beetroot are also known as multi-germ seeds, as one seed can give rise to more than one seedling. The corky exterior of seed contains phenolic compounds and inhibits germination as a physical barrier (Chhikara et al., 2019a). Several bioactive compounds of beetroot are reported to significantly exhibit antioxidant, antimicrobial, anti-inflammatory, antiproliferative, and antiviral activities. Similarly, beetroot extracts have been reported positively to modulate gut microbiota composition, thereby enhancing probiotic metabolism (Rehman et al., 2024).
6. **Borage seeds (*Borago officinalis* L.):** The oil from seeds of Borage is used as a nutritional supplement, and it is known for its anti-inflammatory properties, particularly in the treatment of rheumatoid arthritis. It is also used for various skin conditions and other health concerns such as diabetes and heart diseases. Borage seed oil is rich in gamma-linolenic acid (GLA), an omega-6 fatty acid believed to contribute to its medicinal effects (Pieszak et al., 2012).
7. **Broccoli seeds (*Brassica. oleracea* var. italica)**: Broccoli seeds contain a range of glucosinolates, including SFN and glucoraphanin. These compounds show anticancer properties that help to protect against colon, breast, and prostate cancers (Torres-Contreras et al., 2017).
8. **Chia seeds (*Salvia hispanica* L.)**: The chia seeds have an extensive range of biologically active compounds, which are nutritionally rich including phenolic compounds, carotenoids, sterols, stanols, and phytoestrogens (Kris-Etherton et al., 2002). Chia seeds were traditionally used to lower the risks of obesity (including diabetes) and cardiovascular issues (Mohd Ali et al., 2012). Chia seeds positively influence the cardiac system (i.e. cardio-protective); control hypertension and dyslipidaemia; are antidepressant, antiviral, and anti-carcinogenic; lower LDL cholesterol; increase HDL cholesterol; and repair damaged tissues (Ullah et al., 2016).
9. **Cauliflower seeds (*B. oleracea* var. botrytis)**: Cauliflower seeds contain a range of glucosinolates, including SFN and glucoraphanin, and these compounds have anticancer properties and may help to protect against breast, colon, and prostate cancers (Ayadi et al., 2022).
10. **Cabbage seeds (*B. oleracea* var. capitata)**: Cabbage seeds contain a range of glucosinolates, including sinigrin and gluconapin. These compounds have anti-inflammatory properties and may help to protect against arthritis and other inflammatory conditions (Ayadi et al., 2022).
11. **Coriander seeds (*Coriandrum sativum* L.)**: Coriander seeds are utilized to treat digestive issues such as indigestion and bloating. Essential oils derived from Coriander seeds and herbs are rich in bioactive compounds and have been actively investigated for their chemical composition and biological activities, such as antimicrobial, antioxidant, hypoglycaemic, hypolipidemic, anxiolytic, analgesic, anti-inflammatory, anti-convulsant, and anticancer activities (Laribi et al., 2015a). They also have antimicrobial properties and are thought to reduce blood sugar levels (Nadeem et al., 2013).
12. **Carom Seeds (*Trachyspermum ammi*)**: Carom seed is also known as *ajwain*; it is very valuable for its oil content. Because of its detoxifying, antibacterial, antiviral, and calming properties, it is used as a natural, safe, and cost-effective therapy for several health concerns. Ajwain seeds are most commonly used as a household remedy for diarrhoea, asthma, colic, and dyspepsia, and they also have antifungal, antibacterial, antihelminthic, and antioxidant properties. The active components of ajwain essential oils are phenolic

carvacrol and mainly thymol. Both are responsible for antitussive and antiseptic properties. The phenolic compound thymol is a germicide, antifungal, and antispasmodic agent (Shahida et al., 2019).

13. **Cumin (*Cuminum cyminum* L.) seeds**: Cumin is known as jeera, and black jeera seeds (*Nigella sativa*) are unrelated to *C. cyminum*; however, the name 'Black cumin' was taken from it and used in medicine for many disorders in traditional systems. The cumin seeds have usually been described as its bioactive constituents such as phenols, terpenes, and flavonoids, which are known for their ability to improve digestion and relieve gas and bloating (Srinivasan, 2018). They are also used to treat anaemia, as they contain iron (Allaq et al., 2020)
14. **Caraway (*Carum carvi*) seeds**: Caraway seeds are used to aid digestion, reduce bloating, and treat indigestion. They are also believed to have antimicrobial and antioxidant properties (Rasooli & Allameh, 2016).
15. **Flaxseeds (*Linum usitatissimum* L.)**: Flaxseeds have been used in ancient cultures for their healing properties, particularly for digestive issues and inflammation. It has been shown that consuming flaxseed or flaxseed meal, or cake, can reduce proliferation and nuclear transformation of epithelial cells in mammary glands and that diets with these ingredients are associated with reduced cancer growth. They are considered a natural remedy for constipation and are widely used to promote gastrointestinal health (Shim et al., 2024).
16. **Fenugreek (*Trigonella foenum-graecum* L.)**: Fenugreek seeds have been used as a traditional remedy for the treatment of diabetes and anti-inflammatory, hypercholesterolemia, uterine, and lactation-stimulating properties. They also display immunomodulatory effects as well as gastro- and hepatoprotective and antioxidative properties. They also normalize blood circulation, making the systems active and energetic and maintaining cardiovascular health (Srinivasa & Naidu, 2021). Fenugreek seeds are also used to treat digestive issues, lower blood sugar levels, and manage cholesterol levels. They are also believed to have anti-inflammatory and antioxidant properties (Faisal et al., 2024).
17. **Fennel (*Foeniculum vulgare*) seed**: it is antioxidant-rich, aids digestion, alleviates menstrual cramps, is effective to treat kidney stones due to its diuretic properties, and may help manage appetite and promote breast milk production (Barakat et al., 2022). It demonstrates antioxidant, hepatoprotective, anti-thrombotic, anti-inflammatory, antibacterial, and antifungal qualities (Noreen et al., 2023a).
18. **Lotus (*Nelumbo nucifera*) seeds**: Lotus seeds are edible as fresh or dry, matured, with supplementary medicinal properties as they contain bioactive compounds like alkaloids, flavonoids, and certain micronutrients showing bioactivity such as antioxidant activity and hypoglycaemic, immunomodulatory, antibacterial, anti-inflammatory, analgesic effects as well as gastrointestinal regulation (Zhang et al., 2015). Lotus seeds are in high demand in Ayurvedic medicinal preparations and are widely used in folk medicines to treat tissue inflammation, cancer, diuretics (C.-P. Liu et al., 2004).
19. **Moringa (*Moringa oleifera*) seeds**: Moringa seeds are utilized in traditional medicine for their antimicrobial, antidiabetic, and anti-inflammatory properties (Abd Rani et al., 2018a). They are also thought to support liver and kidney health. Moringa seed works magically in lowering cholesterol levels, and also have antidiabetic, antiseptic, and antimalarial properties. Study also suggests that moringa seeds help in protection from gastric ulcers, hypotension, and anti-inflammation and regulate thyroid hormones (N. Kumar et al., 2021).
20. **Mustard (*Brassica juncea*) seeds**: Mustard seeds are used to treat cold, cough, muscle pain, and skin diseases. They are also used as a stimulant and for improving circulation (Rahman et al., 2024). The seed of *Brassica juncea* aqueous extract, when administered orally showed a beneficial effect on diabetic rats (Thirumalai et al., 2011). Mustard seed

powder significantly decreases the fasting serum glucose, so the plant can be used in the management of pre-diabetic state (Yadav et al., 2004).

21. **Onion (*Allium cepa* L.) seeds**: Onion seeds are the inedible part, but one finding indicated that onion seed extract showed antidiabetic activities in STZ-induced diabetic rats. It was also suggested that the onion seeds might be useful as a supplementary protective agent against the adverse effects of diabetes on the reproductive system in diabetic male rats (Fallah Huseini et al., 2017).
22. **Poppy (*Papaver somniferum* L.) seeds:** The poppy seeds are well known for their health benefits. They are nutritionally rich, including proteins, dietary fibre, oil content, antioxidants, tocopherols, and other micronutrients, and are famous as *khas-khas* (Muhammad et al., 2021).
23. **Peas (*Pisum sativum* L.) seeds:** Wu et al. (2023) reviewed the pea seeds for their chemical composition, physicochemical properties, health benefits, and potential applications. Whole peas are rich in macronutrients, including proteins, starches, dietary fibre, and non-starch polysaccharides. In addition, polyphenols, especially flavonoids and phenolic acids, are an important bioactive ingredient that is distributed mainly in the pea coats. Due to their richness in a variety of nutritional and bioactive ingredients, the consumption of peas has been suggested to be associated with a wide range of health benefits, and there has been increasing focus on their potential as a functional food.
24. **Pumpkin Seeds**: The pumpkin belongs to the Cucurbitaceae family. The edible seeds of Cucurbitaceae are oval, flat, and of diverse colours. Among others, pumpkin (*Cucurbita maxima*, *C. pepo*, and *C. moschata*), winter melon (*Benincasa hispida*), and watermelon (*Citrullus lanatus*) seeds hold the highest food value (Rezig et al., 2019a). The seed flesh is nutty in taste and is rich in protein, PUFAs, minerals (magnesium, phosphorus, copper and potassium, iron, zinc, manganese), carotenoids, β-carotene, and γ-tocopherol (A. Singh & Kumar, 2024). Vitamin E or γ-tocopherol content is particularly high (Hernández-Santos et al., 2016). In traditional medicine, pumpkin seeds have been used to treat bladder and urinary tract infections and even intestinal parasites. They are considered to have diuretic properties and are also used to support prostate health (Dotto & Chacha, 2020a). Watermelon seeds are believed to have diuretic properties, helping with kidney function and detoxification. The seeds are rich in proteins, fats, vitamins, and minerals, and they also contain bioactive compounds such as carotenoids, phenolic acids, and flavonoids, which are used to treat hypertension and improve circulation (Nissar et al., 2025a).
25. **Tomato (*Solanum lycopersicum* L.) seeds**: Tomato seeds contain a range of flavonoids, including lycopene and beta-carotene. These compounds have been shown to have antioxidant and anti-inflammatory properties, which may help to protect against heart disease and certain types of cancer (Kumar et al., 2022a).
26. **Sesame (*Sesamum indicum* L.) Seeds**: In Ayurvedic medicine, sesame seeds are prized for their ability to strengthen bones, improve digestion, and detoxify the body. Sesame oil, extracted from the seeds, is commonly used for skin and hair care due to its moisturizing and antioxidant properties. Sesamin, a kind of lignan found in the oil and seeds of sesame, has been shown to have a variety of biological activities and is beneficial to the human body (Pang et al., 2021).
27. **Sweet corn (*Zea mays* L.) seed:** One of the major corn proteins, α-zein, has therapeutic effects against chronic non-communicable diseases, such as type 2 diabetes and hypertension. Enzymatic hydrolysis of proteins found in foods can produce a wide variety of peptides with various structures and potential bioactivities (Cagnin et al., 2024a). Further, the antioxidant abilities of the phenolic compounds isolated from maize kernel by DPPH (2,2-diphenyl-1-picryl hydrazine) radical, OH radical scavenging activity, and reducing ability, along with α-glucosidase and xanthine oxidase (XO) inhibition, were assessed in *in vitro* conditions (Nile & Park, 2014a).

20.4 MODERN SCIENTIFIC RESEARCH ON PHARMACOLOGICAL ACTIVITIES OF VEGETABLE SEEDS

Modern science has started validation of traditional uses of many vegetable seeds through rigorous studies. Research has confirmed that vegetable seeds contain numerous bioactive compounds with potential therapeutic effects. For instance a review by Bayar and Akkoç (2023) stated that flaxseed supplementation significantly reduced the occurrence of breast cancer in postmenopausal females, supporting its traditional use as a cancer-preventive food. Similarly, Micha et al., (2017) suggested that incorporating omega-3-rich seeds like flax and chia into the diet can improve lipid profiles and lower blood pressure, hence reduce the risk of CVDs. This chapter focuses on some of the important pharmacological activities of vegetable seeds, which can be used as a preventive medicine.

20.4.1 Antioxidant Activities

The antioxidant content in vegetable seeds is another reason for their medicinal value. Seeds like sunflower seeds and sesame seeds contain high concentrations of antioxidants such as vitamin E and phenolic compounds, which help in neutralizing free radicals and reduce oxidative stress. Oxidative stress has been linked to the development of various diseases, including neurodegenerative disorders, cancer, and CVDs, which lead to the ageing process. By scavenging free radicals, antioxidants in vegetable seeds play a vital role in maintaining cellular health and preventing disease progression. Matthäus and Musazcan Özcan (2015) have found rich sources of vitamin-E-active compounds in fennel seed oil (γ-tocotrienol), and peanut oil contains almost equal amounts of α- and γ-tocopherol, with more than 70 mg/100 g of vitamin E in apricot seed oil, apple seed oil, cottonseed oil, and quince seed oil. Cevallos-Casals and Cisneros-Zevallos (2010) showed that germinated edible seeds such as alfalfa, fava, broccoli, fenugreek, kale, lentil, mungbean, mustard, onion, sunflower, soybean, radish, and wheat are a rich source of dietary phenolic antioxidants. Bhat et al. (2014) has reported the high antioxidant activity of sesamin. It enhanced the tocotrienol content, lowering the risk sunburn and tumour development in the skin (Yamada et al., 2008), reduced the UVB-triggered skin erythema, decreased skin inflammation, and prevented skin from light damage and formation of wrinkle (Lin et al., 2019).

20.4.2 Anti-Inflammatory Activities

One of the most well-documented medicinal properties of vegetable seeds is their anti-inflammatory action. In many diseases, chronic inflammation is a common underlying factor, including CVDs, diabetes, and autoimmune disorders. Vegetable seeds such as flaxseed, pumpkin, and chia seeds are rich in omega-3 fatty acids, which have been shown to reduce markers of inflammation in the body. Jiang et al. (2006) and Sumara et al. (2023) reviewed the biomedical functions of various plant seed oils extracted from sesame, black cumin, flax, and pumpkin seeds as they contain bioactive components, which include carotenoids, fatty acids, phytosterols, tocopherols, and phenols. They also discussed the research findings on the medical uses, advantages, and drawbacks of using plant seed oils to treat a variety of illnesses, including cancer, metabolic syndrome, diabetes, premenstrual syndrome, menopause, mastalgia, liver and urinary tract diseases, as well as their use in dermatology and chemotherapy. These seeds modulate the expression of pro-inflammatory cytokines, potentially reducing the risk of chronic inflammatory conditions. Prashanth et al. (2024) reviewed several phytochemicals (arbutin, β-sitosterol, kaemferol, trigonelline, and uridine) from lentils (*Lens culinaris*) which have many advantages, ranging from anti-inflammatory properties, potent antioxidant effects, cholesterol-lowering capabilities, and even potential anticancer properties. Sesame seeds having sesamin have anti-inflammatory and anti-nociceptive activity (Jeng et al., 2005).

20.4.3 Anticancer Properties

Research has indicated that certain vegetable seeds may have potential anticancer properties attributed to their rich content of bioactive compounds such as phytoestrogens (lignans, isoflavones, stilbenes, coumestans, etc.); flavonoids; and saponins. Flaxseeds, in particular, are eminent for their high lignan content, which has demonstrated potential in reduction of the risk of hormone-related cancers such as breast and prostate cancer (Shim et al., 2024). Isoflavones are found in soybeans, soy products, and other legumes such as beans and lentils (Iriti & Varoni, 2017a). Lignans are found in fruits, vegetables, nuts, flaxseed, and whole grains. Coumestans are present in spinach, alfalfa, broccoli, clover, mung bean, and Brussels sprouts, and stilbenes found in peanuts are dietary sources of phytoestrogen, which has anticancer properties (Basu & Maier, 2018). Beneficial bioactivities of glucosinolates found in rapeseed showed antifungal, antibacterial, bioherbicidal, antioxidant, antimutagenic, and anti-carcinogenic activities (Vig et al., 2009).

20.4.4 Antidiabetic Properties

Most of the studies linked the antidiabetic effect of some vegetable seeds to their high phytonutrient content and low glycaemic index. Coriander is one of the traditional herbs, and its various parts can be used, such as seed, stem, leaf, and root. Rajeshwari and Andallu (2011) reported that the seed of coriander contains phytochemicals such as petroselinic acid, linoleic acid, oleic acid, caffeic acid, chlorogenic acid, and palmitic acid, having antidiabetic and antioxidant properties. Flavonoids and phenolic compounds, two types of antioxidants present in high concentrations in watermelon seeds, help prevent oxidative stress and free radical damage, which are linked to chronic diseases like cancer, CVD, and diabetes (Nissar et al., 2025a). Additionally, it has been found that mustard seeds inhibit gluconeogenic and glycolytic enzymes, such as α-glucosidase, which catabolizes the breakdown of complex carbohydrates to glucose; consequently, their inhibition would lower blood glucose levels (Tewari et al., 2003). *Portulaca oleracea* L. is an annual succulent herb, known as the 'vegetable for long life', which is widely distributed worldwide. Hou et al. (2021) showed that purslane seeds can lower blood glucose levels and potentially affect phosphoinositide-3-kinase regulatory subunit-1, peroxisome proliferator-activated receptor gamma (PPAR-γ), tumour necrosis factor (TNF), and other targets associated with inflammation and insulin resistance through the actions of different chemical components such as daidzein, levodopa, and genistein. These potential candidate targets play a critical role in anti-hyperglycaemia, anti-hyperlipidaemia, anti-inflammatory, and immunity regulatory pathways involved in diabetes mellitus, such as Fc gamma R-mediated phagocytosis and the vascular endothelial growth factor (VEGF) signalling pathway, which indicated that purslane seed showed antidiabetic activity through multifaceted actions.

20.4.5 Cardiovascular Effects

Vegetable seeds are particularly useful for heart health due to the presence of omega-3 fatty acids, antioxidants, and a high amount of fibre. Pumpkin seeds, sunflower seeds, and flaxseeds are known to reduce blood pressure, improve lipid profiles, and help decrease the risk of atherosclerosis. Omega-3 fatty acids help in the primary and secondary prevention of cardiovascular (CV) diseases. The DHA and eicosapentaenoic acid (EPA) are relevant to omega-3 fatty acids, which are obtained from alpha-linolenic acid (ALA) from plant sources such as flaxseed oil. Research has proven their cardio-protective actions, anti-inflammatory actions, improving endothelial function, inhibition of platelet aggregation, triglyceride-lowering effects, and regulation of influx and efflux of sodium and calcium channels. The antioxidant activities of polyphenolics may have significant health benefits. Polyphenols are used in the protection against chronic cardioactivity due to their role in transferring electron chelation of ferrous ions, and they also act as scavengers of ROS (Henriksen et al., 1981). Peter and Jacob (2016) reviewed the cardiac vascular (CV) benefits of omega-3 fatty acids, describing their underlying mechanisms of action in the prevention of CV mortality and sudden cardiac death. Omega-3 fatty acids in these seeds help reduce triglyceride levels, prevent blood clot formation, and improve arterial health, thereby lowering the risk of CVDs. Some of the acids found in coriander, such as ascorbic acid (vitamin C), linoleic acid, oleic acid, palmitic acid, and stearic acid, are very effective in reducing the cholesterol level in the blood. They also reduce the cholesterol deposition along the inner walls of the arteries and veins, strengthening the cardiovascular system (Rajeshwari & Andallu, 2011). Watermelon seeds may help lower blood pressure and other cardiovascular illnesses, including hypertension, due to their lower Na/K ratio (Anang et al., 2018a). Additionally, it was found that watermelon seed oil included phytosterols such as stigmasterol, campesterol, and β-sitosterol, which are essential for cardiovascular properties (Rezig et al., 2019a). Cardio protection is accomplished by the activity of quercetin and quercetin glycosides, which protect the LDL from oxidative modification.

20.4.6 Neuroprotective Effects

In various neurodegeneration disorders, such as Alzheimer's disease (AD), Parkinson's disease, and others, medicinal seeds play a significant role in their prevention and treatment, as these are rich in different nutritional phytochemicals. Among these, omega-3 fatty acids and polyphenolic antioxidants may be used as a cure to delay neurodegeneration, improve memory, and cerebral function of the frontal lobes (Arias-Sánchez et al., 2023; Tavan et al., 2024). Seeds such as basil, chia, hemp, pumpkin, sesame, and flax have also been confirmed to have neuroprotective effects against AD due to the interaction between the different signalling pathways initiated by the phytonutrients, resulting in proper folding, leading to neuroinflammation (Duarte et al., 2023).

20.4.6.1 Challenges and Future Directions

Despite the promising therapeutic potential of vegetable seeds, there are still several challenges that need to be addressed. These include:

1. **Standardization:** There is a need for standardization of vegetable seed extracts to ensure consistency and quality. Additionally, exploring the synergistic effects of combining different vegetable seeds may offer new insights into their potential as multifunctional therapeutics.
2. **Regulation:** There is a need for the regulation of vegetable seed extracts to ensure safety and efficacy.
3. **Research:** There is a need for further research into the medicinal properties of vegetable seeds to fully understand their potential health benefits. The bioavailability of some active compounds in seeds can be limited by factors such as processing methods and the presence of anti-nutritional factors (e.g. phytates). Furthermore, extensive clinical trials are necessary to validate the efficacy of vegetable seeds in treating specific health conditions. Future research should focus on understanding the mechanisms involved by which vegetable seeds exert their medicinal effects and how to maximize their bioactive properties through processing and formulation strategies.

20.5 CONCLUSION

Vegetable seeds, often considered humble by-products of agriculture, hold great promise as natural medicine. Their rich nutritional profile is responsible for anti-inflammatory, antioxidant, and anticancer activities, offering a sustainable and accessible approach to health management. While much has been learnt, substantial research is still required to explore the detailed therapeutic capabilities of vegetable seeds and to establish them as integral components of preventive and restorative medicine.

REFERENCES

Abd Rani, N. Z., Husain, K., & Kumolosasi, E. (2018a). Moringa Genus: A Review of Phytochemistry and Pharmacology. *Frontiers in Pharmacology*, *9*, 108. https://doi.org/10.3389/fphar.2018.00108

Abd Rani, N. Z., Husain, K., & Kumolosasi, E. (2018b). Moringa Genus: A Review of Phytochemistry and Pharmacology. *Frontiers in Pharmacology*, *9*, 108. https://doi.org/10.3389/fphar.2018.00108

Alexandrino, T. D., Da Silva, M. G., Ferrari, R. A., Ruiz, A. L. T. G., Duarte, R. M. T., Simabuco, F. M., Bezerra, R. M. N., & Pacheco, M. T. B. (2021). Evaluation of Some in Vitro Bioactivities of Sunflower Phenolic Compounds. *Current Research in Food Science*, *4*, 662–669. https://doi.org/10 1016/j.crfs.2021.09.007

Allai, F. M., Nissar, J., Shafi, A., Bashir, M., Mehreen, Zahoor, I., Ganaie, T. A., Gul, K., & Ahad, T. (2025). Horse Chestnut Starch: Structure, Properties, Modification and Its Applications-A Review. *Carbohydrate Polymer Technologies and Applications*, *10*, 100755. https://doi.org/10.1016/j.carpta.2025.100755

Allaq, A. A., Sidik, N. J., Abdul-Aziz, A., & Ahmed, I. A. (2020). Cumin (Cuminum cyminum L.): A Review of Its Ethnopharmacology, Phytochemistry. *Biomedical Research and Therapy*, *7*(9), 4016–4021. https://doi.org/10.15419/bmrat.v7i9.634

Allaqaband, S., Dar, A. H., Patel, U., Kumar, N., Nayik, G. A., Khan, S. A., Ansari, M. J., Alabdallah, N. M., Kumar, P., Pandey, V. K., Kovács, B., & Shaikh, A. M. (2022). Utilization of Fruit Seed-Based Bioactive Compounds for Formulating the Nutraceuticals and Functional Food: A Review. *Frontiers in Nutrition*, *9*, 902554. https://doi.org/10.3389/fnut.2022.902554

Anang, D. A., Pobee, R. A., Antwi, E., Obeng, E. M., Atter, A., Ayittey, F. K., & Boateng, J. T. (2018a). Nutritional, Microbial and Sensory Attributes of Bread Fortified with Defatted Watermelon Seed Flour. *International Journal of Food Science & Technology*, *53*(6), 1468–1475. https://doi.org/10.1111/ijfs.13727

Anang, D. A., Pobee, R. A., Antwi, E., Obeng, E. M., Atter, A., Ayittey, F. K., & Boateng, J. T. (2018b). Nutritional, Microbial and Sensory Attributes of Bread Fortified with Defatted Watermelon Seed Flour. *International Journal of Food Science & Technology*, *53*(6), 1468–1475. https://doi.org/10.1111/ijfs.13727

Arias-Sánchez, R. A., Torner, L., & Fenton Navarro, B. (2023). Polyphenols and Neurodegenerative Diseases: Potential Effects and Mechanisms of Neuroprotection. *Molecules*, *28*(14), 5415. https://doi.org/10.3390/molecules28145415

Ayadi, J., Debouba, M., Rahmani, R., & Bouajila, J. (2022). Brassica Genus Seeds: A Review on Phytochemical Screening and Pharmacological Properties. *Molecules*, *27*(18), 6008. https://doi.org/10.3390/molecules27186008

Barakat, H., Alkabeer, I. A., Aljutaily, T., Almujaydil, M. S., Algheshairy, R. M., Alhomaid, R. M., Almutairi, A. S., & Mohamed, A. (2022). Phenolics and Volatile Compounds of Fennel (Foeniculum vulgare) Seeds and Their Sprouts Prevent Oxidative DNA Damage and Ameliorates CCl4-Induced Hepatotoxicity and Oxidative Stress in Rats. *Antioxidants*, *11*(12), 2318. https://doi.org/10.3390/antiox11122318

Barot, M., Kumar, N. J. I., & Kumar, R. N. (2016). Bioactive Compounds and Antifungal Activity of Three Different Seaweed Species Ulva Lactuca, Sargassum Tenerrimum and Laurencia Obtusa Collected from Okha Coast, Western India. *Journal of Coastal Life Medicine*, *4*(4), 284–289.

Basu, P., & Maier, C. (2018). Phytoestrogens and Breast Cancer: In Vitro Anticancer Activities of Isoflavones, Lignans, Coumestans, Stilbenes and Their Analogs and Derivatives. *Biomedicine & Pharmacotherapy*, *107*, 1648–1666. https://doi.org/10.1016/j.biopha.2018.08.100

Bayar, İ., & Akkoç, S. (2023). A Mini Review on Components of Flax Seed and Their Effects on Breast Cancer. *Tarım Bilimleri Dergisi*. https://doi.org/10.15832/ankutbd.1349777

Ben Yakoub, A. R., Abdehedi, O., Jridi, M., Elfalleh, W., Nasri, M., & Ferchichi, A. (2018). Flavonoids, phenols, antioxidant, and antimicrobial activities in various extracts from Tossa jute leave (Corchorus olitorus L.). *Industrial Crops and Products*, *118*, 206–213. https://doi.org/10.1016/j.indcrop.2018.03.047

Bettaieb Rebey, I., Aidi Wannes, W., Kaab, S. B., Bourgou, S., Tounsi, M. S., Ksouri, R., & Fauconnier, M. L. (2019a). Bioactive Compounds and Antioxidant Activity of Pimpinella Anisum L. Accessions at Different Ripening Stages. *Scientia Horticulturae*, *246*, 453–461. https://doi.org/10.1016/j.scienta.2018.11.016

Bettaieb Rebey, I., Aidi Wannes, W., Kaab, S. B., Bourgou, S., Tounsi, M. S., Ksouri, R., & Fauconnier, M. L. (2019b). Bioactive Compounds and Antioxidant Activity of Pimpinella anisum L. Accessions at Different Ripening Stages. *Scientia Horticulturae*, *246*, 453–461. https://doi.org/10.1016/j.scienta.2018.11.016

Bhat, K. V., Kumari, R., Pathak, N., & Rai, A. K. (2014). Value addition in sesame: A perspective on bioactive components for enhancing utility and profitability. *Pharmacognosy Reviews*, *8*(16), 147. https://doi.org/10.4103/0973-7847.134249

Bhattacharya, S. (2020). Seeds as Herbal Drugs. In *Nuts and Seeds in Health and Disease Prevention* (pp. 471–483). Elsevier. https://doi.org/10.1016/B978-0-12-818553-7.00033-4

Bonin, E., Avila, V. D., Carvalho, V. M., Cardoso, M. A. P., Matos, A. M., Ramos, A. V. G., Cabral, M. R. P., Baldoqui, D. C., Sarragiotto, M. H., Filho, B. A. D. A., & Prado, I. N. D. (2023). Study of Chemical Constituents, Antioxidants and Antimicrobial Activities of *Tamarindus indica* L. Seed. *Journal of Food Science*, *88*(11), 4639–4652. https://doi.org/10.1111/1750-3841.16739

Cagnin, C., Garcia, B. D. F., Rocha, T. D. S., & Prudencio, S. H. (2024a). Bioactive Peptides from Corn (Zea mays L.) with the Potential to Decrease the Risk of Developing Non-Communicable Chronic Diseases: In Silico Evaluation. *Biology*, *13*(10), 772. https://doi.org/10.3390/biology13100772

Cagnin, C., Garcia, B. D. F., Rocha, T. D. S., & Prudencio, S. H. (2024b). Bioactive Peptides from Corn (Zea mays L.) with the Potential to Decrease the Risk of Developing Non-Communicable Chronic Diseases: In Silico Evaluation. *Biology*, *13*(10), 772. https://doi.org/10.3390/biology13100772

Casalvara, R. F. A., Ferreira, B. M. R., Gonçalves, J. E., Yamaguchi, N. U., Bracht, A., Bracht, L., Comar, J. F., De Sá-Nakanishi, A. B., De Souza, C. G. M., Castoldi, R., Corrêa, R. C. G., & Peralta, R. M. (2024). Biotechnological, Nutritional, and Therapeutic Applications of Quinoa (Chenopodium quinoa Willd.) and Its By-Products: A Review of the Past Five-Year Findings. *Nutrients*, *16*(6), 840. https://doi.org/10.3390/nu16060840

Cevallos-Casals, B. A., & Cisneros-Zevallos, L. (2010). Impact of Germination on Phenolic Content and Antioxidant Activity of 13 Edible Seed Species. *Food Chemistry*, *119*(4), 1485–1490. https://doi.org/10.1016/j.foodchem.2009.09.030

Chhikara, N., Kushwaha, K., Sharma, P., Gat, Y., & Panghal, A. (2019a). Bioactive Compounds of Beetroot and Utilization in Food Processing Industry: A Critical Review. *Food Chemistry*, *272*, 192–200. https://doi.org/10.1016/j.foodchem.2018.08.022

Chhikara, N., Kushwaha, K., Sharma, P., Gat, Y., & Panghal, A. (2019b). Bioactive Compounds of Beetroot and Utilization in Food Processing Industry: A Critical Review. *Food Chemistry*, *272*, 192–200. https://doi.org/10.1016/j.foodchem.2018.08.022

Chinta, M. L., Gandam, P. K., Sivasankar, M. V., & Parcha, S. R. (2025). Tamarind (Tamarindus indica L.) Seed Polysaccharide: A promising biopolymer for drug delivery, wound healing, tissue engineering and beyond. *Carbohydrate Research*, *552*, 109454. https://doi.org/10.1016/j.carres.2025.109454

Dai, Y.-L., Li, Y., Wang, Q., Niu, F.-J., Li, K.-W., Wang, Y.-Y., Wang, J., Zhou, C.-Z., & Gao, L.-N. (2022). Chamomile: A Review of Its Traditional Uses, Chemical Constituents, Pharmacological Activities and Quality Control Studies. *Molecules*, *28*(1), 133. https://doi.org/10.3390/molecules28010133

Dhama, K., Latheef, S. K., Dadar, M., Samad, H. A., Munjal, A., Khandia, R., Karthik, K., Tiwari, R., Yatoo, Mohd. I., Bhatt, P., Chakraborty, S., Singh, K. P., Iqbal, H. M. N., Chaicumpa, W., & Joshi, S. K. (2019). Biomarkers in Stress Related Diseases/Disorders: Diagnostic, Prognostic, and Therapeutic Values. *Frontiers in Molecular Biosciences*, *6*, 91. https://doi.org/10.3389/fmolb.2019.00091

Dhani, S., Hamid, Ngobese, N. Z., Sharma, S., & Jaiswal, A. K. (2025). A Comprehensive Review on Nutritional Composition, Health Benefits, and Industrial Applications of Jackfruit Seeds. *Journal of Agriculture and Food Research*, *19*, 101692. https://doi.org/10.1016/j.jafr.2025.101692

Dotto, J. M., & Chacha, J. S. (2020a). The Potential of Pumpkin Seeds as a Functional Food Ingredient: A Review. *Scientific African*, *10*, e00575. https://doi.org/10.1016/j.sciaf.2020.e00575

Dotto, J. M., & Chacha, J. S. (2020b). The Potential of Pumpkin Seeds as a Functional Food Ingredient: A Review. *Scientific African*, *10*, e00575. https://doi.org/10.1016/j.sciaf.2020.e00575

Duarte, G. M., de Araújo, F. E. A., da Rocha, J. M. C., Idalina Neta, F., do Rego, A. C. M., Araújo Filho, I., . . . & Guzen, F. P. (2023). Neuroprotective potential of seed extracts: Review of in vitro and in vivo studies. *Nutrients*, *15*(11), 2502.

Duda, Ł., Kłosiński, K. K., Budryn, G., JaŚkiewicz, A., Kołat, D., Kałuzińska-Kołat, Ż., & Pasieka, Z. W. (2024). Medicinal Use of Chicory (Cichorium intybus L.). *Scientia Pharmaceutica*, *92*(2), 31. https://doi.org/10.3390/scipharm92020031

Faisal, Z., Irfan, R., Akram, N., Manzoor, H. M. I., Aabdi, M. A., Anwar, M. J., Khawar, S., Saif, A., Shah, Y. A., Afzaal, M., & Desta, D. T. (2024). The multifaceted potential of fenugreek seeds: From health benefits to food and nanotechnology applications. *Food Science & Nutrition*, *12*(4), 2294–2310. https://doi.org/10.1002/fsn3.3959

Fallah Huseini, H., et al. (2017). Antidiabetic effect of *Allium cepa* seed extract in streptozotocin-induced diabetic rats. *Iranian Journal of Pharmaceutical Research*, *16*(3), 1230–1238.

Fiorito, S., Epifano, F., Palmisano, R., Genovese, S., & Taddeo, V. A. (2017). A re-investigation of the phytochemical composition of the edible herb Amaranthus retroflexus L. *Journal of Pharmaceutical and Biomedical Analysis*, *143*, 183–187. https://doi.org/10.1016/j.jpba.2017.05.051

Gautam, B., Vadivel, V., Stuetz, W., & Biesalski, H. K. (2012). Bioactive Compounds Extracted from Indian Wild Legume Seeds: Antioxidant and Type II Diabetes–Related Enzyme Inhibition Properties. *International Journal of Food Sciences and Nutrition*, *63*(2), 242–245. https://doi.org/10.3109/09637486.2011.621413

Ghanbari, R., Anwar, F., Alkharfy, K. M., Gilani, A.-H., & Saari, N. (2012). Valuable Nutrients and Functional Bioactives in Different Parts of Olive (Olea europaea L.) – A Review. *International Journal of Molecular Sciences*, *13*(3), 3291–3340. https://doi.org/10.3390/ijms13033291

Henriksen, T., Mahoney, E. M., & Steinberg, D. (1981). Enhanced Macrophage Degradation of Low Density Lipoprotein Previously Incubated with Cultured Endothelial Cells: Recognition by Receptors for Acetylated Low Density Lipoproteins. *Proceedings of the National Academy of Sciences*, *78*(10), 6499–6503. https://doi.org/10.1073/pnas.78.10.6499

Hernández-Santos, B., Rodríguez-Miranda, J., Herman-Lara, E., Torruco-Uco, J. G., Carmona-García, R., Juárez-Barrientos, J. M., Chávez-Zamudio, R., & Martínez-Sánchez, C. E. (2016). Effect of Oil Extraction Assisted by Ultrasound on the Physicochemical Properties and Fatty Acid Profile of Pumpkin Seed Oil (Cucurbita pepo). *Ultrasonics Sonochemistry*, *31*, 429–436. https://doi.org/10.1016/j.ultsonch.2016.01.029

Hou, J., Zhou, X., Wang, P., Zhao, C., Qin, Y., Liu, F., Yu, L., & Xu, H. (2021). An Integrative Pharmacology-Based Approach for Evaluating the Potential Effects of Purslane Seed in Diabetes Mellitus Treatment Using UHPLC-LTQ-Orbitrap and TCMIP V2.0. *Frontiers in Pharmacology*, *11*, 593693. https://doi.org/10.3389/fphar.2020.593693

Iriti, M., & Varoni, E. (2017a). Pulses, Healthy, and Sustainable Food Sources for Feeding the Planet. *International Journal of Molecular Sciences*, *18*(2), 255. https://doi.org/10.3390/ijms18020255

Iriti, M., & Varoni, E. (2017b). Pulses, Healthy, and Sustainable Food Sources for Feeding the Planet. *International Journal of Molecular Sciences*, *18*(2), 255. https://doi.org/10.3390/ijms18020255

Jeng, K.-C. G., Hou, Rolis. C. W., Wang, J.-C., & Ping, L.-I. (2005). Sesamin Inhibits Lipopolysaccharide-Induced Cytokine Production by Suppression of p38 Mitogen-Activated Protein Kinase and Nuclear Factor-κB. *Immunology Letters*, *97*(1), 101–106. https://doi.org/10.1016/j.imlet.2004.10.004

Jiang, R., Jacobs, D. R., Mayer-Davis, E., Szklo, M., Herrington, D., Jenny, N. S., Kronmal, R., & Barr, R. G. (2006). Nut and Seed Consumption and Inflammatory Markers in the Multi-Ethnic Study of Atherosclerosis. *American Journal of Epidemiology*, *163*(3), 222–231. https://doi.org/10.1093/aje/kwj033

Joshi, P. S., & Singh, S. P. (2024). Comprehensive Review of the Nutritional Composition and Health Benefits of Horse Chestnut (Aesculus hippocastanum) Seeds and Nuts: A Review. *International Journal of Advanced Biochemistry Research*, *8*(3), 01–16. https://doi.org/10.33545/26174693.2024.v8.i3a.668

Khalid, L., Jabeen, I., Inam-ur-Raheem, M., Khan, U. M., Mousavi Khaneghah, A., & Aadil, R. M. (2025). Garden Cress Seed: A Review of Its Functional and Medicinal Properties. *Journal of Agriculture and Food Research*, *21*, 101894. https://doi.org/10.1016/j.jafr.2025.101894

Khalid, W., Arshad, M. S., Aziz, A., Rahim, M. A., Qaisrani, T. B., Afzal, F., Ali, A., Ranjha, M. M. A. N., Khalid, M. Z., & Anjum, F. M. (2023). Chia Seeds (*Salvia hispanica* L.): A Therapeutic Weapon in Metabolic Disorders. *Food Science & Nutrition*, *11*(1), 3–16. https://doi.org/10.1002/fsn3.3035

Kris-Etherton, P. M., Hecker, K. D., Bonanome, A., Coval, S. M., Binkoski, A. E., Hilpert, K. F., Griel, A. E., & Etherton, T. D. (2002). Bioactive Compounds in Foods: Their Role in the Prevention of Cardiovascular Disease and Cancer. *The American Journal of Medicine*, *113*(9), 71–88. https://doi.org/10.1016/S0002-9343(01)00995-0

Kumar, M., Chandran, D., Tomar, M., Bhuyan, D. J., Grasso, S., Sá, A. G. A., Carciofi, B. A. M., Radha, Dhumal, S., Singh, S., Senapathy, M., Changan, S., Dey, A., Pandiselvam, R., Mahato, D. K., Amarowicz, R., Rajalingam, S., Vishvanathan, M., Saleena, L. A. K., & Mekhemar, M. (2022a). Valorization Potential of Tomato (Solanum lycopersicum L.) Seed: Nutraceutical Quality, Food Properties, Safety Aspects, and Application as a Health-Promoting Ingredient in Foods. *Horticulturae*, *8*(3), 265. https://doi.org/10.3390/horticulturae8030265

Kumar, M., Chandran, D., Tomar, M., Bhuyan, D. J., Grasso, S., Sá, A. G. A., Carciofi, B. A. M., Radha, Dhumal, S., Singh, S., Senapathy, M., Changan, S., Dey, A., Pandiselvam, R., Mahato, D. K., Amarowicz, R., Rajalingam, S., Vishvanathan, M., Saleena, L. A. K., & Mekhemar, M. (2022b). Valorization Potential of Tomato (Solanum lycopersicum L.) Seed: Nutraceutical Quality, Food Properties, Safety Aspects, and Application as a Health-Promoting Ingredient in Foods. *Horticulturae*, *8*(3), 265. https://doi.org/10.3390/horticulturae8030265

Kumar, N., Pratibha, & Pareek, S. (2021). Bioactive Compounds of Moringa (Moringa Species). In H. N. Murthy & K. Y. Paek (Eds.), *Bioactive Compounds in Underutilized Vegetables and Legumes* (pp. 503–524). Springer International Publishing. https://doi.org/10.1007/978-3-030-57415-4_28

Laribi, B., Kouki, K., M'Hamdi, M., & Bettaieb, T. (2015a). Coriander (Coriandrum sativum L.) and Its Bioactive Constituents. *Fitoterapia*, *103*, 9–26. https://doi.org/10.1016/j.fitote.2015.03.012

Laribi, B., Kouki, K., M'Hamdi, M., & Bettaieb, T. (2015b). Coriander (Coriandrum sativum L.) and Its Bioactive Constituents. *Fitoterapia*, *103*, 9–26. https://doi.org/10.1016/j.fitote.2015.03.012

Li, J., Wang, J., Zhang, N., Li, Y., Cai, Z., Li, G., Liu, Z., Liu, Z., Wang, Y., Shao, X., & Chen, J. (2023). Anti-Aging Activity and Their Mechanisms of Natural Food-Derived Peptides: Current Advancements. *Food Innovation and Advances*, *2*(4), 272–290. https://doi.org/10.48130/FIA-2023-0028

Lin, T.-Y., Wu, P.-Y., Hou, C.-W., Chien, T.-Y., Chang, Q.-X., Wen, K.-C., Lin, C.-Y., & Chiang, H.-M. (2019). Protective Effects of Sesamin against UVB-Induced Skin Inflammation and Photodamage In Vitro and In Vivo. *Biomolecules*, *9*(9), 479. https://doi.org/10.3390/biom9090479

Liu, C.-P., Tsai, W.-J., Lin, Y.-L., Liao, J.-F., Chen, C.-F., & Kuo, Y.-C. (2004). The Extracts from Nelumbo Nucifera Suppress Cell Cycle Progression, Cytokine Genes Expression, and Cell Proliferation in Human Peripheral Blood Mononuclear Cells. *Life Sciences*, *75*(6), 699–716. https://doi.org/10.1016/j.lfs.2004.01.019

Liu, X.-S., Gao, B., Li, X.-L., Li, W.-N., Qiao, Z.-A., & Han, L. (2020). Chemical Composition and Antimicrobial and Antioxidant Activities of Essential Oil of Sunflower (Helianthus annuus L.) Receptacle. *Molecules*, *25*(22), 5244. https://doi.org/10.3390/molecules25225244

Maestri, D., Barrionuevo, D., Bodoira, R., Zafra, A., Jiménez-López, J., & Alché, J. D. D. (2019). Nutritional Profile and Nutraceutical Components of Olive (Olea europaea L.) Seeds. *Journal of Food Science and Technology*, *56*(9), 4359–4370. https://doi.org/10.1007/s13197-019-03904-5

Maria De Jesus Benevides, C., Almeida Trindade, B., & Vieira Lopes, M. (2018a). Potentialities of Legumes in the Pharmaceutical Industry. *Journal of Analytical & Pharmaceutical Research*, *7*(3). https://doi.org/10.15406/japlr.2018.07.00253

Maria De Jesus Benevides, C., Almeida Trindade, B., & Vieira Lopes, M. (2018b). Potentialities of Legumes in the Pharmaceutical Industry. *Journal of Analytical & Pharmaceutical Research*, *7*(3). https://doi.org/10.15406/japlr.2018.07.00253

Matthäus, B., & Musazcan Özcan, M. (2015). Oil Content, Fatty Acid Composition and Distributions of Vitamin-E-Active Compounds of Some Fruit Seed Oils. *Antioxidants*, *4*(1), 124–133. https://doi.org/10.3390/antiox4010124

Micha, R., Peñalvo, J. L., Cudhea, F., Imamura, F., Rehm, C. D., & Mozaffarian, D. (2017). Association between Dietary Factors and Mortality From Heart Disease, Stroke, and Type 2 Diabetes in the United States. *JAMA*, *317*(9), 912. https://doi.org/10.1001/jama.2017.0947

Mohd Ali, N., Yeap, S. K., Ho, W. Y., Beh, B. K., Tan, S. W., & Tan, S. G. (2012). The Promising Future of Chia, *Salvia hispanica* L. *Journal of Biomedicine and Biotechnology*, *2012*, 1–9. https://doi.org/10.1155/2012/171956

Motyka, S., Skała, E., Ekiert, H., & Szopa, A. (2023). Health-Promoting Approaches of the Use of Chia Seeds. *Journal of Functional Foods*, *103*, 105480. https://doi.org/10.1016/j.jff.2023.105480

Muhammad, A., Akhtar, A., Aslam, S., Khan, R. S., Ahmed, Z., & Khalid, N. (2021). Review on Physicochemical, Medicinal and Nutraceutical Properties of Poppy Seeds: A Potential Functional Food Ingredient. *Functional Foods in Health and Disease*, *11*(10). https://doi.org/10.31989/ffhd.v11i10.836

Nadeem, M., Muhammad Anjum, F., Issa Khan, M., Tehseen, S., El-Ghorab, A., & Iqbal Sultan, J. (2013). Nutritional and Medicinal Aspects of Coriander (*Coriandrum sativum* L.): A Review. *British Food Journal*, *115*(5), 743–755. https://doi.org/10.1108/00070701311331526

Nile, S. H., & Park, S. W. (2014a). Antioxidant, α-Glucosidase and Xanthine Oxidase Inhibitory Activity of Bioactive Compounds from Maize (*Zea mays* L.). *Chemical Biology & Drug Design*, *83*(1), 119–125. https://doi.org/10.1111/cbdd.12205

Nile, S. H., & Park, S. W. (2014b). Antioxidant, α-Glucosidase and Xanthine Oxidase Inhibitory Activity of Bioactive Compounds From Maize (*Zea mays* L.). *Chemical Biology & Drug Design*, *83*(1), 119–125. https://doi.org/10.1111/cbdd.12205

Nissar, J., Sidiqi, U. S., Dar, A. H., & Akbar, U. (2025a). Nutritional Composition and Bioactive Potential of Watermelon Seeds: A Pathway to Sustainable Food and Health Innovation. *Sustainable Food Technology*, *3*(2), 375–395. https://doi.org/10.1039/D4FB00335G

Nissar, J., Sidiqi, U. S., Dar, A. H., & Akbar, U. (2025b). Nutritional Composition and Bioactive Potential of Watermelon Seeds: A Pathway to Sustainable Food and Health Innovation. *Sustainable Food Technology*, *3*(2), 375–395. https://doi.org/10.1039/D4FB00335G

Noreen, S., Tufail, T., Badar Ul Ain, H., & Awuchi, C. G. (2023a). Pharmacological, Nutraceutical, Functional and Therapeutic Properties of Fennel (*foeniculum vulgare*). *International Journal of Food Properties*, *26*(1), 915–927. https://doi.org/10.1080/10942912.2023.2192436

Noreen, S., Tufail, T., Badar Ul Ain, H., & Awuchi, C. G. (2023b). Pharmacological, Nutraceutical, Functional and Therapeutic Properties of Fennel (*foeniculum vulgare*). *International Journal of Food Properties*, *26*(1), 915–927. https://doi.org/10.1080/10942912.2023.2192436

Otunola, G. A. (2022). Culinary Spices in Food and Medicine: An Overview of Syzygium aromaticum (L.) Merr. and L. M. Perry [Myrtaceae]. *Frontiers in Pharmacology*, *12*, 793200. https://doi.org/10.3389/fphar.2021.793200

Pang, Y., Wu, S., He, Y., Nian, Q., Lei, J., Yao, Y., Guo, J., & Zeng, J. (2021). Plant-Derived Compounds as Promising Therapeutics for Vitiligo. *Frontiers in Pharmacology*, *12*, 685116. https://doi.org/10.3389/fphar.2021.685116

Peter, S., & Jacob, J. J. (2016). Role of Omega-3 Fatty Acids in Cardiovascular Disorders. In M. V. Hegde, A. A. Zanwar, & S. P. Adekar (Eds.), *Omega-3 Fatty Acids* (pp. 513–530). Springer International Publishing. https://doi.org/10.1007/978-3-319-40458-5_37

Petrovska, B. (2012). Historical Review of Medicinal Plants' Usage. *Pharmacognosy Reviews*, *6*(11), 1. https://doi.org/10.4103/0973-7847.95849

Prashanth, C. B., Geetha, K. M., Wilson, B., & Banu, S. (2024). Traditional Uses, Bioactive Composition and Pharmacological Activities of Lens Culinaris. *South African Journal of Botany*, *168*, 542–561. https://doi.org/10.1016/j.sajb.2024.03.045

Pieszak, M., Mikolajczak, P. L., & Manikowska, K. (2012). Borage (*Borago officinalis* L.) – A valuable medicinal plant used in herbal medicine. *Herba Polonica*, *58*(4).

Rahman, M., Khatun, A., Liu, L., & Barkla, B. J. (2024). Brassicaceae Mustards: Phytochemical Constituents, Pharmacological Effects, and Mechanisms of Action against Human Disease. *International Journal of Molecular Sciences*, *25*(16), 9039. https://doi.org/10.3390/ijms25169039

Rahul, J., Jain, M. K., Singh, S. P., Kamal, R. K., Anuradha, Naz, A., Gupta, A. K., & Mrityunjay, S. K. (2015). Adansonia digitata L. (baobab): A Review of Traditional Information and Taxonomic Description. *Asian Pacific Journal of Tropical Biomedicine*, *5*(1), 79–84. https://doi.org/10.1016/S2221-1691(15)30174-X

Rajeshwari, U., & Andallu, B. (2011). Medicinal Benefits of Coriander (*Coriandrum sativum* L). *Spatula DD – Peer Reviewed Journal on Complementary Medicine and Drug Discovery*, *1*(1), 51. https://doi.org/10.5455/spatula.20110106123153

Rasooli, I., & Allameh, A. (2016). Caraway (Carum carvi L.) Essential Oils. In *Essential Oils in Food Preservation, Flavor and Safety* (pp. 287–293). Elsevier. https://doi.org/10.1016/B978-0-12-416641-7.00032-8

Rehman, S., Mufti, I. U., Ain, Q. U., & Ijaz, B. (2024). Bioactive Compounds and Biological Activities of Red Beetroot (Beta vulgaris L.). In H. N. Murthy, K. Y. Paek, & S.-Y. Park (Eds.), *Bioactive Compounds in the Storage Organs of Plants* (pp. 845–875). Springer Nature Switzerland. https://doi.org/10.1007/978-3-031-44746-4_42

Rezig, L., Chouaibi, M., Meddeb, W., Msaada, K., & Hamdi, S. (2019a). Chemical Composition and Bioactive Compounds of Cucurbitaceae Seeds: Potential Sources for New Trends of Plant Oils. *Process Safety and Environmental Protection*, *127*, 73–81. https://doi.org/10.1016/j.psep.2019.05.005

Rezig, L., Chouaibi, M., Meddeb, W., Msaada, K., & Hamdi, S. (2019b). Chemical Composition and Bioactive Compounds of Cucurbitaceae Seeds: Potential Sources for New Trends of Plant Oils. *Process Safety and Environmental Protection*, *127*, 73–81. https://doi.org/10.1016/j.psep.2019.05.005

Roy, A., Khan, A., Ahmad, I., Alghamdi, S., Rajab, B. S., Babalghith, A. O., Alshahrani, M. Y., Islam, S., & Islam, Md. R. (2022). Flavonoids a Bioactive Compound from Medicinal Plants and Its Therapeutic Applications. *BioMed Research International*, *2022*, 1–9. https://doi.org/10.1155/2022/5445291

Shahida, S., BA M, H. S., & Fuad, S. H. (2019). Does MSPO certification matter for profitability of Malaysian palm oil companies?. *International Journal of Economics & Management*, *13*(2).

Shahrajabian, M. H., & Sun, W. (2023). Five Important Seeds in Traditional Medicine, and Pharmacological Benefits. *Seeds*, *2*(3), 290–308. https://doi.org/10.3390/seeds2030022

Shahrajabian, M. H., Sun, W., & Cheng, Q. (2020a). Chemical Components and Pharmacological Benefits of Basil (*Ocimum basilicum*): A Review. *International Journal of Food Properties*, *23*(1), 1961–1970. https://doi.org/10.1080/10942912.2020.1828456

Shahrajabian, M. H., Sun, W., & Cheng, Q. (2020b). Chemical Components and Pharmacological Benefits of Basil (*Ocimum basilicum*): A Review. *International Journal of Food Properties*, *23*(1), 1961–1970. https://doi.org/10.1080/10942912.2020.1828456

Shim, Y. Y., Kim, J. H., Cho, J. Y., & Reaney, M. J. T. (2024). Health Benefits of Flaxseed and Its Peptides (linusorbs). *Critical Reviews in Food Science and Nutrition*, *64*(7), 1845–1864. https://doi.org/10.1080/10408398.2022.2119363

Singh, A., & Kumar, V. (2024). Pumpkin Seeds as Nutraceutical and Functional Food Ingredient for Future: A Review. *Grain & Oil Science and Technology*, *7*(1), 12–29. https://doi.org/10.1016/j.gaost.2023.12.002

Singh, B., Singh, J. P., Shevkani, K., Singh, N., & Kaur, A. (2017). Bioactive Constituents in Pulses and Their Health Benefits. *Journal of Food Science and Technology*, *54*(4), 858–870. https://doi.org/10.1007/s13197-016-2391-9

Singh, M. P., Goel, B., Kumar, R., & Rathor, S. (2024). Phytochemical and Pharmacological Aspects of Genus Amaranthus. *Fitoterapia*, *176*, 106036. https://doi.org/10.1016/j.fitote.2024.106036

Siyuan, S., Tong, L., & Liu, R. (2018). Corn Phytochemicals and Their Health Benefits. *Food Science and Human Wellness*, *7*(3), 185–195. https://doi.org/10.1016/j.fshw.2018.09.003

Srinivasa, U. M., & Naidu, M. M. (2021). Fenugreek (Trigonella foenum-graecum L.) Seed: Promising Source of Nutraceutical. In *Studies in Natural Products Chemistry* (Vol. 71, pp. 141–184). Elsevier. https://doi.org/10.1016/B978-0-323-91095-8.00014-3

Srinivasan, K. (2018). Cumin (Cuminum cyminum) and Black Cumin (Nigella sativa) Seeds: Traditional Uses, Chemical Constituents, and Nutraceutical Effects. *Food Quality and Safety*, *2*(1), 1–16. https://doi.org/10.1093/fqsafe/fyx031

Sumara, A., Stachniuk, A., Montowska, M., Kotecka-Majchrzak, K., Grywalska, E., Mitura, P., Saftić Martinović, L., Kraljević Pavelić, S., & Fornal, E. (2023). Comprehensive Review of Seven Plant Seed Oils: Chemical Composition, Nutritional Properties, and Biomedical Functions. *Food Reviews International*, *39*(8), 5402–5422. https://doi.org/10.1080/87559129.2022.2067560

Sun, W., Shahrajabian, M. H., & Cheng, Q. (2019). Anise (*Pimpinella anisum L.*), a Dominant Spice and Traditional Medicinal Herb for Both Food and Medicinal Purposes. *Cogent Biology*, *5*(1), 1673688. https://doi.org/10.1080/23312025.2019.1673688

Tavan, M., Hanachi, P., De La Luz Cádiz-Gurrea, M., Segura Carretero, A., & Mirjalili, M. H. (2024). Natural Phenolic Compounds with Neuroprotective Effects. *Neurochemical Research*, *49*(2), 306–326. https://doi.org/10.1007/s11064-023-04046-z

Tewari, N., Tiwari, V. K., Mishra, R. C., Tripathi, R. P., Srivastava, A. K., Ahmad, R., Srivastava, R., & Srivastava, B. S. (2003). Synthesis and Bioevaluation of Glycosyl Ureas as α-Glucosidase Inhibitors and Their Effect on Mycobacterium. *Bioorganic & Medicinal Chemistry*, *11*(13), 2911–2922. https://doi.org/10.1016/S0968-0896(03)00214-1

Thirumalai, T., Therasa, S. V., Elumalai, E., & David, E. (2011). Hypoglycemic effect of Brassica juncea (seeds) on streptozotocin induced diabetic male albino rat. *Asian Pacific Journal of Tropical Biomedicine*, *1*(4), 323–325. https://doi.org/10.1016/S2221-1691(11)60052-X

Tileuberdi, N., Turgumbayeva, A., Yeskaliyeva, B., Sarsenova, L., & Issayeva, R. (2022). Extraction, Isolation of Bioactive Compounds and Therapeutic Potential of Rapeseed (Brassica napus L.). *Molecules*, *27*(24), 8824. https://doi.org/10.3390/molecules27248824

Torres-Contreras, A. M., Nair, V., Cisneros-Zevallos, L., & Jacobo-Velázquez, D. A. (2017). Stability of Bioactive Compounds in Broccoli as Affected by Cutting Styles and Storage Time. *Molecules*, *22*(4), 636. https://doi.org/10.3390/molecules22040636

Tran, T. T., Bui, P. H., Phan, N. T. T., Hoang, M. D., & Minh Tam Nguyen Kim. (2023). Bioactive Compounds from Portulaca oleracea L. Extract. *Chemical Engineering Transactions*, *106*, 637–642. https://doi.org/10.3303/CET23106107

Ullah, R., Nadeem, M., Khalique, A., Imran, M., Mehmood, S., Javid, A., & Hussain, J. (2016). Nutritional and Therapeutic Perspectives of Chia (Salvia hispanica L.): A Review. *Journal of Food Science and Technology*, *53*(4), 1750–1758. https://doi.org/10.1007/s13197-015-1967-0

Vassilina, G., Sabitova, A., Idrisheva, Z., Zhumabekova, A., Kanapiyeva, F., Orynbassar, R., Zhamanbayeva, M., Kamalova, M., Assilbayeva, J., Turgumbayeva, A., & Abilkassymova, A. (2025a). Bio-Active Compounds and Major Biomedical Properties of Basil (*Ocimum basilicum*, lamiaceae). *Natural Product Research*, *39*(5), 1326–1344. https://doi.org/10.1080/14786419.2024.2357662

Vassilina, G., Sabitova, A., Idrisheva, Z., Zhumabekova, A., Kanapiyeva, F., Orynbassar, R., Zhamanbayeva, M., Kamalova, M., Assilbayeva, J., Turgumbayeva, A., & Abilkassymova, A. (2025b). Bio-Active Compounds and Major Biomedical Properties of Basil (*Ocimum basilicum*, lamiaceae). *Natural Product Research*, *39*(5), 1326–1344. https://doi.org/10.1080/14786419.2024.2357662

Vergara-Jimenez, M., Almatrafi, M., & Fernandez, M. (2017). Bioactive Components in Moringa Oleifera Leaves Protect against Chronic Disease. *Antioxidants*, *6*(4), 91. https://doi.org/10.3390/antiox6040091

Vig, A. P., Rampal, G., Thind, T. S., & Arora, S. (2009). Bio-Protective Effects of Glucosinolates – A Review. *LWT – Food Science and Technology*, *42*(10), 1561–1572. https://doi.org/10.1016/j.lwt.2009.05.023

Wandee, R., Sutthanut, K., Songsri, J., Sonsena, S., Krongyut, O., Tippayawat, P., Tukummee, W., & Rittirod, T. (2022). Tamarind Seed Coat: A Catechin-Rich Source with Anti-Oxidation, Anti-Melanogenesis, Anti-Adipogenesis and Anti-Microbial Activities. *Molecules*, *27*(16), 5319. https://doi.org/10.3390/molecules27165319

Wu, D.-T., Li, W.-X., Wan, J.-J., Hu, Y.-C., Gan, R.-Y., & Zou, L. (2023). A Comprehensive Review of Pea (Pisum sativum L.): Chemical Composition, Processing, Health Benefits, and Food Applications. *Foods*, *12*(13), 2527. https://doi.org/10.3390/foods12132527

Xi, X., Fan, G., Xue, H., Peng, S., Huang, W., & Zhan, J. (2024). Harnessing the Potential of Quinoa: Nutritional Profiling, Bioactive Components, and Implications for Health Promotion. *Antioxidants*, *13*(7), 829. https://doi.org/10.3390/antiox13070829

Yadav, S. P., Vats, V., Ammini, A. C., & Grover, J. K. (2004). Brassica juncea (Rai) Significantly Prevented the Development of Insulin Resistance in Rats Fed Fructose-Enriched Diet. *Journal of Ethnopharmacology*, *93*(1), 113–116. https://doi.org/10.1016/j.jep.2004.03.034

Yamada, Y., Obayashi, M., Ishikawa, T., Kiso, Y., Ono, Y., & Yamashita, K. (2008). Dietary Tocotrienol Reduces UVB-induced Skin Damage and Sesamin Enhances Tocotrienol Effects in Hairless Mice. *Journal of Nutritional Science and Vitaminology*, *54*(2), 117–123. https://doi.org/10.3177/jnsv.54.117

Yan, J., Yang, X., He, L., Huang, Z., Zhu, M., Fan, L., Li, H., Wu, L., Yu, L., & Zhu, W. (2022). Comprehensive Quality and Bioactive Constituent Analysis of Celery Juice Made from Different Cultivars. *Foods*, *11*(18), 2719. https://doi.org/10.3390/foods11182719

Yawadio Nsimba, R., Kikuzaki, H., & Konishi, Y. (2008). Antioxidant Activity of Various Extracts and Fractions of Chenopodium quinoa and Amaranthus spp. Seeds. *Food Chemistry*, *106*(2), 760–766. https://doi.org/10.1016/j.foodchem.2007.06.004

Zhang, Y., Lu, X., Zeng, S., Huang, X., Guo, Z., Zheng, Y., Tian, Y., & Zheng, B. (2015). Nutritional Composition, Physiological Functions and Processing of Lotus (Nelumbo nucifera Gaertn.) Seeds: A Review. *Phytochemistry Reviews*, *14*(3), 321–334. https://doi.org/10.1007/s11101-015-9401-9

Zhao, X.-X., Lin, F.-J., Li, H., Li, H.-B., Wu, D.-T., Geng, F., Ma, W., Wang, Y., Miao, B.-H., & Gan, R.-Y. (2021). Recent Advances in Bioactive Compounds, Health Functions, and Safety Concerns of Onion (Allium cepa L.). *Frontiers in Nutrition*, *8*, 669805. https://doi.org/10.3389/fnut.2021.669805

Index

Note: Page numbers in *italics* indicate a figure and page numbers in **bold** indicate a table on the corresponding page.

D

E

F

G

H

O